W9-ATL-110

Human Life Cycle

edited by
William C. Sze, Ph.D.

foreword by
Theodore Lidz, M.D.
Yale University

Jason Aronson, Inc. New York

Third printing, September 1979

ISBN: 0-87668-199-2

Library of Congress Catalog Number: 75-4783

To the memory of
my parents

Foreword

The preparation of a good book of readings, whatever its purpose, is a difficult undertaking. The compiler needs not simply to have read extensively in the field, but to have mastered it in order to evaluate articles with fine discrimination and to provide a proper descriptive framework for them. Confronted by a wealth of material that arouses his enthusiasm, he painfully suffers the need to exclude favorite articles and even greatly admired authors. When, as with this volume, the purpose is the compilation of articles to augment a course or a textbook that, no matter how good, cannot fully encompass the subject, he must not only know what the students can assimilate but also what will stimulate their curiosities and interests. While he properly seeks to include differing approaches to the topic and a variety of opinions to stimulate the student to puzzle and cogitate, the articles as a group must be reasonably cohesive to avoid fostering hopeless confusion by their disparity. Though the compiler's task is arduous, it is also beneficent and creative. He relieves the student of a laborious search through the mass of literature, directs him to the significant, and presents it in a sequence that guides learning. He provides a solid foundation upon which further reading can rest. He has the pleasure of introducing the student to the gems that delight and excite, and occasionally even dazzle.

Professor Sze set himself the task of providing social work students with supplementary readings for a course or text on the psychosocial approach to the human life cycle, a topic that must everywhere form one of the cornerstones of the education of a social worker. Confronted by countless contributions of pertinence from a variety of disciplines, he has succeeded admirably. He has provided a feast of unusual quality, for it is substantial and still epicurean. Here are many of the names of those who shaped our current concepts of man as a psychosocial being: Anna Freud, Karl Menninger, Hans Selye, H. S. Sullivan, Erik Erikson, Robert Coles, Kenneth Keniston, Lee Rainwater, Alice Rossi, Rene Spitz, Nathan Ackerman. . . . And here are articles that have had an enduring influence upon the way we think: Spiegel's "The Resolution of Role Conflict in the Family," Masters and Johnson's "Counseling with Sexually Incompatible

Marital Partners," Erikson's "Memorandum on Youth," Clark Vincent's "Familia Spongia." The material is excellent — not just informative, but inclusive of some of the seminal articles in the field which are often a pleasure to read for their style as well as for their wisdom.

The articles have been selected to clarify four major segments of the life cycle — childhood, adolescence, early and middle adulthood, and old age — with a series of selections that focus on the interpersonal, intrapersonal, social, and cultural influences affecting each period. An introduction to each of the four major sections provides guidance to understanding the period of life, the divergent views expressed in the articles, and the reasons for their inclusion. In shaping his book to the needs of the social work student, Professor Sze has wisely limited the contributions concerned with the biological and physiological foundations of human adaptation which have only limited relevance to the students' interest and for which they are likely to have meager preparation. However, the omission of such considerations prevents the formulation of a properly integrated holistic understanding of human functioning and the clarification of the particular importance of social and cultural forces in human adaptation. To round out the picture, it may be useful to note, as I have done in my book *The Person,* that it was the evolutionary development of a neuromuscular system that permitted our hominid precursors to use tools, and particularly that tool of tools, language, which eventually enabled man to spread out over this globe and become its master, changing nature to meet his needs. Through the acquisition of language, man could fragment his past and draw converging lines through the present to project an imaginary future. Through creating a future toward which he could plan, he could consider alternatives and go through imagined trial and error before committing himself to irrevocable actions. He became freed from motivation by immediate impulsions and drives alone, and from learning only through conditioning. Because he could use language, his capacities for collaboration with others increased enormously, and he could even collaborate across generations and centuries. Whereas all animals learn and teach by direct example, he could transmit what he learned to subsequent generations, so that each grouping of mankind gradually built up techniques for coping with the environment, as well as codes for living together, which we term a culture and its instrumentalities. Even though each human comes into the world as naked as all other animals, and with a physical makeup virtually unchanged in the past 30,000 years, he is heir to a cultural heritage that he assimilates from those who rear him, and without which he would be closer to his pre-Stone Age ancestors than to an astronaut.

Man's physical makeup establishes many other determinants of the human condition which we cannot afford to forget or neglect as we focus on the environmental forces that help shape his behavior and provoke conflicts within him. His prolonged helplessness and his even longer dependency require that he be raised by others to whom his well-being is a primary concern. He grows up in relation to parental figures, and he will never be free of them, even when in isolation. He requires a family, or some planned substitute for it, to nurture him, structure his personality, and enculture him. Everywhere the process of differentiation from the mother to become a boy or girl member of the family constitutes a cardinal developmental task; the relatively late onset of puberty with its upsurge of sexual impulsions sets other requisites and helps pattern all lives. Then, man alone is confronted by the cognizance that he will age and die, and that he can choose whether he will live or die. Such knowledge, together with his awareness of his inability to control the contingencies of existence, leads him to use ritual and religion as a means of gaining security for himself and those who depend upon him as well as those on whom he depends. These are but a few of the countless ways in which man's physical makeup influences his psychosocial life. It is easy to overlook some of the most essential determinants of behavior, because as they are essential they are omnipresent, and like the air we breathe can simply be taken for granted.

I am indeed pleased to have been asked to help introduce this fine selection of articles, which not only will enhance and lighten the teaching of social workers, but will surely find its place in augmenting the education of persons in many other fields.

Theodore Lidz

Preface

In recent years, students of the behavioral sciences have increasingly come to believe that human behavior can no longer be adequately explained by dichotomization into two exclusive intellectual camps, psychological and sociological. It is further realized that an adequate understanding of the dynamics of human behavior requires that it be viewed in a broad framework. This trend of thought stimulates many applied behavior fields to a re-examination of their knowledge base and theoretical foundations, and spurs the search for a unified theory of human behavior.

To work toward a unified theory of human behavior is a critical objective of all students of the behavioral sciences. Such a high level of theoretical attainment requires rigorous conceptualization and experimentation. In fact, before such an endeavor can be undertaken, there is a great need for cross-fertilization among various behavioral disciplines.

In many universities there is an increase in interdisciplinary teaching for the purpose of filling some of the knowledge gaps within the traditionally defined academic disciplines. It is especially important for social work students, who must be prepared to enhance the well-being of individuals by means of individual counseling, environmental change, or both. The objective of social work is human service, and the means by which this objective is accomplished are rather broad in scope. It is therefore imperative to be well acquainted with as many social constraints and individual determinants affecting the individual's functioning and well-being as possible. Such a broad base of theoretical understanding will lead to better formulation of comprehensive social policy, more effective service to the people, and more accountability in psychosocial counseling.

In light of the above, this book attempts to provide a general theoretical scheme for pursuing a broad knowledge base in human functioning. The scheme suggests that a person who lives in an urban-industrial society is inevitably confronted by various social constraints from his environment and psychological forces within himself. These variables will be viewed as forming a continuum. Considering first the horizontal dimension, we see that at one extreme the individual is struggling with his physical and psychosocial growth and development tasks, and at the other extreme he is enmeshed in and dependent upon the societal structure. In between, there is on the one hand the element of interpersonal relationship, which is vitally important to the individual's functioning, and on the other hand there is cultural variation, which exerts equal impact on the individual. The general theoretical scheme is shown, in part, as follows:

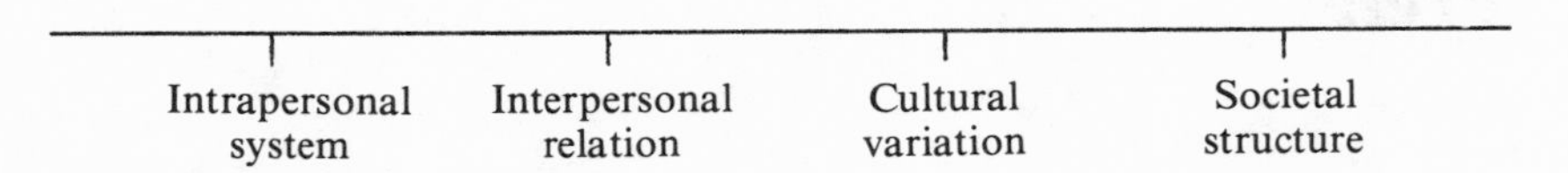

Obviously these constraints and determinants do not operate equally for everyone at any given time, or in the same degree at all times for a given individual. They vary with the individual and with circumstances. This dynamic nature of human behavior makes its study both interesting and difficult. Hence a student searching for determinants of an individual's behavior will have greater success if he has a full range of knowledge of those variables that we know exercise great effect on an individual's functioning.

Since these variables are basic determinants of human behavior, and since each of them exerts its influence on the individual at various points and in varying degrees throughout his life span, the general theoretical scheme further suggests that there is reason to believe some determinants or constraints have greater effect at certain stages of the life cycle than at others. When we view the following diagram on the vertical dimension, it suggests that specific tasks vary in each life stage. That is, (1) psychosocial development tasks are dominant during the periods of childhood and adolescence; (2) family life and work are the major tasks of early and middle adulthood; and (3) behavioral adaptation is the focal task of old age. The broken lines on the diagram illustrate the kind of interconnectedness between the horizontal dimension variables and the vertical dimension variables.

This scheme not only illustrates the interrelationships among variables, but also makes it possible to select readings in the most relevant areas. Thus information from an immensely rich and diversified field of knowledge can be presented in a logical sequence, enabling the student to grasp its essence quickly and form a meaningful conceptual linkage for his professional application.

On the basis of this scheme as an organizing priciple, the life stages are utilized as a backdrop against which various theoretical explanations of individual functioning can be played out. This enables us to illustrate the relationships of the four variables affecting human functioning.

In keeping with an interdisciplinary approach, many theoretical orientations and knowledges that traditionally belong to various discrete academic disciplines have been woven into the framework of this book. Therefore, it is expected that readers will have some basic background in the fields of cultural anthropology, psychology, and sociology in order to reap the full benefit of various points of view.

This book is organized into four major parts, each representing a distinct period of the life cycle. Discussion of each period is organized around four behavioral determinants: intrapersonal system, interpersonal relations, cultural variations, and societal structure. At the beginning of each major part an introductory essay gives a general overview and discusses the issues related to the articles that follow.

Each article included has individual merit, and when several articles with their trenchant viewpoints are synthesized through the cybernetic human

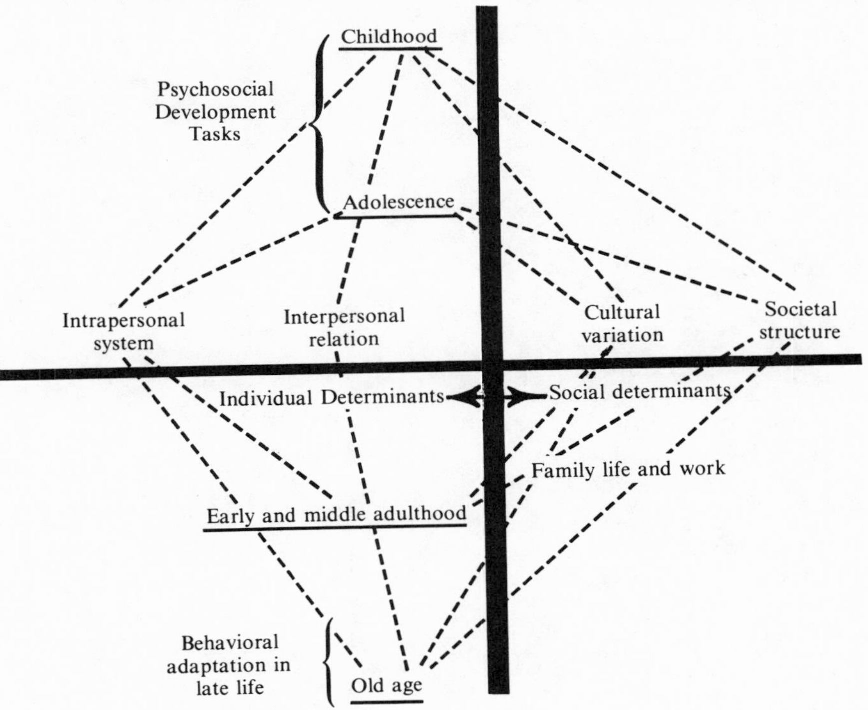

brain of the reader, a sense of new insights and appreciation of the synergetic knowledge of human dynamics emerges. If this is evident to the reader, this volume has then accomplished its purpose, and the credit for whatever virtues it has is due to the contributors for their outstanding work, which has been the seed of the fruit that we all share.

Contents

Human Life Cycle

PART I

Childhood: An Introduction to the Determinates Affecting the Childhood Stage of the Life Cycle

Many theories of child development and behavior have been generated in recent decades. Most child development theories can be generally grouped into four major theoretical orientations: cognitive, psychological, sociological, and developmental; the last approach includes aspects of the other three. Each theoretical orientation naturally has its own focus. The cognitive theorist investigates the capacities of the newborn child — how much intellectual capacity he has and by what processes his intellectual functions are developed. The psychological theorist, especially from the psychoanalytical school, asks how one's psychic apparatus is formed and what are the adaptive mechanisms that affect his behavior. Social theorists focus on the extent to which the child's behavior is influenced by his milieu and social systems — the socialization process, peer influence, social structure, and so on — and how these phenomena come about and are experienced. Developmental theorists are concerned with all these questions, in addition to the child's physical, perceptual, and motor growth and development. This inclusiveness of the developmental orientation has enhanced the general understanding of child development and behavior, but a really comprehensive theory is still in the future. As Paul Mussen and his associates state: "A complete theory would have to include explanatory concepts accounting for the origins, as well as the mechanisms of development and change, of all aspects of psychological functioning — motor, cognitive, emotional and social. It will be impossible to construct such an ideal theory; certainly no one has accomplished it yet" (1).

Nevertheless, each theoretical orientation has contributed its unique perspective and relational dimension to a comprehensive view of child development. The sociological theorist is more concerned with the relationship of man to environment, and of man to man, than is the cognitive theory which concentrates on the relationship of man to his cognitive development. The psychoanalytic theory originally rested on biological foundations but the modern analytical school has moved to interpersonal dimensions as well. In this book, for the best understanding of human behavior, especially the childhood phase, I believe that it would be most appropriate to bring all three relational dimensions into view, so that the intrapersonal, interpersonal, and societal factors can be better understood.

Despite differing viewpoints as to the relative emphasis to be placed on personal, interpersonal, cultural, and societal factors at each developmental phase, and on their interrelationships and contributions to the outcome of human functioning, the various theoretical approaches all agree on the vital importance of early experiences to the formation of personality. Cognitive theorists (2) feel that to understand adult human behavior fully one must know the developmental perspective during childhood. Psychoanalytic theorists such as Erik H. Erikson contend that the time sequence for accomplishment of developmental tasks is a vitally important factor for healthy personality growth. Erikson borrows the epigenetic principle from biological science to illustrate human growth and development. The principle states that the "organ which misses its time of ascendancy is not only doomed as an entity; it endangers at the same time the whole hierarchy of organs" (3). This means that each child has to follow a developmental timetable, and each time period has a predetermined task to be accomplished and crisis to be resolved. Any failure will make the subsequent task more difficult. The sociological study of personality begins, to a large extent, with study of the social development or social learning of the child. James Bossard and Eleanor Boll (4) suggest the importance in the formative years of social influence on a person's later life. The assumption is that a newborn infant is not a social being, and that real human characteristics are developed through years of socialization and social learning (5). Developmental psychologists hold the same view as to the significance of early phases of development. Mussen, Conger, and Kagan state that "a man may be shy and withdrawn or friendly and outgoing, generous or stingy, independent in his actions or dependent upon others, lazy or ambitious, tense or relaxed, passive or aggressive. All these characteristics are the outcomes of his unique personality development, but particularly they are the result of the intimate experiences of his childhood" (6).

In the absence of a comprehensive theory, the selected articles that follow attempt to provide a knowledge base that incorporates the four major orientations in child development and some of the major factors that affect the individual's functioning in childhood and later life. When these readings are considered as a whole, we will have a much clearer conception and understanding of the most complex and uncertain science of human behavior and child development.

Anxiety is a key element in psychodynamics and is sometimes described as a pathological force in human functioning. In Chapter 1, Sylvia Brody and Sidney Axelrad view anxiety as conducive to ego formation in infancy. The experience of anxiety induces the infant to develop the capacity to anticipate, to prepare for, and finally to discharge dangerous stimuli. The authors suggest several phases in ego development. The earliest phase of response to learning is based on the concept of imprinting; the next phase is the process of socialization through which a sequence of awareness of self and non-self and the mother and non-mother emerge. As a result of this phase of socialization, the infant may prepare for losing his love object, and preparedness for anxiety ensues. Brody and Axelrad further contend that anxiety preparedness lays a base for cognition, and that the infant's need to deal with unavoidable quantities of psychological anxiety is nuclear to the development of the ego.

In Chapter 2, a frequently discussed classic treatise on early childhood development, René Spitz convincingly demonstrates through research findings the importance of constant emotional involvement between mother and infant, especially during the first year. Spitz has found that infants need much human stimulation and contact. Interference with the development of a normal object relationship with a mother figure as a significant, trustworthy person is extremely detrimental, resulting in the retardation of the infant's intrapsychic systems such as ego capacity and perceptual growth, and their locomotor radius.

Cognitive development has always been an interesting topic to behavioral scientists. No one has surpassed Jean Piaget's contribution to this subject, and few have translated and developed Piaget's work as well as John Flavell. In Chapter 3 Flavell discusses the genesis of cognitive functioning on the basis of Piaget's work. He defines intelligence and distinguishes between biological structure and cognitive structure, which comes into being only in the course of development. The cognitive structures are the foundation of intelligence and are described as organization and adaptation. Further, adaptation has two subproperties: assimilation and accommodation. An act of intelligence in which assimilation and accommodation are in equilibrium constitutes an intellectual adaptation or intellectual functioning. This article provides a clear understanding of a difficult but basic subject of cognitive development.

Interpersonal interaction during infancy is a critical process for the child's growth and development. Not only does it affect his later interpersonal relationships and socialization; it also affects the infant's intrapersonal system development. Brody and Axelrad deal with the effect of interpersonal interaction on ego formation; Spitz treats the need for object relationship; and Flavell deals with the effect on cognitive growth of adaptation to the environment. These authors all point out the need for functional interpersonal relations even in the very early stages of development, and the necessity for establishing a healthy intrapersonal structure. Therefore, interpersonal and intrapersonal development may be thought of as two sides to the same coin.

Harry S. Sullivan, the eminent interpersonal process theorist, has made a significant contribution to the theory of interpersonal relations with his concept of personification, which is central to his theoretical development. In Chapter 4 Sullivan suggests that interpersonal relations between the mothering figure and the infant result in a permanent personality structure by affecting both the infant's view of himself and his view of the world outside himself. The relationship of the infant to his mothering person at this critical stage of development, her tenderness and rewarding behavior to him, alleviate his anxiety and increase satisfactions and pleasure. These rewards are first experienced in the presence of the parent figure, and form the basis of future relationships to other significant persons by extension. Just as important is the effect of mothering on the infant's view of himself. High-quality mothering, giving the infant the experience of increasing satisfaction and lessening anxiety, will lead to what Sullivan refers to as a "good me" personification. More negative mothering, characterized by experiences that raise anxiety and tension, result in a "bad me" personification. Mothering characterized by anxiety-laden experience and emotional turmoil far beyond normal gives rise to the "not me" personification. During development, the infant's view of the world outside himself and his view of himself are equally important, and their interaction gives rise to the personification of the infant.

In Chapter 5, Anneliese Korner raises a provocative question that has been neglected in the past: Is mother-child interaction a one- or two-way street? Since most Western societies are child-centered, the parents are generally held responsible if their offspring do not develop normally and adequately. Korner attempts to point up the fallacy of this assumption by drawing attention to the variations that can be observed among children, their innate differences in drive organization, in ego functions, and in maturational rates. Therefore, Korner concludes one cannot expect to have a single mode of child rearing.

What are the optimal conditions for the achievement of maximal cognitive and psychosocial development in the child? The importance of the type and quality of interpersonal relations experienced by the child is universally recognized, but there is room for much speculation as to the best means of nurturing interpersonal relationships of high quality. Bettye

Caldwell, in Chapter 6, discusses some of the most basic assumptions of child care principles. Is mother care better than institutional care? Can other types of mothering be substituted for mother-child interaction? Is a two-parent family structure essential for providing optimal child growth and development? Caldwell believes that such traditional child care principles deserve further examination and that alternative models should be considered in view of modern living conditions and social change.

The melting pot concept has been highly regarded in America's social development. Recent social movements, such as black militance and the Indian revolt, make full realization of this concept appear more remote than ever before. Although immigration has slowed to a trickle, cultural divergence remains, with some types of cultural groupings emphasized by economic and social forces. For some who think certain types of culture are backward and unsuited to contemporary social needs, the terms "culture of poverty" and "culturally disadvantaged" become a conceptual reference in analyzing social problems. For instance, Daniel Moynihan (7) attributes the existing disadvantageous position of the Negro in the United States in part to the Negro family structure. One significant feature of the Negro family, according to Moynihan's controversial statement, is its matriarchal system, which he feels has a detrimental effect on the children's ability to adapt in a patriarchal society. Although the validity of Moynihan's claim needs further investigation, the question of cultural influence on human functioning is still very much alive. In Chapter 7, Lee Rainwater's rigorous analysis of the subculture of the Negro lower-class family has yielded illuminating facts about the unfortunate consequences of subcultural influences upon this segment of the American population. He shows that cultural disadvantages impinge upon those who live through their childhood in an environment that provides neither meaningful identity nor rewarding social experience. American blacks are boxed into a subcultural environment that is vastly different from predominant middle-class culture and exhibits much hostility and destructiveness in everyday life.

Catherine Chilman, in Chapter 8, points to the need for studies that will improve the knowledge of behavioral characteristics of lower-class children. She has compiled a list of child-rearing values and behaviors that differ between very poor families and middle-class families. On the basis of her findings, she further suggests that intervention steps can be taken so that certain advantageous middle-class attitudes and behaviors can be incorporated into lower-class child-rearing practices.

After decades of segregation in the deep South, these practices are finally undergoing great change. To what extent this tradition in the midst of change has affected children is little known. Racial struggle has been intensified, especially in the last few years, as school desegregation has come into effect under the federal civil rights law. Chapter 9, written by

Robert Coles during the turmoil of school desegregation, illustrates some means of psychological adaptation used by black and white younger children and their parents to adjust to the racial tension. Coles has also discussed how those children, black and white, face desegregation in their respective lives and especially their coping capacities and defense mechanisms. Coles observed that in general most black children can tolerate the stresses without signs of emotional disturbance. Perhaps, as the author asserts, it is largely a result of their life experience, in which stress has played a large part. The white child's reaction to this stress, on the other hand, depends on his family background, his age, his own nature, and so on. Throughout the study, Coles notes deep feelings of doubt and mistrust on the part of black children toward their white counterparts, feelings much less often noted in white children toward their black counterparts.

Each type of society has its unique advantages and disadvantages for child growth and development. A primitive society has a simple structure and its demands and expectations are few. For instance, Margaret Mead's (8) observations on the lack of neuroses among Samoan children is relevant here. On the other hand, disadvantages are also apparent. In some primitive societies, the infant mortality and morbidity rates are high and the individual's future social mobility and opportunities are very limited. The highly industrialized society also has benefits and problems. Every human group is established within a very complex social system. The functioning of an individual is interwoven with various subsystems under which he lives, such as the family system, educational system, vocational system, and political system. Moreover, each system is in flux, requiring great human adaptability.

In Chapter 10, Urie Bronfenbrenner has eloquently presented this point in his discussion of the differences he perceives in Russian and American child-rearing practices. He discusses the various ways in which children in modern society develop their characteristics. He feels parental influence still exists, but to a lesser degree than in the past. The school environment is important to intellectual achievement not because of good school facilities or teacher-student ratio, but because of the importance of peer group influence. Bronfenbrenner points out that when a moral principle is involved — classroom cheating, for instance — peer influence is the significant factor for American children but not for Russian children. He further explains that a major goal of the Soviet educational process, beginning in the nursery, is to forge a healthy, self-sufficient collective, which in turn has the task of developing the child into a responsible, altruistic, and loyal member of a socialist society. That is certainly not the objective of the American educational system. He further suggests, on the basis of Muzafer Sherif's recent experimentation on group influence, that children and adults should become increasingly involved with each other through some meaningful superordinate goals, regardless of age, sex, and racial differances, thus turning the social system away from its trend toward alienation.

Chapter 11 illustrates one of the social systems which has had great impact on some children, those who for various reasons have been placed under the management of the child welfare system. Henry Maas points out the deficiency of current prevailing child welfare policies and practices directed toward ensuring the well-being of underprivileged children. He calls for rethinking current child welfare policies and urges prevention with a better social environment. There are immediate (or proximal) environments, including the mothering person, and more remote (or distal) environments, including social classes and the culture of poverty. Maas is concerned with both types of environment and their relationship.

What is a psychiatric emergency? According to the most commonly used definition, a psychiatric emergency is the situation of "any individual who develops a sudden or rapid disorganization in his capacity to control his behavior or to carry out his usual personal, vocational and social activities" (9). This definition emphasizes the suddenness or rapidity of changing symptomatic behavior, but on further examination it proves to be inadequate, especially in psychiatric emergency situations among children. In Chapter 12 I examine children's psychiatric emergencies in relation to five variables: behavioral adaptation, alleviating process in a social system, tolerance level in a social system, and labeling and control processes. I conclude that studying these variables and analyzing the related processes are of value in understanding children's psychiatric emergency situations, which often evolve with the interplay of social processes. The complex forces and processes resulting in the child's psychiatric emergency are far more complex than one normally assumes. Any solution, therefore, can be arrived at only after careful consideration of such variables as those examined in Chapter 12.

References

1. Paul H. Mussen, John J. Conger, and Jerome Kagan, *Child Development and Personality*. New York: Harper & Row, 1956, p. 16.

2. John H. Flavell, *The Developmental Psychology of Jean Piaget*. New York: Van Nostrand, 1963, p. 16.

3. Erik H. Erikson, *Childhood and Society*. New York: Norton, 1950, p. 61.

4. James H. S. Bossard and Eleanor S. Boll, *The Sociology of Child Development*. New York: Harper & Row, 1966, pp. 5-6.

5. Frederick Elkin and Gerald Handel, *The Child and Society*. New York: Random House, 1972, pp. 5-6.

6. Paul H. Mussen et al., *Child Development and Personality*, New York: Harper & Row, 1956, p. 6.

7. Lee Rainwater and William L. Yancey, *The Moynihan Report and the Politics of Controversy*. Cambridge: M.I.T. Press, 1967, pp. 51-60.

8. Margaret Mead, *Coming of Age in Samoa.* New York: Morrow, 1928, p. 122.
9. Raymond M. Gasscote, *The Psychiatric Emergency: A Study of Pattern of Service.* Washington, D.C.: American Psychiatric Association and National Association for Mental Health, 1966.

A. INTRAPERSONAL SYSTEM

CHAPTER 1

Ego Formation in Infancy

SYLVIA BRODY and SIDNEY AXELRAD

In recent times the ego has been said, more or less loosely, to be defined by its functions. The statement is useful for descriptive purposes, but it leads away from an understanding of processes involved in the development of the ego as a psychic entity. To us it is crucial to make use of the stricter definition, and to say that perception, motility, memory, reality testing, abstract reasoning are forms of behavior which come under the jurisdiction of the ego. Despite all the advances of modern ego psychology, the ego itself is most usefully to be understood, according to Freud's definition of 1923, as a coherent organization of mental processes, i.e., as a state of psychic organization which makes possible mediation between the demands made upon the organism by the instinctual drives, by reality, and by ethical values. We understand ego *formation,* then, as the development of the cumulative process by which the control of functions necessary for that intrapsychic mediation takes place. Since the quality of this control is highly variable within the individual, especially during periods of change and of heightened conflict, the quality of ego functioning in any person will depend upon the strength and the maturity of specific organic structures as well.

It is generally agreed that anxiety proper can be experienced only at a certain point of ego organization. This statement is, however, too vague. It leaves out any description of the perceptual conditions necessary for recognizing a danger and for recognizing the affective experience of anxiety as happening within a self. We still need to clarify by what means the inner recognition takes place, and what distributions of energy may be required for the affective and the cognitive experiences *per se.* An understanding of those distributions demands a prior insight into certain more elementary functions governed by the ego: object recognition, at least a vague body consciousness, or a vague awareness of sensory responsiveness and motility, and perhaps even some archaic and amorphous notions of cause and effect — some precursors of a secondary process. To a large extent this paper is devoted to an explication of these processes.

Our main proposition is that the emergence of the affect of anxiety and the beginning of ego formation take place in conjunction with one another, and that the two events flow out of a joint process. It appears to us reasonable to suggest that physiological *arousals* promote the exercise of perceptual and motoric structures, and that perception of the tension states thus aroused is reacted to with affect. Cognitive and affective recognition continue, in concert, to promote normal development of ego functions. This proposition does not merely construe for infancy a situation analogous to phases of childhood, when anxiety is frequently a spur to ego development or to mastery of instinctual impulses. The proposition is concerned rather with the inherent quality of anxiety, and with the distribution of cathectic energy which gives it prescience and truly seats it in the ego. A tentative summary description of these psychic events of early infancy would be: Anxiety is one of the principal affects with which the ego is ushered into being; as the infant organism perceives some dystonic condition in his own body and perceives as well his own immediate, involuntary response to the condition, the ego is born and exercises a primary function.

A number of psychoanalysts have made use of empirical findings from physiology and psychology about maturation and development during infancy and transposed the findings into a framework of psychoanalytic theory concerning, for example, ego-id differentiation, early object relations, and anxiety. The psychoanalytic statements resulting from the transposition have been expressed globally, for the most part. But as more and more findings relevant to psychoanalysis appear, because a number of behavioral sciences have concentrated on aspects of early development, we need to examine those that may have effects upon metapsychology.

Carmichael (1951) assembled and reviewed evidence showing that the responsiveness of the foetal organism changes with changing maturity. One same stimulus may evoke different though not necessarily maturer reactions at different points of development, but each response, when it occurs, is specific and not chance. The "disappearance" of certain foetal responses is readily observable in normal neonates, whose changing levels of consciousness and states of need or of alertness affect their immediate responsiveness. Carmichael states clearly that each response that does occur is specific, and is traceable to specific qualities in the stimulus and in the foetal organism; that at any point of growth the non-appearance of a particular response may only indicate a quiescence, not a non-existence, of the related functions; that functional capacities may be demonstrated well before the given function can take a useful role; and that development of many kinds of response proceed with a varying rhythm.

To state the existence of a relationship between individual behavior and specific elements in the total situation in which behavior occurs would appear to be a truism, especially when in the total situation we include both internal and external environments. Applied to the behavior of an infant, this means that while an infant's response would be restricted to what he is so far capable of, organically and psychologically, the response is not to be considered random or inconsistent. It is very difficult to believe this if one makes only short or random or inconsistent observations of neonates, and it is extremely frustrating to measure responsiveness on short order. But an assumption of randomness in infant behavior would imply that determinism operates only after an initial developmental period, and that events of early infancy are as insignificant for later development as the events of childhood were long thought to be. Nissen (1951) has put the problem aptly in his comprehensive discussion of phylogenetic comparisons; where the situation is a relatively open one and provides opportunities for varied activity, the behavior may be called random, but it is never truly so, for it is always determined by environmental conditions plus the perception of them by the infant.

From Pratt's studies (1954), probably the most comprehensive ones available on the subject of neonatal development, we take one finding to be of main relevance here. It is that the greater the duration and intensity of an external stimulus, the more it may absorb the attention of the infant. This may appear too obvious to be of special importance, but it bears major implications. These are: that a series of long-lasting and intense stimuli are likely to have significantly different effects from those of short and weak ones; that long-lasting and intense inner stimuli may at a certain point of duration and strength be sufficient to block out responses to outer stimuli, and vice versa; and that the whole balance of pressure from inner and outer sources is likely to be experienced with wide individual variation at successive stages, and especially at the earliest stages of greatest vulnerability to stimuli of any kind. An assessment of what constitutes long or short duration, with lesser or greater intensity of the stimulus, involves examination of patterns of individual capacities for discrete forms of activity, according to organic possibilities and according to the conditions under which various functions can be evoked.

The literature describing experimental observations of neonatal behavior has been gathering rapidly, and we select for reference only those most immediately relevant to our theme. The best examples differ from those in older literature in two main ways: they report experiments with more subtle and more complex stimuli, and they take greater account of something called "state." The significance of "state" in evaluating neonatal responsiveness has been made explicit by a large group of investigators,[1]

[1] Brazelton (1962), Brown (1964), Escalona (1962), Fish (1963), Graham, Matarazzo and Caldwell (1956), Paine (1965), Prechtl (1964), Wolff (1959).

who have offered evidence that differences in "state," observable in the first days of life, affect the neonate's freedom to distinguish external stimuli and to adapt to them.

Considerations of "state" involve complex issues. The term has been used with different meanings, for different purposes, and with varying assumptions in different contexts. Fries (1935, 1944), when she spoke of activity level, anticipated in behavioral terms the phenomenon referred to by most investigators as "state," but she did not mean quite the same. If one uses "state" or a series of "states" in her sense of congenital activity type, or if one explicates a total set of conditions, possibly even including maturational level, as Escalona (1962) does, many more assumptions have to be made than if one uses the concept of "state," as we do, to describe the gestalt of the infant's physiological equilibrium at a given moment. The gestalt is most readily observable in the infant's degree of arousal at a specific time, or in the degree of tension that accompanies his responses to a specific stimulus at a specific time.

In studies of recent years in which "state" has been taken into account increasingly, infants have been observed to respond to external stimuli much earlier than had been supposed, with varying degrees and kinds of sensorimotor activity, and according to immediate conditions of hunger or satiation, sleepiness or wakefulness, as well as according to native strength and level of irritability.

Without offering details of the reported studies of neonatal and early infant behavior, we summarize some of the main understandings acquired in the last decade.

All findings indicate that once close scrutiny is applied to infant responsiveness, broad descriptions or their behavior as "diffuse," "non-differentiated," or as merely "tension reducing" or "pleasure seeking," all fall short of the mark as too hasty generalizations. In sum, the findings derived from empirical studies of early infant responsiveness are that: (i) Foetal and neonatal responses are never merely random; they always have pattern and direction and effect. (ii) The greater the duration and intensity of the stimulus, the more it may absorb the attention of the infant. (iii) Specific inter-individual and intra-individual variations of sensorimotor responses, in a number of modalities, are observable from the first days of life. (iv) Patterns of sleep in which neurophysiological discharge occurs rhythmically begin in the neonatal period and appear to occur less frequently as life proceeds; but the occurrence must be accounted for in evaluating the effects — possibly reciprocal effects — upon internal arousals of the neonate.

States of alertness and drowsiness are well known to merge and interchange in the first weeks of life. It has seemed to us reasonable, especially in view of many experimental and clinical findings such as have been alluded to here, to consider that the original, transient sensory experiences of the infant of this age, and specifically the first occurrences of the sights and sounds that follow birth, are sharp physiological "experiences." Even though the capacity to respond to separable elements of these "experiences" may be traced back to foetal life, they impinge upon the organism under absolutely unfamiliar conditions.

The conclusion follows that there are structures available to the very young infant for the recognition of sensations, for the assimilation of sensations into perceptions, and for the organization of percepts into engrams. All of the latter processes would bear gradually upon the infant's ultimate capacity to bind and to organize the continuing flow of tension mobilized in growth and development. This implies that psychic determinism can be demonstrated from the moment of birth. The gaps in our present knowledge about the beginnings of psychic structure make it at least premature, we think, to assume that elements of an entity which later on we call psychic structure do not exist in the first days of life. We really do not know the duration of a totally undifferentiated phase.

We conclude that it is reasonable to deduce that the larval elements of an archaic form of mentation are present in the neonate and young infant. These would be the elements requiring the smallest degree of inhibition of energy. They may be conceived of as forming the prestages of primary process. They are far distant from the kind of mentation present in even the older infant of six months; but they constitute the precursors of psychic structure. They are stimulus specific, hence internally uncontrolled, but later they enter into the control system.

A degree of tension is obviously necessary to motivate any behavior. An infant who has experienced optimal degrees of tension, sufficient to motivate activity through which tension may be discharged economically and pleasurably, may be expected to have optimal amounts of energy at his disposal for sensorimotor explorations, for the attainment of gratification from his alertness, and for the development of his curiosities. In contrast, an infant who has experienced unusual stress, with hypercathexis of pain, sensory discomforts, or sensory confusion, may be expected to be left with a reduced readiness to perceive external stimuli and to organize perceptions. We should not be surprised to find an inverse relationship between cumulative physiological stress and the capacity for object perception leading to object cathexis.

But for the present we are limited to dealing with one question: the conditions under which sensations in the infant's body, arising internally or aroused by external stimuli, begin to be perceived and to leave memory traces that may affect the immediate progress of his general psychic development and of his anxiety potential.

Human Imprinting Processes

A number of experiments in imprinting during the last decade touch upon specific connections between sensory and emotional arousal and adaptive behavior in animal infancy. While we anticipate only indirect analogies and comparisons between animal and human infant development or behavior, not alone because of the obviously complex differences between the species, but also because of the additionally complex variables in the imprinting response itself, a number of findings are to be considered because they suggest comparable processes of socialization among human infants.

Among the salient studies are those of Hess (1959a, 1959c), who, like Lorenz, distinguished imprinting from later learning in that it can occur only during a physiologically critical period in early life, and in that it is acquired through an innate releasing mechanism rather than through training. Hess describes it (1959a) as a rigid form of learning of the rough, generalized characteristics of the imprinting object. Furthermore, Hess concluded that the onset of fear ends the critical age for imprinting, and that association learning begins immediately after the peak of imprintability; and he deduced a Law of Effort (1959a, 1959b), i.e., the strength of imprinting equals the logarithm of the effort expended by the animal during following.

Pitz and Ross (1961) saw imprinting as a function of a degree of arousal, and defined arousal as the total amount of stimuli impinging upon the organism. In view of the fact that the period in which the heightened arousal was momentary, Pitz and Ross suggested that maximal imprinting, like other kinds of learning, requires an optimal level of arousal beyond which performance declines; and that in addition to the expenditure of energy which according to Hess facilitates imprinting, something more general is necessary, such as a certain amount of central nervous system activation. We anticipate that the same condition holds for an imprinting process that takes place in human infants.

Imprinting is of obvious survival value for the species as well as for the individual. It makes mothering possible in species where the infant organism must take an active role in being mothered. It also sets one of the conditions for the choice of a sexual partner. The fact that extra-species objects can be imprinted does not speak against the survival value of imprinting. The fact that imprinting is a special form of learning does not speak against its biological function, for statistically, during the critical period and in a state of nature, the imprinting object will belong to the animal's own species. And as various observers have noted, there is no reason to believe that the human species should be exempt from this kind of process.

We should now like to suggest that a number of events believed to be part of a unique ontogenetic development of the human are actually phylogenetic. Imprinting, socialization, and critical periods provide the clearest links.

Gray (1958), defining imprinting as "an innate disposition to learn the parent or parent-surrogate," proposed that the smiling response of infants is the motor equivalent of the following response that occurs below the level of the infra-human primates. He places the critical period for human imprinting from about six weeks to about six months, beginning with the onset of learning, continuing with the smiling response, and ending with discrimination of strangers. Ambrose (1963a), probably correctly, does not accept Gray's reasoning, but agrees that smiling may be a human form of imprinting, and he equates following with smiling. According to Ambrose the infant's smile, along with the mother's loving maternal behavior, makes for greater closeness between mother and infant. He suggests that the smile is dependent upon the ability to fixate upon the face of the mother. Earlier Ambrose (1961) had found that the ability to fixate the mother's eyes begins at approximately the fifth week of life; and that the onset of discrimination of strangers begins at about 12 weeks for family-reared infants, and at about eighteen weeks for institutionally reared infants. This span of weeks appears to Ambrose to be one of many sensitive periods for the development of social responsiveness of the human species.

If, as seems likely, an imprinting-like phenomenon exists among human infants, the infant's smile appears to us to be too mature a response to be regarded as that phenomenon. Our contention would rather be that there are two major critical periods of socialization among human infants: the first, in which there is imprinting to the human species in general, may be signified by the ability to fixate, visually or auditorily, as observed by sharp increases or decreases of body movement (limbs, head, facial features) or of activity level, and by visual pursuit of the human object. The ability to fixate is prerequisite to the ability to smile, in this first period. The second critical period, in which there is imprinting to the mother or her surrogate, may be signified by the ability to discriminate the mother or her surrogate from other human objects. It would be shown by the immediacy off the smiling response to her above others.

The appearance of the smile may be a result of a kind of practice, or work, i.e., an expenditure of effort for an end other than the activity itself. If so, it would be experienced by the infant as a relief of tension. The relief would lie in the achievement of familiarity of a percept: it occurs when no further effort is necessary to attain the percept. It affects a habituation to the hitherto unfamiliar. It is as if the organization of the hitherto amorphous world into patterns and gestalten permits an easy and pleasurable discharge of tension. Smiling proceeds from perception of a mechanically patterned object to a discriminatory response to more and more specific stimuli; i.e., from the most general configuration of the human facial features to the face of the specific mother. The fact that subjectively one has the impression that the first smiles of the infant are the culmination of a process of work provides the analogy to Hess's Law of Effort.

Referring back to the phenomenon of imprinting, we propose that in its initial phase imprinting for the human species in general is likely to be as swift and irreversible as Scott (1963) reports it to be for birds. We suggest that visual fixation upon the human face presented at very close range and especially when reinforced by the human voice, plus visual pursuit movements, may be the human equivalent of following among animals. Human infants can locomote, as it were, only with their eyes in the first weeks of life. Among older infants, the second phase of imprinting, marked by recognition of and smiling approach toward the mother, may indicate a process far more complex, and reversible. A six-week to six-month period of primary socialization, as suggested by Scott (1963), seems too gross. It is in accordance with our observations rather to divide the primary socialization period into at least the two major phases we have alluded to, the first beginning with fixation and pursuit movement, the second beginning with discrimination of the mother *qua* object. Many mothers have reported the latter to occur in the fourth month.

It follows from these considerations that imprinting, or its counterpart in the human species, is a special form of learning, and may involve a hypercathexis of primary percepts; that is, it may involve a perception of objects as not yet apart from the self, and a perception which succumbs to a truer object cathexis at a later point of maturation, when objects come to be recognized as separate from the self and as "interfering" with the object-self unity. Imprinting in the human would thus be seen as a form of object cathexis, perhaps the simplest kind of affective bond, first to the species and then to the mother, subsequently to other parental objects. The "interference" of percepts of external objects, the rise of secondary narcissism, might be viewed in relation to the onset of anxiety arousal and early conflict — a process in some ways analogous to the onset of fear and the ending of imprinting in lower species.

Our purpose in construing these links is to make as discrete as possible the relationships that may inhere among the earliest forms of object cathexis, the dystonic affect of anxiety, and the development of the ego.

Anxiety and the Ego Formation

When in 1926 Freud referred to a certain "preparedness for anxiety," which he said was undoubtedly present in the infant, he did not try to describe what its typical and normative development might be, nor to delineate the point or points at which the preparedness for anxiety became an affective experience. That requires hypotheses by means of which to assess the intensity of an infant's cathexis of the mother or her surrogate, and his threshold for stress when alone and when in need. And it is toward this end, toward an understanding of the elements of anxiety preparedness, that our discussion has been proceeding.

By common psychoanalytic definition it would appear that this affect could emerge only when object perception exists, as only then can a sense of danger from object loss be experienced. Now if the development of object perception follows either developmental lines of other forms of behavior, or the development of the concept of self, then it lacks stability in the period of its initial development in infancy and for variable periods during infancy. We know how far beyond early childhood the image of self and object shift, separate, and merge again. The two discrete percepts may lose their separate unitary qualities under conditions of stress or conflict. Our observations have indicated that the barest elements of that slow process of self-object differentiation can begin to be discerned in the rise and fall of sensorimotor behavior as early as the neonatal period. Either in the infant's own spontaneous or reactive behavior or through the observable medium of the maternal activity, the infant may be observed to increase or to diminish many kinds of activity. Sometimes his movements are arrested entirely, sometimes accelerated, sometimes they are in prolonged flux. At unknown points in the continuum of early experience, differentiations between sensations registered from inside and from outside, vague feelings of the source of stimuli, must begin to occur. The totally satisfied infant, if we can conceive of such a one, would have little need to perceive "reality," and little need to develop mental discrimination between states of tension and states of rest. Freud (1911) referred to the function of attention as instituted to search the external world, and as meeting the sense impressions halfway; and in a footnote he considered the infant's reactivity to distress as signs of the infant's move from the pleasure principle to the reality principle. The unconscious cathexis of particular sensations and the laying down of earliest memory traces would therefore come to serve as instigators of more and more differentiated perceptions, of more and more specific motor discharges and of the capacity for delay, and the capacity to anticipate delay. As Freud (1911) then indicated, all of these would allow at the same time for the further revival of the original gratifying sensations and for the temporary withdrawal from the stimuli of reality.

That there can be withdrawal from the stimuli of reality implies the existence of a protective shield (Freud, 1920). We have further assumed that in earliest infancy a maturation of this shield must take place. All discussions of this shield up to now have devolved upon its protective function. Our purpose is to emphasize, in this context, the receptive function. It does appear true that the infant's physiological defenses *against* distress, his demand for protection, may be dramatically observable; and that his reactions to acceptable, positive stimuli are typically silent and subdued. But although the latter, somato-syntonic, stimuli presumably would play a relatively minor role in the preparedness for anxiety, their registration upon the memory appears to us to be no less significant than the registration of dystonic stimuli.

In view of the endless dovetailing of inner and outer excitations proceeding from the onset of postnatal life, we should expect that the sensory confusion of the infant would be immense before any discrete percepts or discrete memory traces are formed, and that the accomplishment of the "first discrimination" (Freud, 1915) and the "first orientation" would be impressive kinds of infant work. So we should expect that the quality and quantity of the infant's perceptual work leads him toward the psychological experience of being a unit and being alone, and toward the realization that an object that shortly has become cathected is missing. This realization would presumably be thoroughly intertwined with the development of anxiety preparedness.

Considering the readiness of an adult organism in severe physical stress to lose the perception of inner boundaries, and to summon all available energies for physiological defense reactions, we imagine that something very similar may happen to a young infant who is not yet clearly aware of what sensations derive from inside his own body; though with different sequelae, of course. The infant is incomparably less able to locate or to relieve pain when he feels it, and he has not yet developed a psychological attitude toward it. He may be all the more readily overwhelmed by internal sensations, and all the more ready for his energies to be drawn away from perception of the margins between external and internal stimuli.

To the extent that adequate and timely rescues from painful experiences do come, and to the extent that external excitations impinge at a rate that allows for gradually increased cathexis of bodily organs, the protective shield could serve a harmonious adaptation to stress. We observe that some infants are readily irritated to the point of restlessness, squirming, and fretting, and some accept with too little protest the handling to which they are necessarily subjected. We might say metaphorically that in the former the shield is not solid enough or has defects, and that in the latter the shield is too impenetrable and leaves the infant wanting in capacity to receive stimuli and to adapt to them at a moderate pace. The concept of a too thin or a too thick "protective barrier" (as it was called in the translation then available) was proposed by Bergman and Escalona (1949), but solely with reference to over- or underprotection *against* stimuli. We should say that over and above a general degree of solidity in this hypothecated shield, no doubt there would be significant individual variability, structurally and functionally, particularly in sensorimotor areas, making for an individual infant's kind or degree of vulnerability.

The essential nature of the protective shield with which we are dealing is not whether the shield can be pierced by stimuli of a certain intensity — that is a matter of neurophysiological responsiveness. We are concerned rather with the psychological integration and organization of the manifold stimuli which impinge upon the organism — an active, integrative process, rather than a mechanical registration: a process advanced by both the protective *and* the receptive functions of the protective shield. Hypercathexis, which is an active phenomenon but one that can be achieved by rudimentary structures, is the key to this process. It is conceivable that the receptive function of the protective shield is connected to the onset of the secondary process.

These considerations lead us to propose that where *either* internal or external stimuli are too exclusive in their impact upon a young infant, and the protective shield is *either* organically unsound or for dynamic reasons fails in *either* its receptive or its protective functions, there will be an impairment in the infant's ability to differentiate between inner and outer excitations, and in his capacity to experience a balance of passive and active accommodation to stimuli. Consequently, confusion of both sensation and response would aggravate the pain or stress invoked by tension, would increase narcissistic cathexes, would reduce the infant's anxiety preparedness, and would eventually block a fluid cathexis of external objects.

So far we have tried to trace psychic paths by which stimuli may affect the postnatal infant and may reinforce states of either comfortable alertness or discomfort and stress. A consciousness of danger impending from any stress could only result from some disruption or from some noticeable change in the ideal efficacy of the protective shield — when too many discomforts or too few comforts are received. An enlarging consciousness of undiminishing physiological stress (conceivably lasting only fractions of a minute) can be observed to arouse something akin to psychic pain as we know it, i.e., a dystonic sensing of object loss or, more precisely, comfort loss. Full consciousness of the organismic danger entailed by the comfort loss would be achieved in very many small steps, deriving from minimal perceptual cues and gradually giving rise to an affect of anxiety. We should assume that this affect would in turn remain *in statu nascendi* for many weeks. Seen thus, the affect of anxiety comes at the end of a process of a multitude of increases and decreases of stress, and a multitude of perceptions, recognitions, and reactions, all of these psychophysiological experiences becoming crystallized in a capacity to experience signal anxiety.

There would seem to be a direct relationship between the preparedness to deal with physiological stress and the capacity for object perception and reality testing. The longer or more frequently the infant's needs remain unsatisfied, the more intense will the cathexis of inner sensations become. The less physiological tolerance the infant has for stress, the more he may be impelled to fend off new perceptions and to remember instead, albeit with a vague mélange of sensory images, the earlier wish fulfilments. Similar propositions have been brought forward by others (Greenacre, 1945; Hoffer, 1952; Mahler, 1952).

The fact that anxiety is a felt response of the ego generally has been construed to mean that a nascent ego must exist before the affect of anxiety can be experienced.[2] This would seem to be correct if by the *affect* of anxiety we restrict ourselves to the idea of a full-fledged emotion, which we have defined as an awareness of a stirred-up state of the organism. Our construction is that a *process* of anxiety *development* takes place, in which stages of preparedness for anxiety lead through physiological paths to a concomitant emergence of nascent ego functioning *and* of an affect — an emotion — of anxiety.

Phases of Socialization

The propositions that there are discrete stages of preparedness for anxiety and that there is a joint emergence of the affect of anxiety and an ego must be stated in verifiable terms if it is to have more than speculative significance. The developmental steps toward this joint emergence may, we submit, be encompassed in two phases of socialization which influence and are influenced by the preparedness for anxiety.

We hypothesize that the psychological events occur in two early phases of socialization, as follows: the first phase begins with the sensory awakening of the neonate during approximately the first six weeks of life. In these early weeks one can observe that where the infant's attention to environmental arousals is gradual and pleasing, there are correspondingly pleasing arrests, or suspensions, of the infant's motor activity, suspensions that are like little latencies of the impulse life, during which perceptual activity grows apace, and immediately after which motor activity may be spontaneously resumed — producing a rhythm of sensory and motor activities.

Auditory response can quite regularly be elicited. There are now also numerous studies which show that visual fixation and visual pursuit movements occur in the first days, and may be regularly elicited by the

[2]Arlow (1963), Greenacre (1945), Brenner (1953), Kubie (1941), Schur (1953), Spitz (1950), and others.

fourth or fifth week. These reactions depend in part upon the infant's neurological readiness to receive and to be excited by external stimuli. They obviously require no *consciousness* of stimulation or response.

The intensity with which visual pursuit is practiced by any normal infant in the next months is easy to observe. By the third month he spontaneously turns his head to gaze at more distant objects, notices when objects close by are removed, and, what may be of supreme importance for object discrimination, he begins to keep visual attention upon his own hands and seems able to attend to the sound of his own voice. The observable increase of intentional visual searching, inspecting, watching, coordinated with hand and hand-to-mouth movements, occurring in concert with auditory communications, suggest that it is during the third and fourth months of life that consçiousness is attained, and that the first phase of socialization which is most patently marked by visual fixation and pursuit of the human object *qua* species is completed.

An initial hypercathexis or *consciousness* of satisfaction or of its absence, during rhythmic alternations of sensation, perception and response, would appear to constitute the greatest psychological moment in mental development. It seems reasonable to consider that this hypercathexis of sensations and perceptions leads directly to an awareness of a difference between self and object, to object cathexis, even at a most primitive level. The first phase of socialization ends in approximately the third month, and is marked by imprinting of the mother as a species-specific object.

The second phase of socialization, which we estimate begins by the fourth month of life, may, we think, be marked by the infant's discrimination of the mother as an individual, rather than as a member of the species: the mother is the specific imprinting object. Although high social responsiveness may easily be extended to other persons as well, the mother is sought out by eye, ear, hand and voice, and is desired as the infant's primary companion. The ready response to the mother is most observable in the smiling response to her overtures, and the smile as a predictable response to the mother may indicate the onset of a process of discrimination of her from others, a process often evident in an infant's notable curiosity about non-mothers, new persons. Attention to all human objects may contstitute prime positive evidence that specific affects are taking shape, that the overwhelmingly negative reactions are receding, and that hitherto less conspicuous positive reactions are emerging more definitely or more frequently. Consciousness of sensation, of affect, of self, and of object may at this time be attained.

Early in this second phase of socialization, which we may estimate to set in as the fourth month approaches, some capacity for anticipation and delay has already appeared: it has become observable in the quickening of attention to sources of gratification. The infant now seems able to differentiate quickly and sharply between comforting and discomforting stimuli and he shows a "proud" urge for sensorimotor explorations. He does not mind brief disappearances of his mother as long as he finds that he can retrieve her at will, much as he can retrieve the sight of other objects that disappear temporarily — his own hands, or familiar inanimate objects in his environment. Clearly he does not prefer to be alone. Even a most contented infant of about five months will be reported to cry when his mother, after being pleasantly attentive, puts him down in his bed and walks away. He can be comforted by others, but he functions most smoothly in the care of his most familiar mother.

It is usually beginning in his fifth and sixth months that the infant clearly recognizes his mother as the chief object through which gratifications are provided and stresses are relieved. Even a relatively harsh mother appears to acquire the infant's intense emotional investment, and to receive his most immediate shows of delight, anger, protest, or joy. And, in general, the infant who is adequately stimulated by the mother is also quick to reach toward inanimate objects and to explore them visually, manually, orally. He may easily become stimulus-bound, enthralled by life, so to speak, and noisily and physically responsive to it. He has moreover become aware that objects have functions — they can be banged, shaken, dropped, held tight; and as the cathexis of inanimate objects increases, the infant finds himself less threatened by the mother's absence. Now her voice coming from another room may content him for brief periods.

In short, the mother has been endowed with the infant's ambivalent drive derivatives. Her failure to gratify him may now occur even when she is present, and her gratification may be given even while she is not visible (e.g., by her voice). And now other objects, non-mothers, threaten by their very appearance to come between infant and mother: they are regarded with a new kind of caution before they can be accepted. At this point, in which socialization to the mother as an individual has taken place, the preparedness for anxiety may reach a first peak of maturity: the danger of comfort loss is consciously perceived.

The foregoing formulations have of course taken for granted, as a theoretical example, an infant who achieves a preparedness for anxiety through optimal degrees of experience in pleasure and unpleasure. The balance that any infant can achieve in such experiences may be regarded as a significant determinant of the degree of arousal and the duration of physiological stress, and of the further arousals of perceptions. Consciousness of the experiences may set into motion the development of rudimentary ego functions, may inaugurate or literally substantiate anxiety preparedness, and may be the essential component that prepares for the transition from the first to the second phase of socialization. One might think of the presence of this preparedness for anxiety metaphorically, as in an organism's being poised for flight, yet not necessarily taking flight.

We propose that two kinds of sensory arousals, one dominant in each of the two phases of socialization, may occur. The first arousal is automatic, close to an unconditioned reflex, and touches off an innate releasing mechanism. It promotes the onset of the critical period for the first phase of socialization and strengthens the imprinting response to the species. The second arousal occurs when there is a perception of inner tension, and a perception of motor capacity to discharge tension or to summon relief from tension. This second kind of arousal promotes the onset of the critical period for the second phase of socialization, and so for the perceptual consciousness of the mother. It also sets in motion the beginnings of association learning within the second phase of socialization. An essential quality of the cathexis for the mother would include awareness of the possibility of her absence, and a normal preparedness for comfort loss and the affect of anxiety. As a result, the infant might respond actively to felt danger by an impulse to take flight, and this impulse and its discharge would be consciously experienced. It would follow that subsequent to infancy, the degree to which the preparedness for anxiety is augmented or diminished directly affects perceptual consciousness and ego functioning.

This hypothetical description implies that ego rudiments are formed in a course of alternating processes or rhythms, oscillating at first between irritability and reflex response, later between physiological stress reactions and fragmentary sensory arousals, and still later, between psychological reactions or dawning affect and perceptual consciousness. The full first cycle is likely to be attained in approximately the first six weeks of life. It comprises a preparedness for anxiety which remains necessary for survival, and which reaches a maturity in approximately the first six months of life.

Further evidence exists of a parallel between the fear which in the infancy of animals arrests imprintability and begins association learning, and that level of anxiety in human infants in which primarily physiological stress narrows to a psychological stress. Thus, in human infants, perceptual consciousness sets in when physiological stress is superseded by consciousness of unpleasure and of an object that can relieve it. We submit that this accompaniment of stress by awareness leads the infant to a recognition of objects who can be expected regularly to provide comforts and to relieve discomforts. Needless to say, the primary objects may be part objects.

One would expect that among human infants the increasing exercise of perception, responsiveness, and memory would build into the psychic organization strivings toward the further advancement of ego functions, beyond the function of providing signal anxiety. The initial emergence of anxiety as a response to danger would itself be a precipitate of a combined cognitive and affective process, developing hand in hand, and at no point being discrete or separable. Among these processes we should include the capacity for pursuit movements of head, mouth, eye, ear, and hand; of the limbs *en masse* and eventually of one hand; of the vocal apparatus; of the responsive smile.

The readiness for both anxiety with strangers and separation anxiety may be formed in infantile phases of preparedness; i.e., the infant gradually becomes capable of psychological concomitants of stress, first relevant to the absence of the mother *qua* mother. Normally, he retains this capacity to feel a signal that an anxiety-provoking experience is impending, and at best he will develop an increasing capacity to halt the full access of anxiety. The degree to which he attains the capacity in later infancy to receive a signal and to act accordingly will then be determined by a long complementary series of experiences on every psychic level. And the degree and form which the anxiety takes in any infant in the later part of the second phase of socialization, as we have described it, will vary in many ways, for example, as to whether the infant has felt sufficiently or insufficiently gratified by the mother prior to the access of anxiety proper.

It will have become evident that we regard ego rudiments and the affect of anxiety as prerequisite to and part of the process of socialization, and as taking form in developmental steps which, in highly abbreviated form, may be stated in the following propositions:

We submit that for *a first phase of socialization* during the first three months of life:

(i) Sensorimotor responses of a larval nature build recognitions of sensations.

(ii) Consciousness of body functions begins with an awareness of the sensations, the recognitions, the sensorimotor responses, that are taking place and that from time to time may be executed intentionally. The attempt to make intentional movement arouses an element of the capacity for anticipation and delay, and impels the infant toward *further* sensorimotor explorations.

(iii) Advance and control of sensorimotor skills is encouraged by positive stimuli which afford pleasurable affect, and by negative stimuli which bring unpleasure and, optimally, intensify attempts toward mastery. Both kinds of stimuli thus serve the pleasure principle and serve as well the broadening of awareness that sets a base for reality testing.

(iv) Imprinting to the mother as species-specific object is achieved.

For the second phase of socialization, which may begin as early as the third month, and lasts approximately until the seventh month, we submit that:

(i) Rises in physiological tension provide the most visible interferences with the maintenance of pleasure, and promote as well the most visible struggle (optimally, a productive struggle) with real stimuli.

(ii) The central attempt to maximize pleasure and minimize pain brings together an awareness of comfort loss (object loss), and efforts toward autonomous control of sensorimotor skills (partial ego functions, or ego rudiments). In this manner, advancement of ego functions proceeds in coordination with, and is inaugurated with, the formation of anxiety affect.

(iii) Imprinting to the mother as specific object of instinctual cathexis is achieved.

In the course of ego development, the infant faces series upon series of objects that demand the hypercathexis of attention. As a result, and on the basis of biological givens, he must make tiny isolations and at first infinitesimal acts of delay. He must learn to go from the sensations that crowd upon him and the tensions that beset him to the object of the mother. He must learn to transform gross dystonic stimuli to signal anxiety, and move from signal anxiety to the making of appropriate signals to the relieving object, until he can take appropriate action by himself. The first advance requires perception of the mother as a unit of being separate from himself. Later the progression of cathexis is from the mother to the toy (A. Freud, 1960), and still later, from verbalization of perception to verbalization of affect (Katan, 1961). Stated more generally, the capacity to deal with affect and to have normal emotional attachments is prerequisite for abstract thinking even during early childhood (Provence and Lipton, 1962).

Critical periods of development may without doubt be marked off in numerous ways, according to neurophysiological maturation, or to instinctual drive development and affect derivatives, or with respect to object relationships or other metapsychological advances. We have chosen to consider the ways in which ego structure itself may be propelled by the development of the preparedness for anxiety.

Adaptationally, one of the nascent functions of the infantile ego is to externalize sensations so that the infant can summon help from the maturer ego of his mother, or later can take up the task of mastering dystonic stimuli by himself. The accomplishment of delay, of achieving a state in which sensations are held at bay and in which attention can be given to both tension and signal, comes about when both are followed by attention to the means by which, or the object through whom, the tension will be diminished. This constitutes the infantile model of abstract thinking, in which more of the cathexis is upon an object than upon the sensations impinging upon the self.

This discussion of the emergence of the affect of anxiety has led to a re-emphasis of the complex relationships that inhere among the phenomena of stress, tension, defense, and rudimentary ego functions, and that enter into personality formation in successive phases of growth, from the beginning of life.

We have proposed that the affect of anxiety is not only a state or a signal in itself but that it serves as a pilot for the development of other ego functions. In short, we say that anxiety preparedness lays a base for cognition; and that the infant's need to deal with unavoidable quantities of physiological anxiety is nuclear to the development of the ego. Should this set of statements be verified, it is implied that early infantile experiences are of far-reaching and permanent significance for the structuring of the affective as well as the intellectual destiny of the individual.

References

Ambrose, J. A. (1961). "The development of the smiling response in early infancy." *Determinants of Infant Behaviour,* vol. I, ed. B. M. Foss. (New York: Wiley; London: Methuen.)

——— (1963a). "The concept of a critical period for the development of social responsiveness in early human infancy." *Determinants of Infant Behavior,* vol. II, ed. B. M. Foss. (New York: Wiley; London: Methuen.)

——— (1963b). "The age of onset of ambivalence in early infancy: indications from the study of laughing." *J. Child Psychol. Psychiat.,* 4.

Arlow, J. A. (1963). "Conflict, regression, and symptom formation." *Int. J. Psycho-Anal.,* 44.

Bergman, P., and Escalona, S. (1949). "Unusual sensitivities in young children." *Psychoanal. Study Child,* 3-4.

Brazelton, T. B. (1962). "Observations of the neonate." *J. Child Psychiat.,* 1.

Brenner, C. (1953). "An addendum to Freud's theory of anxiety." *Int. J. Psycho-Anal.,* 34.

Brown, J. L. (1964). "States in newborn infants." *Merrill-Palmer Quart.,* 10.

Carmichael, L. (1946). "The onset and early development of behavior." *Manual of Child Psychology,* ed. L. Carmichael. (New York: Wiley.)

—M—— (1951). "Ontogenetic development." In *Handbook of Experimental Psychology,* ed. S. S. Stevens. (New York: Wiley.)

Escalona, S. (1962). "The study of individual differences and the problem of state." *J. Child Psychiat.,* 1.

Fish, B. (1963). "The maturation of arousal and attention in the first months of life: a study of variation in ego development." *J. Child Psychiat.,* 2.

Freud, A. (1960). Four contributions to the psychoanalytic study of the child. Lectures given in New York City, 15 September, 1960.

Freud, S. (1911). "Formulations regarding the two principles of mental functioning." *S.E.,* 12.

——— (1915). "The instincts and their vicissitudes." *S.E.,* 14.

——— (1920). "Beyond the pleasure principle." *S.E.,* 18

——— (1926). "Inhibitions, symptoms and anxiety." *S.E.,* 20.

Fries, M. E. (1935). "The interrelationship of the physical, mental and emotional life of a child from birth to four years of age." *Amer. J. Dis. Child,* 49.

——— (1944). "Psychosomatic relations between mother and infant." *Psychosom. Med.,* 6.

Graham, F. K., Matarazzo, R. G., and Caldwell, B. M. (1956). "Behavioral differences between normal and traumatized newborns." *Psychol. Monographs,* 70.

Gray, P. H. (1958). "Theory and evidence in imprinting in human infants." *J. Psychol.,* 46.

Greenacre, P. (1941). "The predisposition to anxiety." In *Trauma, Growth and Personality.* (New York: Norton, 1952.)

——— (1945). "The biological economy of birth." *Ibid.*

Hess, E. H. (1959a). "Imprinting." *Science,* **130.**

——— (1959b). "Two conditions limiting critical age for imprinting." *J. Comp. and Physiol. Psychol.,* 52.

——— (1959c). "The relationship between imprinting and motivation." *Nebraska Symposium on Motivation,* ed. M. J. Jones. (Lincoln: Univ. Nebraska Press.)

Hoffer, W. (1952). "The mutual influences on the development of ego and id: earliest stages." *Psychoanal. Study Child,* 7.

Katan, A. (1961). "Some thoughts about the role of verbalization in early childhood." *Psychoanal. Study Child,* 16.

Kubie, L. S. (1941). "A physiological approach to the concept of anxiety." *Psychosom. Med.,* 3.

Mahler, M. (1952). "On child psychosis and schizophrenia autistic and symbiotic infantile psychosis." *Psychoanal. Study Child,* 7.

Paine, R. S. (1965). "The contribution of developmental neurology to child psychiatry." *J. Child Psychiat.,* 4.

Pitz, G. F., and Ross, R. B. (1961). "Imprinting as a function of arousal." *J. Comp Physiol. Pyychol.,* 54.

Pratt, K. C. (1954). "The neonate." In *Manual of Child Psychology,* ed. L. Carmichael. (New York: Wiley.)

Prechtl, H., and Beintema, D. (1964). *The Neurological Examination of the Full Term Newborn Infant.* (London: The Spastic Society Medical Education and Information Unit with Heinemann Medical Books.)

Provence, S., and Lipton, R. (1962). *Infants in Institutions.* (New York: Int. Univ. Press.)

Schur, M. (1953). "The ego in anxiety." In *Drives, Affects, Behavior,* ed. R. M. Loewenstein. (New York: Int. Univ. Press.)

Scott, J. P. (1963). "The process of primary socialization in canine and human infants." Monograph. Society for Research in Child Development, 28.

Spitz, R. A. (1950). "Anxiety in infancy." *Int. J. Psycho-Anal.,* 31.

Wolff, P. H. (1959). "Observations on newborn infants." *Psychosom. Med.,* 21.

CHAPTER 2

Hospitalism: The Genesis of Psychiatric Conditions in Early Childhood

RENE A. SPITZ

The aim of my research is to isolate and investigate the pathogenic factors responsible for the favorable or unfavorable outcome of infantile development. A psychiatric approach might seem desirable; however, infant psychiatry is a discipline not yet existent: its advancement is one of the aims of the present study.

The Procedures

With this purpose in mind, a long-term study of 164 children was undertaken.[1] In view of the findings of previous investigations, this study was largely limited to the first year of life, and confined to two institutions, in order to embrace the total population of both (130 infants). Since the two institutions were situated in different countries of the Western Hemisphere, a basis of comparison was established by investigating non-institutionalized children of the same age group in their parents' homes in both countries. A total of 34 of these were observed. We thus have four environments:

Table I

Environment	Institution no. 1[a]	Corresponding private background[b]	Institution no. 2	Corresponding private background
Number of children	69	11	61	23

[1] I wish to thank K. Wolf, Ph.D., for her help in the experiments carried out in "Nursery" and in private homes, and for her collaboration in the statistical evaluation of the results.

In each case an anamnesis was made which whenever possible included data on the child's mother; and in each case the Hetzer-Wolf baby tests were administered. Problems cropping up in the course of our investigations for which the test situation did not provide answers were subjected to special experiments elaborated for the purpose. Such problems referred, for instance, to attitude and behavior in response to stimuli offered by inanimate objects, by social situations, etc. All observations of unusual or unexpected behavior of a child were carefully protocoled and studied.

A large number of tests, all the experiments and some of the special situations were filmed on 16/mm film. A total of 31,500 feet of film preserve the results of our investigation to date. In the analysis of the movies the following method was applied: Behavior was filmed at sound speed, i.e., 24 frames per second. This makes it possible to slow action down during projection to nearly one-third of the original speed so that it can be studied in slow motion. A projector with additional hand drive also permits study of the films frame by frame, if necessary, to reverse action, and to repeat projection of every detail as often as required. Simultaneously the written protocols of the experiments are studied and the two observations compared.

a Institution no. 1 will from here on be called "Nursery;" institution no. 2, "Foundling Home."

b The small number of children observed in this particular environment was justified by the fact that it has been previously studied extensively by other workers; our only aim was to correlate our results with theirs. However, during the course of one year each child was tested at least at regular monthly intervals.

Table II

Type of environment	Cultural and social background	Developmental quotients	
		Average of first four months	Average of last four months
Parental home	Professional	133	131
	Village population	107	108
Institution	"Nursery"	101.5	105
	"Foundling home"	124	72

For the purpose of orientation we established the average of the developmental quotients for the first third of the first year of life for each of the environments investigated. We contrasted these averages with those for the last third of the first year. This comparison gives us a first hint of the significance of environmental influences for development.

Children of the first category come from professional homes in a large city; their developmental quotient, high from the start, remains high in the course of development.

Children in the second category come from an isolated fishing village of 499 inhabitants, where conditions of nutrition, housing, hygienic and medical care are very poor indeed; their developmental quotient in the first four months is much lower and remains at a lower level than that of the previous category.

In the third category, "Nursery," the children were handicapped from birth by the circumstances of their origin, which will be discussed below. At the outset their developmental quotient is even somewhat lower than that of the village babies; in the course of their development they gain slightly.

In the fourth category, "Foundling Home," the children are of an unselected urban (Latin) background. Their developmental quotient on admission is below that of our best category but much higher than that of the other two. The picture changes completely by the end of the first year, when their developmental quotient sinks to the astonishingly low level of 72.

Thus the children in the first three environments were at the end of their first year on the whole well developed and normal, whether they were raised in their progressive middle-class family homes (where obviously optimal circumstances prevailed and the children were well in advance of average development) or in an institution or a village home, where the development was not brilliant but still reached a perfectly normal and satisfactory average. The children in the fourth environment, though starting at almost as high a level as the best of the others, had spectacularly deteriorated.

The children in Foundling Home showed all the manifestations of hospitalism, both physical and mental. In spite of the fact that hygiene and precautions against contagion were impeccable, the children showed, from the third month on, extreme susceptibility to infection and illness of any kind. There was hardly a child in whose case history we did not find reference to otitis media, or morbilli, or varicella, or eczema, or intestinal disease of one kind or another. No figures could be elicited on general mortality; but during my stay an epidemic of measles swept the institution, with staggeringly high mortality figures, notwithstanding liberal administration of convalescent serum and globulins, as well as excellent hygienic conditions. Of a total of eighty-eight children up to the age of 2½, twenty-three died. It is striking to compare the mortality among the forty-five children up to 1½ years to that of the forty-three children ranging from 1½ to 2½ years: usually the *incidence* of measles is low in the younger age group, but among those infected the mortality is higher than that in the older age group. Since in the case of Foundling Home every child was infected, the question of incidence does not enter; however, contrary to expectation, the mortality was much higher in the older age group. In the younger group, six died, i.e., approximately 13%. In the older group, seventeen died, i.e., close to 40%. The significance of these figures becomes apparent when we realize that the mortality from measles during the first year of life in the community in question, outside the institution, was less than ½%.

In view of the damage sustained in all personality sectors of the children during their stay in this institution, we believe it licit to assume that their vitality (whatever that may be), their resistance to disease, was also progressively sapped. In the ward of the children ranging from eighteen months to 2½ years only two of the twenty-six surviving children speak a couple of words. The same two are able to walk. A third child is beginning to walk. Hardly any of them can eat alone. Cleanliness habits have not been acquired and all are incontinent.

In sharp contrast to this is the picture offered by the oldest inmates in Nursery, ranging from eight to twelve months. The problem here is not whether the children walk or talk by the end of the first year; the problem with these ten-month olds is how to tame the healthy toddlers' curiosity and enterprise. They climb up the bars of the cots after the manner of South Sea islanders climbing palms. Special measures to guard them from harm have had to be taken after one ten-month old actually succeeded in diving right over the more than two-foot railing of the cot. They vocalize freely and some of them actually speak a word or two. And all of them understand the significance of simple social gestures. When released from their cots, all walk with support and a number walk without it.

What are the differences between the two institutions that result in the one turning out normally acceptable children and the other showing such appalling effects?

The Effect of Institutional Care on Infants

Similarities[2]

1. Background of the children

Nursery is a penal institution in which delinquent girls are sequestered.

[2] Under this heading we enumerate not only actual similarities but also differences that are of no etiological significance for the deterioration in Foundling Home. These differences comprise two groups: differences of no importance whatever, and differences that actually favor the development of children in Foundling Home.

When, as is often the case, they are pregnant on admission, they are delivered in a neighboring maternity hospital and after the lying-in period their children are cared for in Nursery from birth to the end of their first year. The background of these children provides for a markedly negative selection since the mothers are mostly delinquent minors as a result of social maladjustment or feeble-mindedness, or because they are psychically defective, psychopathic, or criminal. Psychic normalcy and adequate social adjustment are almost excluded.

The other institution is a foundling home pure and simple. A certain number of the children housed have a background not much better than that of the Nursery children; but a sufficiently relevant number come from socially well-adjusted, normal mothers whose only handicap is inability to support themselves and their children (which is no sign of maladjustment in women of Latin background). This is expressed in the average of the developmental quotients of the two institutions during the first 4 months, as shown in Table II.

The background of the children in the two institutions does therefore not favor Nursery; on the contrary, it shows a very marked advantage for Foundling Home.

2. Housing conditions

Both institutions are situated outside the city, in large spacious gardens. In both hygienic conditions are carefully maintained. In both infants at birth and during the first six weeks are segregated from the older babies in a special newborns' ward, to which admittance is permitted only in a freshly sterilized smock after hands are washed. In both institutions infants are transferred from the newborns' ward after two or three months to the older babies' wards, where they are placed in individual cubicles which in Nursery are completely glass enclosed, in Foundling Home glass enclosed on three sides and open at the end. In Foundling Home the children remain in their cubicles up to fifteen to eighteen months; in Nursery they are transferred after the sixth month to rooms containing four to five cots each.

One-half of the children in Foundling Home are located in a dimly lighted part of the ward; the other half, in the full light of large windows facing southeast, with plenty of sun coming in. In Nursery, all the children have well-lighted cubicles. In both institutions the walls are painted in a light neutral color, giving a white impression in Nursery, a gray-green impression in Foundling Home. In both, the children are placed in white painted cots. Nursery is financially the far better provided one: we usually find here a small metal table with the paraphernalia of child care, as well as a chair, in each cubicle; whereas in Foundling Home it is the exception if a low stool is to be found in the cubicles, which usually contain nothing but the child's cot.

3. Food

In both institutions adequate food is excellently prepared and varied according to the needs of the individual child at each age; bottles from which children are fed are sterilized. In both institutions a large percentage of the younger children are breast-fed. In Nursery this percentage is smaller, so that in most cases a formula is soon added, and in many cases weaning takes place early. In Foundling Home all children are breast-fed as a matter of principle as long as they are under three months unless disease makes a deviation from this rule necessary.

4. Clothing

Clothing is practically the same in both institutions. The children have adequate pastel-colored dresses and blankets. The temperature in the rooms is appropriate. We have not seen any shivering child in either setup.

5. Medical care

Foundling Home is visited by the head physician and the medical staff at least once a day, often twice, and during these rounds the chart of each child is inspected as well as the child itself. For special ailments a laryngologist and other specialists are available; they also make daily rounds. In Nursery no daily rounds are made, as they are not necessary. The physician sees the children when called.

Up to this point it appears that there is very little significant difference between the children of the two institutions. Foundling Home shows, if anything, a slight advantage over Nursery in the matter of selection of admitted children, of breast feeding and of medical care. It is in the items that now follow that fundamental differences become visible.

Differences

1. Toys

In Nursery it is the exception when a child is without one or several toys. In Foundling Home my first impression was that not a single child had a toy. This impression was later corrected. In the course of time, possibly in reaction to our presence, more and more toys appeared, some of them quite intelligently fastened by a string above the baby's head so that he could reach it. By the time we left, a large percentage of the children in Foundling Home had a toy.

2. Visual radius

In Nursery the corridor running between the cubicles, though rigorously white and without particular adornment, gives a friendly impression of warmth. This is probably because trees, landscape and sky are visible from both sides and because a bustling activity of mothers carrying their children, tending them, feeding them, playing with them, chatting with each other with babies in their arms, is usually present. The cubicles of the children are enclosed but the glass panes of the partitions reach low enough for every child to be able at any time to observe everything going on all around. He can see into the corridor as soon as he lifts himself on his elbows. He can look out of the windows, and can see babies in the other cubicles by just turning his head; witness the fact that whenever the experimenter plays with a baby in one of the cubicles the babies in the two adjoining cubicles look on fascinated, try to participate in the game, knock at the panes of the partition, and often begin to cry if no attention is paid to them. Most of the cots are provided with widely spaced bars that are no obstacle to vision. After the age of six months, when the child is transferred to a ward of older babies, the visual field is enriched as a number of babies are then together in the same room, and accordingly play with each other.

In Foundling Home the corridor into which the cubicles open, though full of light on one side at least, is bleak and deserted, except at feeding time, when five to eight nurses file in and look after the children's needs. Most of the time nothing goes on to attract the babies' attention. A special routine of Foundling Home consists in hanging bedsheets over the foot and the side railing of each cot. The cot itself is approximately eighteen inches high. The side railings are about twenty inches high; the foot and head railings are approximately twenty-eight inches high. Thus, when bedsheets are hung over the railings, the child lying in the cot is effectively screened from the world. He is completely separated from the other cubicles, since the glass panes of the wooden partitions begin six to eight inches higher than even the head railing of the cot. The result of this system is that each baby lies in solitary confinement up to the time when he is able to stand up in his bed, and that the only object he can see is the ceiling.

3. Radius of locomotion

In Nursery the radius of locomotion is circumscribed by the space available in the cot, which up to about ten months provides a fairly satisfactory range.

Theoretically the same would apply to Foundling Home. But in practice this is not the case, for, probably owing to the lack of stimulation, the babies lie supine in their cots for many months and a hollow is worn into their mattresses. By the time they reach the age when they might turn from back to side (approximately the seventh month) this hollow confines their activity to such a degree that they are effectively prevented from turning in any direction. As a result we find most babies, even at ten and twelve months, lying on their backs and playing with the only object at their disposal, their own hands and feet.

4. Personnel

In Foundling Home there is a head nurse and five assistant nurses for a total of forty-five babies. These nurses have the *entire* care of the children on their hands, except for the babies so young that they are breast-fed. The latter are cared for to a certain extent by their own mothers or by wet nurses; but after a few months they are removed to the single cubicles of the general ward, where they share with at least seven other children the ministrations of *one* nurse. It is obvious that the amount of care one nurse can give to an individual child when she has eight children to manage is small indeed. These nurses are unusually motherly, baby-loving women; but of course the babies of Foundling Home nevertheless lack all human contact for most of the day.

Nursery is run by a head nurse and her three assistants, whose duties do not include the care of the children, but consist mainly in teaching the children's mothers in child care, and in supervising them. The children are fed, nursed and cared for by their own mothers or, in those cases where the mother is separated from her child for any reason, by the mother of another child, or by a pregnant girl who in this way acquires the necessary experience for the care of her own future baby. Thus in Nursery each child has the full-time care of his own mother, or at least that of the substitute, which the very able head nurse tries to change about until she finds someone who really likes the child.

Discussion

To say that every child in Nursery has a full-time mother is an overstatement, from a psychological point of view. However modern a penal institution may be, and however constructive and permissive its reeducative policies, the deprivation it imposes upon delinquent girls is extensive. Their opportunities for an outlet for their interests, ambitions, activity are very much impoverished. The former sexual satisfactions as well as the satisfactions of competitive activity in the sexual field are suddenly stopped: regulations prohibit flashy dresses, vivid nail polish, or extravagant hair-do's. The kind of social life in which the girls could show off has vanished. This is especially traumatic as these girls become delinquent because they have not been able to sublimate their sexual drives, to find substitute gratifications, and therefore do not possess a pattern for relinquishing pleasure when frustrated. In addition, they do not have compensation in relations with family and friends, as formerly they had. These factors, combined with the loss of personal liberty, the deprivation of private property and the regimentation of the penal institution, all add up to a severe narcissistic trauma from the time of admission; and they continue to affect the narcissistic and libidinal sectors during the whole period of confinement.

Luckily there remain a few safety valves for their emotions: (1) the relationship with wardens, matrons and nurses; (2) with fellow prisoners; (3) with the child. In the relationship with the wardens, matrons and nurses, who obviously represent parent figures, much of the prisoner's aggression and resentment is bound. Much of it finds an outlet in the love and hate relationship to fellow prisoners, where all the phenomena of sibling rivalry are revived.

The child, however, becomes for them the representative of their sexuality, a product created by them, an object they own, which they can dress up and adorn, on which they can lavish their tenderness and pride, and of whose accomplishments, performance and appearance they can boast. This is manifested in the constant competition among them as to who has the better dressed, more advanced, more intelligent, better looking, the heavier, bigger, more active — in a word, the better baby.[3] For their own persons they have more or less given up the competition for love, but they are intensely jealous of the attention given to their children by the matrons, wardens, and fellow prisoners.

It would take an exacting experimenter to invent an experiment with conditions as diametrically opposed in regard to the mother-child relationship as they are in these two institutions. Nursery provides each child with a mother to the nth degree, a mother who gives the child everything a good mother does and, beyond that, everything else she has.[4] Foundling Home does not give the child a mother, nor even a substitute mother, but only an eighth of a nurse.

We are now in a position to approach more closely and with better understanding the results obtained by each of the two institutions. We have already cited a few: we mentioned that the developmental quotient of Nursery achieves a normal average of about 105 at the end of the first year, whereas that of Foundling Home sinks to 72; and we mentioned the striking difference of the children in the two institutions at first sight. Let us first consider the point at which the developments in the two institutions deviate.

On admission the children of Foundling Home have a much better average than the children of Nursery; their hereditary equipment is better than that of the children of delinquent minors. But while Foundling Home shows a rapid fall of the developmental index, Nursery shows a steady rise.

[3] The psychoanalytically oriented reader of course realizes that for these girls in prison the child has become a hardly disguised phallic substitute. However, for the purposes of this article I have carefully avoided any extensive psychoanalytic interpretation, be it ever so tempting, and limited myself as closely as possible to results of direct observations of behavior. At numerous other points it would be not only possible but natural to apply analytic concepts; that is reserved for future publication.

[4] For the non-psychoanalytically oriented reader we note that this intense mother-child relationship is not equivalent to a relationship based on love of the child. The mere fact that the child is used as a phallic substitute implies what a large part unconscious hostility plays in the picture.

They cross between the fourth and fifth months, and from that point on the curve of the average developmental quotient of Foundling Home drops downward with increasing rapidity, never again to rise (Curve I).

The point where the two curves cross is significant. The time when the children in Foundling Home are weaned is the beginning of the fourth month. The time lag of one month in the sinking of the index below normal is explained by the fact that the quotient represents a cross-section including all sectors of development, and that attempts at compensation are made in some of the other sectors.

However, when we consider the sector of body mastery (Curve II), which according to Wolf[5] is most indicative for the mother-child relationship, we find that the curves of the children in Nursery cross the body mastery curve of the Foundling Home children between the third and fourth month. The inference is obvious. As soon as the babies in Foundling Home are weaned, the modest human contacts which they have had during nursing at the breast stop, and their development falls below normal.

One might be inclined to speculate as to whether the further deterioration of the children in Foundling Home is not due to other factors also, such as the perceptual and motor deprivations from which they suffer. It might be argued that the better achievement of the Nursery children is due to the fact that they were better provided for in regard to toys and other perceptual stimuli. We shall therefore analyze somewhat more closely the nature of deprivations in perceptual and locomotor stimulation.

First of all it should be kept in mind that the nature of the inanimate perceptual stimulus, whether it is a toy or any other object, has only a very minor importance for a child under twelve months. At this age the child is not yet capable of distinguishing the real purpose of an object. He is only able to use it in a manner adequate to his own functional needs (1). Our thesis is that perception is a function of libidinal cathexis and therefore the result of the intervention of an emotion of one kind or another.[6] Emotions are provided for the child through the intervention of a human partner, i.e., by the mother or her substitute. A progressive development of emotional interchange with the mother provides the child with perceptive experiences of its environment. The child learns to grasp by nursing at the mother's breast and by combining the emotional satisfaction of that experience with tactile perceptions. He learns to distinguish animate objects from inanimate ones by the spectacle provided by his mother's face (6) in

[5] K. Wolf, "Body Mastery of the Child as an Index for the Emotional Relationship Between Mother and Child" (in preparation).

[6] This is stating in psychoanalytic terms the conviction of most modern psychologists, beginning with Compayré (2) and shared by such familiar authorities in child psychology as Stern (3) and Buhler (4), and in animal psychology, Tolman (5).

situations fraught with emotional satisfaction. The interchange between mother and child is loaded with emotional factors and it is in this interchange that the child learns to play. He becomes acquainted with his surroundings through the mother's carrying him around; through her help he learns security in locomotion as well as in every other respect. This security is reinforced by her being at his beck and call. In these emotional relations with the mother the child is introduced to learning, and later to imitation. We have previously mentioned that the motherless children in Foundling Home are unable to speak, to feed themselves, or to acquire habits of cleanliness: it is the security provided by the mother in the field of locomotion, the emotional bait offered by the mother calling her child, that "teaches" him to walk. When this is lacking, even children two to three years old cannot walk.

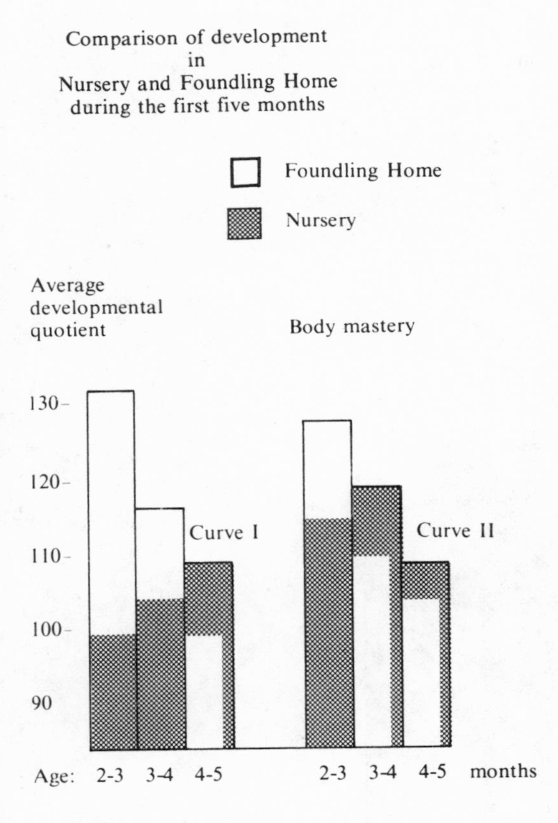

The children in Foundling Home have, theoretically, as much radius of locomotion as the children in Nursery. They did not at first have toys, but they could have exerted their grasping and tactile activity on the blankets, on their clothes, even on the bars of the cots. We have seen children in Nursery without toys; they are the exception — but the lack of material is not enough to hamper them in the acquisition of locomotor and grasping skills. The presence of a mother or her substitute is sufficient to compensate for all the other deprivations.

It is true that the children in Foundling Home are condemned to solitary confinement in their cots. But we do not think that it is the lack of perceptual stimulation *in general* that counts in their deprivation. We believe that they suffer because their perceptual world is emptied of human partners, that their isolation cuts them off from any stimulation by any

Comparison of development
in
Nursery and Foundling Home

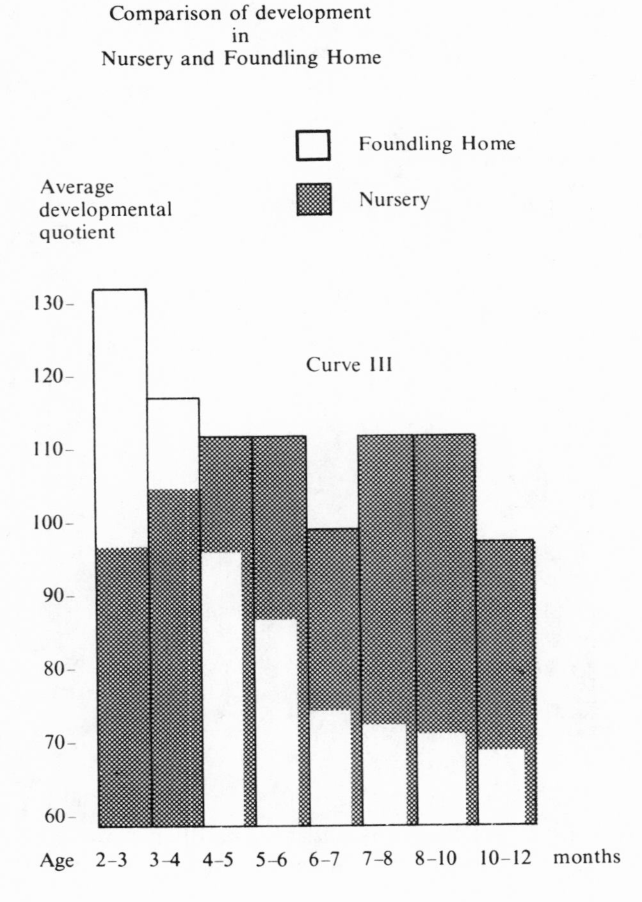

persons who could signify mother representatives for the child at this age.[7] The result, as Curve III shows, is a complete restriction of psychic capacity by the end of the first year.

This restriction of psychic capacity is not a temporary phenomenon. It is, as can be seen from the curve, a progressive process. How much this deterioration could have been arrested if the children were taken out of the institution at the end of the first year is an open question. The fact that they remain in Foundling Home probably furthers this progressive process. By the end of the second year the developmental quotient sinks to 45, which corresponds to a mental age of approximately ten months, and would qualify these children as imbeciles.

The curve of the children in Nursery does not deviate significantly from the normal. The curve sinks at two points, between the sixth and seventh and between the tenth and twelfth months. These deviations are within the normal range; their significance will be discussed in a separate article. It has nothing to do with the influence of institutions, for the curve of the village group is nearly identical.

The contrasting pictures of these two institutions show the significance of the mother-child relationship for the development of the child during the first year. Deprivations in other fields, such as perceptual and locomotor radius, can all be compensated by adequate mother-child relations. "Adequate" is not here a vague general term. The examples chosen represent the two extremes of the scale.

The children in Foundling Home do have a mother — for a time, in the beginning — but they must share her immediately with at least one other child, and from three months on, with seven other children. The quantitative factor here is evident. There is a point under which the mother-child relations cannot be restricted during the child's first year without inflicting irreparable damage. On the other hand, the exaggerated mother-child relationship in Nursery introduces a different quantitative factor. To anyone familiar with the field it is surprising that Nursery should achieve such excellent results, for we know that institutional care is destructive for children during their first year; but in Nursery the destructive factors have been compensated by the increased intensity of the mother-child relationship.

These findings should not be construed as a recommendation for overprotection of children. In principle the libidinal situation of Nursery is almost as undesirable as the other extreme in Foundling Home. Neither in the nursery of a penal institution nor in a foundling home for parentless

[7] This statement is to be developed further in a forthcoming article on "The Beginning of the Social Relations of the Child."

children can the normal libidinal situation that obtains in a family home be expected. The two institutions have here been chosen as experimental setups for the purpose of examining variations in libidinal factors ranging from extreme frustration to extreme gratification. That the extreme frustration practiced in Foundling Home has deplorable consequences has been shown; the extreme gratification in Nursery can be tolerated by the children housed there for two reasons:

(1) The mothers have the benefit of the intelligent guidance of the head nurse and her assistants, and the worst exaggerations are thus corrected.

(2) Children during their first year of life can stand the ill effects of such a situation much better than at a later age. In this respect Nursery has wisely limited the duration of the children's stay to the first twelve months. For children older than this we should consider a libidinal setup such as that in Nursery very dangerous indeed.

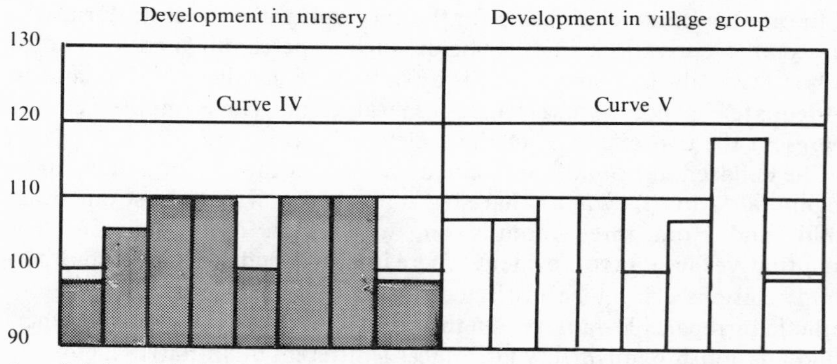

Apart from the severe developmental retardation, the most striking single factor observed in Foundling Home was the change in the pattern of the reaction to strangers in the last third of the first year (7). The usual behavior was replaced by something that could vary from extreme friendliness to any human partner combined with anxious avoidance of inanimate objects, to a generalized anxiety expressed in blood-curdling screams which could go on indefinitely. It is evident that these deviant behavior patterns require a more thorough and extensive discussion than our present study would have permitted.

We also observed extraordinary deviations from the normal in the time of appearance and disappearance of familiar developmental patterns; and certain phenomena unknown in the normal child, such as bizarre stereotyped motor patterns distinctly reminiscent of the stereotypy in catatonic motility. These and other phenomena observed in Foundling Home require an extensive discussion in order to determine which are to be classified as maturation phenomena (which appear even under the most unfavorable circumstances, and which appear with commensurate retardation when retardation is general) and which can be considered as the first symptoms of the development of serious psychiatric disturbances. In connection with this problem a more thorough discussion of the rapidity with which the developmental quotients recede in Foundling Home is intended.

Another study is to deal with the problems created by the enormous overprotection practiced in Nursery.

And finally the rationale of the one institutional routine as against that of the other will have to be discussed in greater detail. This study will offer the possibility of deciding how to compensate for unavoidable changes in the environment of children orphaned at an early age. It will also shed some light on the social consequences of the progressive disruption of home life caused by the increase of female labor and by the demands of war; we might state that we foresee in the course of events a corresponding increase in asociality, in the number of problem and delinquent children, of mental defectives, and of psychotics.

It will be necessary to take into consideration in our institutions, in our charitable activities, in our social legislation, the overwhelming and unique importance of adequate and satisfactory mother-child relationship during the first year, if we want to decrease the unavoidable and irreparable psychiatric consequences deriving from neglect during this period.

References

1. Buhler, Ch., *Kindheit und Jugend,* Leipzig, 1931, p. 67.
2. Compayré, G., *L'évolution intellectuelle et morale de l'enfant,* Paris, 1893.
3. Stern, Wm., *Psychology of Early Childhood,* London, 1930.
4. Buhler, K., *Die geistige Entwicklung des Kindes,* 4th ed., Jena, 1942, p. 106 and p. 116.
5. Tolman, E. C., *Purposive Behavior,* New York, 1932, p. 27 ff.
6. Gesell, A., and Ilg, F., *Feeding Behavior of Infants,* Philadelphia, 1937, p. 21.
7. Gesell, A., and Thompson, H., *Infant Behavior, Its Genesis and Growth,* New York, 1934, p. 208.

CHAPTER 3

Basic Properties of Cognitive Functioning

JOHN H. FLAVELL

Biology and Intelligence

Every theory of intelligence, Piaget argues, ought to begin with some basic conception of its subject matter. What sort of thing is this intelligence we study? What relationship, if any, does it bear to other processes not ordinarily called by that name? In Piaget's view these are related questions. The search for the defining and fundamental characteristics of intelligence must begin by a search for even more fundamental processes from which intelligence derives and to which, in its essentials, it remains similar. For Piaget, the key to *ipse intellectus* lies in a careful examination of these "even more fundamental processes." What are they?

They are biological in nature. For Piaget, the one-time biologist, intelligence can be meaningfully considered only as an extension of certain fundamental biological characteristics, fundamental in the sense that they obtain wherever life obtains. (It is indicative of Piaget's biological orientation toward matters intellectual that he sometimes refers to cognitive development as "mental embryology," e.g., 1947, p. 143.) Intellectual functioning is a special form of biological activity and, as such, possesses important attributes in common with the parent activities from which it derives. In other words, intelligence bears a biological imprint, and this imprint defines its essential characteristics. But to say that intelligence is founded upon a biological substrate can imply two quite different things (1952c, pp. 1-3). Let us examine each of these in turn.

Specific Heredity

Intelligence is first of all allied to biology in the sense that inherited biological structures condition what we may directly perceive. For example, our nervous and sensory system is such that only certain wavelengths give rise to color sensations, and we are unable to perceive space in more than three dimensions. Our perceptions constitute only a selected segment within a totality of conceivable perceptions. There can be no doubt that these biological limitations influence the construction of our most fundamental concepts. In this sense, there is certainly an intimate relation between basic physiological and anatomical fundaments and intelligence.

However, this relation is not the most important kind of liaison between biology and cognition. As a matter of fact, it is characteristic of intelligence that it will eventually transcend the limitations imposed upon it by these structural properties, this *specific heredity,* as Piaget call it (ibid., p.2). We come to be able to *cognize* wavelengths which we never *see*. We *hypothesize* spatial dimensions we can never *experience* directly. In short, the neurological and sensory structures which constitute our species-specific inheritance can be said to impede or facilitate intellectual functioning, but they can hardly be said to *account for* functioning itself. For this we must look to a second kind of connection between biology and intelligence.

General Heredity

This second kind of relation is more subtle and elusive than the first. Most simply, it is this. We inherit as biological *anlagen* not only structural limitations but also something else, which, as we have seen, permits these limitations to be overcome. That is, our biological endowment consists not only of inborn structures which can be thought of as obstacles to intellectual progress, but it also consists of that which makes intellectual progress possible at all, that something which lies behind intellectual achievement. Two questions immediately arise. First, what is the nature of this something? And second, what is its relation to biological processes at large?

The positive, constructive something which we inherit, Piaget argues, is a *mode of intellectual functioning*. We do not inherit cognitive structures as such; these come into being only in the course of development. What we do inherit is a *modus operandi,* a specific manner in which we transact business with the environment. There are two important general characteristics of this mode of functioning. First, it generates cognitive structures. Structures come into being in the course of intellectual functioning; it is through functioning, and only through functioning, that cognitive structures get formed. Second, and this is a most important point, the mode of functioning which Piaget says constitutes our biological heritage remains essentially constant throughout life. That is, the fundamental properties of intellectual functioning are always and everywhere the same, despite the wide varieties of cognitive structures which this functioning creates. It is because of this constancy of functioning in the face of changing structures that its fundamental properties, soon to be described, are referred to as *functional invariants.*

Let us review what has been said so far. The really important biological endowment, so far as intelligence is concerned, is a set of functional characteristics rather than a set of inborn structural limitations. These functional characteristics are at the very heart and soul of intelligence because they constitute unvarying common elements amid a panorama of structural changes and because it is precisely through functioning that the succession of structures get constituted. The sought-for *ipse intellectus* is to be found in intellectual functioning itself, nowhere else.

But it remains to be shown in what sense intellectual functioning can be considered a biological endowment. In order to do this, it is necessary to take a preliminary look at the fundamental characteristics of intellectual functioning, those defining attributes which were said to be invariant over the whole developmental span. There are two principal ones. The first is *organization;* the second is *adaptation* and comprises two intimately related but conceptually distinct subproperties: *assimilation* and *accommodation.* The nature of these functional invariants is the subject of the next section. The important thing which needs to be understood in advance of their definitions is this. *These invariant characteristics, which define the essence of intellectual functioning and hence the essence of intelligence, are also the very characteristics which hold for biological functioning in general.* All living matter adapts to its environment and possesses organizational properties which make the adaptation possible. Intellectual functioning is only a special case, a special extension of biological functioning at large; and its fundamental and invariant properties are the same as those found in biological activity. This is the second and more important sense in which it could be said that a biological substrate underlies intelligence. In addition to a *specific heredity* of inborn and limiting anatomical *structures,* we have a *general heredity,* of a functional kind, upon which all positive cognitive acquisitions are built (ibid., p. 2). By virtue of the fact that we are living organisms, we begin life with certain irreducible properties held in common by all organisms, and these fundaments are a set of peculiarly functional characteristics. It is these characteristics which supply the continuity between biology in general and intelligence in particular. And it is these which, despite their lowly origins, make possible the most sublime of intellectual accomplishments. Piaget sums it up in this way:

> Now, this second type of hereditary psychological reality is of primary

> importance for the development of intelligence. If there truly in fact exists a functional nucleus of the intellectual organization which comes from biological organization in its most general aspect, it is apparent that this invariant will orient the whole of the successive structures which the mind will then work out in its contact with reality. It will thus play the role that philosophers assigned to the *a priori;* that is to say, it will impose on the structures certain necessary and irreducible conditions (ibid., pp. 2-3).

And again the continuity with biological functioning:

> In fact there exist, in mental development, elements which are variable and others which are invariant. . . . Just as the main functions of the living being are identical in all organisms but correspond to organs which are very different in different groups, so also between the child and the adult a continuous creation of varied structures may be observed although the main functions of thought remain constant (ibid, p. 4).

The Functional Invariants

Of the two basic invariants of functioning, *organization* and *adaptation,* the latter is subdivided into two interrelated components, *assimilation* and *accommodation.* These invariants provide the crucial link between biology and intelligence because they hold equally for both. This isomorphism permits us to see intelligence in its proper context, as an interesting and highly developed extension of more primitive activities whose most general characteristics — the functional invariants — it shares. Let us then begin our analysis of the functional invariants by first seeing how they characterize an elementary biological process; their application to intelligence will follow later. There are two advantages to doing this. First, basic physiological events are more palpable for most of us than psychological processes are, and a preliminary understanding of these concepts may be facilitated if they are first demonstrated in their biological context. Indeed, Piaget himself introduces them in this way (ibid., pp. 5-6). Second, the isomorphism between biology and intelligence as regards these invariants needs to be documented, not simply asserted. An illustrative biological example would help to serve that need.

A Biological Example

A very fundamental — probably the most fundamental — function of living matter is that of incorporating into its structure nutrition-providing elements from the outside. The organism sustains itself and grows by means of such transactions with its milieu. The invariant attributes of this kind of functioning are the following.

First of all, the process is one of *adaptation* to the environment.

Adaptation is said to occur whenever a given organism-environment interchange has the effect of modifying the organism in such a way that further interchanges, favorable to its preservation, are enhanced (ibid., p. 5). Not everything that an organism does is adaptive in this sense, of course, but the incorporation of nutritive substances ordinarily is. Now this particular form of adaptation (and all adaptations generally) involves two conceptually distinguishable components.

First, the organism must and will transform the substances it takes in in order to incorporate their food values into its system. An initial transformation occurs when the substance is ingested by chewing. Thus, hard and sharply contoured objects become pulpy and formless. Still more drastic changes occur as the substance is slowly digested, and eventually it will lose its original identity entirely by becoming part of the structure of the organism. The process of changing elements in the milieu in such a way that they can become incorporated into the structure of the organism is called *assimilation,* i.e., the elements are assimilated to the system. The manner in which the incorporation is carried out and the structures into which elements are incorporated are extremely variable. But the process itself, *qua* process, always obtains whenever and wherever adaptation takes place. In this sense Piaget speaks of assimilation as a functional invariant.

In the process of assimilating foodstuffs to itself, the organism is also doing something else. It is also adjusting itself to them. This it does in a variety of ways and at all stages in the adaptation process. The mouth (or whatever corresponds to it in a given species) must open or the substance cannot enter the system at all. The object must be chewed if its structure demands chewing. And finally, the digestive processes must adapt themselves to the object's specific chemical and physical properties or no digestion can take place. Just as objects must be adjusted to the peculiar structure of the organism in any adaptational process, so also must the organism adjust itself to the idiosyncratic demands of the object. The first aspect of adaptation has been called *assimilation.* The second aspect, the adjustment to the object, Piaget labels *accommodation* — i.e., the organism must accommodate its functioning to the specific contours of the object it is trying to assimilate. As was the case for assimilation, the details of the accommodatory process are highly variable. What is invariant is its existence, as a process, in all adaptation.

Although assimilation and accommodation are distinguished conceptually, they are obviously indissociable in the concrete reality of any adaptational act. As will become clear when intellectual adaptation is discussed, every assimilation of an object to the organism simultaneously involves an accommodation of the organism to the object; conversely, every accommodation is at the same time an assimilatory modification of the object accommodated to. Taken together, they make up the constant attributes of even the most elementary adaptational act.

Adaptation, through its twin components assimilation and accommodation, expresses the dynamic, outer aspect of biological functioning. But an adaptive act always presupposes an underlying *organization,* and this is the second major functional invariant (ibid., p. 5). Actions are coordinated affairs, governed by laws of totality — this is Piaget's now familiar holism again. The assimilation of foodstuffs to the organism and the simultaneous accommodation of the organism to these nutritive substances are organized activities carried out by an organized being. Adaptive, directed behavior cannot proceed from a chaotic and completely undifferentiated source. There are subordinating structures and subordinated structures, and so on. Once again, the specific nature of the organization which lies behind an adaptive act will vary, but organization of some kind there must be.

It is Piaget's position, as we have said, that intellectual functioning can be characterized in terms of the same invariants that hold for more elementary biological processes.[1] Let us begin with *organization.*

Cognitive Organization

Cognition, like digestion, is an organized affair. Every act of intelligence presumes some kind of intellectual structure, some sort of organization, within which it proceeds. The apprehension of reality always involves multiple interrelationships among cognitive actions and among the concepts and meanings which these actions express.

As to the nature of this organization, its specific characteristics, like those of biological organizations, differ markedly from stage to stage in development. Though structural change is what Piaget studies, there are stage-independent properties which the very fact of organization always implies. All intellectual organizations can be conceived of as *totalities,* systems of *relationships* among elements, to use Piaget's terms (ibid., p.10). An act of intelligence, be it a crude motor movement in infancy or a complex and abstract judgment in adulthood, is always related to a system or totality of such acts of which it is a part.

The relation of part to whole need not be simply static and

[1] Our treatment of the functional invariants of cognition will contain some deliberate (although minor) deviations from Piaget's own presentation for the sake of clarity and consistency. Piaget initially uses the expressions *regulative function, implicative function,* and *explicative function* when referring to the intellectual counterparts of the biological invariants organization, assimilation, and accommodation, respectively (ibid., p. 9). However, since the former terms tend to drop out of usage in his subsequent discussions of intellectual functioning, it seems superfluous to define and discuss them here. We also refrain from systematic presentation of the *categories of reason* (ibid., p. 9) under the same rationale.

configurational, as the proverbial trees are to the forest. Acts are also organized directionally in terms of means to ends, or *values* to *ideals* in Piaget's phraseology (ibid., pp. 10-11). Moreover, the finalism which may characterize individual sets of actions — an infant bangs his rattle (means) in order to hear a noise (end) — also holds, in the large, for cognitive development itself. The ontogenetic development of structures can be thought of as a process of successive approximations to a kind of ideal equilibrium, an end state never completely achieved. Development itself, then, constitutes a totality with a goal or ideal subordinating means.

Cognitive Adaptation: Assimilation and Accommodation

Intellectual functioning, in its dynamic aspect, is also characterized by the invariant processes of *assimilation* and *accommodation*. An act of intelligence in which assimilation and accommodation are in balance or equilibrium constitutes an intellectual *adaptation*.[2] Adaptation and organization are two sides of the same coin, since adaptation presupposes an underlying coherence, on the one hand, and organizations are created through adaptations, on the other. In Piaget's words:

> . . . Organization is inseparable from adaptation: They are two complementary processes of a single mechanism, the first being the internal aspect of the cycle of which adaptation constitutes the external aspect. . . . The "accord of thought with things" and the "accord of thought with itself" express this dual functional invariant of adaptation and organization. These two aspects of thought are indissociable: It is by adapting to things that thought organizes itself and it is by organizing itself that it structures things (1952c, pp. 7-8).

[2] This restriction of the meaning of the term is likely to puzzle the reader, since adaptation is supposed to be an invariant of all intellectual functioning. Although the matter will become clearer in the course of subsequent reading in this chapter, a few things may be said at this point. Organization, assimilation, and accommodation are truly invariant; every instance of cognitive functioning presupposes these three characteristics. However, the *relationships between* assimilation and accommodation are quite variable, both across development and within any developmental period. In the most narrow meaning Piaget ever gives it, adaptation refers to those organism-environment exchanges in which assimilation and accommodation are in equilibrium, neither one predominating (1951a, 1952c). This implies that some intelligent actions are more truly adaptations than others. Except when specific arguments are made which hinge on these distinctions, e.g., Piaget's analysis of imitation and play (1951a), the term tends to become denotatively broader than this and even appears to be synonymous with intellectual functioning itself at times. Taken in its broadest meaning, then, adaptation is certainly a functional invariant.

What is the nature of cognitive as opposed to physiological assimilation and accommodation? Assimilation here refers to the fact that every cognitive encounter with an environmental object necessarily involves some kind of cognitive structuring (or restructuring) of that object in accord with the nature of the organism's existing intellectual organization. As Piaget says: "Assimilation is hence the very functioning of the system of which organization is the structural aspect" (ibid., p. 410). Every act of intelligence, however rudimentary and concrete, presupposes an interpretation of something in external reality, that is, an assimilation of that something to some kind of meaning system in the subject's cognitive organization. To use a happy phrase of Kelly's (1955), to adapt intellectually to reality is to *construe* that reality, and to construe it in terms of some enduring *construct* within oneself. Piaget's epistemological position is essentially the same on this point, requiring only the substitution of *assimilate* for *construe* and *structure* or *organization* for *construct.* And it is Piaget's argument that intellectual assimilation is not different in principle from a more primary biological assimilation: in both cases the essential process is that of bending a reality event to the templet of one's ongoing structure.

If intellectual adaptation is always and essentially an assimilatory act, it is no less an accommodatory one. In even the most elemental cognition there has to be some coming to grips with the special properties of the thing apprehended. Reality can never be infinitely malleable, even for the most autistic of cognizers, and certainly no intellectual development can occur unless the organism in some sense adjusts his intellectual receptors to the shapes reality presents him. The essence of accommodation is precisely this process of adapting oneself to the variegated requirements or demands which the world of objects imposes upon one. And once again, Piaget underscores the essential continuity between biological accommodation, on the one hand, and cognitive accommodation, on the other: a receptive and accommodating mouth and digestive system are not really different in principle from a receptive and accommodating cognitive system.

However necessary it may be to describe assimilation and accommodation separately and sequentially, they should be thought of as simultaneous and indissociable as they operate in a living cognition. Adaptation is a unitary event, and assimilation and accommodation are merely abstractions from this unitary reality. As in the case of food ingestion, the cognitive incorporation of reality always implies both an assimilation *to* structure and an accommodation *of* structure. To assimilate an event it is necessary at the same time to accommodate to it and vice versa. As subsequent pages will make clear, the balance between the two invariants can and does vary, both from stage to stage and within a given stage. Some cognitive acts show a relative preponderance of the assimilative component; others seem heavily weighted towards accommodation. However, "pure" assimilation and "pure" accommodation nowhere obtain in cognitive life; intellectual acts always presuppose each in some measure:

> . . . From the beginning assimilation and accommodation are indissociable from each other. Accommodation of mental structures to reality implies the existence of assimilatory schemata[3] apart from which any structure would be impossible. Inversely, the formation of schemata through assimilation entails the utilization of external realities to which the former must accommodate, however crudely . . . (1954a, pp. 352-353).
>
> Assimilation can never be pure because by incorporating new elements into its earlier schemata the intelligence constantly modifies the latter in order to adjust them to new elements. Conversely, things are never known by themselves, since this work of accommodation is only possible as a function of the inverse process of assimilation (1952c, pp. 6-7).

Having endowed the organism with these twin mechanisms of intellectual adaptation, two problems remain. First, how does the action of assimilation and accommodation permit the organism to make cognitive progress as opposed to remaining fixated at the level of familiar and habitual cognitions? That is, how is the organism able to do something other than repeat past accommodations and assimilate the results of these accommodations to the same old system of meanings? Secondly, assuming that cognitive progress or cognitive development can somehow result from assimilatory and accommodatory operations, what prevents it from occurring all at once and of a piece? That is, why is intellectual development the slow and gradual process we know it to be? To indulge in metaphor, we need to know both what makes the cognitive engine progress at all and what limits its velocity and acceleration, assuming the possibility of movement.

Cognitive progress, in Piaget's system, is possible for several reasons. First of all, accommodatory acts are continually being extended to new and different features of the surround. To the extent that a newly accommodated-to feature can fit somewhere in the existing meaning structure, it will be assimilated to that structure. Once assimilated, however, it tends to change the structure in some degree and, through this change, make possible further accommodatory extensions. Also, as discussion of schemas will show, assimilatory structures are not static and

[3] The plural of *scheme* is sometimes rendered *schemas* and sometimes *schemata* in translations of Piaget's works; usage varies. We use *schemas* in this book except when quotations (such as the above) demand *schemata*. The meaning and significance of the schema concept itself is taken up in the next section of this chapter.

unchanging, even in the absence of environmental stimulation. Systems of meanings are constantly becoming reorganized internally and integrated with other systems. This continuous process of internal renovation is itself, in Piaget's system, a very potent source of cognitive progress (1952c, p. 414). Thus, both kinds of changes — reorganizations of purely endogenous origin and reorganizations induced more or less directly by new accommodatory attempts — make possible a progressive intellectual penetration into the nature of things. Once again the twin invariants innervate each other in reciprocal fashion: changes in assimilatory structure direct new accommodations, and new accommodatory attempts stimulate structural reorganizations.

If cognitive progress is insured under this interpretation of the invariants, it is certainly well established empirically that this progress is typically slow and gradual. It is not immediately clear why this should be so. What prevents the organism from mastering, in one fell swoop, all that is cognizable in a given terrain? The answer is that the organism can assimilate only those things which past assimilations have prepared it to assimilate. There must already be a system of meanings, an existing organization, sufficiently advanced that it can be modified to admit the candidates for assimilation which accommodation places before it. There can never be a radical rupture between the new and the old; events whose interpretation requires a complete extension or reorganization of the existing structure simply cannot be accommodated to and thence assimilated. As Piaget states (1954a, pp. 352-354), assimilation is by its very nature conservative, in the sense that its primary function is to make the unfamiliar familiar, to reduce the new to the old. A new assimilatory structure must always be some variate of the last one acquired, and it is this which insures both the gradualness and continuity of intellectual development.

In summary, the functional characteristics of the assimilatory and accommodatory mechanisms are such that the possibility of cognitive change is insured, but the magnitude of any given change is always limited. The organism adapts repeatedly, and each adaptation necessarily paves the way for its successor. Structures are not infinitely modifiable, however, and not everything which is potentially assimilable can in fact be assimilated by organism A at point X in his development. On the contrary, the subject can incorporate only those components of reality which its ongoing structure can assimilate without drastic change.

A Concrete Example

Piaget's concepts tend to become more meaningful when examined in behavioral context. Let us consider a sample of cognitive activity and see how it would be described in the terms we have been discussing. An infant comes in contact for the first time with a ring suspended from a string. He makes a series of exploratory accommodations: he looks at it, touches it, causes it to swing back and forth, grasps it, and so on. These accommodatory acts of course do not take place *in vacuo;* through past interactions with various other objects the child already possesses assimilatory structures (schemas) which set in motion and direct those accommodations. Piaget would say here that the ring is assimilated to concepts of touching, moving, seeing, etc., concepts which are already part of the child's cognitive organization. The child's actions with respect to the ring are at once accommodations of these concepts or structures to the reality contours of the ring and assimilations of this new object to these concepts.

But the infant does more than simply repeat behaviors acquired earlier. The structures which are defined by grasping, seeing, touching, etc., are themselves modified in a number of ways as they accommodate themselves to the ring and assimilate it. The varieties of structural modification constitute the subject of the following section on the schema concept, but we may anticipate two of them here. First, the structures are *generalized* to assimilate the new object. In ordinary language, the child learns that rings *too* may be sucked, pulled at, visually scanned, etc.; his cognitive structures are modified in the sense of being extended to fit one more object. Second, they are changed in so far as the structure of the new object necessitates some *variation* in the way one sucks it, pulls at it, scans it, etc. In other words, cognitive structures are not only *generalized* to the new object but are also *differentiated* as a consequence of its idiosyncratic structural demands. Thus, the child learns that one sucks ringlike objects a little differently from other objects sucked in the past, and that ringlike objects look and feel somewhat different from objects seen and touched in the past. The important consequence of the structural changes wrought by this generalization and differentiation is, of course, the fact that this change makes possible new and different accommodations to future objects encountered. These new accommodations engender further changes in intellectual organization, and so the cycle repeats itself.

The example of the child and the ring can also serve to illustrate the limitations to which structural change is subject in a single organism-reality transaction. In the first place, there are a number of potentially cognizable features of the ring which we may be sure the infant will not accommodate to and will not assimilate. There is nothing in the infant's present structural repertoire which will permit him to accommodate to the ring as an exemplar of the abstract class of circles, for example. Similarly, he cannot apprehend the ring as an object which can be rolled like a hoop, as an object which can be worn as a bracelet, and so forth. In Piaget's terms, the organism cannot accommodate itself to those object potentialities which it is unable to assimilate to something in its present system of meanings. The hiatus between the new and the old cannot be too great. This fact, that new structures must arise almost imperceptibly from the foundations provided by present ones, is what always insures the gradualness of cognitive development.

The Concept of Schema

We have said that assimilatory and accommodatory functioning always presupposes some sort of quasi-enduring organization or structural system within the organism. Objects are always assimilated *to* something. Although the differing properties of cognitive structures at various ontogenetic levels constitute the subject matter of subsequent chapters, the over-all character — the stage-free properties — of structures in general can be discussed here. Piaget makes extensive use of a certain structural concept which, although more specific than the noncommittal term *structure* itself, is nonetheless transdevelopmental, not bound to any particular stage. The concept in question is the one Piaget would invariably insert in sentences of the following type: "The infant assimilated the nipple to the — of sucking." In Piaget's system the missing term could only be *schema.*

Basic Properties

The notion of *schema* needs careful examination in any explication of Piaget's system. First, it occupies a very prominent place in Piaget's account of cognitive development, especially cognitive development in infancy.[4] Second, a thorough explanation of what schemas are and how they function sheds further light on the functional invariants of organization and adaptation to which they are so closely related in the theory.

What is a schema? As we have seen to be the case for other theoretical

[4] Although the concept of schema is definitely stage-free and is invoked by Piaget at all age levels, it is used most frequently and elaborated most extensively in connection with the sensory-motor period of infancy, probably because Piaget has available more specific and delimited structural concepts (*groupings,* etc.) to describe postinfancy developments. Nonetheless, schemas of one kind or another do continue to be introduced in connection with these later periods, often in textual contiguity with the more specific structural concepts. The most important of these postinfancy schemas, perhaps, are the *operational schemas* of adolescence (Inhelder and Piaget, 1958).

concepts, Piaget does not give a careful and exhaustive definition of the term in any single place; rather, its full meaning is developed in successive fragments of definition spanning several volumes (1951a, 1952c, 1954a, 1958a). It is, despite its vagueness, a rich and subtle notion, full of shifting nuances and most thoroughly bound up with Piaget's whole conception of cognitive development. A preliminary and somewhat inadequate rendering may be the following. A *schema* is a cognitive structure which has reference to a class of similar action sequences, these sequences of necessity being strong, bounded totalities in which the constituent behavioral elements are tightly interrelated. In actuality, it is easier to get at least a global image of what schemas are and how they operate than this rather forbidding definition might suggest. A word of caution, however. Just as the concepts of organization, assimilation and accommodation become enriched when schema is explained, so also will a really adequate grasp of the latter probably have to await a detailed description of infantile development, the period in which schemas figure so prominently.

The first and most obvious thing to be said about schemas is that they are labeled by the behavior sequences to which they refer. Thus, in discussing sensory-motor development, Piaget speaks of the *schema of sucking,* the *schema of prehension,* the *schema of sight,* and so on (1952c). Similarly, there is in middle childhood a *schema of intuitive qualitative correspondence* which refers to a strategy by which the child tries to assess whether or not two sets of elements are numerically equivalent (1952b, p. 88). And, as mentioned in footnote 4, adolescents possess a number of *operational schemas* also defined, ultimately, in terms of observable behavior in the face of certain tasks.

But if schemas are named by their referent action sequences, it is not completely accurate to say that they *are* these sequences and nothing more. To be sure, Piaget would certainly say that an infant who performs an organized sequence of grasping behaviors is in fact applying a grasping schema to reality, and that the behavior itself does constitute the schema (1952c, pp. 405-407; 1957c, pp. 46, 74). However, and the point is a rather subtle one, to say that a grasping sequence forms a schema is to imply more than the simple fact that the infant shows organized grasping behavior. It implies that assimilatory functioning has generated a specific cognitive *structure,* an organized *disposition* to grasp objects on repeated occasions. It implies that there has been a change in over-all cognitive organization such that a new behavioral totality has become part of the child's intellectual repertoire. Finally, it implies that a psychological "organ" has been created, *functionally* (but not, of course, *structurally)* equivalent to a physiological digestive organ in that it constitutes an instrument for incorporating reality "aliments" (1952c, pp. 13, 359). In brief, a schema is the organized overt behavior content which names it, but with important structural connotations not indigenous to the concrete content itself.[5] There are certain characteristics that a behavior sequence must possess in order to be conceptualized in schematic terms. To be sure, it is clear that schemas subsume behavior sequences of widely differing magnitude and complexity: compare the brief and simple sucking sequence of the neonate with the complex problem-solving strategies of a bright adult. Schemas come in all sizes and shapes. However, they all possess one general characteristic in common: the constituent behavior sequence is an organized totality. Thus, an action sequence, if it is to constitute a schema, must have a certain cohesiveness and must maintain its identity as a quasi-stable, repeatable unit. It must possess component actions which are tightly interconnected and governed by a core meaning. However elementary the schema, it is a schema precisely by virtue of the fact that the behavior components which it sets into motion form a strong whole, a recurrent and identifiable figure against a background of less tightly organized behaviors. Piaget states:

> As far as "totality" is concerned, we have already emphasized that every schema of assimilation constitutes a true totality, that is to say, an ensemble of sensorimotor elements mutually dependent or unable to function without each other. It is due to the fact that schemata present this kind of structure that mental assimilation is possible and any object whatever can be incorporated or serve as aliment to a given schema (1952c, p. 244).

One sees this "ensemble of sensorimotor elements mutually dependent or unable to function without each other" in any behavior series governed by a schema. For example, an elementary schema of prehension or grasping consists of interconnected reaching, finger-curling, and retracting subsequences which together make up an identifiable and repeatable unit. At certain phases of infant development, this particular schema, as a unit, tends to run itself off whenever an object is placed near the child. In Piaget's terms, all reachable objects become "aliments" which nourish this schema.

Another characteristic of schemas is hinted at by the phrase "class of similar action sequences" in our initial definition. A schema is a kind of concept, category, or underlying strategy which subsumes a whole collection of distinct but similar action sequences. For example, it is clear that no two grasping sequences are ever going to be exactly alike; a grasping schema — a "concept" of grasping — is nonetheless said to be

[5] The cognitive-structure connotations of the concept of schema become more apparent in the postinfancy periods. There, where so much of intelligent behavior is internalized and symbolic, Piaget's use of the term more unequivocally implies a plan of action, a strategy, or literally, a *scheme*. Note, for example, the *schema of intuitive qualitative correspondence alluded to above*.

operative when any such sequence is seen to emerge. Schemas therefore refer to *classes* of total acts, acts which are distinct from one another and yet share common features. Although the terms *schema* and *concept* are not completely interchangeable, Piaget has recognized a certain similarity between them: "The schema, as it appeared to us, constitutes a sort of sensorimotor concept, or more broadly, the motor equivalent of a system of relations and classes" (ibid., p. 385).

The earlier discussion of functional invariants suggests two more important general properties of schemas. A schema, being a cognitive structure, is a more or less fluid form or plastic organization to which actions and objects are assimilated during cognitive functioning. As Piaget expresses it, schemas are "mobile frames" successively applied to various contents (ibid., pp. 385-386). The fact that schemas accommodate to things (adapt and change their structure to fit reality) while assimilating them attests to their dynamic, supple quality. In short, they are the very antithesis of congealed molds into which reality is poured.

Again schemas, being structures, are both created and modified by intellectual functioning. They may be envisaged as the structural precipitates of a recurrent assimilatory activity. Not all the connotations of the term *precipitate* hold, however; schemas, unlike chemical precipitates, are far from being static and inert residues. The fact that schemas are created by functioning and are highly mobile and plastic is repeatedly stressed by Piaget:

> From the psychological point of view, the assimilatory activity . . . is consequently the primary fact. Now this activity, precisely to the extent that it leads to repetition, engenders an elementary schema . . . (ibid., p. 389).
>
> In effect, the schemata have always seemed to us to be not autonomous entities but the products of a continuous activity which is immanent in them and of which they constitute the sequential moments of crystallization (ibid., p. 388).

Operation and Development of Schemas

We have so far described schemas from a more or less static, attributive standpoint: what schemas are and what characteristics describe them. It remains to describe them in their dynamic aspect: how they function and change with development, how they relate to one another, and so forth.

One of the most important single characteristics of an assimilatory schema is its tendency toward repeated application. In fact, only behavior patterns which recur again and again in the course of cognitive functioning are conceptualized in terms of schemas. It is fundamental to Piaget's conception of development that organized behavior totalities are in evidence from birth onward (organization is a functional invariant, as we have seen) and that these totalities are set into motion again and again. Piaget speaks of *reproductive* or *functional assimilation* in referring to this ubiquitous tendency towards repetition (1951a, 1952c, 1954a). This concept of reproductive assimilation — the almost repetition-compulsion character of assimilation — will be shown to have important bearing on Piaget's conception of intellectual motivation. For the moment, suffice it to say it is indigenous to schemas that, once constituted, they apply themselves again and again to assimilable aspects of the environment.

In the course of this repeated exercise, individual schemas are of course transformed in several important ways (1952c, pp. 33-36); functioning not only creates structures but, as we have seen, changes them continually. First of all, schemas are forever extending their field of application so as to assimilate new and different objects. Piaget speaks of *generalizing assimilation* to indicate this important characteristic of assimilatory activity. In discussing reflex schemas in early infancy, he states:

> This need for repetition is only one aspect of a more general process which we can qualify as assimilation. The tendency of the reflex being to reproduce itself, it incorporates into itself every object capable of fulfilling the function of excitant. Two distinct phenomena must be mentioned here. . . . The first is what we may call "generalizing assimilation," that is to say, the incorporation of increasingly varied objects into the reflex schema. . . . Thus, according to chance contacts, the child, from the first two weeks of life, sucks his fingers, the fingers extended to him, his pillow, quilt, bed clothes, etc.; consequently he assimilates these objects to the activity of the reflex . . . the newborn child at once incorporates into the global schema of sucking a number of increasingly varied objects, whence the generalizing aspect of this process of assimilation (ibid., pp. 33-34).

Discrimination is the complement of generalization, as students of learning have long known, and the second important kind of change which schemas are said to undergo is that of internal differentiation. In a rudimentary way, the infant gradually discriminates objects which are to be sucked from those which are not to be sucked, or, at least, not to be sucked when one is very hungry. An elementary "recognition" of certain objects is the consequence of this differentiation within an initially undifferentiated schema, and Piaget speaks here of *recognitory assimilation*:

> This search and this selectivity seem to us to imply the beginning of differentiation in the global schema of sucking, and consequently a beginning of recognition, a completely practical and motor recognition, needless to say, but sufficient to be called recognitory assimilation (ibid., p. 36).
>
> More precisely, repetition of the reflex leads to a general and generalizing assimilation of objects to its activity, but, due to the varieties which gradually enter this activity (sucking for its own sake, to stave off hunger, to eat, etc.), the schema of assimilation becomes differentiated and, in the most important differentiated cases, assimilation becomes recognitory (ibid., p. 37).

Thus we see the three basic functional and developmental characteristics of all assimilatory schemas: repetition, generalization, and differentiation-recognition. These three are naturally contemporaneous in the stream of intellectual functioning, as Piaget points out. The following summary, although stated in connection with neonatal reflex schemas, holds true for the operation of schemas at any developmental level:

> In conclusion, assimilation . . . appears in three forms: cumulative repetition, generalization of the activity with incorporation of new objects to it, and finally, motor recognition. But in the last analysis, these three forms are but one: The reflex must be conceived as an organized totality whose nature it is to preserve itself by functioning and consequently to function sooner or later for its own sake (repetition) while incorporating into itself objects propitious to this functioning (generalized assimilation) and discerning situations necessary to certain special modes of its activity (motor recognition) (ibid., pp. 37-38).

We have so far seen how cognitive development proceeds through the vicissitudes of a single schema. Repetition consolidates and stabilizes it, as well as providing the necessary condition for change. Generalization enlarges it by extending its domain of application. And differentiation has the consequence of dividing the originally global schema into several new schemas, each with a sharper, more discriminating focus on reality. But it is characteristic of schemas not only to undergo individual changes of this kind but also to form ever more complex and interlocking relationships with other schemas. Two schemas may undergo separate developments up to a point, e.g., generalization to new objects, differentiation, etc., and then unite to form a single, supraordinate schema. The principal uniting relationship between two hitherto separate schemas is called *reciprocal assimilation,* that is, each schema assimilates the other:

> Organization exists within each schema of assimilation since . . . each one constitutes a real whole, bestowing on each element a meaning relating to this totality. But there is above all total organization; that is to say, coordination among the various schemata of assimilation. Now, as we have seen, this coordination is not formed differently from the single schemata, except only that each one comprises the other, in a reciprocal assimilation. . . . In short, the conjunction of two cycles or of two schemata is to be conceived as a new totality, self-enclosed (ibid., pp. 142-143).

A concrete illustration of the coalescence of originally separate and independent structures can be seen in Piaget's account of the development of visual schemas:

Thus it may be said that, independently of any coordination between vision and other schemata (prehension, touch, etc.), the visual schemata are organized among themselves and constitute more or less well-coordinated totalities. But the essential thing for this immediate question is the coordination of the visual schemata, no longer among themselves, but with the other schemata. Observation shows that very early, perhaps from the very beginnings of orientation in looking, coordinations exist between vision and hearing. . . . Subsequently the relationships between vision and sucking appear . . . then between vision and prehension, touch, kinesthetic impressions, etc. These intersensorial coordinations, this organization of heterogeneous schemata will give the visual images increasingly rich meanings and make visual assimilation no longer an end in itself but an instrument at the service of vaster assimilations. When the child seven or eight months old looks at unknown objects for the first time before swinging, rubbing, throwing and catching them, etc., he no longer tries to look for the sake of looking (pure visual assimilation in which the object is a simple aliment for looking), nor even for the sake of seeing (generalizing or recognitory visual assimilation in which the object is incorporated without adding anything to the already elaborated visual schemata), but he looks in order to act, that is to say, in order to assimilate the new object to the schemata of weighing, friction, falling, etc. There is therefore no longer only organization inside the visual schemata but between those and all the others. It is this progressive organization which endows the visual images with their meanings and solidifies them in inserting them in a total universe (ibid., pp. 75-76).

Assimilation-Accommodation Relationships

Assimilation and accommodation constitute the most fundamental ingredients of intellectual functioning. Both functions are present in every intellectual act, of whatever type and developmental level. However, while their co-occurrence may be said to be strictly invariant, this is not the case for the relationship between them. On the contrary, it will be shown that this relationship changes drastically, both within and between developmental stages. Because the functional invariants themselves make up the core of intelligence in Piaget's system, alterations in the relationship between them must necessarily have important consequences for the kind of intellectual functioning which takes place. For this reason an analysis of the vicissitudes of this relationship is as necessary to an account of Piaget's theory as was the basic description of the invariants themselves.

The first evolution of this kind occurs during the period of sensory-motor development in infancy and carries with it momentous changes in the relationship between cognizer and cognized. The basic characteristics of this evolution are then repeated again, in vertical décalage in the course of subsequent, postinfantile ontogenesis — and again with similar alterations in the organism-environment relationship. Finally, beginning early in the sensory-motor period but becoming increasingly important just subsequent to it, one sees changes of an essentially nondevelopmental kind, that is, momentary, short-lived variations in the assimilation-accommodation relationship within a given stage.

Developmental Changes in Infancy: The Basic Paradigm

The fundamental transformation in the assimilation-accommodation relationship which occurs during the first two years of life can be broadly stated as follows: development proceeds from an initial state of profound egocentrism, in which assimilation and accommodation are undifferentiated from each other and yet mutually antagonistic or opposed in their functioning, to a final state of objectivity and equilibrium in which the two functions are relatively separate and distinct, on the one hand, and coordinated and complementary, on the other (1954a, Conclusion). Let us now examine the constituent terms of this "law of evolution," as Piaget refers to it (ibid., p. 352).

The young infant begins life with certain elementary schemas which, in accord with the schema development we have already described, soon begin to stabilize, differentiate, and generalize through repeated applications to the surround (1952c). For example, the newborn tends to assimilate objects placed in his mouth to a slowly forming sucking schema, making the necessary crude accommodatory adjustments to object structure as he does so. Now, according to Piaget, it is characteristic of this early period that assimilation and accommodation are both undifferentiated one from the other and yet — paradoxically — antagonistic or opposed to each other in their action (1952c, pp. 19, 364; 1954a, pp. 350-354).

Early assimilation and accommodation are undifferentiated in that an object and the activity to which the object is assimilated constitute for the young infant a single, indivisible experience. Thus, the act of assimilating an object to a schema (the sensory-motor equivalent of making sense out of the object) is hopelessly confused with and undifferentiated from the accommodatory adjustments intrinsic to this act. It is not that the infant fails to take account of objects, i.e., accommodate his movements to their specific contours. This he does, and these clumsy accommodations produce changes in the assimilatory schemas. Rather, the infant has no way of distinguishing his acts from the reality events which these acts produce or the reality objects upon which they bear. In short, agent and object, ego and outside world are inextricably linked together in every infantile action, and the distinction between assimilation of objects to the self and accommodation of the self to objects simply does not exist. Piaget describes this pervasive undifferentiation as follows:

> In its beginnings, assimilation is essentially the utilization of the external environment by the subject to nourish his hereditary or acquired schemata. It goes without saying that schemata such as those of sucking, sight, prehension, etc., constantly need to be accommodated to things, and that the necessities of this accommodation often thwart the assimilatory effort. But this accommodation remains so undifferentiated from the assimilatory processes that it does not give rise to any special active behavior pattern but merely consists in an adjustment of the pattern to the details of the things assimilated. Hence it is natural that at this developmental level the external world does not seem formed by permanent objects, that neither space nor time is yet organized in groups and objective series, and that causality is not spatialized or located in things. In other words, at first the universe consists in mobile and plastic perceptual images centered about personal activity. But it is self-evident that to the extent that this activity is undifferentiated from the things it constantly assimilates to itself it remains unaware of its own subjectivity; the external world therefore begins by being confused with the sensations of a self unaware of itself, before the two factors become detached from one another and are organized correlatively (1954a, p. 351).

The opposition between assimilation and accommodation stems from this very undifferentiatedness. Since the infant cannot distinguish his actions from their environmental consequences, the necessity to make new and difficult accommodations in order to assimilate novel objects to already established schemas can only be experienced as frustrating. There is a fundamental antagonism, in this developmental period, between assimilation to the familiar, which is essentially "conservative," and accommodation to the novel, which is inherently "progressive." It is in the nature of assimilatory schemas, as we have seen, to apply and reapply themselves to any reality "aliments" which have the capacity to nourish and sustain them. A new, exploratory accommodation, instead of constituting a welcome foray into the unknown which will result in a differentiation of existing schemas, is at first experienced simply as a troublesome obstacle to habitual assimilation and is performed, as it were, only under duress:

> In their initial directions, assimilation and accommodation are obviously opposed to one another, since assimilation is conservative and tends to subordinate the environment to the organism as it is, whereas accommodation is the source of changes and bends the organism to the successive constraints of the environment (ibid., p. 352).

This initial state of undifferentiation and antagonism between the functional invariants essentially defines what is perhaps the most widely known (although perhaps also the most widely misunderstood) of Piagetian concepts: *egocentrism.* The concept of egocentrism is a most important one in Piaget's thinking and has been from the very earliest writings (e.g., 1926). It denotes a cognitive state in which the cognizer sees the world from a single point of view only — his own — but without knowledge of the existence of viewpoints or perspectives and, *a fortiori,* without awareness that he is the prisoner of his own. Thus Piaget's egocentrism is by definition an egocentrism of which the subject cannot be aware; it might be said that the egocentric subject is a kind of solipsist aware of neither self nor solipsism (1927a, 1954a). To quote Piaget again:

> Through an apparently paradoxical mechanism whose parallel we have described a propos of the egocentrism of thought of the older child, it is precisely when the subject is most self-centered that he knows himself the least, and it is to the extent that he discovers himself that he places himself in the universe and constructs it by virtue of that fact. In other words, egocentrism signifies the absence of both self-perception and objectivity, whereas acquiring possession of the object as such is on a par with the acquisition of self-perception (1954a, p. xii).

If assimilation and accommodation are undifferentiated and opposed in the radical egocentrism of the neonate, one of the most important fruits of sensory-motor development is their growing articulation and complementation. With a gradual separation between self and world, fine-grained accommodations to the nuances of things come to be experienced as interesting pursuits in and of themselves, pursuits now distinguished from the assimilations that these new discoveries make possible. The network of assimilatory schemas is now so rich and complex that it can with relative effortlessness extend itself to encompass and interpret the reality products which accommodation presents to it; reflexively, this very richness and complexity of schemas provides a guiding framework of meanings which can explicitly direct accommodatory exploration further and further into the unknown. If the invariants may be momentarily reified for the sake of exposition, one might say that schemas not only interpret better what accommodation presents, they also tell it what to look for on the next sortie. In summary, assimilation and accommodation are now at once articulated and in a state of complementary balance or relative equilibrium, one with the other:

> To the extent that new accommodations multiply because of the demands of the environment on the one hand and of the coordinations between schemata on the other, accommodation is differentiated from assimilation, and by virtue of that very fact becomes complementary to it. It is differentiated, because, in addition to the accommodation necessary for the usual circumstances, the subject becomes interested in novelty and pursues it for its own sake. The more the schemata are differentiated, the smaller the gap between the new and the familiar becomes, so that novelty, instead of constituting an annoyance avoided by the subject, becomes a problem and

invites searching. Thereafter and to the same extent, assimilation and accommodation enter into relations of mutual dependence. On the one hand, the reciprocal assimilation of the schemata and the multiple accommodations which stem from them favor their differentiation and consequently their accommodation; on the other hand, the accommodation to novelties is extended sooner or later into assimilation, because, interest in the new being simultaneously the function of resemblances and of differences in relation to the familiar, it is a matter of conserving new acquisitions and of reconciling them with the old ones (ibid., pp. 353-354).

Consequences of the Changing Assimilation-Accommodation Relationship During Sensory-Motor Development

With increasing differentiation and equilibration of the two functions during the sensory-motor period comes a development of great significance for intelligence: there is simultaneously a centrifugal process of gradual objectification of external reality and a centripetal process of burgeoning self-awareness — the self comes to be seen as an object among objects. Initially, as we have said, the infant knows neither self nor world as distinct and differentiated entities; he experiences only a mélange of feelings and perceptions concomitant with what an adult observer would label as contacts between his actions and outside objects. Cognition really begins at the boundary between self and object and with development invades both self and object from this initial "zone of undifferentiation" (ibid., p. 356). Knowledge of self and knowledge of objects are thus the dual resultants of the successive differentiation and equilibration of the invariant functions which characterize sensory-motor development. Piaget describes this highly significant consequence in the following way:

> Thus it may be seen that intellectual activity begins with confusion of experience and of awareness of the self, by virtue of the chaotic undifferentiation of accommodation and assimilation. In other words, knowledge of the external world begins with an immediate utilization of things, whereas knowledge of self is stopped by this purely practical and utilitarian contact. Hence there is simply interaction between the most superficial zone of external reality and the wholly corporal periphery of the self. On the contrary, gradually as the differentiation and coordination of assimilation and accommodation occur, experimental and accommodative activity penetrates to the interior of things, while assimilatory activity becomes enriched and organized. Hence there is a progressive formation of relationships between zones that are increasingly deep and removed from reality and the increasingly intimate operations of personal activity. Intelligence thus begins neither with knowledge of the self nor of things as such but with knowledge of their interaction, and it is by orienting itself simultaneously toward the two poles of that interaction that intelligence organizes the world by organizing itself.

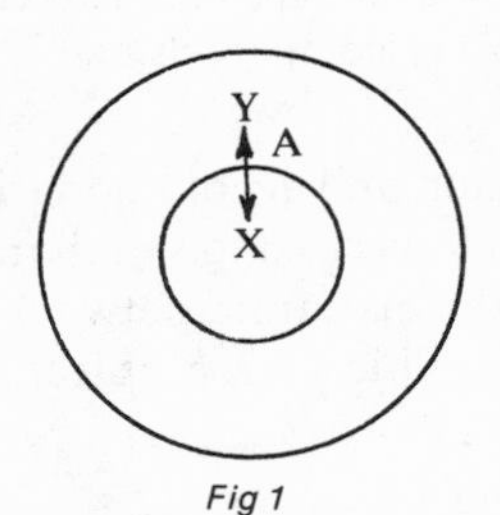

Fig 1

> A diagram will make the thing comprehensible. Let the organism be represented by a small circle inscribed in a large circle which corresponds to the surrounding universe. The meeting between the organism and the environment takes place at point A and at all analogous points, which are simultaneously the most external to the organism and to the environment itself. In other words, the first knowledge of the universe or of himself that the subject can acquire is knowledge relating to the most immediate appearance of things or to the most external and material aspect of his being. From the point of view of consciousness, this primitive relation between subject and object is a relation of undifferentiation, corresponding to the protoplasmic consciousness of the first weeks of life when no distinction is made between the self and the non-self. From the point of view of behavior this relation constitutes the morphologic-reflex organization, in so far as it is a necessary condition of primitive consciousness. But from this point of junction and undifferentiation A, knowledge proceeds along two complementary roads. By virtue of the very fact that all knowledge is simultaneously accommodation to the object and assimilation to the subject, the progress of intelligence works in the dual direction of externalization and internalization, and its two poles will be the acquisition of physical experience (➔Y) and the acquisition of consciousness of the intellectual operation itself (➔X) (ibid., pp. 354-356).

The centrifugal process, that of objectification and solidification of things in the milieu, merits particular attention for the relation it bears to developmental changes in assimilatory schemas. In brief, Piaget's position is that the objectification of sensory-motor reality is a consequence of the increasingly rich network of interrelated assimilatory schemas which the child constructs during this period (1952c, pp. 413-415). In early infancy, as we have seen, external objects are cognitively indissociated from the few simple action schemas (sucking, grasping, etc.) through which the baby comes in contact with them. There are no objects as such, only undifferentiated object-action amalgams. Objectification of reality — the population of the external milieu with things recognized as independent of a self which cognizes them — can come about only when objects come to be inserted into a whole network of intercoordinated schemas. For the adult, the object *chair* may be said to have status as an independent entity with discriminated properties as a function of the network of interconnected concepts or classes in which it can be inserted: "wooden," "four-legged," "to-sit-on," etc. So it is with the infant. The rattle gradually emerges as a thing distinct from his action only when he can insert it in multiple sensory-motor schemas, e.g., when he can apprehend it as something which can be visually fixated (schema 1), in order to grasp (schema 2), in order to shake and listen to the sound (schemas 3 and 4), and so forth. As Piaget puts it:

> It is coordination itself, that is to say, the multiple assimilation constructing an increasing number of relationships between the compounds "action × object" which explains the objectification (ibid., p. 415).

In summary, the constant working of assimilation and accommodation gives rise during sensory-motor development to an increasingly elaborate and complex schematic organization. In turn, the elaborate network of interrelated schemas so constituted makes it possible to see objects as things-out-there, independent of one's activity.

Developmental Changes After Infancy

It has been asserted that the fundamental change in the assimilation-accommodation relationship during infant development is one from undifferentiation and antagonism to differentiation and balance or equilibrium. The important cognitive consequences of this development have been shown to include a gradual change from an initial and profound egocentrism in which subject and object are indissociable to an articulation and objectification of outside reality and a parallel differentiation and objectification of the subject himself. The discussion of the concept of vertical décalage may have led the reader to anticipate that the ontogenetic scheme just described repeats itself again in post-infantile developmental periods.

Consider first the case for the general paradigm: evolution from an assimilation-accommodation undifferentiation and antagonism to differentiation and equilibrium. The concept of equilibrium is of central importance in Piaget's recent thinking about development. What is important is that Piaget sees the whole of ontogenetic development as a series of differing equilibrium states, or perhaps as a succession of phases or nodes in a grand equilibration process (1955e, 1957c, 1957f). The immediate and concrete significance of this rather complicated thesis for the assimilation-accommodation problem is as follows. Whereas we have said that assimilation and accommodation are in relative equilibrium or balance with respect to the two-year-old's overt, *sensory-motor* acts, this is not at all the case for his *symbolic* or *representational* manipulations of the world. Quite the contrary, the toddler's first attempts at conceptual or symbolic rapport with reality bear all the earmarks of relative disequilibrium, of relative undifferentiation and antagonism between assimilation and accommodation, seen in the neonate's first traffic with the sensory-motor world (1954a, Conclusion).

The fact that, through vertical décalage, the undifferentiation and opposition now concern symbolic-representational rather than sensory-motor cognitions does not mean the consequences are any less drastic. The preschool child shows every bit of the egocentrism which this undifferentiation and antagonism ordinarily imply. In fact, Piaget first introduced the concept of egocentrism, not in connection with sensory-motor intelligence, but to describe the character of the child's conceptual thought in the preschool and early school years (e.g., 1923, 1926). As we shall see, the egocentric preschooler, analogous to the egocentric three-month-old, is unaware of the fact that his representations of reality are in various ways distorted as a consequence of his failure to see things from points of view other than his own. As with the infant, his representational accommodations to reality are both confused with and antagonistic to one-perspective assimilations by which he is forced to interpret it. Since it is always a subject-object undifferentiation relative to a differentiation and equilibrium yet to be achieved, egocentrism of course reappears in attenuated form at genetic levels beyond those of neonate and preschooler. As Piaget points out, the subject in middle childhood and early adolescence can also be considered egocentric and in relative assimilation-accommodation disequilibrium with respect to certain more abstract symbolic manipulations which he tries to perform (Inhelder and Piaget, 1958, ch. 18).

Just as egocentrism reappears in its various forms in post-infantile development, so also do its opposite terms get repeatedly reconstituted at ever higher levels: objectification of external reality and undistorted knowledge of self. As in sensory-motor development, this dual progression forms an indissoluble whole: it is by conceptualizing the self as a distinct and separate center which perceives reality from a particular viewpoint that it becomes possible to correct egocentric distortions about reality; it is by penetrating deeper into the fabric of reality that self-knowledge becomes possible. Thus, with the various developmental levels of symbolic construction as well as with elemental sensory-motor behavior, cognition always begins on the margins of both self and milieu and works its way simultaneously into the inner regions of each.

Nondevelopmental Changes

The subject of the preceding pages has been *developmental* changes in the assimilation-accommodation relationship: namely, an ontogenetic movement towards differentiation and equilibrium between the invariant functions, first during infancy for sensory-motor cognitions and later, through vertical décalage, for cognitions of a conceptual-symbolic nature. In each case the ideal norm towards which intelligence moves is, to repeat, one or another form of equilibrium between the twin invariants. Although the concept of equilibrium will be shown to mean much more than this in Piaget's system, so far as assimilation and accommodation are concerned it connotes a kind of balance, a functional state in which potentially slavish and naively realistic (in the epistemological sense) accommodations to reality are effectively held in check by an assimilatory process which can organize and direct accommodations, and in which assimilation is kept from being riotously autistic by a sufficiency of continuing accommodatory adjustments to the real world. In short, intelligent functioning, when equilibrium obtains, is made up of a balanced recipe of about equal parts of assimilation and accommodation. Through this fine balance, a both realistic (accommodation) and meaningful (assimilation) rapport between subject and object is secured.

However, there are two important kinds of cognition which do not manifest this delicate balance between the functions. The first of these is *play* in the broad sense, including all forms of dream and dreamlike activity as well as the various kinds of play and make-believe. The second is termed *imitation* and includes all copying or imitative behavior, either in overt behavior or internally. Piaget has devoted a separate book (1951a) to the complexities of play and imitation. For the matter at hand, play and imitation are significant primarily as cognitive activities in which assimilation and accommodation are decidedly not in balance. In play the primary object is to mold reality to the whim of the cognizer, in other words, to assimilate reality to various schemas with little concern for precise accommodation to that reality. Thus, as Piaget puts it (ibid., p. 87), in play there is "primacy of assimilation over accommodation." In imitation, on the other hand, it is accommodation which reigns supreme. All energy is focused on taking exact account of the structural niceties of the reality one is imitating and in precisely dovetailing one's schematic repertoire to these details. In other words, as in play the primary concern is to adapt reality to the self (assimilation), in imitation the paramount object is to adapt the self to reality (accommodation). When referring to the case in which neither function dominates the other — that is, the case of equilibrium or balance — Piaget uses variously the terms *adapted intelligence, intelligent adaptation, adaptation,* or simply *intelligence* (ibid.). The distinction between the three can best be expressed by concrete example:

> . . . Intelligence tends towards permanent equilibrium between assimilation and accommodation. For instance, in order to draw an objective towards him by means of a stick, the child must assimilate both stick and objective to the schema of prehension and that of movement through contact, and he must also accommodate these schemas to the objects, their length, distance, etc., in accordance with the causal order hand-stick-objective. Imitation, on the contrary, is the continuation of accommodation . . . to which it subordinates assimilation. For instance, imitation will reproduce the motion made by the stick in reaching the objective, the movement of the hand thus being determined by those of the stick and the objective (which is by definition

> accommodation), without the hand actually affecting the objects (which would be assimilation). There is, however, a third possibility, that of assimilation *per se.* Let us assume, for instance, that the stick does not reach its objective and that the child consoles himself by hitting something else, or that he suddenly becomes interested in moving the stick for its own sake, or that when he has no stick he takes a piece of paper and applies the schema of the stick to it for fun. In such cases there is a kind of free assimilation, without accommodation to spatial conditions or to the significance of the objects. This is simply play, in which reality is subordinated to assimilation which is distorting, since there is no accommodation. Intelligent adaptation, imitation and play are thus the three possibilities, and they result according as there is stable equilibrium between assimilation and accommodation or primacy of one of these two tendencies over the other (ibid., pp. 85-86).

It remains to be shown in what sense the alterations in assimilation-accommodation relationship indigenous to play and imitation can be considered nondevelopmental in contrast to the developmental changes described earlier. It is certainly not the case that play and imitation themselves are nondevelopmental phenomena; on the contrary, from their indistinct beginnings in the sensory-motor period, both play and imitation are said to undergo a whole series of important genetic changes (ibid.). Rather, it is that within any given developmental stage (except possibly early sensory-motor development) cognition may show *either* a relative balance between assimilation and accommodation, and hence some kind of adapted intelligence, *or* an imbalance in one direction or the other, and hence play or imitation in any of their myriad forms. In other words, within any developmental period one sees momentary, essentially agenetic fluctuations in the assimilation-accommodation recipe — now play, now imitation, now intelligent adaptation. Thus, one might think of the relationship between the functions as changing simultaneously along two dimensions. There are the "horizontal" genetic changes from undifferentiation and antagonism towards equilibrium and mutual complementation; these are cyclic developments, and recur at ever higher genetic levels. Then there are the "vertical" nondevelopmental shifts in assimilation-accommodation balance which are moment-to-moment affairs superimposed on the developmental changes.

References

Kelly, G. A., *The Psychology of Personal Constructs.* vol. 1, *A Theory of Personality.* New York: Norton, 1955.

Piaget, J., "La pensée symbolique et la pensée de l'enfant," *Arch. Psychol.,* Geneva, 1923, 18, pp. 273-304.

———, *The Language and Thought of the Child.* New York: Harcourt, Brace, 1926.

———, "La Premiere Année de l'enfant," *Brit. J. Psychol.,* 1927, 18, pp. 97-120 (a).

———, "Du rapport des sciences avec la philosophie," *Synthese,* 1947, 6, pp. 130-150.

———, *Play, Dreams, and Imitation in Childhood.* New York: Norton, 1952(a).
———, *The Child's Conception of Number.* New York: Humanities Press, 1952(b).
———, *The Origins of Intelligence in Childhood.* New York: International Universities Press, 1952(c).
———, *The Construction of Reality in the Child.* New York: Basic Books, 1954.
———, "Rapport." In A. Michotte et al., *La perception.* Paris: Presses Univer. France, 1955, pp. 2-30.
———, "Logique et équilibre dans les comportements du sujet." In L. Apostel, B. Mandelbrot, and J. Piaget, *Logique et équilibre. Etudes d'épistémologie génétique,* vol. 2, pp. 27-117. Paris: Presses Univer. France, 1957(a).
———, "Le rolé de la notion d'équilibre dans l'explication en psychologie proc." 15th Int. Congr. Pyychol., 1957, pp. 51-62(b).
———' "Les étapes du développement mental," *Bull. Psychol.,* Paris, 1957-1958, 11, pp. 217-19, 347-51, 438-40, 520-22, 678-85, 878-82.
———, "La lecture de l'expérience," *Etudes d'épistémologie génétique,* vol. 5. Paris: Presses Univer. France, 1958, pp. 49-108(a)

B. INTERPERSONAL RELATIONS

CHAPTER 4

Beginnings of the Self-System

HARRY S. SULLIVAN

Three Aspects of Interpersonal Cooperation

We have got our human animal as far, in the process of becoming a person, as the latter part of infancy, and we find him being subjected more and more to the social responsibilities of the parent. As the infant comes to be recognized as educable, capable of learning, the mothering one modifies more and more the exhibition of tenderness, or the giving of tenderness, to the infant. The earlier feeling that the infant must have unqualified cooperation is now modified to the feeling that the infant should be learning certain things, and this implies a restriction, on the part of the mothering one, of her tender cooperation under certain circumstances.

Successful training of the functional activity of the anal zone of interaction accentuates a new aspect of tenderness — namely, the additive role of tenderness as a sequel to what the mothering one regards as good behavior. Now this is, in effect — however it may be prehended by the infant — a *reward,* which, once the approved social ritual connected with defecating has worked out well, is added to the satisfaction of the anal zone. Here is tenderness taking on the attribute of a reward for having learned something, or for behaving right.

Thus the mother, or the parent responsible for acculturation or socialization, now adds tenderness to her increasingly neutral behavior in a way that can be called rewarding. I think that very, very often the parent does this with no thought of rewarding the infant. Very often the rewarding tenderness merely arises from the pleasure of the mothering one in the skill which the infant has learned — the success which has attended a venture on the toilet chair, or something of that kind. But since tenderness in general is becoming more restricted by the parental necessity to train, these incidents of straightforward tenderness, following the satisfaction of a need like that to defecate, are really an addition — a case of getting something extra for good behavior — and this is, in its generic pattern, a reward. This type of learning can take place when the training procedure has been well adjusted to the learning capacity of the infant. The friendly response, the pleasure which the mother takes in something having worked out well, comes more and more to be something special in the very last months of infancy, whereas earlier, tenderness was universal when the mothering one was around, if she was a comfortable mothering one. Thus, to a certain extent, this type of learning can be called learning under the influence of reward — the reward being nothing more or less than tender behavior on the part of the acculturating or socializing mothering one.

Training in the functional activity of the oral-manual behavior — that is, conveying things by the hand to the mouth and so on — begins to accentuate the differentiation of anxiety-colored situations in contrast to approved situations. The training in this particular field is probably, in almost all cases, the area in which *grades of anxiety* first become of great importance in learning; as I have already stressed, behavior of a certain unsatisfactory type provokes increasing anxiety, and the infant learns to keep a distance from, or to veer away from, activities which are attended by increasing anxiety, just as the amoebae avoid high temperatures.

This is the great way of learning in infancy, and later in childhood — by the grading of anxiety, so that the infant learns to chart his course by mild forbidding gestures, or by mild states of worry, concern, or disapproval mixed with some degree of anxiety on the part of the mothering one. The infant plays, one might say, the old game of getting hotter or colder, in charting a selection of behavioral units which are not attended by an increase in anxiety. Anxiety in its most severe form is a rare experience after infancy, in the more fortunate courses of personality development, and anxiety as it is a function in chronologically adult life, in a highly civilized community confronted by no particular crisis, is never very severe for most people. And yet it is necessary to appreciate that it is anxiety which is responsible for a great part of the inadequate, inefficient, unduly rigid, or otherwise unfortunate performances of people; that anxiety is responsible in a basic sense for a great deal of what comes to a psychiatrist for attention. Only when this is understood, can one realize that this business of whether one is getting more or less anxious is in a large sense the basic influence which determines interpersonal relations — that is, it is not the motor, it does not call interpersonal relations into being, but it more or less directs the course of their development. And even in late infancy there is a good deal of learning by the anxiety gradient, particularly where there is a mothering one who is untroubled, but still intensely interested in producing the right kind of child; and this learning is apt to first manifest itself when the baby is discouraged from putting the wrong things in the mouth, and the like. This kind of learning applies over a vast area of behavior. But in this discussion I am looking for where things are apt to start.

Training of the manual-exploratory function — which I have discussed in connection with the infant's getting his hands near the anus, or into the feces, or, perhaps, in contact with the external genitals — almost always begins the discrimination of situations which are marked by what we shall later discuss as *uncanny emotion.* This uncanny feeling can be described as the abrupt supervention of *severe anxiety,* with the arrest of anything like the learning process, and with only gradual informative recall of the noted circumstances which preceded the extremely unpleasant incident.

Early in infancy, when situations approach the "all-or-nothing" character, the induction of anxiety is apt to be the sudden translation from a condition of moderate euphoria to one of very severe anxiety. And this severe anxiety, as I have said before, has a little bit the effect of a blow on the head, in that later one is not clear at all as to just what was going on at the time anxiety became intense. The educative effect is not by any means as simple and useful as is the educative effect in the other two situations which we have discussed, because the sudden occurrence of severe anxiety practically prohibits any clear prehension, or understanding, of the immediate situation. It does not, however, preclude recall, and as recall develops sufficiently so that one recalls what was about to occur when severe anxiety intervened — in other words, when one has a sense of what one's action was addressed to at the time when everything was disorganized by severe anxiety — then there comes to be in all of us certain areas of "uncanny taboo," which I think is a perfectly good way of characterizing those things which one stops doing, once one has caught himself doing them. This type of training is much less immediately useful, and, shall I say, is productive of much less healthy acquaintance with reality, than are the other two.

Good-Me, Bad-Me, and Not-Me

Now here I have set up three aspects of interpersonal cooperation which are necessary for the infant's survival, and which dictate learning. That is, these aspects of interpersonal cooperation require acculturation or socialization of the infant. Infants are customarily exposed to all of these before the era of infancy is finished. From experience of these three sorts — with rewards, with the anxiety gradient, and with practically obliterative sudden severe anxiety — there comes an initial personification of three phases of what presently will be *me,* that which is invariably connected with the sentience of *my body* — and you will remember that *my body* as an organization of experience has come to be distinguished from everything else by its self-sentient character. These beginning personifications of three different kinds, which have in common elements of the prehended body, are organized in about mid-infancy — I can't say exactly when. I have already spoken of the infant's very early double personification of the actual mothering one as the good mother and the bad mother. Now, at this time, the beginning personifications of *me* are *good-me, bad-me* and *not-me.* So far as I can see, in practically every instance of being trained for life, in this or another culture, it is rather inevitable that there shall be this tripartite cleavage in personifications, which have as their central tie — the thing that binds them ultimately into one, that always keeps them in very close relation — their relatedness to the growing conception of "my body."

Good-me is the beginning personification which organizes experience in which satisfactions have been enhanced by rewarding increments of tenderness, which come to the infant because the mothering one is pleased with the way things are going; therefore, and to that extent, she is free, and moves toward expressing tender appreciation of the infant. Good-me, as it ultimately develops, is the ordinary topic of discussion about "I."

Bad-me, on the other hand, is the beginning personification which organizes experience in which increasing degrees of anxiety are associated with behavior involving the mothering one in its more-or-less clearly prehended interpersonal setting. That is to say, bad-me is based on this increasing gradient of anxiety and that, in turn, is dependent, at this stage of life, on the observation, if misinterpretation, of the infant's behavior by someone who can induce anxiety.[1] The frequent coincidence of certain behavior on the part of the infant with increasing tenseness and increasingly evident forbidding on the part of the mother is the source of the type of experience which is organized as a rudimentary personification to which we may apply the term bad-me.

So far, the two personifications I have mentioned may sound like a sort of laboring of reality. However, these personifications are a part of the communicated thinking of the child, a year or so later, and therefore it is not an unwarranted use of inference to presume that they exist at this earlier stage. When we come to the third of these beginning personifications, *not-me,* we are in a different field — one which we know about only through certain very special circumstances. And these special circumstances are not outside the experience of any of us. The personification of not-me is most conspicuously encountered by most of us

[1]Incidentally, for all I know, anybody can induce anxiety in an infant, but there is no use cluttering up our thought by considering that, because frequency of events is of very considerable significance in all learning processes; and at this stage of life, when the infant is perhaps nine or ten months old, it is likely to be the mother who is frequently involved in interpersonal situations with the infant.

in an occasional dream while we are asleep; but it is very emphatically encountered by people who are having a severe schizophrenic episode, in aspects that are to them most spectacularly real. As a matter of fact, it is always manifest — not every minute, but every day, in every life — in certain peculiar absences of phenomena where there should be phenomena; and in a good many people — I know not what proportion — it is very striking in its indirect manifestations (dissociated behavior), in which people do and say things of which they do not and could not have knowledge, things which may be quite meaningful to other people but are unknown to them. The special circumstances which we encounter in grave mental disorders may be, so far as you know, outside your experience; but they were not once upon a time. It is from the evidence of these special circumstances — including both those encountered in everybody and those encountered in grave disturbances of personality, all of which we shall presently touch upon — that I choose to set up this third beginning personification which is tangled up with the growing acquaintance of "my body," the personification of *not-me.* This is a very gradually evolving personification of an always relatively primitive character — that is, organized in unusually simple signs in the parataxic mode of experience, and made up of poorly grasped aspects of living which will presently be regarded as "dreadful," and which still later will be differentiated into incidents which are attended by awe, horror, loathing, or dread.

This rudimentary personification of not-me evolves very gradually, since it comes from the experience of intense anxiety — a very poor method of education. Such a complex and relatively inefficient method of getting acquainted with reality would naturally lead to relatively slow evolution of an organization of experiences; furthermore, these experiences are largely truncated, so that what they are really about is not clearly known. Thus organizations of these experiences marked by uncanny emotion — which means experiences which, when observed, have led to intense forbidding gestures on the part of the mother, and induced intense anxiety in the infant — are not nearly as clear and useful guides to anything as the other two types of organizations have been. Because experiences marked by uncanny emotion, which are organized in the personification of not-me, cannot be clearly connected with cause and effect — cannot be dealt with in all the impressive ways by which we explain our referential processes later — they persist throughout life as relatively primitive, unelaborated, parataxic symbols. Now that does not mean that the not-me component in adults is infantile; but it does mean that the not-me component is, in all essential respects, practically beyond discussion in communicative terms. Not-me is part of the very "private mode" of living. But, as I have said, it manifests itself at various times in the life of everyone after childhood — or of nearly everyone, I can't swear to the statistics — by the eruption of certain exceedingly unpleasant emotions in what are called nightmares.

These three rudimentary personifications of *me* are, I believe, just as distinct as the two personifications of the objectively same mother were earlier. But while the personifications of me are getting under way, there is some change going on with respect to the personification of mother. In the latter part of infancy, there is some evidence that the rudimentary personality, as it were, it already fusing the previously disparate personifications of the good and the bad mother; and within a year and a half after the end of infancy we find evidence of this duplex personification of the mothering one as the good mother and the bad mother clearly manifested only in relatively obscure mental processes, such as these dreamings while asleep. But, as I have suggested, when we come to consider the question of the peculiarly inefficient and inappropriate interpersonal relations which constitute problems of mental disorder, there again we discover that the trend in organizing experience which began with this duplex affair has not in any sense utterly disappeared.

The Dynamism of the Self-System

From the essential desirability of being good-me, and from the increasing ability to be warned by slight increases of anxiety — that is, slight diminutions in euphoria — in situations involving the increasingly significant other person, there comes into being the start of an exceedingly important, as it were, secondary dynamism, which is purely the product of interpersonal experience arising from anxiety encountered in the pursuit of the satisfaction of general and zonal needs. This secondary dynamism I call the *self-system.* As a dynamism it is secondary in that it does not have any particular zones of interaction, any particular physiological apparatus, behind it; but it literally uses all zones of interaction and all physiological apparatus which is integrative and meaningful from the interpersonal standpoint. And we ordinarily find its ramifications spreading throughout interpersonal relations in every area where there is any chance that anxiety may be encountered.

The essential desirability of being good-me is just another way of commenting on the essential undesirability of being anxious. Since the beginning personification of good-me is based on experience in which satisfactions are enhanced by tenderness, then naturally there is an essential desirability of living good-me. And since sensory and other abilities of the infant are well matured by now — perhaps even space perception, one of the slowest to come along, is a little in evidence — it is only natural that along with this essential desirability there goes increasing ability to be warned by slight forbidding — in other words, by slight anxiety. Both these situations, for the purpose now under discussion, are situations involving another person — the mothering one, or the congeries of mothering ones — and she is becoming increasingly significant because, as I have already said, the manifestation of tender cooperation by her is now complicated by her attempting to teach, to socialize the infant; and this makes the relationship more complex, so that it requires better, more effective differentiation by the infant of forbidding gestures, and so on. For all these reasons, there comes into being in late infancy an organization of experience which will ultimately be of nothing less than stupendous importance in personality, and which comes entirely from the interpersonal relations in which the infant is now involved — and these interpersonal relations have their motives (or their motors, to use a less troublesome word) in the infant's general and zonal needs for satisfaction. But out of the social responsibility of the mothering one, which gets involved in the satisfaction of the infant's needs, there comes the organization in the infant of what might be said to be a dynamism directed at how to live with this significant other person. The self-system thus is an organization of educative experience called into being by the necessity to avoid or to minimize incidents of anxiety.[2] The functional activity of the self-system — I am now speaking of it from the general standpoint of a dynamism — is primarily directed to avoiding and minimizing this disjunctive tension of anxiety, and thus indirectly to protecting the infant from this evil eventuality in connection with the pursuit of satisfactions — the relief of general or of zonal tensions.

Thus we may expect, at least until well along in life, that the components of the self-system will exist and manifest functional activity in relation to every general need that a person has, and to every zonal need that the excess supply of energy to the various zones of interaction gives rise to. How conspicuous the "sector" of the self-system connected with any particular general need or zonal need will be, or how frequent its manifestations, is purely a function of the past experience of the person concerned.

I have said that the self-system begins in the organizing of experience with the mothering one's forbidding gestures, and that these forbidding gestures are refinements in the personification of the bad mother; this might seem to suggest that the self-system comes into being by the *incorporation* or *introjection* of the bad mother, or simply by the introjection of the mother. These terms, incorporation or introjection, have been used in this way, not in speaking of the self-system, but in

[2]Since *minimize* in this sense can be ambiguous, I should make it clear that I refer, by minimizing, to moving, in behavior, in the direction which is marked by diminishing anxiety. I do not mean, by minimize, to "make little of," because so far as I know, human ingenuity cannot make little of anxiety.

speaking of the psychoanalytic superego, which is quite different from any conception of the self-system. But, if I have been at all adequate in discussing even what I have presented thus far, it will be clear that the use of such terms in connection with the development of the self-system is a rather reckless oversimplification, if not also a great magic verbal gesture the meaning of which cannot be made explicit. I have said that the self-system comes into being because the pursuit of general and zonal needs for satisfaction is increasingly interfered with by the good offices of the mothering one in attempting to train the young. And so the self-system, far from being anything like a function of or an identity with the mothering one, is an organization of experience for avoiding increasing degrees of anxiety which are connected with the educative process. But these degrees of anxiety cannot conceivably, in late infancy (and the situation is similar in most instances at any time in life), mean to the infant what the mothering one, the socializing person, believes she means, or what she actually represents, from the standpoint of the culture being inculcated in the infant. This idea that one can, in some way, take in another person to become a part of one's personality is one of the evils that comes from overlooking the fact that between a doubtless real "external object" and a doubtless real "my mind" there is a group of processes — the act of perceiving, understanding, and what not — which is intercalated, which is highly subject to past experience and increasingly subject to foresight of the neighboring future. Therefore, it would in fact be one of the great miracles of all time if our perception of another person were, in any greatly significant number of respects, accurate or exact. Thus I take some pains at this point to urge you to keep your mind free from the notion that I am dealing with something like the taking over of standards of value and the like from another person. Instead, I am talking about the organization of experience connected with relatively successful education in becoming a human being, which begins to be manifest late in infancy.

When I talk about the self-system, I want it clearly understood that I am talking about a *dynamism* which comes to be enormously important in understanding interpersonal relations. This dynamism is an explanatory conception; it is not a thing, a region, or what not, such as superegos, egos, ids, and so on.[3] Among the things this conception explains is something that can be described as a quasi-entity, the personification of the self.

[3]Please do not bog down unnecessarily on the problem of whether my self-system ought to be called the superego or the ego. I surmise that there is some noticeable relationship, perhaps in the realm of cousins or closer, between what I describe as the personification of the self and what is often considered to be the psychoanalytic ego. But if you are wise, you will dismiss that as facetious, because I am not at all sure of it; it has been so many years since I found anything but headaches in trying to discover parallels between various theoretical systems that I have left that for the diligent and scholarly, neither of which includes me.

The personification of the self is what you are talking about when you talk about yourself as "I," and what you are often, if not invariably, referring to when you talk about "me" and "my." But I would like to make it forever clear that *the relation of personifications to that which is personified is always complex and sometimes multiple;* and that *personifications are not adequate descriptions of that which is personified.* In my effort to make that clear, I have gradually been compelled, in my teaching, to push the beginnings of things further and further back in the history of the development of the person, to try to reach the point where the critical deviations from convenient ideas become more apparent. Thus I am now discussing the beginning of the terrifically important self-dynamism as the time when — far from there being a personification of the self — there are only rudimentary personifications of good-me and bad-me, and the much more rudimentary personification of not-me. These rudimentary personifications constitute anything but a personification of the self such as you all believe you manifest, and which you believe serves its purpose, when you talk about yourselves one to another in adult life.

The Necessary and Unfortunate Aspects of the Self-System

The origin of the self-system can be said to rest on the irrational character of culture or, more specifically, society. Were it not for the fact that a great many prescribed ways of doing things have to be lived up to, in order that one shall maintain workable, profitable, satisfactory relations with his fellows; or, were the prescriptions for the types of behavior in carrying on relations with one's fellows perfectly rational — then, for all I know, there would not be evolved, in the course of becoming a person, anything like the sort of self-system that we always encounter. If the cultural prescriptions which characterize any particular society were better adapted to human life, the notions that have grown up about incorporating or introjecting a punitive, critical person would not have arisen.

But even at that, I believe that a human being without a self-system is beyond imagination. It is highly probable that the type of education which we have discussed, even probably the inclusion of certain uncanny experience that tends to organize in the personification of not-me, would be inevitable in the process of the human animal's becoming a human being. I say this because the enormous capacity of the human animal which underlies human personality is bound to lead to exceedingly intricate specializations — differentiations of living, function, and one thing and another; to maintain a workable, profitable, appropriate, and adequate type of relationship among the great numbers of people that can become involved in a growing society, the young have to be taught a vast amount before they begin to be significantly involved in society outside the home group. Therefore, the special secondary elaboration of the sundry types of learning — which I call the self-system — would, I believe, be a ubiquitous aspect of all really human beings in any case. But in an ideal culture, which has never been approximated and at the present moment looks as if it never will be, the proper function of the self-system would be conspicuously different from its actual function in the denizens of our civilization. In our civilization, no parental group actually reflects the essence of the social organization for which the young are being trained in living; and after childhood, when the family influence in acculturation and socialization begins to be attenuated and augmented by other influences, the discrete excerpts, you might say, of the culture which each family has produced as its children come into collision with other discrete excerpts of the culture — all of them more or less belonging to the same cultural system, but having very different accents and importances mixed up in them. As a result of this, the self-system in its actual functioning in life in civilized societies, as they now exist, is often very unfortunate. But do not overlook the fact that the self-system comes into being because of, and can be said to have as its goal, the securing of necessary satisfaction without incurring much anxiety. And however unfortunate the manifestations of the self-system in many contexts may seem, always keep in mind that, if one had no protection against very severe anxiety, one would do practically nothing — or, if one still had to do something, it would take an intolerably long time to get it done.

So you see, however truly the self-system is the principal stumbling block to favorable changes in personality — a point which I shall develop later on — that does not alter the fact that it is also the principal influence that stands in the way of unfavorable changes in personality. And while the psychiatrist is skillful, in large measure, in his ability to formulate the self-system of another person with whom he is integrated, and to, shall I say, "intuit" the self-system aspects of his patient which tend to perpetuate the type of morbid living that the patient is showing, that still, in no sense, makes the self-system something merely to be regretted. In any event, it is always before us, whether we regret or praise it. This idea of the self-system is simply tremendously important in understanding the vicissitudes of interpersonal relations from here on. If we understand how the self-system begins, then perhaps we will be able to follow even the most difficult idea connected with its function.

The self-system is a product of educative experience, part of which is of the character of reward, and a very important part of which has the graded anxiety element that we have spoken of. But quite early in life, anxiety is also a very conspicuous aspect of the self-dynamism *function.* This is another way of saying that experience functions in both recall and foresight. Since troublesome experience, organized in the self-system, has been experience connected with increasing grades of anxiety, it is not astounding that this element of recall, functioning on a broad scale, makes the intervention of the self-dynamism in living tantamount to the warning, or foresight, of anxiety. And warning of anxiety means noticeable anxiety, really a warning that anxiety will get worse.

There are two things which I would like to mention briefly at this point. One is the infant's discovery of the unobtainable, his discovery of situations in which he is powerless, regardless of all the cooperation of the mothering one. The infant's crying for the full moon is an illustration of this. Now even before the end of infancy, it is observable that these unattainable objects gradually come to be treated *as if* they did not exist; that is, they do not call out the expression of zonal needs. This is possibly the simplest example of a very important process manifested in living which I call *selective inattention.*

The other thing I would like to mention is this: Where the parental influence is peculiarly incongruous to the actual possibilities and needs of the infant — before speech has become anything except a source of marvel in the family, before it has any communicative function whatever, before alleged words have any meaning — there can be inculcated in this growing personification of bad-me and not-me disastrous distortions which will manifest themselves, barring very fortunate experience, in the whole subsequent development of personality.

CHAPTER 5

Mother-Child Interaction

ANNELIESE F. KORNER

In the last twenty years, an increasing number of investigations have demonstrated the crucial importance of adequate mothering for early child development and sound ego growth. Maternal deprivation studies, scrutiny of the early ego development of institutionalized children, and the cumulative evidence from clinical case histories all point to the inescapable relationship between inadequate or insufficient maternal care and a variety of ego deficits in the child.

In more recent years numerous investigations have addressed themselves to the opposite side of this question: To what extent do characteristics in the child himself influence parent-child interaction? This search for factors in the child that may jeopardize a mother's capacity to relate to him was largely stimulated by work with childhood psychoses and infantile autism. From this concern with the extreme deviations has sprung an interest in the individual differences among normal children.

Research findings clearly demonstrate that individual differences in disposition and temperament do exist among newborn babies, but there is a lag in incorporating this fact in child-rearing practices and in considering these basic differences in diagnostic and therapeutic formulations. It is always experience that is held to account for any deviation.

Mother Is Held Responsible

Often when parent-child interaction is discussed, reference is really being made to what a mother does *with* or *to* a child. Unwittingly, the interaction is seen as a one-way street rather than as a true reciprocal exchange. Individual differences among children are rarely accorded sufficient importance in explaining variations in development and the differences in ease of mastering various developmental tasks. It is always the mother's handling, *her* personality, and *her* difficulties that are considered to be at the root of the child's problems. While the mental health professions, in their recognition of the tremendous importance of adequate mothering, have done a great deal to promote enlightened child care practices, they also have helped create a generation of guilt-ridden parents. In some quarters, to understand any difficulty in any phase of any child's development the dictum is *"cherchez la mere."* In the process, the child — as a separate organism with his own propensities and vulnerabilities —is overlooked.

The notion that the parent is at the root of his child's problems and is also responsible for the normality of his development has become especially popular in the United States. Undoubtedly this faith has deep roots in our cultural heritage. The extensive migration to this country was motivated by the wish to leave behind all the Old World's inequalities of class and birth. Our culture, more than most, has become child and youth centered, partly, no doubt, from the hope that through the new generation the frustrations and limitations of the Old World generation will be offset. The idea that there might be basic genetic and biological differences among people had to be rejected as too close to Old World values, too close to being undemocratic and unequalitarian. Furthermore, since it was felt that there was nothing that could be done about these differences, recognition of them would only lead to therapeutic and educational pessimism. Therefore, it was better to proceed as if they did not exist.

Clearly, by making the parent totally responsible for the normality or inadequacy of his offspring's development, the baffling and complex job of child rearing is reduced by some important variables —those presented by the individuality of the child. There are obvious advantages to taking the position that environment and experience are the sole determinants of how development will proceed — all that is necessary is to learn to do the "correct thing," whatever that might be. If things do not work out, we tell ourselves we must have mishandled the situation in some way. This position, although fraught with guilt-producing elements, at least gives us the illusion that, at least potentially, we are in control of our destiny and the destiny of our children (Korner, 1961, pp. 339-42).

Yet, in the long run, by not taking into account the child's individual makeup and by concentrating purely on what the mother did or did not do, the mental health professions may become helpless, reaching the limits of their effectiveness through use of this approach. What should emerge, after a good deal of research, are many different kinds of child-rearing methods, methods that are geared to deal variably with the strengths and vulnerabilities of a given child's endowment.

Individual Differences

The importance of variations in ego and drive endowment for the genesis and type of neurosis has long been recognized by psychoanalysts, but, probably for the cultural reasons mentioned earlier, these variations have been neglected as a field of investigation. Freud, in one of his last contributions, "Analysis Terminable and Interminable," stated his conviction that "each individual ego is endowed from the beginning with his own peculiar dispositions and tendencies" (Freud, 1950, pp. 316-57). In an earlier paper, which discussed a case of obsessional neurosis, he linked these early dispositions to later type of illness (Freud, 1946, pp. 122-32). Hartmann outlined how the primary ego apparatuses may influence later choice of defense (Hartmann, 1950, pp. 94-96). Benjamin, who has done a great deal of longitudinal work with children, stressed how primary endowment factors may mold the way a child responds selectively to environmental influences. As he put it:

> Not only can innate differences in drive organization, in ego functions and in maturational rates determine different responses to objectively identical experiences, but they can also help determine what experiences will be experienced and how they will be perceived. (Benjamin, 1961, pp. 19-42).

What are some of these individual differences at birth that may influence the way a mother can relate to her infant and that may throw an individual cast on the manner in which a child will experience and master each developmental step?

Margaret Fries was one of the earliest research workers to seize on what she termed an infant's *congenital activity type,* relating it to later personality development through longitudinal follow-up (Fries and Woolf, 1953, pp. 48-62). A host of workers are currently investigating variations in *autonomic reactivity* among newborns, hoping to detect the physiological antecedents to different types of affect management and the precursors to choices in psychosomatic disease. Escalona has been studying individual differences in *sensory responsiveness* among infants after her clinical observations alerted her to the fact that many autistic children have a history of unusual sensory sensitivity. From their histories it was suspected that these children may have had to erect an unusually high stimulus barrier to protect themselves against their sensory vulnerability (Bergman and Escalona, 1949, pp. 333-52). Thomas et al. in New York and Meili in Switzerland have been studying variations in *responses to novel situations* among infants and have found this to be an enduring characterological trait that emerges long before accommodation to and identification with parental attitudes seem plausible (Thomas et al., 1963; Meili; 1957). Mary Shirley (1933) noted irritability to be such an enduring trait. Peter Wolff found that, up to a point, hunger tension improves hand-mouth co-ordination in the newborn, but with increased hunger tension coordination breaks down. He noted individual variations in the point of time between feedings when *dedifferentiation* (reversion to a more generalized or primitive condition) of the baby's behavior sets in. This has implications not only for the optimal time to feed a given baby, but may also have ramifications for later development. For example, if some babies are more readily and more intensely vulnerable to dedifferentiation of behavior under the pressures of drive tension, does this represent a general vulnerability to regression? If so, this is an extremely early manifestation of one of the most lasting and potentially pathogenic ego qualities.

There are many other characteristics observable at birth that differentiate one baby from another. The author's as yet incomplete observations suggest that while sucking on a nipple babies' *response to massive stimulation* can be freezing, sucking more vigorously, or global motor discharge.[1] What a prototype for possible later defense choice! The baby's reaction, particularly if consistent, may herald later tendencies to motor discharge or displacement behavior aimed at tension reduction; it may foreshadow the need for warding-off mechanisms or propensities toward flooding of the ego apparatuses in the face of massive stimulation.

In addition, there are differences in the *quantity and mode of mouthing behavior* among newborn infants. Some babies mouth much more than others, irrespective of hunger. Some suck vigorously, others passively. Some are primarily chewers. Some drool, others spit. In Erikson's terms the quality of the infant's mouthing may be primarily incorporative, retentive, eliminative, or intrusive in character (Erikson, 1950). Might not these variations be the earliest manifestations of both the strength and the direction of instinctual drive endowment that may predispose to greater or lesser difficulties in resolving the developmental tasks posed by later psychosexual stages? They may then be one nonexperiential factor in making some babies more susceptible than others to certain zonal and modal fixations in their psychosexual development. (For further elaboration of these issues, see Korner, 1964.)

"Reading" a Baby's Needs

As babies differ, so do their specific needs for care. Some babies are easier to "read" than others, and some mothers are better "readers." For mutuality to be established the communications must go both ways. What makes a baby easy to "read"? Certainly the clarity with which he can express hunger, discomfort, and sleepiness, and the predictability and periodicity of these events. Babies differ markedly in this capacity to

[1] These, and the observations made later in this paper, are a part of an investigation supported by a Public Health Service Research Grant from the National Institute of Child Health and Human Development.

communicate their needs. Their expressions range from fuzzy and vague to clear-cut. What makes a mother a good "reader" at this stage of development? An almost symbiotic capacity to tune in and to make her ego an extension of the baby's almost nonexistent one.

The formation of consistent and trustful internal sets of expectations may be hampered not only by a mother's lack of dependability, but also by the baby's unpredictable and indistinct experience of various internal states. And yet, it is more usually suspected that the mother's care was erratic and that this solely explains a child's difficulty in establishing reliable sets of internal expectations.

There are many examples in which either internal factors or maternal handling may account for a given development. A good illustration is Rubinfine's hypothesis that severe pathology may emerge from a much too early perception by the baby of separateness from the mother owing to a high level of need tension or pain from which the mother cannot or does not protect the child (Rubinfine, 1962, pp. 265-82). This expresses the whole duality of the problem — that the mother *cannot* or *does not* protect the child. An early and painful feeling of separateness may thus emerge from the mother's aloofness or insensitivity or from the strength of internal conditions over which the mother, in spite of her best efforts to comfort the child, may have little control.

Furthermore, there appear to be capacities within the newborn himself that make early perception of separateness from the mother a more or less likely development. Ongoing research by Birns and Bridger at Albert Einstein College of Medicine in New York and in the author's own laboratory suggests that there may be a wide variation in the degree to which babies can be soothed and in the extent to which they are capable of comforting themselves (Bridger and Birns, 1963, pp. 1-6). Even casual observation confirms this, such as the fact that some babies begin to suck their thumbs the first day of life.

More Knowledge Is Needed

The implications of these research findings are obvious: we need to know a great deal more. We have just begun to get a glimpse of the vicissitudes and variations of the developmental process and to understand the crucial import of adequate mothering.

Adequate mothering has to take into account not only individual variations of the growth process, but also the infant's phase-specific needs. In a short eighteen months, to meet her infant's needs, a mother has to — in sequence — be a symbiotic object, a shielder and mediator of excessive stimulation (during the first three months), be a provider of stimulation and reciprocator of the child's first play (between age 2½ and 5 months), accept the severance of the symbiotic tie by welcoming his first initiative (5-

9 months), tolerate the baby's demands that she alone fulfill his needs (9-15 months), and tolerate his self-assertiveness, which usually develops between 12 and 18 months.[2] By 18 months, life has just begun as far as diversification of a child's needs is concerned. What an artist the mother must be to meet adequately all these different needs!

Obviously, the mother is going through a recapitulation of her own development and, at best, she has an opportunity to master residual difficulties of her own. This is what Erikson must have meant when he talked about children bringing up their parents. Under less favorable circumstances, the mother's own unresolved difficulty with regard to a particular developmental issue may cause her to falter in dealing with it. This is where therapeutic intervention can be most helpful. By this is meant helping the mother over a specific hump and not demanding of her, as is so often done, unconditional surrender as a long-term patient.

What has been said has obvious practical implications for child rearing, parental guidance, and casework. It is clear that we cannot work by rote or formula. There is no *one* correct way to deal with children or to be a parent. What may be correct for one child may be misguided for another, what may be good practice for one period of development may be entirely out of place in another. A flexible frame of reference is imperative if we wish to understand and help resolve a specific difficulty. It is equally imperative that such a flexible frame of reference be conveyed to parents. A healthy respect for individual differences and for the changing requirements of child care will reduce uniform expectations of what is normal and result in less zeal regarding the efficacy of many child care practices. Most important, by learning to appreciate the separateness and individuality of his child, the parent may come to feel less guilty about the crises in his development. In turn, this should free him for more intelligent action.

References

Benjamin, John D., "The Innate and the Experiential in Development," in *Lectures in Experimental Psychiatry.* Pittsburgh: University of Pittsburgh Press, 1961, pp. 19-42.

Bergman, Paul, and Sybille Escalona, "Unusual Sensitivity in Very Young Children," in *Psychoanalytic Study of the Child,* vols. 3/4, pp. 333–52. New York: International Universities Press, 1949.

Bridger, Wagner H., and Beverly Birns, "Neonates, Behavioral and Autonomic Responses During Soothing," in *Recent Advances in Biological Psychiatry,* vol. 5, pp. 1-6. New York: Plenum Press, 1963.

Erikson, Erik H., "The Theory of Infantile Sexuality," in *Childhood and Society.* New York: Norton, 1950.

Freud, Sigmund, "The Predisposition of Obsessional Neurosis: A Contribution to the

[2]This sequence of the infant's changing needs emerged from a longitudinal study conducted by Eleanor Pavenstedt and Louis W. Sander. See Louis W. Sander, 1962, pp. 141-66.

Problem of the Option of Neurosis," in *Collected Papers,* vol. 2, pp. 122-32. London: Hogarth Press, 1946.

———, "Analysis Terminable and Interminable," in *Collected Papers,* vol. 5, pp. 316-57. London: Hogarth Press, 1950.

Fries, Margaret E., and Paul Y. Woolf, "Some Hypotheses on the Role of the Congenital Activity Type in Personality Development," in *Psychoanalytic Study of the Child,* vol. 8, pp. 48-62. New York: International Universities Press, 1953.

Hartman, Heinz, "Comments on the Psychoanalytic Theory of the Ego," in *Psychoanalytic Study of the Child,* vol. 5, pp. 94-96. New York: International Universities Press, 1950.

Korner, Anneliese F., "The Parent Takes the Blame," *Social Casework,* vol. 42, no. 7 (July 1961), pp. 339-42.

———, "Some Hypotheses Regarding the Significance of Individual Differences at Birth for Later Development," in *Psychoanalytic Study of the Child,* vol. 19. New York: International Universities Press, 1964.

Meili, Richard, "Anfange der Charakterentwicklung," *Beitrage zur genetischen Characterologic,* no. 1. Bern, Switzerland: Hans Huber, 1957.

Rubinfine, David L., "Maternal Stimulation, Psychic Structure and Early Object Relations," in *Psychoanalytic Study of the Child,* vol. 17, pp. 265-82. New York: International Universities Press, 1962.

Sander, Louis W., "Issues in Early Mother-Child Interaction," *Journal of the American Academy of Child Psychiatry,* vol. 1, no. 1 (January 1962), pp. 141-66.

Shirley, Mary M., *The First Two Years: A Study of Twenty-five Babies,* Institute of Child Welfare Monograph Series No. 8, vol. 3, *Personality Manifestation.* Minneapolis: University of Minnesota Press, 1933.

Thomas, Alexander, et al., *Behavioral Individuality in Early Childhood.* New York: New York University Press, 1963.

Wolff, Peter H., *The Causes, Controls and Organization of Behavior in the Newborn,* in the monograph series *Psychological Issues.* New York: International Universities Press, in press.

CHAPTER 6

What Is the Optimal Learning Environment for the Young Child?

BETTYE M. CALDWELL

A truism in the field of child development is that the milieu in which development occurs influences that development. As a means of validating the principle, considerable scientific effort has gone into the Linnaean task of describing and classifying milieus and examining developmental consequences associated with different types. Thus we know something about what it is like to come of age in New Guinea,[29] in a small Midwestern town,[4] in villages and cities in Mexico[25] in families of different social-class level in Chicago[12] or Boston,[27,31] in a New York slum,[46] in Russian collectives,[9] in Israeli Kibbutzim,[23, 34, 41] in the eastern part of the United States,[33] and in a Republican community in Central New York.[10] Most of these milieu descriptions have placed great stress on the fact that they were just that and nothing more, i.e., they have expressed the customary scientific viewpoint that to describe is not to judge or criticize. However, in some of the more recent milieu descriptions which have contrasted middle- and lower-class family environments or highlighted conditions in extreme lower-class settings,[31, 46] often more than a slight suggestion has crept in that things could be better for the young child from the deprived segment of the culture. Even so, there remains a justifiable wariness about recommending or arranging any environment for the very young child other than the type regarded as its natural habitat, viz., within its own family.

Of course, optimizing environments are arranged all the time under one guise or another. For example, for disturbed children whose family environments seem effectively to reinforce rather than extinguish psychopathology, drastic alterations of milieu often are attempted. This may take the form of psychotherapy for one or both parents as well as the disturbed child, or it may involve total removal of the child from the offending environment with temporary or prolonged placement of the child in a milieu presumably more conducive to normal development. Then there is the massive milieu arrangement formalized and legalized as "education" which profoundly affects the lives of all children once they reach the age of five or six. This type of arrangement is not only tolerated but fervently endorsed by our culture as a whole. In fact, any subculture (such as the Amish) which resists the universalization of this pattern of milieu arrangement is regarded as unacceptably deviant and as justifying legal action to enforce conformity.

For very young children, however, there has been a great deal of timidity about conscious and planned arrangement of the developmental milieu, as though the implicit assumption has been made that any environment which sustains life is adequate during this period. This is analogous to suggesting that the intrauterine environment during the period of maximal cellular proliferation is less important than it is later, a suggestion that patently disregards evidence from epidemiology and experimental embryology. The rate of proliferation of new behavioral skills during the first three years of life and the increasing accumulation of data pointing to the relative permanence of deficit acquired when the environment is inadequate during this period make it mandatory that careful attention be given to the preparation of the developmental environment during the first three years of life.

Conclusions from Inadequate Environments

It is, of course, an exaggeration to imply that no one has given attention to the type of environment which can nourish early and sustained growth and development. For a good three decades now infants who are developing in different milieus have been observed and examined, and data relating to their development have made it possible to identify certain strengths and deficiencies of the different types of environments. Of all types described, the one most consistently indicted by the data is the institution. A number of years ago Goldfarb[19] published an excellent series of studies contrasting patterns of intellectual functioning shown by a group of adopted adolescents who had been reared in institutions up to age three and then transferred to foster homes or else placed shortly after birth in foster homes. The development of the group that had spent time in the institution was deficient in many ways compared to the group that had gone directly into foster homes. Provence and Lipton[33] recently published a revealing description of the early social and intellectual development of infants in institutions, contrasting their development with that of home-reared children. On almost every measured variable the institutional infants were found wanting — less socially alert and outgoing, less curious, less responsive, less interested in objects, and generally less advanced. The findings of this study are almost prototypic of the literature in the field, as pointed out in excellent reviews by Yarrow[47] and Ainsworth.[1]

Although there are many attributes in combination that comprise the institutional environment, the two most obvious elements are (1) absence of a mother and (2) the presence of a group. These basic characteristics have thus been identified as the major carriers of the institutional influence and have been generalized into an explicit principle guiding our recommendations for optimal environments — learning or otherwise — for young children whenever any type of milieu arrangement is necessary. This principle may be stated simply as: the optimal environment for the young child is one in which the child is cared for in his own home in the context of a warm, continuous emotional relationship with his own mother under conditions of varied sensory input. Implicit in this principle is the conviction that the child's mother is the person best qualified to provide a stable and warm interpersonal relationship as well as the necessary pattern of sensory stimulation. Implicit also is the assumption that socio-emotional development has priority during the first three years and that if this occurs normally, cognitive development, which is of minor importance during this period anyway, will take care of itself. At a still deeper level lurks the assumption that attempts to foster cognitive development will interfere with socio-emotional development. Advocacy of the principle also implies endorsement of the idea that most homes are adequate during this early period and that no formal training (other than possibly some occasional supervisory support) for mothering is necessary. Such an operating principle places quite an onus on mothers and assumes that they will possess or quickly acquire all the talents necessary to create an optimal learning environment. And this author, at least, is convinced that a majority of mothers have such talents or proclivities and that they are willing to try to do all they can to create for their children the proper developmental milieu.

But there are always large numbers of children for whom family resources are not available and for whom some type of substitute milieu arrangement must be made. On the whole, such attempts have followed the entirely logical and perhaps evolutionary approach to milieu development — they have sought to create substitute families. The same is usually true when parents themselves seek to work out an alternate child-care arrangement because of less drastic conditions, such as maternal employment. The most typical maneuver is to try to obtain a motherly person who will "substitute" for her (not supplement her) during her hours away from her young child.

Our nation has become self-consciously concerned with social evolution, and in the past decade a serious attempt has been made to assimilate valid data from the behavioral and social sciences into planning for social action. In this context it would be meaningful to examine and question some of the hidden assumptions upon which our operating principle about the optimal environment for the young child rests.

Examining the Hidden Assumptions

1. *Do intermittent, short-term separations of the child from the mother impair the mother-child relationship or the development of the child?* Once having become sensitized to the consequences of institutionalization, and suspicious that the chief missing ingredient was the continued presence of the mother, the scientific and professional community went on the *qui vive* to the possibly deleterious consequences of any type of separation of an infant from its mother. Accordingly, a number of studies[10, 18, 21, 35, 39] investigated the consequences of short-term intermittent separation and were unable to demonstrate in the children the classical syndrome of the "institutional child." In reviewing the literature, Yarrow[47] stressed the point that available data do not support the tendency to assume that maternal deprivation, such as exists in the institutional environment, and maternal separation are the same thing. Apparently short cyclic interruptions culminated by reunions do not have the same effect as prolonged interruptions, even though quantitatively at the end of a designated period the amount of time spent in a mother-absent situation might be equal for the two experiences. Also in this context it is well to be reminded that in the institutional situation there is likely to be no stable mother-child relationship to interrupt. These are often never-mothered rather than ever-mothered children, a fact which must be kept in mind in generalizing from data on institutional groups. Thus until we have data to indicate that such intermittent separation-reunion cycles have similar effects on young children as prolonged separations, we are probably unjustified in assuming that an "uninterrupted" relationship is an essential ingredient of the optimal environment.

2. *Is group upbringing invariably damaging?* In studies done in West European and American settings, social and cognitive deficits associated with continuous group care during infancy have been frequently demonstrated. Enough exceptions have been reported, however, to warrant an intensification of the search for the "true" ingredient in the group situation associated with the observed deficits. For example, Freud and Dann[17] described the adjustment of a group of six children reared in a concentration camp orphanage for approximately three years, where they were cared for by overworked and impersonal inmates of the camp, and then transported to a residence for children in England. The children, who had never known their own mothers but who had been together as a group

for approximately three years, were intensely attached to one another. Although their adjustment to their new environment was slow and differed from the pattern one would expect from home-reared children, it was significant that they eventually did make a reasonably good adjustment. That the children were able to learn a new language while making this emotional transition was offered as evidence that many of the basic cognitive and personality attributes remained unimpaired in spite of the pattern of group upbringing. The accumulation of data showing that Kibbutz-reared children [34] do not have cognitive deficits also reinforces the premise that it is not necessarily group care *per se* that produces the frequently reported deficit and that it is possible to retain the advantages of group care while systematically eliminating its negative features. Grounds for reasonable optimism also have been found in retrospective studies by Maas [26] and Beres and Obers,[6] although in both cases the authors found evidence of pathology in some members of the follow-up sample. Similarly Dennis and Najarian [14] concluded from their data that the magnitude of the deficit varied as a function of the type of instrument used to measure deficit, and Dennis[13] showed that in institutions featuring better adult-child ratios and a conscious effort to meet the psychological needs of the infants the development of the children was much less retarded than was the case in a group of children residing in institutions with limited and unsophisticated staff. It is not appropriate to go into details of limitations of methodology in any of these studies; however, from the standpoint of an examination of the validity of a principle, it is important to take note of any exceptions to the generality of that principle.

In this context it is worth considering a point made by Gula.[20] He recently has suggested that some of the apparent consistency in studies comparing institutionalized infants with those cared for in their own homes and in foster homes might disappear if it were possible to equate the comparison groups on the variable of environmental adequacy. That is, one could classify all three types of environments as good, marginal, or inadequate on a number of dimensions. Most of the studies have compared children from palpably "inadequate" institutions with children from "good" foster and own homes. He suggests that, merely because most institutions studied have been inadequate in terms of such variables as adult-child ratio, staff turnover, and personal characteristics of some of the caretakers, etc., one is not justified in concluding *ipso facto* that group care is invariably inferior or damaging.

3. *Is healthy socio-emotional development the most important task of the first three years? Do attempts to foster cognitive growth interfere with social and emotional development?* These paired assumptions, which one finds stated in one variety or another in many pamphlets and books dealing with early child development, represent acceptance of a closed system

model of human development. They seem to conceptualize development as compartmentalized and with a finite limit. If the child progresses too much in one area he automatically restricts the amount of development that can occur in another area. Thus one often encounters such expressions as "cognitive development at the *expense* of socio-emotional development." It is perhaps of interest to reflect that, until our children reach somewhere around high school age, we seldom seem to worry that the reverse might occur. But, of course, life is an open system, and on the whole it is accurate to suggest that development feeds upon development. Cognitive and socio-emotional advances tend on the whole to be positively, not negatively correlated.

The definition of intelligence as *adaptivity* has not been adequately stressed by modern authors. It is, of course, the essence of Piaget's definition[32] as it was earlier of Binet.[7] Unfortunately, however, for the last generation or so in America we have been more concerned with how to measure intelligent behavior than how to interpret and understand it. Acceptance of the premise that intelligent behavior is adaptive behavior should help to break the set of many persons in the field of early child development that to encourage cognitive advance is to discourage healthy socio-emotional development. Ample data are available to suggest that quite the reverse is true either for intellectually advanced persons[42,43] or an unselected sample. In a large sample of young adults from an urban area in Minnesota, Anderson[3] and associates found that the best single predictor of post-high school adjustment contained in a large assessment battery was a humble little group intelligence test. Prediction based on intelligence plus teacher's ratings did somewhat better, but nothing exceeded the intelligence test for single measure efficiency.

It is relevant here to mention White's[45] concept of competence or effectance as a major stabilizing force in personality development. The emotional reinforcement accompanying the old "I can do it myself" declaration should not be undervalued. In Murphy's report[30] of the coping behavior of preschool children one sees evidence of the adjustive supports gained through cognitive advances. In his excellent review of cognitive stimulation in infancy and early childhood, Fowler[16] raises the question of whether there is any justification for the modern anxiety (and, to be sure, it is a modern phenomenon) over whether cognitive stimulation may damage personality development. He suggests that in the past severe and harmful methods may have been the culprits whenever there was damage and that the generalizations have confused methods of stimulation with the process of stimulation *per se.*

4. *Do cognitive experiences of the first few months and years leave no significant residual?* Any assumption that the learnings of infancy are evanescent appears to be a fairly modern idea. In his *Emile,* first published

in 1762, Rousseau[38] stressed the point that education should begin while the child is still in the cradle. Perhaps any generalization to the contrary received its major modern impetus from a rather unlikely place — from longitudinal studies of development covering the span from infancy to adulthood. From findings of poor prediction of subsequent intellectual status[5] one can legitimately infer that the infant tests measure behavior that is somewhat irrelevant to later intellectual performance. Even though these behaviors predictive of later cognitive behavior elude most investigators, one cannot infer that the early months and years are unimportant for cognitive development.

Some support for this assumption has come from experimental studies in which an attempt has been made to produce a durable effect in human subjects by one or another type of intervention offered during infancy. One cogent example is the work of Rheingold,[36] in which she provided additional social and personal stimulation to a small group of approximately six-month-old, institutionalized infants for a total of eight weeks. At the end of the experimental period, differences in social responsiveness between her stimulated group and a control group composed of other babies in the institution could be observed. There were also slight but nonsignificant advances in postural and motor behavior on a test of infant development. However, when the babies were followed up approximately a year later, by which time all but one were in either adoptive or boarding homes or in their own natural homes, the increased social responsiveness formerly shown by the stimulated babies was no longer observed. Nor were there differences in level of intellectual functioning. Rheingold and Bayley[37] concluded that the extra mothering provided during the experimental period was enough to produce an effect at the time but not enough to sustain this effect after such a time as the two groups were no longer differentially stimulated. However, in spite of their conservative conclusion, it is worth noting that the experimentally stimulated babies were found to vocalize more during the follow-up assessments than the control babies. Thus there may have been enough of an effect to sustain a developmental advance in at least this one extremely important area.

Some very impressive recent unpublished data obtained by Skeels offer a profound challenge to the assumption of the unimportance of the first three years for cognitive growth. This investigator has followed up after approximately 25 years most of the subjects described in a paper by Skeels and Dye.[40] Thirteen infants had been transferred from an orphanage because of evidence of mental retardation and placed in an institution for the retarded under the care of adolescent retardates who gave them a great deal of loving care and as much cognitive stimulation as they could. The 13 subjects showed a marked acceleration in development after this transfer.

In contrast a group of reasonably well matched infants left on the wards of the orphanage continued to develop poorly. In a recent follow-up of these cases, Skeels discovered that the gains made by the transferred infants were sustained into their adult years, whereas all but one of the control subjects developed the classic syndrome of mental retardation.

The fact that development and experience are cumulative makes it difficult ever to isolate any one antecedent period and assert that its influence was or was not influential in a subsequent developmental period. Thus even though it might be difficult to demonstrate an effect of some experience in an adjacent time period, delayed effects may well be of even greater developmental consequence. In a recent review of data from a number of longitudinal studies, Bloom[8] has concluded that during the first three to four years (the noncognitive years, if you will) approximately 50 per cent of the development of intelligence that is ever to occur in the life cycle takes place. During this period a particular environment may be either abundant or deprived in terms of the ingredients essential for providing opportunities for the development of intelligence and problem solving. Bloom[8] states:

> The effects of the environments, especially of the extreme environments, appear to be greatest in the early (and more rapid) periods of intelligence development and least in the later (and less rapid) periods of development. Although there is relatively little evidence of the effects of changing the environment on the changes in intelligence, the evidence so far available suggests that marked changes in the environment in the early years can produce greater changes in intelligence than will equally marked changes in the environment at later periods of development (pp. 88-89).

5. *Can one expect that, without formal planning, all the necessary learning experiences will occur?* There is an old legend that if you put six chimpanzees in front of six typewriters and leave them there long enough they eventually will produce all the works in the British Museum. One could paraphrase this for early childhood by suggesting that six children with good eyes and ears and hands and brains would, if left alone in nature, arrive at a number system, discover the laws of conservation of matter and energy, comprehend gravity and the motions of the planets, and perhaps arrive at the theory of relativity. All the "facts" necessary to discern these relationships are readily available. Perhaps a more realistic example would be to suggest that, if we surround a group of young children with a carefully selected set of play materials, they would eventually discover for themselves the laws of color mixture, of form and contour, of perspective, of formal rhythm and tonal relationships, and biological growth. And, to be sure, all this *could* occur. But whether this will necessarily occur with any frequency is quite another matter. We also assume that at a still earlier period a child will learn body control, eye-hand coordination, the

rudiments of language, and styles of problem solving in an entirely incidental and unplanned way. In an article in a recent issue of a popular woman's magazine, an author[22] fervently urges parents to stop trying to teach their young children in order that the children may learn. And, to be sure, there is always something to be said for this caution; it is all too easy to have planned learning experiences become didactic and regimented rather than subtle and opportunistic.

As more people gain experience in operating nursery school programs for children with an early history deficient in many categories of experience, the conviction appears to be gaining momentum that such children often are not able to avail themselves of the educational opportunities and must be guided into meaningful learning encounters. In a recent paper dealing with the pre-school behavior of a group of 21 children from multiproblem families, Malone[28] describes the inability of the children to carry out self-directed exploratory maneuvers with the toys and equipment as follows:

> When the children first came to nursery school they lacked interest in learning the names and properties of objects. Colors, numbers, sizes, shapes, locations, all seemed interchangeable. Nothing in the room seemed to have meaning for a child apart from the fact that another child had approached or handled it or that the teacher's attention was turned toward it. Even brief play depended on the teacher's involvement and support (p. 5).

When one reflects on the number of carefully arranged reinforcement contingencies necessary to help a young child learn to decode the simple message, "No," it is difficult to support the position that in early learning, as in anything else, nature should just take its course.

6. *Is formal training for child-care during the first three years unnecessary?* This assumption is obviously quite ridiculous, and yet it is one logical derivative of the hypothesis that the only adequate place for a young child is with his mother or a permanent mother substitute. There is, perhaps unfortunately, no literacy test for motherhood. This again is one of our interesting scientific paradoxes. That is, proclaiming in one breath that mothering is essential for the healthy development of a child, we have in the very next breath implied that just any mothering will do. It is interesting in this connection that from the elementary school level forward we have rigid certification statutes in most states that regulate the training requirements for persons who would qualify as teachers of our children. (The same degree of control over the qualifications and training of a nursery school teacher has not prevailed in the past, but we are moving into an era when it will.) So again, our pattern of social action appears to support the implicit belief in the lack of importance of the first three years of life.

In 1928, John B. Watson[44] wrote a controversial little trade book called

The Psychological Care of Infant and Child. He included one chapter heretically entitled, "The Dangers of Too Much Mother Love." In this chapter he suggested that child training was too important to be left in the hands of mothers, apparently not because he felt them intellectually inadequate but because of their sentimentality. In his typical "nondirective" style Watson[44] wrote:

> Six months' training in the actual handling of children from two to six under the eye of competent instructors should make a fairly satisfactory child's nurse. To keep them we should let the position of nurse or governess in the home be a respected one. Where the mother herself must be the nurse — which is the case in the vast majority of American homes — she must look upon herself while performing the functions of a nurse as a professional woman and not as a sentimentalist masquerading under the name of "Mother" (p. 149).

At present in this country a number of training programs are currently being formulated which would attempt to give this kind of professional training called for by Watson and many others. It is perhaps not possible to advance on all fronts at the same time, and the pressing health needs of the young child demanded and received top priority in earlier decades. Perhaps it will now be possible to extend our efforts at social intervention to encompass a broader range of health, education, and welfare activities.

7. *Are most homes and most parents adequate for at least the first three years?* Enough has been presented in discussing other implicit assumptions to make it unnecessary to amplify this point at length. The clinical literature, and much of the research literature of the last decade dealing with social-class differences, has made abundantly clear that all parents are not qualified to provide even the basic essentials of physical and psychological care to their children. Such reports as those describing the incidence of battered children[15, 24] capture our attention, but reports concerned with subtler and yet perhaps more long-standing patterns of parental deficit also fill the literature. In her description of the child-rearing environments provided by low lower-class families, Pavenstedt[31] has described them as impulse determined with very little evidence of clear planfulness for activities that would benefit either parent or child. Similarly, Wortis and associates[46] have described the extent to which the problems of the low-income mother so overwhelm her with reactions of depression and inadequacy that behavior toward the child is largely determined by the needs of the moment rather than by any clear plan about how to bring up children and how to train them to engage in the kind of behavior that the parents regard as acceptable or desirable. No social class and no cultural or ethnic group has exclusive rights to the domain of inadequate parentage; all conscientious parents must strive constantly for improvement on this score. However, relatively little attention has been

paid to the possibly deleterious consequences of inadequacies during the first three years of life. Parents have been blamed for so many problems of their children in later age periods that a moderate reaction formation appears to have set in. But again, judging by the type of social action taken by the responsible professional community, parental inadequacy during the first three years is seldom considered as a major menace. Perhaps, when the various alternatives are weighed, it appears by comparison to be the least of multiple evils; but parental behavior of the first three years should not be regarded as any more sacrosanct or beyond the domain of social concern than that of the later years.

Planning Alternatives

At this point the exposition of this paper must come to an abrupt halt, for insufficient data about possible alternative models are available to warrant recommendation of any major pattern of change. At present there are no completed research projects that have developed and evaluated alternative approximations of optimal learning environments for young children in our culture. One apparent limitation on ideas for alternative models appears to be the tendency to think in terms of binary choices. That is, we speak of individual care *versus* group care, foster home *versus* institution, foster home *versus* own home, and so on. But environments for the very young child do not need to be any more mutually exclusive than they are for the older children. After all, what is our public education system but a coordination of the efforts of home plus an institution? Most of us probably would agree that the optimal learning environment for the older child is neither of these alone but rather a combination of both. Some of this same pattern of combined effort also may represent the optimal arrangement for the very young child.

A number of programs suggesting alternatives possibly worth considering are currently in the early field trial stage. One such program is the one described by Caldwell and Richmond.[11] This program offers educationally oriented day care for culturally deprived children between six months and three years of age. The children spend the better part of five days a week in a group care setting (with an adult-child ratio never lower than 1:4) but return home each evening and maintain primary emotional relationships with their own families. Well child care, social and psychological services, and parent education activities are available for participating families. The educational program is carefully planned to try to help the child develop the personal-social and cognitive attributes conducive to learning and to provide experiences which can partially compensate for inadequacies which may have existed in the home environment. The strategy involved in offering the enrichment experience to children in this very young age group is to maximize their potential and hopefully prevent the deceleration in rate of development which seems to occur in many deprived children around the age of two to three years. It is thus an exercise in circumvention rather than remediation. Effectiveness of the endeavor is being determined by a comparision of the participating children with a control group of children from similar backgrounds who are not enrolled in the enrichment program. Unfortunately at this juncture it is too early for such projects to do more than suggest alternatives. The degree of confidence which comes only from research evidence plus replicated experience will have to wait a little longer.

Effective social action, however, can seldom await definitive data. And in the area of child care the most clamorous demand for innovative action appears to be coming from a rather unlikely source — not from any of the professional groups, not particularly from social planners who try to incorporate research data into plans for social action, but from *mothers*. From mothers themselves is coming the demand that professionals in the field look at some of the alternatives. We need not be reminded here that in America at the present time there are more than three million working mothers with children under six years of age.[2] And these mothers are looking for professional leadership to design and provide child-care facilities that help prepare their children for today's achievement-oriented culture. The challenge which has been offered is inevitable. After almost two decades of bombarding women with the importance of their mothering role, we might have predicted the weakening of their defenses and their waving the flag of truce as though to say, "I am not good enough to do all that you are saying I must do."

It is a characteristic of social evolution that an increased recognition of the importance of any role leads to the professionalization of that role, and there can be no doubt but that we are currently witnessing the early stages of professionalization of the mother-substitute role — or, as I would prefer to say, the mother-supplement. It is interesting to note that no one has as yet provided a satisfactory label for this role. The term "baby-sitter" is odious, reminding us of just about all some of the "less well trained" professionals do — sit with babies. If English were a masculine-feminine language, there is little doubt that the word would be used in the feminine gender, for we always speak of this person as a "she" (while emphasizing that young children need more contact with males). We cannot borrow any of the terms from already professionalized roles, such as "nurse" or "teacher," although such persons must be to a great extent both nurse and teacher. Awkward designations such as "child-care worker," or hybridized terms such as "nurse-teacher" do not quite seem to fill the bill; and there appears to be some reluctance to accept an untranslated foreign word like the Hebrew "metapelet" or the Russian "Nyanya." When such a word does appear, let us hope that it rhymes well and has a strong trochaic rhythm, for it will have to sustain a whole new era of poetry and song. (This author is convinced that the proper verb is *nurture*. It carries the desired connotations, but even to one who is not averse to neologisms such nominative forms as "nuturist," "nurturer," and "nurturizer" sound alien and inadequate.)[1]

Another basis for planning alternatives is becoming available from a less direct but potentially more persuasive source — from increasing knowledge about the process of development. The accumulation of data suggesting that the first few years of life are crucial for the priming or cognitive development call for vigorous and imaginative action programs for those early years. To say that it is premature to try to plan optimal environments because we do not fully understand how learning occurs is unacceptable. Perhaps only by the development of carefully arranged environments will we attain a complete understanding of the learning process. Already a great deal is known which enables us to specify some of the essential ingredients of a growth-fostering milieu. Such an environment must contain warm and responsive people who by their own interests invest objects with value. It must be supportive and as free of disease and pathogenic agents as possibly can be arranged. It also must trace a clear path from where the child is to where he is to go developmentally; objects and events must be similar enough to what the child has experienced to be assimilated by the child and yet novel enough to stimulate and attract. Such an environment must be exquisitely responsive, as a more consistent pattern of response is required to foster the acquisition of new forms of behavior than is required to maintain such behavior once it appears in the child's repertoire. The timing of experiences also must be carefully programmed. The time table for the scheduling of early postnatal events may well be every bit as demanding as that which obtains during the embryological period. For children whose early experiences are know to be deficient and depriving, attempts to program such environments seem mandatory if subsequent learning difficulties are to be circumvented.

Summary

Interpretations of research data and accumulated clinical experience have led over the years to a consensual approximation of an answer to the question: what is the optimal learning environment for the young child? As judged from our scientific and lay literature and from practices in health

[1]In a letter to the author written shortly after the meeting at which this paper was presented, Miss Rena Corman of New York City suggested that the proper term should be "nurcher," a compound of words "nurse" and "teacher." To be sure, a "nurcher" sounds nurturant.

and welfare agencies, one might infer that the optimal learning environment for the young child is that which exists when (a) a young child is cared for in his own home (b) in the context of a warm and nurturant emotional relationship (c) with his mother (or a reasonable facsimile thereof) under conditions of (d) varied sensory and cognitive input. Undoubtedly until a better hypothesis comes along, this is the best one available. This paper has attempted to generate constructive thinking about whether we are justified in overly vigorous support of (a) when (b), (c) or (d), or any combination thereof, might not obtain. Support for the main hypothesis comes primarily from other hypotheses (implicit assumptions) rather than from research of experimental data. When these assumptions are carefully examined they are found to be difficult if not impossible to verify with existing data.

The conservatism inherent in our present avoidance of carefully designed social action programs for the very young child needs to be re-examined. Such a re-examination conducted in the light of research evidence available about the effects of different patterns of care forces consideration of whether formalized intervention programs should not receive more attention than they have in the past and whether attention should be given to a professional training sequence for child-care workers. The careful preparation of the learning environment calls for a degree of training and commitment and personal control not always to be found in natural caretakers and a degree of richness of experience by no means always available in natural environments.

References

1. Ainsworth, Mary. 1962. Reversible and irreversible effects of maternal depriviation on intellectual development. Child Welfare League of America, 42-62

2. American Women. 1963. Report of the President's Commission on the Status of Women. (Order from Supt. of Documents, Washington, D.C.)

3. Anderson, J. E., et al. 1959. A survey of children's adjustment over time. Minneapolis, Minn.: University of Minnesota.

4. Barker, R. G., and H. F. Wright. 1955. Midwest and its Children: The Psychological Ecology of an American Town. New York: Harper & Row.

5. Bayley, Nancy. 1949. Consistency and variability in the growth of intelligence from birth to eighteen years. J. Genet. Psychol. 75: 165-96.

6. Beres, D., and S. Obers. 1950. The effects of extreme deprivation in infancy on psychic structure in adolescence. Psychoanal. Stud. of the Child. 5: 121-40.

7. Binet, A., and T. Simon. 1916. The Development of Intelligence in Children. Elizabeth S. Kite, trans. Baltimore: Williams and Wilkins.

8. Bloom, B. S. 1964. Stability and Change in Human Characteristics. New York: John Wiley.

9. Bronfenbrenner, Urie. 1962. Soviet studies of personality development and socialization. *In* Some Views on Soviet Psychology. Amer. Psychol. Assoc., Inc. Pp. 63-85.

10. Caldwell, Bettye M., et al. 1963. Mother-infant interaction in monomatric and polymatric families. Amer. J. Orthopsychiat. 33: 653-64.

11. Caldwell, Bettye M., and J. B. Richmond. 1964. Programmed day care for the very young child — a preliminary report. J. Marriage and the Family. 26: 481-88.

12. Davis, A., and R. J. Havighurst. 1946. Social class and color differences in child-rearing. Amer. Sociol. Rev. 11: 698-710.

13. Dennis, W. 1960. Causes of retardation among institutional children. J. Genet. Psychol. 96: 47-59.

14. Dennis, W., and P. Najarian. 1957. Infant development under environmental handicap. Psychol. Monogr. 71: 7 (Whole No. 536).

15. Elmer, Elizabeth. 1963. Identification of abused children. Children. 10: 180-184.

16. Fowler, W. 1962. Cognitive learning in infancy and early childhood. Psychol. Bull. 59: 116-52.

17. Freud, Anna, and Sophie Dann. 1951. An experiment in group upbringing. Psychoanal. Study of the Child. 6: 127-68.

18. Gardner, D. B., G. R. Hawkes, and L. G. Burchinal. 1961. Noncontinuous mothering in infancy and development in later childhood. Child Develpm. 32: 225-34.

19. Goldfarb, W. 1949. Rorschach test differences between family-reared, institution-reared and schizophrenic children. Amer. J. Orthopsychiat. 19: 624-33.

20. Gula, H. January, 1965. Paper given at Conference on Group Care for Children. Children's Bureau.

21. Hoffman, Lois Wladis. 1961. Effects of maternal employment on the child. Child Developm. 32: 187-97.

22. Holt, J. 1965. How to help babies learn — without teaching them. Redbook. 126 (l): 54-55, 134-37.

23. Irvine, Elizabeth E. 1952. Observations on the aims and methods of child-rearing in communal settlements in Israel. Human Relations. 5: 247-75.

24. Kempe, C. H., et al. 1962. The battered-child syndrome. J. Amer. Med. Asso. 181: 17-24.

25. Lewis, O. 1959. Five families. New York: Basic Books.

26. Maas, H. 1963. Long-term effects of early childhood separation and group care. Vita Humana. 6: 34-56.

27. Maccoby, Eleanor, and Patricia K. Gibbs. 1954. Methods of child-rearing in two social classes. *In* Readings in Child Developm. W. E. Martin and Celia B. Stendler, eds. New York: Harcourt, Brace & Co. Pp. 380-96.

28. Malone, C. A. 1966. Safety first: comments on the influence of external danger in the lives of children of disorganized families. Amer. J. Orthopsychiat. 36: 3-12.

29. Mead, Margaret. 1953. Growing up in New Guinea. New York: The New American Library.

30. Murphy, Lois B., et al. 1962. The Widening World of Childhood. Basic Books, Inc., New York.

31. Pavenstedt, E. 1965. A comparison of the child-rearing environment of upper-lower and very low-lower class families. Amer. J. Orthopsychiat. 35: 89-98.

32. Piaget, J. 1952. The Origins of Intelligence in Children. Margaret Cook, trans. New York: International Universities Press.

33. Provence, Sally, and Rose C. Lipton. 1962. Infants in institutions. New York: International Universities Press

34. Rabin, A. I. 1957. Personality maturity of Kibbutz and non-Kibbutz children as reflected in Rorschach findings. J. Proj. Tech. Pp. 148-53.

35. Radke Yarrow, Marian. 1961. Maternal employment and child rearing. Children. 8: 223-28.

36. Rheingold, Harriet. 1956. The modification of social responsiveness in institutional babies. Monogr. Soc. Res. Child Develpm. 21: (63).

37. Rheingold, Harriet L., and Nancy Bayley. 1959. The later effects of an experimental modification of mothering. Child Develpm. 30: 363-72.

38. Rousseau, J. J. 1950. Emile (1762). Great Neck, N. Y.: Barron's Educational Series.

39. Siegel, Alberta E., and Miriam B. Hass. 1963. The working mother: a review of research. Child Develpm. 34: 513-42.

40. Skeels, H. and H. Dye. 1939. A study of the effects of differential stimulation on mentally retarded children. Proc. Amer. Assoc., on Ment. Def. 44: 114-36.

41. Spiro, M. 1958. Children of the Kibbutz. Cambridge, Mass.: Harvard U. Press.

42. Terman, L. M., et al. 1925. Genetic studies or genius: Vol. 1. Mental and physical traits of a thousand gifted children. Stanford University, Calif.: Stanford University Press.

43. Terman, L. M., and Melita H. Oden. 1947. The gifted child grows up: twenty-five years' follow-up of a superior group. Stanford University, Calif.: Stanford University Press.

44. Watson, J. B. 1928. Psychological care of infant and child. London: Allen and Unwin.

45. White, R. W. 1959. Motivation reconsidered: the concept of competence. Psychol. Rev. 66: 297-33.

46. Wortis, H., et al. 1963. Child-rearing practices in a low socio-economic group. Pediatrics, 32: 298-307.

47. Yarrow, L. J. 1961. Maternal deprivation: toward an empirical and conceptual re-evaluation. Psychol. Bull. 58: 459-90.

C. CULTURE VARIATIONS

CHAPTER 7
Crucible of Identity: TheNegro Lower-Class Family

LEE RAINWATER

This paper is concerned with a description and analysis of slum Negro family patterns as these reflect and sustain Negroes' adaptations to the economic, social, and personal situation into which they are born and in which they must live. As such it deals with facts of lower-class life that are usually forgotten or ignored in polite discussion. We have chosen not to ignore these facts in the belief that to do so can lead only to assumptions which would frustrate efforts at social reconstruction, to strategies that are unrealistic in the light of the actual day-to-day reality of slum Negro life. Further, this analysis will deal with family patterns which interfere with the efforts slum Negroes make to attain a stable way of life as working or middle-class individuals and with the effects such failure in turn has on family life. To be sure, many Negro families live *in* the slum ghetto, but are not *of* its culture (though even they, and particularly their children, can be deeply affected by what happens there). However, it is the individuals who succumb to the distinctive family life style of the slum who experience the greatest weight of deprivation and who have the greatest difficulty responding to the few self-improvement resources that make their way into the ghetto. In short, we propose to explore in depth the family's role in the "tangle of pathology" which characterizes the ghetto.

For their part, Negroes creatively adapt to the system in ways that keep them alive and extract what gratification they can find, but in the process of adaptation they are constrained to behave in ways that inflict a great deal of suffering on those with whom they make their lives, and on themselves. The ghetto Negro is constantly confronted by the immediate necessity to suffer in order to get what he wants of those few things he can have, or to make others suffer, or both — for example, he suffers as exploited student and employee, as drug user, as loser in the competitive game of his peer-group society; he inflicts suffering as disloyal spouse, petty thief, knife- or gun-wielder, petty con man.

It is the central thesis of this paper that the caste-facilitated infliction of suffering by Negroes on other Negroes and on themselves appears most poignantly within the confines of the family, and that the victimization process as it operates in families prepares and toughens its members to function in the ghetto world, at the same time that it seriously interferes with their ability to operate in any other world. This, however, is very different from arguing that "the family is to blame" for the deprived situation ghetto Negroes suffer; rather we are looking at the logical outcome of the operation of the widely ramified and interconnecting caste system. In the end we will argue that only palliative results can be expected from attempts to treat directly the disordered family patterns to be described. Only a change in the original "inputs" of the caste system, the structural conditions inimical to basic social adaptation, can change family forms.

Almost thirty years ago, E. Franklin Frazier foresaw that the fate of the Negro family in the city would be a highly destructive one. His readers would have little reason to be surprised at observations of slum ghetto life today:

> . . . As long as the bankrupt system of southern agriculture exists, Negro families will continue to seek a living in the towns and cities. . . . They will crowd the slum areas of southern cities or make their way to northern cities where their families will become disrupted and their poverty will force them to depend upon charity.[1]

The Autonomy of the Slum Ghetto

Just as the deprivations and depredations practiced by white society have had their effect on the personalities and social life of Negroes, so also has the separation from the ongoing social life of the white community had its effect. In a curious way, Negroes have had considerable freedom to fashion their own adaptations within their separate world. The larger society provides them with few resources but also with minimal interference in the Negro community on matters which did not seem to affect white interests. Because Negroes learned early that there were a great many things they could not depend upon whites to provide they developed their own solutions to recurrent human issues. These solutions can often be seen to combine, along with the predominance of elements from white culture, elements that are distinctive to the Negro group. Even more distinctive is the *configuration* which emerges from those elements Negroes share with whites and those which are different.

It is in this sense that we may speak of a Negro subculture, a distinctive *patterning* of existential perspectives, techniques for coping with the problems of social life, views about what is desirable and undesirable in particular situations. This subculture, and particularly that of the lower-class, the slum, Negro, can be seen as his own creation out of the elements available to him in response to (1) the conditions of life set by white society and (2) the selective freedom which that society allows (or must put up with given the pattern of separateness on which it insists).

Out of this kind of "freedom" slum Negroes have built a culture which has some elements of intrinsic value and many more elements that are highly destructive to the people who must live in it. The elements that whites can value they constantly borrow. Negro arts and language have proved so popular that such commentators on American culture as Norman Mailer and Leslie Fiedler have noted processes of Negro-ization of white Americans as a minor theme of the past thirty years.[2] A fairly large proportion of Negroes with national reputations are engaged in the occupation of diffusing to the larger culture these elements of intrinsic value.

On the negative side, this freedom has meant, as social scientists who have studied Negro communities have long commented, that many of the protections offered by white institutions stop at the edge of the Negro ghetto: there are poor police protection and enforcement of civil equities, inadequate schooling and medical service, and more informal indulgences which whites allow Negroes as a small price for feeling superior.

For our purposes, however, the most important thing about the freedom which whites have allowed Negroes within their own world is that it has required them to work out their own ways of making it from day to day, from birth to death. The subculture that Negroes have created may be imperfect but it has been viable for centuries; it behooves both white and Negro leaders and intellectuals to seek to understand it even as they hope to change it.[3]

Negroes have created, again particularly within the lower-class slum group, a range of institutions to structure the tasks of living a victimized life and to minimize the pain it inevitably produces. In the slum ghetto these institutions include prominently those of the social network — the extended kinship system and the "street system" of buddies and broads which tie (although tenuously and unpredictably) the "members" to each other — and the institutions of entertainment (music, dance, folk tales) by which they instruct, explain, and accept themselves. Other institutions function to provide escape from the society of the victimized: the church (Hereafter!) and the civil rights movement (Now!).

The Functional Autonomy of the Negro Family

At the center of the matrix of Negro institutional life lies the family. It is in the family that individuals are trained for participation in the culture and find personal and group identity and continuity. The "freedom" allowed by white society is greatest here, and this freedom has been used to create an institutional variant more distinctive perhaps to the Negro subculture than any other. (Much of the content of Negro art and entertainment derives exactly from the distinctive characteristics of Negro family life.) At each stage in the Negro's experience of American life — slavery, segregation, *de facto* ghettoization — whites have found it less necessary to interfere in the relations between the sexes and between parents and children than in other areas of the Negro's existence. His adaptations in this area, therefore, have been less constrained by whites than in many other areas.

Now that the larger society is becoming increasingly committed to integrating Negroes into the main stream of American life, however, we can expect increasing constraint (benevolent as it may be) to be placed on the autonomy of the Negro family system.[4] These constraints will be designed to pull Negroes into meaningful integration with the larger society, to give up ways which are inimical to successful performance in the larger society, and to adopt new ways that are functional in that society. The strategic questions of the civil rights movement and of the war on poverty are ones that have to do with how one provides functional equivalents for the existing subculture before the capacity to make a life within its confines is destroyed.

The history of the Negro family has been ably documented by historians and sociologists.[5] In slavery, conjugal and family ties were reluctantly and ambivalently recognized by the slave holders, were often violated by them, but proved necessary to the slave system. This necessity stemmed both from the profitable offspring of slave sexual unions and the necessity for their nurture, and from the fact that the slaves' efforts to sustain patterns of sexual and parental relations mollified the men and women whose labor could not simply be commanded. From nature's promptings, the thinning memories of African heritage, and the example and guilt-ridden permission of the slave holders, slaves constructed a partial family system and sets of relations that generated conjugal and familial sentiments. The slave holder's recognition in advertisements for runaway slaves of marital and family sentiments as motivations for absconding provides one indication that strong family ties were possible, though perhaps not common, in the slave quarter. The mother-centered family with its emphasis on the primacy of the mother-child relation and only tenuous ties to a man, then, is the legacy of adaptations worked out by Negroes during slavery.

After emancipation this family design often also served well to cope with the social disorganization of Negro life in the late nineteenth century. Matrifocal families, ambivalence about the desirability of marriage, ready acceptance of illegitimacy, all sustained some kind of family life in situations which often made it difficult to maintain a full nuclear family. Yet in the hundred years since emancipation, Negroes in rural areas have been able to maintain full nuclear families almost as well as similarly situated whites. As we will see, it is the move to the city that results in the very high proportion of the mother-headed households. In the rural system the man continues to have important functions; it is difficult for a woman to make a crop by herself, or even with the help of other women. In the city, however, the woman can earn wages just as a man can, and she can receive welfare payments more easily than he can. In rural areas, although there may be high illegitimacy rates and high rates of marital disruption, men and women have an interest in getting together; families are headed by a husband-wife pair much more often than in the city. That pair may be much less stable than in the more prosperous segments of Negro and white communities but it is more likely to exist among rural Negroes than among urban ones.

The matrifocal character of the Negro lower-class family in the United States has much in common with Caribbean Negro family patterns; research in both areas has done a great deal to increase our understanding of the Negro situation. However, there are important differences in the family forms of the two areas.[6] The impact of white European family models has been much greater in the United States than in the Caribbean both because of the relative population proportions of white and colored peoples and because equalitarian values in the United States have had a great impact on Negroes even when they have not on whites. The typical Caribbean mating pattern is that women go through several visiting and common-law unions but eventually marry; that is, they marry legally only relatively late in their sexual lives. The Caribbean marriage is the crowning of a sexual and procreative career; it is considered a serious and difficult step.

In the United States, in contrast, Negroes marry at only a slightly lower rate and slightly higher age than whites.[7] Most Negro women marry relatively early in their careers; marriage is not regarded as the same kind of crowning choice and achievement that it is in the Caribbean. For lower-class Negroes in the United States marriage ceremonies are rather informal affairs. In the Caribbean, marriage is regarded as quite costly because of the feasting which goes along with it; ideally it is performed in church.

In the United States, unlike the Caribbean, early marriage confers a kind of permanent respectable status upon a woman which she can use to deny any subsequent accusations of immorality or promiscuity once the marriage is broken and she becomes sexually involved in visiting or common-law relations. The relevant effective status for many Negro women is that of "having been married" rather than "being married"; having the right to be called "Mrs." rather than currently being Mrs. Someone-in-Particular.

For Negro lower-class women, then, first marriage has the same kind of importance as having a first child. Both indicate that the girl has become a woman but neither one that this is the last such activity in which she will engage. It seems very likely that only a minority of Negro women in the urban slum go through their child-rearing years with only one man around the house.

Among the Negro urban poor, then, a great many women have the experience of heading a family for part of their mature lives, and a great many children spend some part of their formative years in a household with out a father-mother pair. From Table 1 we see that in 1960, forty-seven percent of the Negro poor urban families with children had a female head. Unfortunately cumulative statistics are hard to come by; but, given this very high level for a cross-sectional sample (and taking into account the fact that the median age of the children in these families is about six years), it seems very likely that as many as two-thirds of Negro urban poor children will not live in families headed by a man and a woman throughout the first eighteen years of their lives.

One of the other distinctive characteristics of Negro families, both poor and not so poor, is the fact that Negro households have a much higher proportion of relatives outside the mother-father-children triangle than is the case with whites. For example, in St. Louis Negro families average 0.8 other relatives per household compared to only 0.4 for white families. In the case of the more prosperous Negro families this is likely to mean that an older relative lives in the home providing baby-sitting services while both the husband and wife work and thus further their climb toward stable working- or middle-class status. In the poor Negro families it is much more likely that the household is headed by an older relative who brings under her wings a daughter and that daughter's children. It is important to note

TABLE 1

Proportion of Female Heads for Families with Children by Race, Income, and Urban-Rural Categories

	Rural	*Urban* (*percent*)	*Total*
Negroes			
under $3000	18	47	36
$3000 and over	5	8	7
Total	14	23	21
Whites			
under $3000	12	38	22
$3000 and over	2	4	3
Total	4	7	6

Source: U. S. Census: 1960, PC (1) D. U. S. Volume, Table 225; State Volume, Table 140.

that the three-generation household with the grandmother at the head exists only when there is no husband present. Thus, despite the high proportion of female-headed households in this group and despite the high proportion of households that contain other relatives, we find that almost all married couples in the St. Louis Negro slum community have their own household. In other words, when a couple marries it establishes its own household; when that couple breaks up the mother either maintains that household or moves back to her parents or grandparents.

Finally we should note that Negro slum families have more children than do either white slum families or stable working- and middle-class Negro families. Mobile Negro families limit their fertility sharply in the interest of bringing the advantages of mobility more fully to the few children that they do have. Since the Negro slum family is both more likely to have the father absent and more likely to have more children in the family, the mother has a more demanding task with fewer resources at her disposal. When we examine the patterns of life of the stem family we shall see that even the presence of several mothers does not necessarily lighten the work load for the principal mother in charge.

The Formation and Maintenance of Families

We will outline below the several stages and forms of Negro lower-class family life. At many points these family forms and the interpersonal relations that exist within them will be seen to have characteristics in common with the life styles of white lower-class families.[8] At other points there are differences, or the Negro pattern will be seen to be more sharply divergent from the family life of stable working and middle-class couples.

It is important to recognize that lower-class Negroes know that their particular family forms are different from those of the rest of the society and that, though they often see these forms as representing the only ways of behaving given their circumstances, they also think of the more stable family forms of the working class as more desirable. That is, lower-class Negroes know what the "normal American family" is supposed to be like, and they consider a stable family-centered way of life superior to the conjugal and familial situations in which they often find themselves. Their conceptions of the good American life include the notion of a father-husband who functions as an adequate provider and interested member of the family, a hard working home-bound mother who is concerned about her children's welfare and her husband's needs, and children who look up to their parents and perform well in school and other outside places to reflect credit on their families. This image of what family life can be like is very real from time to time as lower-class men and women grow up and move through adulthood. Many of them make efforts to establish such families but find it impossible to do so either because of the direct impact of economic disabilities or because they are not able to sustain in their day-to-day lives the ideals which they hold.[9] While these ideals do serve as a meaningful guide to lower-class couples who are mobile out of the group, for a great many others the existence of such ideas about normal family life represents a recurrent source of stress within families as individuals become aware that they are failing to measure up to the ideals, or as others within the family and outside it use the ideals as an aggressive weapon for criticizing each other's performance. It is not at all uncommon for husbands or wives or children to try to hold others in the family to the norms of stable family life while they themselves engage in behaviors which violate these norms. The effect of such criticism in the end is to deepen commitment to the deviant sexual and parental norms of a slum subculture. Unless they are careful, social workers and other professionals exacerbate the tendency to use the norms of "American family life" as weapons by supporting these norms in situations where they are in reality unsupportable, thus aggravating the sense of failing and being failed by others which is chronic for lower-class people.

Going Together

The initial steps toward mating and family formation in the Negro slum take place in a context of highly developed boys' and girls' peer groups. Adolescents tend to become deeply involved in their peer-group societies beginning as early as the age of twelve or thirteen and continue to be involved after first pregnancies and first marriages. Boys and girls are heavily committed both to their same sex peer groups and to the activities that those groups carry out. While classical gang activity does not necessarily characterize Negro slum communities everywhere, loosely-knit peer groups do.

The world of the Negro slum is wide open to exploration by adolescent boys and girls: "Negro communities provide a flow of common experience in which young people and their elders share, and out of which delinquent behavior emerges almost imperceptibly."[10] More than is possible in white slum communities, Negro adolescents have an opportunity to interact with adults in various "high life" activities; their behavior more often represents an identification with the behavior of adults than an attempt to set up group standards and activities that differ from those of adults.

Boys and young men participating in the street system of peer-group activity are much caught up in games of furthering and enhancing their status as significant persons. These games are played out in small and large gatherings through various kinds of verbal contests that go under the names of "sounding," "signifying," and "working game." Very much a part of a boy's or man's status in this group is his ability to win women. The man who has several women "up tight," who is successful in "pimping off" women for sexual favors and material benefits, is much admired. In sharp contrast to white lower-class groups, there is little tendency for males to separate girls into "good" and "bad" categories.[11] Observations of groups of Negro youths suggest that girls and women are much more readily referred to as "that bitch" or "that whore" than they are by their names, and this seems to be a universal tendency carrying no connotation that "that bitch" is morally inferior to or different from other women. Thus, all women are essentially the same, all women are legitimate targets, and no girl or woman is expected to be virginal except for reason of lack of opportunity or immaturity. From their participation in the peer group and according to standards legitimated by the total Negro slum culture, Negro boys and young men are propelled in the direction of girls to test their "strength" as seducers. They are mercilessly rated by both their peers and the opposite sex in their ability to "talk" to girls; a young man will go to great lengths to avoid the reputation of having a "weak" line.[12]

The girls share these definitions of the nature of heterosexual relations; they take for granted that almost any male they deal with will try to seduce them and that given sufficient inducement (social not monetary) they may wish to go along with his line. Although girls have a great deal of ambivalence about participating in sexual relations, this ambivalence is minimally moral and has much more to do with a desire not to be taken advantage of or get in trouble. Girls develop defenses against the exploitative orientations of men by devaluing the significance of sexual relations ("he really didn't do anything bad to me"), and as time goes on by developing their own appreciation of the intrinsic rewards of sexual intercourse.

The informal social relations of slum Negroes begin in adolescence to be highly sexualized. Although parents have many qualms about boys and, particularly, girls entering into this system, they seldom feel there is much they can do to prevent their children's sexual involvement. They usually confine themselves to counseling somewhat hopelessly against girls becoming pregnant or boys being forced into situations where they might have to marry a girl they do not want to marry.

Girls are propelled toward boys and men in order to demonstrate their maturity and attractiveness; in the process they are constantly exposed to pressures for seduction, to boys "rapping" to them. An active girl will "go with" quite a number of boys, but she will generally try to restrict the number with whom she has intercourse to the few to whom she is attracted or (as happens not infrequently) to those whose threats of physical violence she cannot avoid. For their part, the boys move rapidly from girl to girl seeking to have intercourse with as many as they can and thus build up their "reps." The activity of seduction is itself highly cathected; there is gratification in simply "talking to" a girl as long as the boy can feel that he has acquitted himself well.

> At sixteen Joan Bemias enjoys spending time with three or four very close girl friends. She tells us they follow this routine when the girls want to go out and none of the boys they have been seeing lately is available: "Every time we get ready to go someplace we look through all the telephone numbers of boys we'd have and we call them and talk so sweet to them that they'd come on around. All of them had cars you see. (I: What do you do to keep all these fellows interested?) Well nothing. We don't have to make love with all of them. Let's see, Joe, J. B., Albert, and Paul, out of all of them I've been going out with I've only had sex with four boys, that's all." She goes on to say that she and her girl friends resist boys by being unresponsive to their lines and by breaking off relations with them on the ground that they're going out with other girls. It is also clear from her comments that the girl friends support each other in resisting the boys when they are out together in groups.
>
> Joan has had a relationship with a boy which has lasted six months, but she has managed to hold the frequency of intercourse down to four times. Initially she managed to hold this particular boy off for a month but eventually gave in.

Becoming Pregnant

It is clear that the contest elements in relationships between men and women continue even in relationships that become quite steady. Despite the girls' ambivalence about sexual relations and their manifold efforts to reduce its frequency, the operation of chance often eventuates in their becoming pregnant.[13] This was the case with Joan. With this we reach the second stage in the formation of families, that of premarital pregnancy. (We are outlining an ideal-typical sequence and not, of course, implying that all girls in the Negro slum culture become pregnant before they marry but only that a great many of them do.)

Joan was caught despite the fact that she was considerably more sophisticated about contraception than most girls or young women in the group (her mother had both instructed her in contraceptive techniques and constantly warned her to take precautions). No one was particularly surprised at her pregnancy although she, her boy friend, her mother, and others regarded it as unfortunate. For girls in the Negro slum, pregnancy before marriage is expected in much the same way that parents expect their children to catch mumps or chicken pox; if they are lucky it will not happen but if it happens people are not too surprised and everyone knows what to do about it. It was quickly decided that Joan and the baby would stay at home. It seems clear from the preparations that Joan's mother is making that she expects to have the main responsibility for caring for the infant. Joan seems quite indifferent to the baby; she shows little interest in mothering the child although she is not particularly averse to the idea so long as the baby does not interfere too much with her continued participation in her peer group.

Establishing who the father is under these circumstances seems to be important and confers a kind of legitimacy on the birth; not to know who one's father is, on the other hand, seems the ultimate in illegitimacy. Actually Joan had a choice in the imputation of fatherhood; she chose J. B. because he is older than she, and because she may marry him if he can get a divorce from his wife. She could have chosen Paul (with whom she had also had intercourse at about the time she became pregnant), but she would have done this reluctantly since Paul is a year younger than she and somehow this does not seem fitting.

In general, when a girl becomes pregnant while still living at home it seems taken for granted that she will continue to live there and that her parents will take a major responsibility for rearing the children. Since there are usually siblings who can help out and even siblings who will be playmates for the child, the addition of a third generation to the household does not seem to place a great stress on relationships within the family. It seems common for the first pregnancy to have a liberating influence on the mother once the child is born in that she becomes socially and sexually more active than she was before. She no longer has to be concerned with preserving her status as a single girl. Since her mother is usually willing to take care of the child for a few years, the unwed mother has an opportunity to go out with girl friends and with men and thus become more deeply involved in the peer-group society of her culture. As she has more children and perhaps marries she will find it necessary to settle down and spend more time around the house fulfilling the functions of a mother herself.

It would seem that for girls pregnancy is the real measure of maturity, the dividing line between adolescence and womanhood. Perhaps because of this, as well as because of the ready resources for child care, girls in the Negro slum community show much less concern about pregnancy than do girls in the white lower-class community and are less motivated to marry the fathers of their children. When a girl becomes pregnant the question of marriage certainly arises and is considered, but the girl often decides that she would rather not marry the man either because she does not want to settle down yet or because she does not think he would make a good husband.

It is in the easy attitudes toward premarital pregnancy that the matrifocal character of the Negro lower-class family appears most clearly. In order to have and raise a family it is simply not necessary, though it may be desirable, to have a man around the house. While the AFDC program may make it easier to maintain such attitudes in the urban situation, this pattern existed long before the program was initiated and continues in families where support comes from other sources.

Finally it should be noted that fathering a child similarly confers maturity on boys and young men although perhaps it is less salient for them. If the boy has any interest in the girl he will tend to feel that the fact that he has impregnated her gives him an additional claim on her. He will be stricter in seeking to enforce his exclusive rights over her (though not exclusive loyalty to her). This exclusive right does not mean that he expects to marry her but only that there is a new and special bond between them. If the girl is not willing to accept such claims she may find it necessary to break off the relationship rather than tolerate the man's jealousy. Since others in the peer group have a vested interest in not allowing a couple to be too loyal to each other they go out of their way to question and challenge each partner about the loyalty of the other, thus contributing to the deterioration of the relationship. This same kind of questioning and challenging continues if the couple marries and represents one source of the instability of the marital relationship.

Getting Married

As noted earlier, despite the high degree of premarital sexual activity and the rather high proportion of premarital pregnancies, most lower-class Negro men and women eventually do marry and stay together for a shorter or longer period of time. Marriage is an intimidating prospect and is approached ambivalently by both parties. For the girl it means giving up a familiar and comfortable home that, unlike some other lower-class subcultures, places few real restrictions on her behavior. (While marriage can appear to be an escape from interpersonal difficulties at home, these difficulties seldom seem to revolve around effective restrictions placed on her behavior by her parents.) The girl also has good reason to be suspicious of the likelihood that men will be able to perform stably in the role of husband and provider; she is reluctant to be tied down by a man who will not prove to be worth it.

From the man's point of view the fickleness of women makes marriage problematic. It is one thing to have a girl friend step out on you, but it is quite another to have a wife do so. Whereas premarital sexual relations and fatherhood carry almost no connotation of responsibility for the welfare of the partner, marriage is supposed to mean that a man behaves more responsibly, becoming a provider for his wife and children even though he may not be expected to give up all the gratifications of participation in the street system.

For all of these reasons both boys and girls tend to have rather negative views of marriage as well as a low expectation that marriage will prove a stable and gratifying existence. When marriage does take place it tends to represent a tentative commitment on the part of both parties with a strong tendency to seek greater commitment on the part of the partner than on one's own part. Marriage is regarded as a fragile arrangement held together primarily by affectional ties rather than instrumental concerns.

In general, as in white lower-class groups, the decision to marry seems to be taken rather impulsively.[14] Since everyone knows that sooner or later he will get married, in spite of the fact that he may not be sanguine about the prospect, Negro lower-class men and women are alert for clues that the time has arrived. The time may arrive because of a pregnancy in a steady relationship that seems gratifying to both partners, or as a way of getting out of what seems to be an awkward situation, or as a self-indulgence during periods when a boy and a girl are feeling very sorry for themselves. Thus, one girl tells us that when she marries her husband will cook all of her meals for her and she will not have any housework; another girl says that when she marries it will be to a man who has plenty of money and will have to take her out often and really show her a good time.

Boys see in marriage the possibility of regular sexual intercourse without having to fight for it, or a girl safe from veneral disease, or a relationship to a nurturant figure who will fulfill the functions of a mother. For boys, marriage can also be a way of asserting their independence from the peer group if its demands become burdensome. In this case the young man seeks to have the best of both worlds.[15]

Marriage as a way out of an unpleasant situation can be seen in the case of one of our informants, Janet Cowan:

> Janet has been going with two men, one of them married and the other single. The married man's wife took exception to their relationship and killed her husband. Within a week Janet and her single boyfriend, Howard, were married. One way out of the turmoil the murder of her married boyfriend stimulated (they lived in the same building) was to choose marriage as a way of "settling down." However, after marrying the new couple seemed to have little idea how to set themselves up as a family. Janet was reluctant to leave her parents' home because her parents cared for her two illegitimate children. Howard was unemployed and therefore unacceptable in his parents-in-law's home, nor were his own parents willing to have his wife move in with them. Howard was also reluctant to give up another girl friend in another part of town. Although both he and his wife maintained that it was all right for a couple to step out on each other so long as the other partner did not know about it, they were both jealous if they suspected anything of this kind. In the end they gave up on the idea of marriage and went their separate ways.

In general, then, the movement toward marriage is an uncertain and tentative one. Once the couple does settle down together in a household of their own, they have the problem of working out a mutually acceptable organization of rights and duties, expectations and performances, that will meet their needs.

Husband-Wife Relations

Characteristic of both the Negro and white lower class is a high degree of conjugal role segregation.[16] That is, husbands and wives tend to think of themselves as having very separate kinds of functioning in the instrumental organization of family life, and also as pursuing recreational and outside interests separately. The husband is expected to be a provider; he resists assuming functions around the home so long as he feels he is doing his proper job of bringing home a pay check. He feels he has the right to indulge himself in little ways if he is successful at this task. The wife is expected to care for the home and children and make her husband feel welcome and comfortable. Much that is distinctive to Negro family life stems from the fact that husbands often are not stable providers. Even when a particular man is, his wife's conception of men in general is such that she is pessimistic about the likelihood that he will continue to do well in this area. A great many Negro wives work to supplement the family income. When this is so the separate incomes earned by husband and wife tend to be treated not as "family" income but as the individual property of the two persons involved. If their wives work, husbands are likely to feel that they are entitled to retain a larger share of the income they provide; the wives, in turn, feel that the husbands have no right to benefit from the purchases they make out of their own money. There is, then, "my money" and "your money." In this situation the husband may come to feel that the wife should support the children out of her income and that he can retain all of his income for himself.

While white lower-class wives often are very much intimidated by their husbands, Negro lower-class wives come to feel that they have a right to give as good as they get. If the husband indulges himself, they have the right to indulge themselves. If the husband steps out on his wife, she has the right to step out on him. The commitment of husbands and wives to each other seems often a highly instrumental one after the "honeymoon" period. Many wives feel they owe the husband nothing once he fails to perform his provider role. If the husband is unemployed the wife increasingly refuses to perform her usual duties for him. For example one woman, after mentioning that her husband had cooked four eggs for himself, commented, "I cook for him when he's working but right now he's unemployed; he can cook for himself." It is important, however, to understand that the man's status in the home depends not so much on whether he is working as on whether he brings money into the home. Thus, in several of the families we have studied in which the husband receives disability payments his status is as well-recognized as in families in which the husband is working.[17]

Because of the high degree of conjugal role segregation, both white and Negro lower-class families tend to be matrifocal in comparison to middle-class families. They are matrifocal in the sense that the wife makes most of the decisions that keep the family going and has the greatest sense of responsibility to the family. In white as well as in Negro lower-class families women tend to look to their female relatives for support and counsel, and to treat their husbands as essentially uninterested in the day-to-day problems of family living.[18] In the Negro lower-class family these tendencies are all considerably exaggerated so that the matrifocality is much clearer than in white lower-class families.

The fact that both sexes in the Negro slum culture have equal right to the various satisfactions of life (earning an income, sex, drinking, and peer-group activity which conflicts with family responsibilities) means that there is less pretense to patriarchal authority in the Negro than in the white lower class. Since men find the overt debasement of their status very threatening, the Negro family is much more vulnerable to disruption when men are temporarily unable to perform their provider roles. Also, when men are unemployed the temptations for them to engage in street adventures that have repercussions on the marital relationship are much greater. This fact is well-recognized by Negro lower-class wives; they often seem as concerned about what their unemployed husbands will do instead of working as they are about the fact that the husband is no longer bringing money into the home.

It is tempting to cope with the likelihood of disloyalty by denying the usual norms of fidelity, by maintaining instead that extra-marital affairs are acceptable as long as they do not interfere with family functioning. Quite a few informants tell us this, but we have yet to observe a situation in which a couple maintains a stable relationship under these circumstances without a great deal of conflict. Thus one woman in her forties who has been married for many years and has four children first outlined this deviant norm and then illustrated how it did not work out:

> My husband and I, we go out alone and sometimes stay all night. But when I get back my husband doesn't ask me a thing and I don't ask him anything. . . . A couple of years ago I suspected he was going out on me. One day I came home and my daughter was here. I told her to tell me when he left the house. I went into the bedroom and got into bed and then I heard him come in. He left in about ten minutes and my daughter came in and told me he was gone. I got out of bed and put on my clothes and started following him. Soon I saw him walking with a young girl and I began walking after them. They were just laughing and joking right out loud right on the sidewalk. He was carrying a large package of hers. I walked up behind them until I was about a yard from

> them. I had a large dirk which I opened and had decided to take one long slash across the both of them. Just when I decided to swing at them I lost my balance — I have a bad hip. Anyway, I didn't cut them because I lost my balance. Then I called his name and he turned around and stared at me. He didn't move at all. He was shaking all over. That girl just ran away from us. He still had her package so the next day she called on the telephone and said she wanted to come pick it up. My husband washed his face, brushed his teeth, took out his false tooth and started scrubbing it and put on a clean shirt and everything, just for her. We went downstairs together and gave her the package and she left.
>
> So you see my husband does run around on me and it seems like he does it a lot. The thing about it is he's just getting too old to be pulling that kind of stuff. If a young man does it then that's not so bad — but an old man, he just looks foolish. One of these days he'll catch me but I'll just tell him, "Buddy, you owe me one," and that'll be all there is to it. He hasn't caught me yet, though.

In this case, as in others, the wife is not able to leave well enough alone; her jealousy forces her to a confrontation. Actually seeing her husband with another woman stimulates her to violence.

With couples who have managed to stay married for a good many years, these peccadillos are tolerable although they generate a great deal of conflict in the marital relationship. At earlier ages the partners are likely to be both prouder and less inured to the hopelessness of maintaining stable relationships; outside involvements are therefore much more likely to be disruptive of the marriage.

Marital Breakup

The precipitating causes of marital disruption seem to fall mainly into economic or sexual categories. As noted, the husband has little credit with his wife to tide him over periods of unemployment. Wives seem very willing to withdraw commitment from husbands who are not bringing money into the house. They take the point of view that he has no right to take up space around the house, to use its facilities, or to demand loyalty from her. Even where the wife is not inclined to press these claims, the husband tends to be touchy because he knows that such definitions are usual in his group, and he may therefore prove difficult for even a well-meaning wife to deal with. As noted above, if husbands do not work they tend to play around. Since they continue to maintain some contact with their peer groups, whenever they have time on their hands they move back into the world of the street system and are likely to get involved in activities which pose a threat to their family relationships.

Drink is a great enemy of the lower-class housewife, both white and Negro. Lower-class wives fear their husbands' drinking because it costs money, because the husband may become violent and take out his frustrations on his wife, and because drinking may lead to sexual involvements with other women.[19]

, The combination of economic problems and sexual difficulties can be seen in the case of the following couple in their early twenties:

When the field worker first came to know them, the Wilsons seemed to be working hard to establish a stable family life. The couple had been married about three years and had a two-year-old son. Their apartment was very sparsely furnished but also very clean. Within six weeks the couple had acquired several rooms of inexpensive furniture and obviously had gone to a great deal of effort to make a liveable home. Husband and wife worked on different shifts so that the husband could take care of the child while the wife worked. They looked forward to saving enough money to move out of the housing project into a more desirable neighborhood. Six weeks later, however, the husband had lost his job. He and his wife were in great conflict. She made him feel unwelcome at home and he strongly suspected her of going out with other men. A short time later they had separated. It is impossible to disentangle the various factors involved in this separation into a sequence of cause and effect, but we can see something of the impact of the total complex.

First Mr. Wilson loses his job: "I went to work one day and the man told me that I would have to work until 1:00. I asked him if there would be any extra pay for working overtime and he said no. I asked him why and he said, 'If you don't like it you can kiss my ass.' He said that to me. I said, 'Why do I have to do all that?' He said, 'Because I said so.' I wanted to jam [fight] him but I said to myself I don't want to be that ignorant, I don't want to be as ignorant as he is, so I just cut out and left. Later his father called me [it was a family firm] and asked why I left and I told him. He said, 'If you don't want to go along with my son then you're fired.' I said O.K. They had another Negro man come in to help me part time before they fired me. I think they were trying to have him work full time because he worked for them before. He has seven kids and he takes their shit."

The field worker observed that things were not as hard as they could be because his wife had a job, to which he replied, "Yeah, I know, that's just where the trouble is. My wife has become independent since she began working. If I don't get a job pretty soon I'll go crazy. We have a lot of little arguments about nothing since she got so independent." He went on to say that his wife had become a completely different person recently; she was hard to talk to because she felt that now that she was working and he was not there was nothing that he could tell her. On her last pay day his wife did not return home for three days; when she did she had only seven cents left from her pay check. He said that he loved his wife very much and had begged her to quit fooling around. He is pretty sure that she is having an affair with the man with whom she rides to work. To make matters worse his wife's sister counsels her that she does not have to stay home with him as long as he is out of work.

Finally the wife moved most of their furniture out of the apartment so that he came home to find an empty apartment. He moved back to his parents' home (also in the housing project).

One interesting effect of this experience was the radical change in the husband's attitudes toward race relations. When he and his wife were doing well together and had hopes of moving up in the world he was quite critical of Negroes; "Our people are not ready for integration in many cases because they really don't know how to act. You figure if our people don't want to be bothered with whites then why in hell should the white man want to be bothered with them. There are some of us who are ready; there are others who aren't quite ready yet so I don't see why they're doing all of this hollering." A scarce eight months later he addressed white people as he spoke for two hours into a tape recorder, "If we're willing to be with you, why aren't you willing to be with us? Do our color make us look dirty and low down and cheap? Or do you know the real meaning of 'nigger'? Anyone can be a nigger, white, colored, orange or any other color. It's something that you labeled us with. You put us away like you put a can away on the shelf with a label on it. The can is marked 'Poison: stay away from it.' You want us to help build your country but you don't want us to live in it. . . . You give me respect; I'll give you respect. If you threaten to take my life, I'll take yours and believe me I know how to take a life. We do believe that man was put here to live together as human beings; not one that's superior and the one that's a dog, but as human beings. And if you don't want to live this way then you become the dog and we'll become the human beings. There's too much corruption, too much hate, too much one individual trying to step on another. If we don't get together in a hurry we will destroy each other." It was clear from what the respondent said that he had been much influenced by Black Muslim philosophy, yet again and again in his comments one can see the displacement into a public, race relations dialogue of the sense of rage, frustration and victimization that he had experienced in his ill-fated marriage.[20]

Finally, it should be noted that migration plays a part in marital disruption. Sometimes marriages do not break up in the dramatic way described above but rather simply become increasingly unsatisfactory to one or both partners. In such a situation the temptation to move to another city, from South to North, or North to West, is great. Several wives told us that their first marriages were broken when they moved with their children to the North and their husbands stayed behind.

"After we couldn't get along I left the farm and came here and stayed away three or four days. I didn't come here to stay. I came to visit but I liked it and so I said, 'I'm gonna leave!' He said, 'I'll be glad if you do.' Well, maybe he didn't mean it but I thought he did. . . . I miss him sometimes, you know. I think about him, I guess. But just in a small way. That's what I can't understand about life sometimes; you know — how people can go on like that and still break up and meet somebody else. Why couldn't — oh, I don't know!"

The gains and losses in marriage and in the post-marital state often seem quite comparable. Once they have had the experience of marriage, many women in the Negro slum culture see little to recommend it in the future, important as the first marriage may have been in establishing their maturity and respectability.

The House of Mothers

As we have seen, perhaps a majority of mothers in the Negro slum community spend at least part of their mature life as mothers heading a family. The Negro mother may be a working mother or she may be an AFDC mother, but in either case she has the problems of maintaining a household, socializing her children, and achieving for herself some sense of membership in relations with other women and with men. As is apparent from the earlier discussion, she often receives her training in how to run such a household by observing her own mother manage without a husband. Similarly she often learns how to run a three-generation household because she herself brought a third generation into her home with her first, premarital, pregnancy.

Because men are not expected to be much help around the house, having to be head of the household is not particularly intimidating to the Negro mother if she can feel some security about income. She knows it is a hard, hopeless, and often thankless task, but she also knows that it is possible. The maternal household in the slum is generally run with a minimum of organization. The children quickly learn to fend for themselves, to go to the store, to make small purchases, to bring change home, to watch after themselves when the mother has to be out of the home, to amuse themselves, to set their own schedules of sleeping, eating, and going to school. Housekeeping practices may be poor, furniture takes a terrific beating from the children, and emergencies constantly arise. The Negro mother in this situation copes by not setting too high standards for herself, by letting things take their course. Life is most difficult when there are babies and preschool children around because then the mother is confined to the home. If she is a grandmother and the children are her daughter's, she is often confined since it is taken as a matter of course that the mother has the right to continue her outside activities and that the grandmother has the duty to be responsible for the child.

In this culture there is little of the sense of the awesome responsibility of caring for children that is characteristic of the working and middle class. There is not the deep psychological involvement with babies which has been observed with the working-class mother.[21] The baby's needs are cared for on a catch-as-catch-can basis. If there are other children around and they happen to like babies, the baby can be over-stimulated; if this is not the case, the baby is left alone a good deal of the time. As quickly as he can move around he learns to fend for himself.

The three-generation maternal household is a busy place. In contrast to working- and middle-class homes it tends to be open to the world, with many non-family members coming in and out at all times as the children are visited by friends, the teenagers by their boy friends and girl friends, the mother by her friends and perhaps an occasional boy friend, and the grandmother by fewer friends but still by an occasional boy friend.

The openness of the household is, among other things, a reflection of the mother's sense of impotence in the face of the street system. Negro lower-class mothers often indicate that they try very hard to keep their young children at home and away from the streets; they often seem to make the children virtual prisoners in the home. As the children grow and go to school they inevitably do become involved in peer-group activities. The mother gradually gives up, feeling that once the child is lost to this pernicious outside world there is little she can do to continue to control him and direct his development. She will try to limit the types of activities that go on in the home and to restrict the kinds of friends that her children can bring into the home, but even this she must give up as time goes on, as the children become older and less attentive to her direction.

The grandmothers in their late forties, fifties, and sixties tend increasingly to stay at home. The home becomes a kind of court at which other family members gather and to which they bring their friends for sociability, and as a by-product provide amusement and entertainment for the mother. A grandmother may provide a home for her daughters, their children, and sometimes their children's children, and yet receive very little in a material way from them; but one of the things she does receive is a sense of human involvement, a sense that although life may have passed her by she is not completely isolated from it.

The lack of control that mothers have over much that goes on in their households is most dramaticaly apparent in the fact that their older children seem to have the right to return home at any time once they have moved and to stay in the home without contributing to its maintenance. Though the mother may be resentful about being taken advantage of, she does not feel she can turn her children away. For example, sixty-five-year-old Mrs. Washington plays hostess for weeks or months at a time to her forty-year-old daughter and her small children, and to her twenty-three-year-old grandaughter and her children. When these daughters come home with their families the grandmother is expected to take care of the young children and must argue with her daughter and granddaughter to receive contributions to the daily household ration of food and liquor. Or a twenty-year-old son comes home from the Air Force and feels he has the right to live at home without working and to run up an eighty-dollar long-distance telephone bill.

Even aged parents living alone in small apartments sometimes acknowledge such obligations to their children or grandchildren. Again, the only clear return they receive for the hospitality is the reduction of isolation that comes from having people around and interesting activity going on. When in the Washington home the daughter and granddaughter and their children move in with the grandmother, or when they come to visit for shorter periods of time, the occasion has a party atmosphere. The women sit around talking and reminiscing. Though boy friends may be present, they take little part; instead they sit passively, enjoying the stories and drinking along with the women. It would seem that in this kind of party activity the women are defined as the stars. Grandmother, daughter, and granddaughter in turn take the center of the stage telling a story from the family's past, talking about a particularly interesting night out on the town or just making some general observation about life. In the course of these events a good deal of liquor is consumed. In such a household as this little attention is paid to the children since the competition by adults for attention is stiff.

Boy Friends, Not Husbands

It is with an understanding of the problems of isolation which older mothers have that we can obtain the best insight into the role and function of boy friends in the maternal household. The older mothers, surrounded by their own children and grandchildren, are not able to move freely in the outside world, to participate in the high life which they enjoyed when younger and more foot-loose. They are disillusioned with marriage as providing any more secure economic base than they can achieve on their own. They see marriage as involving just another responsibility without a concomitant reward — "It's the greatest thing in the world to come home in the afternoon and not have some curly headed twot in the house yellin' at me and askin' me where supper is, where I've been, what I've been doin', and who I've been seein'."In this situation the woman is tempted to form relationships with men that are not so demanding as marriage but still provide companionship and an opportunity for occasional sexual gratification.

There seem to be two kinds of boy friends. Some boy friends "pimp" off mothers; they extract payment in food or money for their companionship. This leads to the custom sometimes called "Mother's Day," the tenth of the month when the AFDC checks come.[22] On this day one can observe an influx of men into the neighborhood, and much partying. But there is another kind of boy friend, perhaps more numerous than the first, who instead of being paid for his services pays for the right to be a pseudo family member. He may be the father of one of the woman's children and for this reason makes a steady contribution to the family's support, or he may simply be a man whose company the mother enjoys and who makes reasonable gifts to the family for the time he spends with them (and perhaps implicitly for the sexual favors he receives). While the boy friend does not assume fatherly authority within the family, he often is known and liked by the children. The older children appreciate the meaningfulness of their mother's relationship with him — one girl said of her mother's boy friend: "We don't none of us [the children] want her to marry again. It's all right if she wants to live by herself and have a boy friend. It's not because we're afraid we're going to have some more sisters and brothers, which it wouldn't make us much difference, but I think she be too old."

Even when the boy friend contributes ten or twenty dollars a month to the family he is in a certain sense getting a bargain. If he is a well-accepted boy friend he spends considerable time around the house, has a chance to relax in an atmosphere less competitive than that of his peer group, is fed and cared for by the woman, yet has not responsibilities which he cannot renounce when he wishes. When women have stable relationships of this kind with boy friends they often consider marrying them but are reluctant to take such a step. Even the well-liked boy friend has some shortcomings — one woman said of her boy friend: "Well, he works; I know that. He seems to be a nice person, kindhearted. He believes in survival for me and my family. He don't much mind sharing with my youngsters. If I ask him for a helping hand he don't seem to mind that. The only part I dislike is his drinking."

The woman in this situation has worked out a reasonably stable adaptation to the problems of her life; she is fearful of upsetting this adaptation by marrying again. It seems easier to take the "sweet" part of the relationship with a man without the complexities that marriage might involve.

It is in the light of this pattern of women living in families and men living by themselves in rooming houses, odd rooms here and there, that we can understand Daniel Patrick Moynihan's observations that during their mature years men simply disappear; that is, that census data show a very high sex ratio of women to men.[23] In St. Louis, starting at the age range twenty to twenty-four there are only seventy-two men for every one hundred women. This ratio does not climb to ninety until the age range fifty to fifty-four. Men often do not have real homes; they move about from one household where they have kinship or sexual ties to another; they live in flophouses and rooming houses; they spend time in institutions. They are not household members in the only "homes" that they have — the homes of their mothers and of their girl friends.

It is in this kind of world that boys and girls in the Negro slum community learn their sex roles. It is not just, or even mainly, that fathers are often absent but that the male role models around boys are ones which emphasize expressive, affectional techniques for making one's way in the world. The female role models available to girls emphasize an exaggerated self-sufficiency (from the point of view of the middle class) and the danger of allowing oneself to be dependent on men for anything that is crucial. By the time she is mature, the woman learns that she is most secure when she herself manages the family affairs and when she dominates her men. The man learns that he exposes himself to the least risk of failure when he does not assume a husband's and father's responsibilities but instead counts on his ability to court women and to ingratiate himself with them.

Identity Processes in the Family

Up to this point we have been examining the sequential development of family stages in the Negro slum community, paying only incidental attention to the psychological responses family members make to these social forms and not concerning ourselves with the effect the family forms have on the psychosocial development of the children who grow up in them. Now we want to examine the effect that growing up in this kind of a system has in terms of socialization and personality development.

Household groups function for cultures in carrying out the initial phases of socialization and personailty formation. It is in the family that the child learns the most primitive categories of existence and experience, and that he develops his most deeply held beliefs about the world and about himself.[24] From the child's point of view, the household *is* the world; his experiences as he moves out of it into the larger world are always interpreted in terms of his particular experience within the home. The painful experiences which a child in the Negro slum culture has are therefore interpreted as in some sense a reflection of this family world. The impact of the system of victimization is transmitted through the family; the child cannot be expected to have the sophistication an outside observer has for seeing exactly where the villians are. From the child's point of view, if he is hungry it is his parents' fault; if he experiences frustrations in the streets or in the school it is his parents' fault; if that world seems incomprehensible to him it is his parents' fault; if people are aggressive or destructive toward each other it is his parents' fault, not that of a system of race relations. In another culture this might not be the case; if a subculture could exist which provided comfort and security within its limited world and the individual experienced frustration only when he moved out into the larger society, the family might not be thought so much to blame. The effect of the caste system, however, is to bring home through a chain of cause and effect all of the victimization processes, and to bring them home in such a way that it is often very difficult even for adults in the system to see the connection between the pain they feel at the moment and the structured patterns of the caste system.

Let us take as a central question that of identity formation within the Negro slum family. We are concerned with the question of who the individual believes himself to be and to be becoming. For Erikson, identity means a sense of continuity and social sameness which bridges what the individual "*was* as a child and what he is *about to become* and also reconciles his *conception of himself* and his community's recognition of him." Thus identity is a "self-realization coupled with a mutual recognition."[25] In the early childhood years identity is family-bound since the child's identity is his identity vis-a-vis other members of the family. Later he incorporates into his sense of who he is and is becoming his experiences outside the family, but always influenced by the interpretations and evaluations of those experiences that the family gives. As the child tries on identities, *announces* them, the family sits as judge of his pretensions. Family members are both the most important judges and the most critical ones, since who he is allowed to become affects them in their own identity strivings more crucially than it affects anyone else. The child seeks a sense of valid identity, a sense of being a particular person with a satisfactory degree of congruence between who he feels he is, who he announces himself to be, and where he feels his society places him.[26] He is uncomfortable when he experiences disjunction between his own needs and the kinds of needs legitimated by those around him, or when he feels a disjunction between his sense of himself and the image of himself that others play back to him.[27]

"Tell It Like It Is"

When families become involved in important quarrels the psychosocial underpinnings of family life are laid bare. One such quarrel in a family we have been studying brings together in one place many of the themes that seem to dominate identity problems in Negro slum culture. The incident illustrates in a particularly forceful and dramatic way family processes which our field work, and some other contemporary studies of slum family life, suggests unfold more subtly in a great many families at the lower-class level. The family involved, the Johnsons, is certainly not the most disorganized one we have studied; in some respects their way of life represents a realistic adaptation to the hard living of a family nineteen years on AFDC with a monthly income of $202 for nine people. The two oldest daughters, Mary Jane (eighteen years old) and Esther (sixteen) are pregnant; Mary Jane has one illegitimate child. The adolescent sons, Bob and Richard, are much involved in the social and sexual activities of their peer group. The three other children, ranging in age from twelve to fourteen, are apparently also moving into this kind of peer-group society.

When the argument started Bob and Esther were alone in the apartment with Mary Jane's baby. Esther took exception to Bob's playing with the baby because she had been left in charge; the argument quickly progressed to a fight in which Bob cuffed Esther around, and she tried to cut him with a knife. The police were called and subdued Bob with their nightsticks. At this point the rest of the family and the field worker arrived. As the argument continued, these themes relevant to the analysis which follows appeared:
1) The sisters said that Bob was not their brother (he is a half-brother to Esther, and Mary Jane's full brother). Indeed, they said their mother "didn't have no husband. These kids don't even know who their daddies are." The mother defended herself by saying that she had one legal husband, and one common-law husband, no more.
2) The sisters said that their fathers had never done anything for them, nor had their mother. She retorted that she had raised them "to age of womanhood" and now would care for their babies.
3) Esther continued to threaten to cut Bob if she got a chance (a month later they fought again, and she did cut Bob, who requireed twenty-one stitches).
4) The sisters accused their mother of favoring their lazy brothers and asked her to put them out of the house. She retorted that the girls were as lazy, that they made no contribution to maintaining the household, could not get their boy friends to marry them or support their children, that all the support came from her AFDC check. Mary Jane retorted that "the baby has a check of her own."
5) The girls threatened to leave the house if their mother refused to put their brothers out. They said they could force their boy friends to support them by taking them to court, and Esther threatened to cut her boy friend's throat if he did not co-operate.
6) Mrs. Johnson said the girls could leave if they wished but that she would keep their babies; "I'll not have it, not knowing who's taking care of them."
7) When her thirteen-year-old sister laughed at all of this, Esther told her not to laugh because she, too, would be pregnant within a year.
8) When Bob laughed, Esther attacked him and his brother by saying that both were not man enough to make babies, as she and her sister had been able to do.
9) As the field worker left, Mrs. Johnson sought his sympathy. "You see, Joe, how hard it is for me to bring up a family. . . . They sit around and talk to me like I'm some kind of a dog and not their mother."
10) Finally, it is important to note for the analysis which follows that the following labels — "black-assed," "black bastard," "bitch," and other profane terms — were liberally used by Esther and Mary Jane, and rather less liberally by their mother, to refer to each other, to the girls' boy friends, to Bob, and to the thirteen-year-old daughter.

Several of the themes outlined previously appear forcefully in the course of this argument. In the last year and a half the mother has become a grandmother and expects shortly to add two more grandchildren to her household. She takes it for granted that it is her responsibility to care for the grandchildren and that she has the right to decide what will be done with the children since her own daughters are not fully responsible. She makes this very clear to them when they threaten to move out, a threat which they do not really wish to make good nor could they if they wished to.

However, only as an act of will is Mrs. Johnson able to make this a family. She must constantly cope with the tendency of her adolescent children to disrupt the family group and to deny that they are in fact a family — "He ain't no brother of mine"; "The baby has a check of her own." Though we do not know exactly what processes communicate these facts to the children, it is clear that in growing up they have learned to regard themselves as not fully part of a solidary collectivity. During the quarrel this message was reinforced for the twelve-, thirteen-, and fourteen-year-old daughters by the four-way argument among their older sisters, older brother, and their mother.

The argument represents vicious unmasking of the individual members' pretenses to being competent individuals.[28] The efforts of the two girls to present themselves as masters of their own fate are unmasked by the mother. The girls in turn unmask the pretensions of the mother and of their two brothers. When the thirteen-year-old daughter expresses some amusement they turn on her, telling her that it won't be long before she too becomes pregnant. Each member of the family in turn is told that he can expect to be no more than a victim of his world, but that this is somehow inevitably his own fault.

In this argument masculinity is consistently demeaned. Bob has no right to play with his niece, the boys are not really masculine because at fifteen and sixteen years they have yet to father children, their own fathers were no-goods who failed to do anything for their family. These notions probably come originally from the mother, who enjoys recounting the story of having her common-law husband imprisoned for nonsupport, but this comes back to haunt her as her daughters accuse her of being no better than they in ability to force support and nurturance from a man. In contrast, the girls came off somewhat better than the boys, although they must accept the label of stupid girls because they have similarly failed and inconveneintly become pregnant in the first place. At least they can and have had children and therefore have some meaningful connection with the ongoing substance of life. There is something important and dramatic in which they participate, while the boys, despite their sexual activity, "can't get no babies."

In most societies, as children grow and are formed by their elders into suitable members of the society, they gain increasingly a sense of competence and ability to master the behavioral environment their particular world presents. But in Negro slum culture growing up involves an ever-increasing appreciation of one's shortcomings, of the impossibility of finding a self-sufficient and gratifying way of living.[29] It is in the family first and most devastatingly that one learns these lessons. As the child's sense of frustration builds, he too can strike out and unmask the pretensions of others. The result is a peculiar strength and a pervasive weakness. The strength involves the ability to tolerate and defend against degrading verbal and physical aggressions from others and not to give up completely. The weakness involves the inability to embark hopefully on any course of action that might make things better, particularly action which involves cooperating and trusting attitudes toward others. Family members become potential enemies to each other, as the frequency of observing the police being called in to settle family quarrels brings home all too dramatically.

The conceptions parents have of their children are such that they are constantly alert as the child matures to evidence that he is as bad as everyone else. That is, in lower-class culture human nature is conceived of as essentially bad, destructive, immoral.[30] This is the nature of things. Therefore any one child must be inherently bad unless his parents are very lucky indeed. If the mother can keep the child insulated from the outside world, she feels she may be able to prevent his inherent badness from coming out. She feels that once he is let out into the larger world the badness will come to the fore since that is his nature. This means that in the identity development of the child he is constantly exposed to identity labeling by his parents as a bad person. Since as he grows up he does not experience his world as particularly gratifying, it is very easy for him to conclude that this lack of gratification is due to the fact that something is wrong with him. This, in turn, can readily be assimilated to the definitions of being a bad person offered him by those with whom he lives.[31] In this way the Negro slum child learns his culture's conception of being-in-the-world, a conception that emphasizes inherent evil in a chaotic, hostile, destructive world.

Blackness

To a certain extent these same processes operate in white lower-class groups, but added for the Negro is the reality of blackness. "Black-assed" is not an empty pejorative adjective. In the Negro slum culture several distinctive appellations are used to refer to oneself and others. One involves the terms "black" or "nigger." Black is generally a negative way of naming, but nigger can be either negative or positive, depending upon the context. It is important to note that, at least in the urban North, the initial development of racial identity in these terms has very little directly to do with relations with whites. A child experiences these identity placements in the context of the family and in the neighborhood peer group; he probably very seldom hears the same terms used by whites (unlike the situation in the South). In this way, one of the effects of ghettoization is to mask the ultimate enemy so that the understanding of the fact of victimization by a caste system comes as a late acquisition laid over conceptions of self and of other Negroes derived from intimate, and to the child often traumatic, experience within the ghetto community. If, in addition, the child attends a ghetto school where his Negro teachers either overtly or by implication reinforce his community's negative conceptions of what it means to be black, then the child has little opportunity to develop a more realistic image of himself and other Negroes as being damaged by whites and not by themselves. In such a situation, an intelligent man like Mr. Wilson (quoted earlier) can say with all sincerity that he does not feel most Negroes are ready for integration — only under the experience of certain kinds of intense personal threat coupled with exposure to an ideology that places the responsibility on whites did he begin to see through the direct evidence of his daily experience.

To those living in the heart of a ghetto, black comes to mean not just "stay back," but also membership in a community of persons who think poorly of each other, who attack and manipulate each other, who give each other small comfort in a desperate world. Black comes to stand for a sense of identity as no better than these destructive others. The individual feels that he must embrace an unattractive self in order to function at all.

We can hypothesize that in those families that manage to avoid the destructive identity imputations of "black" and that manage to maintain solidarity against such assaults from the world around, it is possible for children to grow up with a sense of both Negro and personal identity that allows them to socialize themselves in an anticipatory way for participation in the larger society.[32] This broader sense of identity, however, will remain a brittle one as long as the individual is vulnerable to attack from within the Negro community as "nothing but a nigger like everybody else" or from the white community as "just a nigger." We can hypothesize further that the vicious unmasking of essential identity as black described above is least likely to occur within families where the parents have some stable sense of security, and where they therefore have less need to protect themselves by disavowing responsibility for their children's behavior and denying the children their patrimony as products of a particular family rather than of an immoral nature and an evil community.

In sum, we are suggesting that Negro slum children as they grow up in their families and in their neighborhoods are exposed to a set of experiences — and a rhetoric which conceptualizes them — that brings home to the child an understanding of his essence as a weak and debased person who can expect only partial gratification of his needs, and who must seek even this level of gratification by less than straight-forward means.

Strategies for Living

In every society complex processes of socialization inculcate in their members strategies for gratifying the needs with which they are born and those which the society itself generates. Inextricably linked to these strategies, both cause and effect of them, are the existential propositions which members of a culture entertain about the nature of their world and of effective action within the world as it is defined for them. In most of American society two grand strategies seem to attract the allegiance of its members and guide their day-to-day actions. I have called these strategies those of *the good life* and of *career success*.[33] A good life strategy involves efforts to get along with others and not to rock the boat, a comfortable familism grounded on a stable work career for husbands in which they perform adequately at the modest jobs that enable them to be good providers. The strategy of career success is the choice of ambitious men and women who see life as providing opportunities to move from a lower to a higher status, to "accomplish something," to achieve greater than ordinary material well-being, prestige, and social recognition. Both of these strategies are predicated on the assumption that the world is inherently rewarding if one behaves properly and does his part. The rewards of the world may come easily or only at the cost of great effort, but at least they are there.

In the white and particularly in the Negro slum worlds little in the experience that individuals have as they grow up sustains a belief in a rewarding world. The strategies that seem appropriate are not those of a good, family-based life or of a career, but rather *strategies for survival*.

Much of what has been said above can be summarized as encouraging three kinds of survival strategies. One is the strategy of the *expressive life style* which I have described elsewhere as an effort to make yourself interesting and attractive to others so that you are better able to manipulate their behavior along lines that will provide some immediate gratification.[34] Negro slum culture provides many examples of techniques for seduction, of persuading others to give you what you want in situations where you have very little that is tangible to offer in return. In order to get what you want you learn to "work game," a strategy which requires a high development of a certain kind of verbal facility, a sophisticated manipulation of promise and interim reward. When the expressive strategy fails or when it is unavailable there is, of course, the great temptation to adopt a *violent strategy* in which you force others to give you what you need once you fail to win it by verbal and other symbolic means.[35] Finally, and increasingly as members of the Negro slum culture grow older, there is the *depressive strategy* in which goals are increasingly constricted to the bare necessities for survival (not as a social being but simply as an organism).[36] This is the strategy of "I don't bother anybody and I hope nobody's gonna bother me; I'm simply going through the motions to keep body (but not soul) together." Most lower-class people follow mixed strategies, as Walter Miller has observed, alternating among the excitement of the expressive style, the desperation of the violent style, and the deadness of the depressed style.[37] Some members of the Negro slum world experiment from time to time with mixed strategies that also incorporate the stable working-class model of the good American life, but this latter strategy is exceedingly vulnerable to the threats of unemployment or a less than adequate pay check, on the one hand, and the seduction and violence of the slum world around them, on the other.

Remedies

Finally, it is clear that we, no less than the inhabitants of the ghetto, are not masters of their fate because we are not masters of our own total society. Despite the battles with poverty on many fronts we can find little evidence to sustain our hope of winning the war given current programs and strategies.

The question of strategy is particularly crucial when one moves from an examination of destructive cultural and interaction patterns in Negro families to the question of how these families might achieve a more stable and gratifying life. It is tempting to see the family as the main villain of the piece, and to seek to develop programs which attack directly this family pathology. Should we not have extensive programs of family therapy, family counseling, family-life education, and the like? Is this not the prerequisite to enabling slum Negro families to take advantage of other opportunities? Yet, how pale such efforts seem compared to the deep-seated problems of self-image and family process described above. Can an army of social workers undo the damage of three hundred years by talking and listening without massive changes in the social and economic situations of the families with whom they are to deal? And, if such changes take place, will the social-worker army be needed?

If we are right that present Negro family patterns have been created as adaptations to a particular socioeconomic situation, it would make more sense to change that socioeconomic situation and then depend upon the people involved to make new adaptations as time goes on. If Negro providers have steady jobs and decent incomes, if Negro children have some realistic expectations of moving toward such a goal, if slum Negroes come to feel that they have the chance to affect their own futures and to receive respect from those around them, then (and only then) the destructive patterns described are likely to change. The change, though slow and uneven from individual to individual, will in a certain sense be automatic because it will represent an adaptation to changed socioeconomic circumstances which have direct and highly valued implications for the person.

It is possible to think of three kinds of extra-family change that are required if family patterns are to change; these are outlined below as pairs of current deprivations and needed remedies:

Deprivation effect of caste vicitimization	*Needed remedy*
I. Poverty	Employment income for men; income maintenance for mothers
II. Trained incapacity to function in a bureaucratized and industrialized world	Meaningful education of the next generation
III. Powerlessness and stigmatization	Organizational participation for aggressive pursuit of Negroes' self-interest
	Strong sanctions against callous or indifferent service to slum Negroes
	Pride in group identity, Negro *and* American

Unless the major effort is to provide these kinds of remedies, there is a very real danger that programs to "better the structure of the Negro family" by direct intervention will serve the unintended functions of distracting the country from the pressing needs for socioeconomic reform and providing an alibi for the failure to embark on the basic institutional changes that are needed to do anything about abolishing both white and Negro poverty. It would be sad indeed if, after the Negro revolt brought to national prominence the continuing problem of poverty, our expertise about Negro slum culture served to deflect the national impulse into symptom-treatment rather than basic reform. If that happens, social scientists will have served those they study poorly indeed.

References

1. E. Franklin Frazier, *The Negro Family in the United States* (Chicago, 1939), p. 487.

2. Norman Mailer, "The White Negro" (City Light Books, San Francisco, Calif., 1957); and Leslie Fiedler, *Waiting for the End* (New York, 1964), pp. 118-37.

3. See Alvin W. Gouldner, "Reciprocity and Autonomy in Functional Theory," in Llewellyn Gross (ed.), *Symposium of Sociological Theory* (Evanston, Ill., 1958), for a discussion of functional autonomy and dependence of structural elements in social systems. We are suggesting here that the lower-class groups have a relatively high degree of functional autonomy *vis a vis* the total social system because that system does little to meet their needs. In general the fewer the rewards a society offers members of a particular group in the society, the more autonomous will that group prove to be with reference to the norms of the society. Only by constructing an elaborate repressive machinery, as in the concentration camps, can the effect be otherwise.

4. For example, the lead sentence in a *St. Louis Post Dispatch* article of July 20, 1965, begins "A White House study group is laying the ground work for an attempt to better the structure of the Negro family."

5. See Kenneth Stampp, *The Peculiar Institution* (New York, 1956); John Hope Franklin, *From Slavery to Freedom* (New York, 1956); Frank Tannenbaum, *Slave and Citizen* (New York, 1946); E. Franklin Frazier, *op. cit.*; and Melville J. Herskovits, *The Myth of the Negro Past* (New York, 1941).

6. See Raymond T. Smith, *The Negro Family in British Guiana* (New York, 1956); J. Mayone Stycos and Kurt W. Back, *The Control of Human Fertility in Jamaica* (Ithaca, N.Y., 1964); F. M. Henriques, *Family and Colour in Jamaica* (London, 1953); Judith Blake, *Family Structure in Jamaica* (Glencoe, Ill., 1961); and Raymond T. Smith, "Culture and Social Structure in The Caribbean," *Comparative Studies in Society and History* (The Hague, The Netherlands, October 1963), vol.6, pp. 24-46. For a broader comparative discussion of the matrifocal family see Peter Kunstadter, "A Survey of the Consanguine or Matrifocal Family," *American Anthropologist*, vol. 65, no. 1 (February 1963), pp. 56-66; and Ruth M. Boyer, "The Matrifocal Family Among the Mescalero: Additional Data," *American Anthropologist*, vol. 66, no. 3 (June 1964), pp. 593-602.

7. Paul C. Glick, *American Families* (New York, 1957), pp. 133 ff.

8. For discussions of white lower-class families, see Lee Rainwater, Richard P. Coleman, and Gerald Handel, *Workingman's Wife* (New York, 1959); Lee Rainwater, *Family Design* (Chicago, 1964); Herbert Gans, *The Urban Villagers* (New York, 1962); Albert K. Cohen and Harold M. Hodges, "Characteristics of the Lower-Blue-Collar-Class," *Social Problems,* vol.10, no. 4 (Spring 1963), pp. 303-34; S. M. Miller, "The American Lower Classes: A Typological Approach," in Arthur B. Shostak and William Gomberg, *Blue Collar World* (Englewood Cliffs, N. J., 1964); and Mirra Komarovsky, *Blue Collar Marriage* (New York, 1964). Discussions of Negro slum life can be found in St. Clair Drake and Horace R. Cayton, *Black Metropolis* (New York, 1962), and Kenneth B. Clark, *Dark Ghetto* (New York, 1965); and of Negro community life in small-town and rural settings in Allison Davis, Burleigh B. Gardner, and Mary Gardner, *Deep South* (Chicago, 1944), and Hylan Lewis, *Blackways of Kent* (Chapel Hill, N.C., 1955).

9. For general discussions of the extent to which lower-class people hold the values of the larger society, see Albert K. Cohen, *Delinquent Boys* (New York, 1955); Hyman Rodman, "The Lower Class Value Stretch," *Social Forces*, vol. 42, no. 2 (December 1963), pp. 205 ff; and William L. Yancey, "The Culture of Poverty: Not So Much Parsimony," unpublished manuscript, Social Science Institute, Washington University.

10. James F. Short, Jr., and Fred L. Strodtbeck, *Group Process and Gang Delinquency* (Chicago, 1965), p. 114. Chapter 5 (pages 102-115) of this book contains a very useful discussion of differences between white and Negro lower-class communities.

11. Discussions of white lower-class attitudes toward sex may be found in Arnold W. Green, "The Cult of Personality and Sexual Relations," *Psychiatry*, vol. 4 (1941), pp. 343-48;

William F. Whyte, "A Slum Sex Code," *American Journal of Sociology,* vol. 49, no. 1 (July 1943), pp. 24-31; and Lee Rainwater, "Marital Sexuality in Four Cultures of Poverty," *Journal of Marriage and the Family*, vol. 26, no. 4 (November 1964), pp. 457-66.

12. See Boone Hammond, "The Contest System: A Survival Technique," Master's Honors paper, Washington University, 1965. See also Ira L. Reiss, "Premarital Sexual Permissiveness Among Negroes and Whites," *American Sociological Review*, vol. 29, no. 5 (October 1964). pp. 688-98.

13. See the discussion of aleatory processes leading to premarital fatherhood in Short and Strodtbeck, *op. cit.*, pp. 44-45.

14. Rainwater, *And the Poor Get Children*, pp. 61-63. See also Carlfred B. Broderick, "Social Heterosexual Development Among Urban Negroes and Whites," *Journal of Marriage and the Family*, vol. 27 (May 1965), pp. 200-212. Broderick finds that although white boys and girls and Negro girls become more interested in marriage as they get older, Negro boys become *less* interested in late adolescence than they were as preadolescents.

15. Walter Miller, "The Corner Gang Boys Get Married," *Trans-action*, vol. 1, no. 1 (November 1963), pp.10-12.

16. Rainwater, *Family Design*, pp. 28-60.

17. Yancey, *op. cit.* The effects of unemployment on the family have been discussed by E. Wright Bakke, *Citizens Without Work* (New Haven, Conn., 1940); Mirra Komarovsky, *The Unemployed Man and His Family* (New York, 1960); and Earl L. Koos, *Families in Trouble* (New York, 1946). What seems distinctive to the Negro slum culture is the short time lapse between the husband's loss of a job and his wife's considering him superfluous.

18. See particularly Komarovsky's discussion of "barriers to marital communications" (Chapter 7) and "confidants outside of marriage" (Chapter 9), in *Blue Collar Marriage*.

19. Rainwater, *Family Design*, pp. 305-308.

20. For a discussion of the relationship between black nationalist ideology and the Negro struggle to achieve a sense of valid personal identity, see Howard Brotz, *The Black Jews of Harlem* (New York, 1963), and E. U. Essien-Udom, *Black Nationalism: A Search for Identity in America* (Chicago, 1962).

21. Rainwater, Coleman, and Handel, *op. cit.*, pp. 88-102.

22. Cf. Michael Schwartz and George Henderson, "The Culture of Unemployment: Some Notes On Negro Children," in Schostak and Gomborg, *op. cit.*

23. Daniel Patrick Moynihan, "Employment, Income, and the Ordeal of the Negro Family," *The Negro American* (Boston, 1966), pp. 149-50.

24. Talcott Parsons concludes his discussion of child socialization, the development of an "internalized family system" and internalized role differentiation by observing, "The internalization of the family collectivity as an object and its value should not be lost sight of. This is crucial with respect to . . . the assumption of representative roles outside the family on behalf of it. Here it is the child's family membership which is decisive, and thus his acting in a role in terms of its values for 'such as he.' " Talcott Parsons and Robert F. Bales, *Family, Socialization and Interaction Process* (Glencoe, Ill., 1955), p.113

25. Erik H. Erikson, "Identity and the Life Cycle," *Psychological Issues,* vol. 1, no. 1 (1959).

26. For discussion of the dynamics of the individual's *announcements* and the society's *placements* in the formation of identity, see Gregory Stone, "Appearance and the Self," in Arnold Rose, *Human Behavior in Social Process* (Boston, 1962), pp. 86-118

27. The importance of identity for social behavior is discussed in detail in Ward Goodenough, *Cooperation and Change* (New York, 1963), pp. 176-251, and in Lee Rainwater, "Work and Identity in the Lower Class," in Sam H. Warner, Jr., *Planning for the Quality of Urban Life* (Cambridge, Mass., forthcoming). The images of self and of other

family members is a crucial variable in Hess and Handel's psychosocial analysis of family life; see Robert D. Hess and Gerald Handel, *Family Worlds* (Chicago, 1959), especially pp. 6-11.

28. See the discussion of "masking" and "unmasking" in relation to disorganization and re-equilibration in families by John P. Spiegel, "The Resolution of Role Conflict within the Family," in Norman W. Bell and Ezra F. Vogel, *A Modern Introduction to the Family* (Glencoe, Ill., 1960), pp. 375-77.

29. See the discussion of self-identity and self-esteem in Thomas F. Pettigrew, *A Profile of the Negro American* (Princeton, N. J., 1964), pp. 6-11.

30. Rainwater, Coleman and Handel, *op. cit.*, pp. 44-51. See also the discussion of the greater level of "anomie" and mistrust among lower-class people in Ephriam Mizruchi, *Success and Opportunity* (New York, 1954). Unpublished research by the author indicates that for one urban lower-class sample (Chicago) Negroes scored about 50 per cent higher on Srole's anomie scale that did comparable whites.

31. For a discussion of the child's propensity from a very early age for speculation and developing explanations, see William V. Silverberg, *Childhood Experience and Personal Destiny* (New York, 1953) pp. 81 ff.

32. See Ralph Ellison's autobiographical descriptions of growing up in Oklahoma City in his *Shadow and Act* (New York, 1964).

33. Rainwater, "Work and Identity in the Lower Class."

34. Ibid.

35. Short and Strodtbeck see violent behavior in juvenile gangs as a kind of last resort strategy in situations where the actor feels he has no other choice. See Short and Strodtbeck, *op. cit.1, pp. 248-64.*

36. Wiltse speaks of a "pseudo depression syndrome" as characteristic of many AFDC mothers. Kermit T. Wiltse, "Orthopsychiatric Programs for Socially Deprived Groups," *American Journal or Orthopsychiatry*, vol. 33, no. 5 (October 1963), pp. 806-813.

37. Walter B. Miller, "Lower Class Culture as a Generating Milieu of Gang Delinquency," *Journal of Social Issues*, vol. 14, no. 3 (1958), pp. 5-19.

CHAPTER 8

Child-Rearing and Family Relationship Patterns of the Very Poor

CATHERINE S. CHILMAN

Poverty that continues from generation to generation is a problem of major concern today. This so-called cycle of poverty has many causes. Among them are certain cultural patterns — the ways in which parents in very poor families rear their children and relate to each other. One of the approaches to an expansion of social and economic opportunities for the very poor consists of a variety of action programs directed, at least in part, toward helping families adopt a life style which is more effective in coping with the demands of an increasingly urban, technical, and complex society.

The knowledge on which such programs must be based includes a delineation of significant cultural patterns of the poor in regard to child-rearing and family relationships — an area in which much investigation has already been carried out. These studies are summarized here and the findings compared with those of studies concerned with "ideal" patterns of child-rearing and family life. Finally, the numerous implications of this body of information for service programs and for further research are indicated.

Limitations of Past Research

The term "subcultures of poverty" is used in this paper rather than the more familiar term, "culture of poverty," because the very poor apparently subscribe at least partially to much of the middle-class culture while adapting it to their own more disadvantaged situation.[1] Also, it is likely that there are a number of subcultures of poverty based on regional, ethnic, religious, and racial variation within the United States, although most of the published studies have failed to concern themselves with considerations of this kind.

Furthermore, the subcultures of poverty, as discussed here, refer to the subcultures of the very poor, as distinct from those of the stable working class. Until recently, research on low-income culture patterns has tended to treat these two groups together. More gross discriminations have generally been made between middle-class and lower class families. The criteria must be precise to differentiate the working class, the employed poor, the unemployed poor, or the chronically dependent poor. It is highly probable that there are more differences between the very poor (the chronically unemployed or casual, unskilled labor groups) and the skilled labor groups than between the latter and the middle class.[2] It is also probable that, with the increasing prosperity and economic security of skilled labor, this group is becoming more and more middle class in orientation.

Studies of lower class culture that have been made so far tend to be lacking in methodological precision. Further attention needs to be paid to a more rigorous definition of the nature of the sample (for instance, size of sample, randomness of sampling, and characteristics of the sample — racial, regional, national, religious, age, sex, and a finer delineation of social class level).

Many studies have been based on interview and observation techniques, with insufficient consideration given to issues of reliability and observer bias. While valuable insights can be gained from such methods, there is a tendency to make global generalizations that lack adequate support.

Little use has been made of standardized tests. While such tests have their limitations, including middle-class cultural bias and questionable applicability for subjects who are illiterate, more explorations might be made in this area, including the use of appropriate projective test techniques.

The current interest in the cultural characteristics of various groups has sometimes led to a tendency to forget that individuals within a group do not necessarily have all the characteristics of the socioeconomic group to which they belong. Moreover, most individuals belong to a number of subcultural groups and quite probably draw attitudes and values from each of them.

It seems appropriate also to bear in mind that individual behavior is a product of multiple causation. Other factors include the individual's physical constitution, his particular life experience and current situation, and his level of intellectual functioning.

Some of the hazards of quick, simple generalizations have been sketched here. While studies to date regarding the subcultures of the poor are limited in number, scope, and method, it is not realistic to wait for all of the required evidence in order to present some of it. Nor is it realistic to defer action-oriented programs until research has pointed the way with clairty and precision. As a stimulant both to further study and to action programs, the following material is presented, with the recognition that many of the findings summarized here may be changed or modified with time and further research.

Child-Rearing Patterns

While each family has its individual style of living, research findings indicate that certain child-rearing patterns tend to be more characteristic of lower class than of middle-class parents, as listed in tables 1-4. (In regard to these tables it should be noted that because a pattern has been found to be more characteristic of poor families than of more affluent ones, this pattern is not necessarily predominant in these families. It is also important to recognize that there is a growing body of research which strongly indicates that parental practices apparently have a differential effect on boys and on girls.)

The tables also summarize many of the major findings of a large number of studies of child-rearing practices that are found to be variously conducive to a child's emotional health; educational achievement; social success; and "good moral character," i.e., ability to resist temptation and to be responsible and honest.

When the relevant child-rearing patterns of the very poor are compared with those which are found to be conducive to educational achievement, the contrast is striking (table 1). While it would be premature and overly simple to conclude that disadvantaged families tend to rear their children for failure in school, evidence accumulated to date points in this direction.

Similarly, the opportunity for very poor children to grow up as emotionally healthy individuals would seem to be seriously jeopardized if their parents employ the child-rearing patterns which research indicates to

Table 1. Child-rearing patterns found to be characteristic of families of children who are educationally achieving compared with relevant patterns found to be characteristic of very poor families.

Conducive (Refs. 3-8)	Low-income (Refs. 2, 6, 9-15)
1. Child given freedom within consistent limits to explore and experiment.	1. Limited freedom for exploration (partly imposed by crowded and dangerous aspects of environment).
2. Wide range of parent-guided experiences, offering visual, auditory, kinesthetic, and tactile stimulation.	2. Constricted lives led by parents: fear and distrust of the unknown.
3. Goal-commitment and belief in long-range success potential.	3. Fatalistic, apathetic attitudes.
4. Gradual training for and value placed on independence.	4. Tendency for abrupt transition to independence: parents tend to "lose control" of children at early age.
5. Educational-occupational success of parents; model as continuing "learners" themselves.	5. Tendency to educational-occupational failure; reliance on personal versus skill attributes of vocational success.
6. Reliance on objective evidence.	6. Magical, rigid thinking.
7. Much verbal communication.	7. Little verbal communication.

be more typical of low-income families than of middle-class ones (table 2). Very poor parents would also seem to have a tendency to rear their children in ways that tend to be prejudicial to their social acceptability in the usual middle-class group, as indicated in table 3.

The higher rates of delinquency and rejection by authority figures of the children of the very poor are probably related in part to the child-rearing methods used by their parents. Such character traits as honesty,

Table 2. Child-rearing patterns found to be characteristic of families of children who are emotionally healthy compared with relevant patterns found to be characteristic of very poor families.

(Refs. 3–7, 9, 16–22)	(Refs. 2, 10–13, 15, 23, 25, 28–38)
1. Respect for child as individual whose behavior has multiple causes.	1. Misbehavior regarded as such in terms of concrete pragmatic outcomes; reasons for behavior not considered.
2. Commitment to slow development of child from infancy to maturity; stresses and pressures of each stage accepted by parent because of perceived worth of ultimate goal of rearing "happy," successful son or daughter.	2. Lack of goal commitment, impulsive gratification, fatalism, lack of belief in long-range success; main object for parent and child is "keep out of trouble."
3. Discipline chiefly verbal, mild, reasonable, consistent, based on needs of child, family, and society; more emphasis on rewarding good behavior than on punishing bad behavior.	3. Discipline harsh, inconsistent, physical, makes use of ridicule; based on whether child's behavior does or does not annoy parent.
4. Open, free, verbal communication between parent and child; control largely verbal.	4. Limited verbal communication; control largely physical.
5. Democratic rather than autocratic or laissez faire methods of rearing, with both parents in equalitarian but not necessarily interchangeable roles.	5. Authoritarian methods; mother chief child-care agent; father, when in home, mainly a punitive figure.
6. Parents view selves as generally competent adults.	6. Low parental self-esteem.
7. Intimate, expressive, warm relationship between parent and child, allowing for gradually increasing independence.	7. Large families; more impulsive, narcissistic parent behavior.
8. Presence of father in home.	8. Father out of home.
9. Free verbal communication about sex, acceptance of child's sex needs, channeling of sex drive through "healthy" psychological defenses, acceptance of slow growth toward impulse control and sex satisfaction in marriage; sex education by both father and mother.	9. Repressive, punitive attitude about sex, sex questioning, and experimentation. Sex viewed as exploitative relationship.
10. Acceptance of child's drive for aggression but channeling it into socially approved outlets.	10 Alternating encouragement and restriction of aggression, primarily related to consequences of aggression for parents.

responsibility, dependability, resistance to temptation, and capacity for impulse control would seem to be consonant with adherence to many of the legal requirements set up by our society. These traits also are favored by such success-relevant figures as school authorities and employers. While evidence from research is rather fragmentary in the area of parental practices which are conducive to producing this "good moral character," that which is available strongly suggests that the children of the very poor have less chance to develop socially desirable character traits than children of other socioeconomic groups (table 4).

While the tables tend to oversimplify the evidence and thus may lead to unwarranted generalizations, they do indicate that lower-class parents tend far more than middle-class ones to bring up their children in ways that are conducive to serious difficulties in the four crucial areas examined. The phenomenon of generation-to-generation poverty may well be associated in part with such parental methods, behavior, and attitudes.

Although it is recognized that, in effect, the dice are loaded against very poor parents in the presentation of evidence such as this, our predominantly middle-class society is also inherently and inevitably loaded against them. Criteria of emotional health, social acceptability, and "good moral character" are all middle-class concepts. Research in these areas is generally built on such criteria. Therefore, it is inevitable that lower class children more often than middle-class ones would fail to meet the standards on which most studies are based. Moreover, schools are basically middle-class institutions; again it is no wonder that lower class children have trouble achieving in such settings.

But this "lack of fit" cannot be tossed aside because it can be explained. The middle-class culture has in part produced our prosperous, technically advanced society; it has also in part been produced by these technical

Table 3. Child-rearing patterns found to be characteristic of families of children who are socially successful compared with relevant patterns found to be characteristic of very poor families.

Conducive (Refs. 6, 7)	**Low-income** (Refs. 2, 13, 27, 32, 39)
1. Social skills in dress, manners, speech, games, etc., according to middle-class norms.	1. Little skill in prevalent middle-class behavior.
2. Sensitivity to feelings and attitudes of others.	2. Slight awareness of subtleties of interpersonal relations.
3. Ability to be flexible and conform to group.	3. Tendency to be rigid and non-conforming to middle-class norms.
4. Good impulse control.	4. Poor impulse control.
5. Cheerful, happy, self-assured attitude.	5. Low self-esteem, distrust, tendency to hostile aggression and/or withdrawal.

advances. There is a cycle of prosperity, just as there is a cycle of poverty. The individual or family that wishes to participate in the first cycle very probably must abandon the second. For instance, fatalism, apathy, impulsiveness, and lack of goal-commitment ill-serve the person who wants to escape from the urban gray areas into the suburban green ones.

It is also recognized that research findings are dependent on what questions are asked. The weakness of child-rearing methods used by the poor may well be overemphasized in this report partly because of what research personnel have looked for — their hypotheses and related questions.

Family Relationship Patterns

Table 5 summarizes the research evidence on attitudes and behavior found to be more characteristic of low-income families as a group than of middle-class ones as a group which concern interpersonal relations between the sexes. It also summarizes the findings that show a strong association with "good" marital adjustment.

While the summary of findings presented in table 5, like those in the preceding tables, may lead to oversimplification and easy generalizations, it shows that the attitudes and behavior of the very poor tend, more than those of the middle class, to be antithetical to a stable, personally fulfilling marriage. (Although the matter of "personal fulfillment" in marriage is a highly subjective one, the research leading to the findings in table 5 includes a number of studies, done at various times with various populations. Self-ratings as to marital happiness have been most generally used as criteria of

Table 4. Child-rearing patterns found to be characteristic of families of children who are of "good character"* compared with relevant patterns found to be characteristic of very poor families.

Conducive (Ref. 40)	Low-income (Refs. 2, 10, 13, 15, 23, 25, 28–30, 32, 35)
1. Democratic child-rearing methods.	1. Authoritarian methods.
2. Mild, reasonable, consistent discipline.	2. Harsh, physical, inconsistent discipline.
3. Child's capacity for moral judgment, according to basic principles, is viewed as a slowly developing ability.	3. Reasons for child's misbehavior tend not to be considered; specific behavioral outcomes rather than principles considered paramount.
4. Moral values are discussed and clarified.	4. Little verbal communication and discussion.
5. Parents set example by their own behavior.	5. Parental behavior more apt to be impulsive and gratification-oriented.

*Defined as including honesty, responsibility, dependability, and ability to resist temptation.

personal fulfillment since there is doubt as to whether better criteria exist. Other criteria have also been used, such as lack of separation, desertion, and divorce.) Such attitudes and behavior probably play an important part in the far higher rate of family instability — separation, divorce, desertion, illegitimacy — found among the very poor. Since a stable, "happy" marital adjustment of the parents is important to the psychological, social, and educational well-being of the child, not only during childhood but also as he grows up and establishes his own family, the implications of the

Table 5. Attitudes and behavior found to be prevalent among low-income men and women compared with attitudes and behavior strongly associated with "good" marital adjustment.*

Good marital adjustment (Refs. 34, 41–51)	Low-income behavior and attitudes (Refs. 2, 10, 23–26, 28, 34, 51–54)
1. Happiness of parents' marriage and lack of divorce in family of origin.	1. High divorce and separation rates of the very poor.
2. High level of self-perceived childhood happiness.	2. Research evidence lacking; professional observation suggests higher levels of childhood unhappiness among very poor.
3. Mild but firm discipline during childhood with only moderate punishment.	3. Severe, harsh, inconsistent punishment in childhood.
4. Adequate sex information, especially from encouraging parents.	4. Repressive, punitive attitudes toward sex.
5. Substantial time of acquaintance and engagement before marriage.	5. Short courtship, marriage often because of pregnancy.
6. Marriage in the middle 20's or older.	6. Teen-age marriages common.
7. Mutual enjoyment of sex relations.	7. Sex regarded as enjoyable only for male, women regarding it as source of economic security; basically exploitative on both sides.
8. Open expression of affection between sexes, confidence in each other, equalitarian, free communication.	8. Little expressed affection, male in dominant role, women aligned against men, little communication, attitudes of mutual exploitation.
9. Joint participation in outside interests, friends in common.	9. Masculine and feminine worlds separate, few friends.
10. Residence stable, in single home.	10. Residence in multiple dwellings, frequent moving.
11. Education at least through high school.	11. Education less than high school.
12. High socioeconomic level, income at least $5,000 per year.	12. Low socioeconomic level, income less than $3,000 per year.

*Insofar as studies have been done concerning differences between Negroes and whites, findings indicate apparent differences largely disappear when data are carefully analyzed in terms of strictly comparable socioeconomic levels. Thus studies reporting racial differences are likely to be reporting class differences.

tendencies of the very poor toward destructive heterosexual attitudes and behavior are serious indeed. As in the case of child-rearing patterns, it is apparent that the interparental life style of this group is more likely to play into the continuing cycle of generation-to-generation disadvantages of the "hard-core" poor than is the interparental life style of the middle class. However, as with the findings on child-rearing patterns, it should be noted that concepts of marital happiness and stability are middle-class in nature and that research findings derive in large part from the questions that are asked.

Associated Outcomes

Some of the major expected outcomes of the child-rearing patterns and family life styles of low-income parents have already been highlighted. The subcultures of poverty, in comparison with other cultural patterns, would seem to be more conducive not only to poor emotional health, school failure, social rejection, inadequate impulse control (lack of "good" character), and family breakdown but also to collateral areas of behavioral malfunctioning. These include difficulties in obtaining and holding a job, in effective participation in the various social systems of the community, in adequate home-management and consumer skills, and in political participation as a citizen in a democracy.

Implications for Service

The cultural patterns that have been outlined present numerous implications for development of services for families living in poverty. A few of these implications are:

1. Since the conditions of poverty are likely to promote the subcultures of poverty, it is obvious that such conditions as insufficient income, lack of equal opportunities, poor housing, inadequate schools, exploitation, etc., must be ameliorated in order to produce a milieu in which new behavior patterns can develop and flourish.

2. Since the subcultures of poverty tend to be defeating to the poor, it seems that remedial efforts planned with and for the poor should provide ways in which some of these culture patterns can be modified. While it is generally inappropriate to impose middle-class ways on the disadvantaged, it may well be appropriate to develop plans and services with them that may lead to the achievement of certain middle-class attitudes and behavior, since these attitudes and behavior are more likely to bring about the social and economic rewards generally associated with middle-class membership to which many of the poor aspire.

3. Perhaps the subcultures of poverty could be tentatively diagnosed as

being (at least, in the American middle-class value orientation) a comparatively immature system of behavior and attitudes since, in the light of available research, some of their outstanding features are magical thinking, subjective judgment, impulse gratification, use of force rather than reason, alienation from authority figures, lack of goal commitment, distrust of heterosexual relationships, projection of problems on to others, and other similar characteristics.

If it is agreed that these characteristics are more typically associated with childhood and youth than with adult maturity in the middle-class American culture,[55] does it follow that many of the poor need, in effect, to be guided with the same kind of mild, firm, consistent discipline that research reveals to be so closely associated with positive mental health, educational achievement, "good moral character," and eventual marital success? Such guidance would require modification, of course, in its use with adults.

4. Since a pragmatic, concrete, personal, physical learning style appears to be characteristic of most low-income persons, how can services be designed that will work with, rather than in opposition to, such a style? Clues from a variety of sources indicate that a direct, assertive, specific approach is more effective in group and individual counseling than the more subtle, abstract, insight-oriented approach often used with middle-class clients.[56] Role-playing and other "learning-by-doing" projects seem to be more effective than strictly verbal approaches to problem-solving.[57]

Considerable expermentation is going on at present in reference to specific structuring of the external environment and direct teaching of individuals to play certain roles in this environment rather than relying on internal personality changes developed from verbalized insights and a therapeutic relationship. By acting differently in a structured situation, some people, at least, seem to begin to feel and think differently.[58]

5. The possible benefits to be gained from cultural enrichment of disadvantaged preschool children presents a direct challenge. Experimentation and research in ways of overcoming the cultural deficits of poor children at the preschool level are being financed by various programs in the Welfare Administration, as well as by other Federal programs and by foundations.

Experimentation that includes opportunities in cultural enrichment for older children, youth, and adults is going forward in a number of communities.[59] Such experimentation will doubtlessly increase under the job-training, educational programs, and community development features of the Economic Opportunity Act.

6. Day care centers and homemaker services not only offer needed direct services to many low-income families but also help parents and children learn new cultural patterns and skills by observation and participation.

7. Further training of public welfare staff members in reference to the subcultures of poverty and their implications seems relevant. Such training might be carried forward by a variety of formal and informal arrangements, including inservice training.

In line with its policy of seeking to develop more effective treatment as well as research and demonstration programs, the Welfare Administration through the Office of the Commissioner, the Children's Bureau, and the Bureau of Family Services recently participated in a 3-day national group consultation on parent and family life education. This consultation was planned by the Interdepartmental Committee on Children and Youth and brought together specialized consultants from many professions and from public and voluntary agencies and universities. (Other participating Federal agencies were the Public Health Service, Department of Agriculture, Office of Education, Housing and Home Finance Agency, Department of Labor, and Selective Services of the Department of Defense). A Federal publication providing guides to services and research in parent education with low-income families will be one of the results of this consultation.

Implications for Research

Descriptive Research

At the descriptive level, it is clear that more precise and rigorous studies are required in order to obtain more exact information about the various ethnic, regional, racial, and religious subcultures of poverty. A refinement of techniques is called for, with:

1. Careful attention paid to random sampling from a defined population.
2. Adequate sample size and description of sample.
3. Development of testing instruments with due consideration given to their applicability to the research questions of the study and to their reliability and validity.
4. Analysis of data through the use of appropriate statistical methods.
5. Comparison of samples from several defined populations (for example, lower-lower class compared with (a) working class and (b) middle class).

Also, at the descriptive level, it is important to know more about the characteristics of groups of individuals who manage to escape from poverty. If the child-rearing and family life patterns of the very poor do tend to block the chances of the individual to move upward, how do the poor who are upwardly mobile differ from those who are not, in reference to the values, attitudes, and practices of their parents and in reference to their own present cultural characteristics? Studies along these lines are in progress, underwritten by Welfare Administration funds.

Since there are usually large differences between the sexes in reference to personality characteristics, interests, aspirations, and values, descriptive studies aimed at differentiating the successful from the nonsuccessful "slum child" should be designed with questions directed towards male and female groups considered separately.

Experimental Research

There is a strong need for action research aimed at finding out whether specific treatment programs actually "make a difference." While descriptive studies have their important place in yielding more sophisticated insights and suggesting programs for change, the crucial question is whether these insights and suggestions add up to effective programs. Whether or not such programs actually represent an improvement must be objectively tested, insofar as this is possible.

For example, in preceding section suggestions were made for various ways of devising services for very poor families which might lead to more effective child-rearing practices and family life style. Programs designed along such lines should be subjected to carefully designed evaluation techniques. Part of this evaluation needs to be focused on behavioral outcomes that can be objectively measured.

Evaluation too frequently is focused on attempts to measure such factors as changes in attitudes and information. One of the outstanding deficiencies of measurements of this kind is that there is no guarantee that changed attitudes and information lead to changed behavior.

For instance, if, as a result of a parent discussion group, parents say that harsh, physical, dictatorial punishment has a bad effect on children, does this mean that they actually abstain from this kind of punishment and substitute in its place mild, consistent, reasoned discipline? To push this point one step further, if parents do manage to behave differently in reference to discipline, does their children's behavior show an objectively observable difference? Theoretically, such a parental change should lead to a reduction in a child's overt aggression in the home, school, and community. Do such changes occur and are they maintained over time?

Long-range behavioral outcomes associated with service programs designed for the very poor include: improved school attendance and academic success, obtaining and holding employment, better handling of family income, Improved housekeeping, better physical health, reduction in deviant behavior, and increase in marital stability.

If goals such as these are reached, which individuals reach them and under what program methodologies and content? For which individuals does the program seem to be a failure? What are their characteristics? We can learn just as much from reported failures as we can from reported successes, although the tendency seems to be to publish reports only of the successes. Does the group served by a program change to a greater extent than a group not served? If behavioral changes are found to occur, do they persist over time?

Experimental programs with an evaluation component need to be repeated with different groups of subjects, with different staff composition, and in different parts of the country in order to find out whether or not the service methodology has general application.

Technical difficulties are present in research of this kind. Measurement of change is one of these difficulties. The Division of Research of the Welfare Administration is currently engaged in an attempt to devise a method of measuring family change that can be used by research personnel as one evaluation method in a number of experimental projects.

Some very difficult research and evaluation questions have been raised here. Finding at least some of the answers will not only add to scientific knowledge and give clues to further research but also contribute to the development of public welfare practice aimed at family and individual rehabilitation and the reduction of economic dependency.

References

1. Irelan, Lola M. "Escape from the Slums: A Focus for Research" *Welfare in Review*, vol. 2, no. 12, December 1964, pp. 19-23.

2. Herzog, Elizabeth: "Some Assumptions About the Poor" *Social Service Review*, vol. 37, no. 4, December 1963, pp. 389-402.

3. Brackbill, Y., and D. Jack. "Discrimination Learning in Children as a Function of Reinforcement Value," *Child Development*, vol. 29, 1959, pp. 185-90.

4. Crandall, W. J., Anne Preston, and Alice Robson. "Maternal Reactions and the Development of Independence and Achievement Behavior in Young Children" *Child Development*, vol. 31, pp. 243-51, 1960, p. 20.

5. Ackerman, Nathan, et al. "Child and Family Psychopathy: Problems of Correlation," in P. H. Hock and J. Juzer, eds., *Psychopathology of Childhood*. Grune & Stratton, New York, 1955.

6. Jersild, Arthur. *Child Psychology*. Prentice-Hall, New York, 1954, pp. 235-49.

7. Watson, Robert. *Psychology of the Child*. Wiley, New York, 1959, pp. 539-47.

8. Siegel, Irving E. "The Attainment of Concepts" In Martin L. Hoffman and Lois W. Hoffman, eds., *Review of Child Development Research*, vol. 1. Russell Sage Foundation, New York, 1964.

9. Yarrow, Leon J. "Separation from Parents During Early Childhood" In Martin L. Hoffman and Lois W. Hoffman, eds., *Review of Child Development Research*, vol. 1. Russell Sage Foundation, New York, 1964.

10. Lewis, Oscar. *The Children of Sánchez*. Random House, New York, 1961, p. xxiv.

11. Riessman, Frank: *The Culturally Deprived Child*. Harper, New York, 1962.

12. Davis, A., and R. J. Havighurst. "Social Class and Color Differences in Child-Rearing" *American Sociological Review*, vol. 1. 1946, pp. 698-710.

13. Sewell, William H. "Some Recent Developments in Socialization Theory and

Research" *Annals of the American Academy of Political and Social Science*, vol. 349, September 1963, pp. 163-81.

14. Bernstein, Basil. "Social Class and Linguistic Development: A Theory of Social Learning" In A. H. Halsey, Jean Flowd, and C. Arnold Anderson, eds., *Education, Economy and Society*. Free Press of Glencoe, New York, 1961.

15. Hunt, J. McV. "Social Class and Parental Values" *American Journal of Sociology*, vol. 64, January 1959, pp. 337-51.

16. Mussen, Paul H., and Luther Distler. "Masculinity, Identity, and Father-Son Relationships" *Journal of Abnormal and Social Psychology*, vol. 59, November 1959; and "Consequences of Masculine Sex Typing in Adolescent Boys" *Psychological Monographs*, vol. 75, 1961.

17. Lynn, David B., and William L. Saurey. "The Effects of Father Absence on Norwegian Boys and Girls" *Journal of Abnormal and Social Psychology*, vol. 59, September 1959, pp. 258-62.

18. Stolz, L. M., et al. *Father Relations of War-Born Children*. Stanford University Press, Stanford, Calif., 1954.

19. Bach, G.R. "Father Fantasies and Father Typing in Father Separated Children" *Child Development*, vol. 17, 1946, pp. 63-80.

20. Glidewell, Thomas, et al. *Parent Attitudes and Child Behavior*. Charles C. Thomas, Springfield, Ill., 1961.

21. Bronfenbrenner, Urie. "Socialization and Social Class Through Time and Space" In E. E. Maccoby, T. M. Newcomb, and E. L. Hartley, eds., *Readings in Social Psychology*. Holt, New York, 1961.

22. Baldwin, A. L., J. Kalhorn, and F. H. Breese. "Patterns of Parent Behavior" *Psychological Monographs*, vol. 48, no. 3, 1945.

23. Meyers, Jerome I., and Bertram H. Roberts. *Family and Class Dynamics in Mental Health*. Wiley, New York, 1959.

24. Miller, Walter B. "Implications of Urban Lower-Class Culture for Social Work" *Social Service Review*, vol. 32, September 1959, pp. 219-36.

25. Lewis, Hylan. "Child-Rearing Practices Among Low-Income Families in the District of Columbia." Presented at the National Conference on Social Welfare, May 16, 1961, Minneapolis, Minn. Health and Welfare Council of the National Capitol Area, Washington, D. C., 1961. Mimeographed.

26. Vincent, Clark. *Unmarried Mothers*. Free Press of Glencoe, New York, 1961.

27. Bandura, A., and R. H. Walters. *Adolescent Aggression*. Ronald Press, New York, 1959.

28. Kohn, Melvin L. "Social Class and the Exercise of Parental Authority" *American Sociological Review*, vol. 24, June 1959.

29. Miller, S. M., and Frank Reissman. "The Working Class Subculture: A New View" *Social Problems*, vol. 9, summer 1961, pp. 86-97.

30. Kohn, Melvin L. "Social Class and Parental Values" *American Journal of Sociology*, vol. 64, January 1959, pp. 337-51.

31. Rodman, Hyman. "On Understanding Lower-Class Behavior" *Social and Economic Studies*, vol. 7, December 1959, pp. 441-50.

32. Caldwell, Bettye M. "The Effects of Infant Care" In Martin L. Hoffman and Lois W. Hoffman, eds., *Review of Child Development Research*, vol. 1. Russell Sage Foundation, New York, 1964.

33. Klatskin, E. H. "Shift in Child Care Practices Under an Infant Care Program of Flexible Methodology" *American Journal of Orthopsychiatry*, vol. 22, 1952, pp. 52-61.

34. Maccoby, E. E., and P. K. Gibbs. "Methods of Child-Rearing in Two Social Classes" In W. E. Martin and C. B. Stendler, eds., *Readings in Child Development*. Harcourt Brace, New York, 1954, pp. 380-96.

35. White, M. S. "Social Class, Child-Rearing Practices, and Child Behavior" *American Sociological Review*, vol. 22, 1957, pp. 704-712.

36. Kohn, M. L., and E. E. Carroll. "Social Class and the Allocation of Parental Responsibilities." *Sociometry*, vol. 23, 1960, pp. 372-92.

37. Bayley, N., and E. S. Shaefer. "Relationships Between Socioeconomic Variables and the Behavior of Mothers Toward Young Children" *J. Genet. Psychol.*, vol. 96, 1960, pp. 61-77.

38. Wortis, H., et al. "Child-Rearing Practices in a Low Socioeconomic Group," *Pediatrics*, vol. 32, 1963, pp. 298-307.

39. Sears, Robert, Eleanor Maccoby, and Harry Levin. *Patterns of Child-Rearing*. Row, Peterson, Evanston, Ill., and White Plains, N.Y., 1957.

40. Kohlberg, Lawrence. "Development of Moral Character and Moral Ideology," In Martin L. Hoffman and Lois W. Hoffman, eds., *Review of Child Development Research*, vol. 1. Russel Sage Foundation, New York, 1964.

41. Kirkpatrick, Clifford. *The Family as Process and Institution*. Ronald Press, New York, 1963, pp. 375-407.

42. Burgess, Ernest W., and Leonard S. Cottrell. *Predicting Success or Failure in Marriage*. Prentice-Hall, Englewood Cliffs, N.J., 1939, pp. 58-74.

43. Terman, L. M. *Psychological Factors in Marital Happiness*. McGraw-Hill, New York, 1938, pp. 48-83.

44. Locke, Harvey J. *Predicting Adjustment in Marriage: A Comparison of a Divorced and a Happily Married Group*. Holt, New York, 1951.

45. Stroup, Atlee L. "Predicting Marital Success or Failure in an Urban Population" *American Sociological Review*, vol. 28, 1953, p. 560.

46. King, Charles E. "The Burgess-Cottrell Method of Measuring Marital Adjustment Applied to a Non-White Southern Urban Population" *Marriage and Family Living*, vol. 14, 1952, p. 284.

47. Locke, Harvey J., and Karl M. Wallace. "Short Marital-Adjustment and Prediction Tests: Their Reliability and Validity" *Marriage and Family Living*, vol. 21, 1959, pp. 251-55.

48. Bowerman, Charles. "Adjustment in Marriage: Over-All and in Specific Areas." *Sociology and Social Research*, vol. 41, 1957, pp. 257-63.

49. Locke, Harvey J., and Robert C. Williamson. "Marital Adjustment: A Factor Analysis Study" *American Sociological Review*, vol. 23, 1958, pp. 562-69.

50. Nye, F. Ivan, and Evelyn MacDougall. "The Dependent Variable in Marital Research" *Pacific Sociological Review*, vol. 2, 1959, pp. 67-70.

51. U.S. Bureau of the Census. "Marital Status, Economic Status, and Family Status: March 1957" *Current Population Reports*, Series P-20, no. 81, March 19, 1958, p. 10.

52. Rohrer, John H., and Munro S. Edmondson. *The Eighth Generation*. Harper, New York, 1960.

53. Rainwater, Lee, and Karol Kane Weinstein. *And the Poor Get Children*. Quadrangle Books, Chicago, 1960.

54. Handel, Gerald, and Lee Rainwater. "Working-Class People and Family Planning" *Social Work*, vol. 6, no. 2, April 1961.

55. Young, Leontine. *Wednesday's Children*. McGraw-Hill, New York, 1964.

56. Weinandy, Janet. "Casework with Multi-Problem Families," *Journal of Marriage and the Family*, November 1964.

57. Riessman, Frank. "New Approaches to Mental Health Treatment for Labor and Low Income Groups," Mental Health report no. 2. National Institute for Labor Education, New York, 1964.

58. Riessman, Frank. *The Revolution in Social Work: The New Non-Professional*. Training Department, Mobilization for Youth and Urban Study Center, Rutgers University, New Brunswick, N.J., October 1962. Mimeographed.

59. Chilman, Catherine, and Ivor Kraft. "Helping Low Income Parents Through Parent Education Groups" *Children*, July-August 1963, pp. 127-32

CHAPTER 9

Southern Children Under Desegregation

M. ROBERT COLES

Confronting certain children in the South today are the tasks of getting an education under unusual conditions of social stress. These children, their families and teachers, have happened upon a moment in historical change, and their behavior in such circumstances deserves our attention. For nearly ten years now the deep South has faced increasing national pressure upon its avowedly segregated system of public education, and slowly, with stated reluctance, it has been forced to begin to permit the entry of a few Negroes into its colleges first, its high and elementary schools later.

What has happened again and again is the explosion of words of hate and deeds of violence by certain white persons, who were prompted by fears and anxieties. Public disorder, organized scorn, or, under conditions of order, carefully patrolled calm, such are the external conditions somehow faced and managed by those scatterings of Negro children who continue to press their admission and attendance upon formerly all white schools; and these dark children are received by their waiting white classmates in various ways. As psychiatrists, we must look further, trying to find out how these several groups think and feel, looking at their responses under tension, evaluating their survival and manner of accommodation with new realities daily urged upon them. In any crisis of society are the individuals who enact it, seek it, are caught in it, attempt to flee it. This particular struggle has many roots in our history and culture, many implications in our current political and economic life, all of them beyond the present concerns of this paper, which must be with the individual child and his psychological development and adjustment. We may only emphasize that such distinctions are arbitrary if necessary. No child can be separated from the history of his people, from their lot and their condition. This is not the place to chronicle the fate of the Negro in America, nor to trace the development of the white man's attitudes in the South, or North likewise, toward him. Such has been done by historians (1-4), sociologists (5), psychologists (6-8), and psychiatrists (9-14), and as we take up our work here we are forced merely to mention these facts of "culture" or "environment," or "socioeconomic context." It must be obvious, however, that at every moment in the life of these children, such facts are firm realities for them.

Nature of This Study

Our Negro children have been selected by others, by themselves and their parents by virtue of application, by federal judges by virtue of decision, by school boards by virtue of compliance and permission. Our white children have been those who attended class with these Negroes, in the beginning simply living their normal lives of growing and being schooled, and suddenly finding themselves with new questions of attitude and behavior. Finally, the parents of all of these children, white and colored, and their teachers, all of them white, have been a part of our study, supplying valuable observations about their children and students, talking a good deal about their own problems in what is for many of them, of both races, a critical adjustment in habit and thought.

We have tried for lengthy acquaintance with children of several ages. They have been as young as five and as old as eighteen. In New Orleans they were the little ones, entering first grade, attending second and third grades. They included the four Negro girls who, at age six, first entered two white schools in 1961, and four of their white classmates; the six additional Negro children who entered the same and other schools the next year, and seven of their white classmates. In Atlanta they were the ten Negro children who were selected for that city's first year of desegregation, five to enter the twelfth grade and five the eleventh of four high schools, and ten of their white classmates.

All of these children have been followed for two full school years. We have seen the children in Atlanta al least once a week, often twice weekly, depending upon the child and the circumstances at work in his personal and school life. We have visited the children in New Orleans every month, seeing them two or three times in a period of a week's stay. We talked with the teenagers and their parents at their homes; the younger children we supplied with toys, crayons, and paints. Though regular home visitors, we have also been to their schools, talked with their teachers separately and in groups. We have known many of them months before schools opened, watched them enter on their first day, and graduate or be promoted on their last. Among white children we have aimed at diverse opinions on segregation. In both cities we have children from families actively in favor of desegration and very earnestly opposed. In Atlanta one family is connected with the Klan, two others are militantly segregationist; in New Orleans four families are of like mind.

In brisker fashion we have journeyed to other cities, other Southern regions where desegregation has begun, hoping in a few interviews with children, parents, teachers for comparative information on adjustment and progress. Little Rock in Arkansas, Clinton and Oak Ridge in Tennessee, Burnsville, Asheville, and Charlotte in North Carolina, have offered rural and urban contrasts, and information from a more northerly South than Georgia and Louisiana.

Roles and Reactions of Negro and White Children and Adults

Our chief concerns have been what color means to a child, how these children, Negro and white, see themselves and respond to one another, how they handle the strains of desegregation, the shock of one another's presence, the anxiety and anger on the streets or the fear inside their homes; finally, how their attitudes and feelings change or persist over the months of these eventful and unusual school years. It would be fatuous indeed to ignore the separate nature of existence for the black and white child in the South. We must therefore tell their fate and our findings separately, starting, in each case, with the young and moving on to the older child.

Negro Children

We may note, initially, that there are no obvious social or economic ways to distinguish these families whose children participated in the first years of desegregation. These children have come from all classes, the very poor to the comfortable, the educated to the barely literate, long-time city dwellers and recent arrivals from fast disappearing farms. A wide variety of motives have pressed upon them or their parents, most especially the desire for better education, and the advancement of their people's struggle for equality. In no case were they cajoled, rewarded, or even asked to apply.[1] In both Atlanta and New Orleans many were eager, but only very few chosen. Selection was based in both cities upon the declared regard of school officials for seeking children able and capable of surviving an obviously

[1] One of the most common beliefs I have met in the South, held not only by segregationists, but by many people very sympathetic to desegregation, often by many kind teachers, is that the NAACP in some way recruits these children, then rewards them, pays them, bribes them, the verb depending on the attitudes of the particular person. I have yet to see evidence of this with any of our children.

special and possibly traumatic experience. In both cities and in others over the South, psychological tests were given, personal interviews held. Four in New Orleans, ten in Atlanta, survived these for first-year desegregation. These children represent, therefore, a residue deemed by educators fit and superior to other aspirants.

We have seen, and we must emphasize, heavy and hard, that each of these Negro children must learn and has learned how to manage in the white world of the South. From their first years they must be taught and slowly begin to realize fully for themselves what they may do, where they may go, how they must behave. These dusky children of six, their yet smaller brothers and sisters of three or four, are fast learning such facts of social life, and their older kind in Atlanta have already mastered these childhood lessons quite well.

Drawings of these young children reveal their emerging sense of shame and worthlessness, their feelings of weakness before white skin and their envy of it. While all children have their troubles with many of these emotions, early sorrows, frights, rivalries and angers, for the Negro child, in contrast with the white child's experience, they have a real basis in the world outside the home, regardless of the pleasure and happiness within.

For instance, a six-year-old boy, several months at a heavily boycotted but desegrated New Orleans school, draws his view of himself and one of his few white classmates, and, in another picture, sketches his idea of his place in school. Another child, a girl of seven, having faced daily mobs, draws her school and herself. The boy leaves some white on his right arm and leg, draws himself smaller, more bound and animal-like than his blond classmate, who, in reality, is smaller and much less active and darting. He puts himself in one part of his school, his white schoolmates in another, although, he assures me, in actual fact they are together in classes. ("They say we're not supposed to, but we are.") The girl pictures herself a lonely blackbird, cautiously winging her way toward the school, hesitant, fragile before white mankind as birds generally are before people, a host of her black kin segregated in distant knots of black upon a blue sky. Another of the four girls, one who saw a whole school empty in the face of her admission, draws her school with bright flowers on one side, a stealthy brown squirrel on another. Again, she draws a white girl friend as large, smiling and blooming with cheer, while a colored child ("It's not me," she hastens to tell me with no prodding) appears in a corner of the paper, dour and droopy, lips puckered, legs looking lamed.

In these homes I have seen white dolls kept separated from colored ones, called, in two instances, King and Queen. On the other hand I have seen our same children eventually feel free enough to hit, crush, or shatter these small white worlds of dolls and toys, looking at first fearfully or guiltily at me, later able to talk rather clearly about their deep feelings of resentment and hurt. ("If the white hits the colored, the colored has to hit him back.")

In her very overworked and overstated affection for whites in paper and in play a five-year-old girl shows other, contrary attitudes at work. She never quite felt free enough to express them directly, but their symbolic presence could not be concealed. She had always given white people features, had drawn carefully and precisely their blue eyes, had marked their fine noses by thin pink lines, had outlined their slim, curved lips, jarred only rarely by all too excessive teeth which loomed larger than the mouth and almost threatened the entire face. Negroes never had eyes, noses, mouths, only large ears. They were anonymous, and would thus not reveal their true feelings, though they were listening all the time. One day she looked furtively at me as she started drawing eyes and a nose for a colored mother and her daughter, then quickly left her crayons for the escape of skipping and dancing; we both knew why. She had always been able to talk about how white children played and felt, telling me stories to go with her drawings, telling me thereby about many of her own normal problems of growing and getting along with her mother, her father, her sister and brothers. What she feared talking about was one problem, how she felt about being a Negro, how she felt about the restrictions and prohibitions which even then she was learning, which even then were being reinforced in their power over her spontaneous urges, in buses, in stores, in movie houses, at lunch counters. At school she could play with white children and study with them, but she knew this condition was special, and precarious at that. She found attendance at a desegregated school much easier, however, than revealing her private, her family's private fears and hates, of those white children, of their parents, of their entire world.

Positive feelings toward this white doctor occur. Apart from an occasional doorstep visit from an insurance salesman or such, these children and their parents have never before acted as friendly host to a white guest in their home. Real trust must therefore be distinguished from early, anxious, fearful ingratiation, just as these must be differentiated in older children and adults, race apart, too. In three children, particularly expressive with crayons and paints, lessened distrust and fright seemed to appear with pictures of me, to which they would hastily add brown skin or coloring. First my tan or dappled complexion would be erased with a new picture started, or an apologetic remark indicating an error voiced. After a while I could be left colored, and they learned that it caused me no pain. Two asked me if I minded. No, I didn't, they found out, then tested my words and tried their freedom to make us seem more alike by embarking upon a spate of such drawings for a few months. Eventually they abandoned this game. One of them was able to tell me once that she was soiling me, for which she was very sorry. I remarked to her that she had painted a large shining sun, and that, perhaps I was being suntanned. Yes, she murmured, that too. Those children who found desegregation a happier experience than expected drew schools progressively less forbidding and impersonal, played games less hectic or guarded. Those whose school experiences had been frought with much tension expressed much of this in our games, trying, oftentimes, to make sense out of the experience by reenacting it. One girl would imitate the crowds, the unfriendly crowds, then turn on them, disperse them in a model of police efficiency painfully in contrast with what had actually happened on the streets. "I'm better than the police," she told me once.

These angry crowds took on a highly individual implications and meanings for each child. Mob reproaches are fitted into the normal problems of growing. A child who feels guilty about her wishes to hurt or eliminate her brothers and sisters can wonder whether the epithets of street-walkers are not really designed to punish her for her secret and sometimes not so hidden fantasies. At five, six, or seven, they are very much involved in the rights and wrongs of the world, how to tell them apart, how to mediate between their evolving conscience and quick wishes.

What the parents intuitively knew was that they must acknowledge their children's perceptions of a sullen or hostile world, all the while assuring them that there were other worlds of affection, and, even among whites, support. One mother put it this way:

> I tells her that she's right and those folks don't like her nor me either just 'cause we're colored. But I say that me and her daddy went against worse in the older days, and we is going to go right through this, and we got a lot of the whites who are helping us all the way.

For many parents this was a forthright chance to talk openly with their children about their problems as Negroes. By no means is it easy for Negro families to talk with one another about such matters. Shame, self-hatred which can express itself in contempt for one another, make such open talk difficult.

Balancing outside threats and anxiety were the inside supports of the school as well as the home. While these children were applauded by their own communities, and many people all over the world, they also received help from their teachers and white classmates, few to many in number, depending upon the year and the school involved. Three of them, learning their letters in an abandoned school, had no white classmates the first year and about fifteen the second. One of them studies largely alone at first, though she always had about five, then about ten white students with her at the beginning, and later, in the second year, over one hundred returned to school with her. Perhaps if we hear two recorded comments, from a Negro and a white mother, we will understand what happened inside these battered schools. A Negro mother told us: "Ruby came home the other day and told me that a new white girl came and she came right up to Ruby and told her in the middle of the recess that she wasn't going to play with her, 'cause her mother done told her to stay clear of niggers, but Ruby said they played together right afterwards." A white mother, who kept her children in school under constant threat and intimidation, commented:

> Once they were in school, it was fine. Why, they practically worshipped that little girl [Negro]. After all, they'd never had it so good . . . so few of them to a teacher, and all that excitement. They loved it. I got scared, all the calls and their faces . . . but the kids kept me calmer than I kept them. They weren't afraid . . . they had a real school spirit . . . they sang soungs like 'Frantz School Will Survive' . . . they became real close together. . . . Yes, it *was* like a family, and I guess the police *were* like their fathers, and they even joined in with the singing. I think the teachers became more determined than ever to stick it out and protect the children after they saw what those people were like. I've never heard such foul language. . . .

We saw evidence of anxiety, nervous loss of appetite, fretful disposition, unusually scrappy behavior with brothers and sisters, rare tearfulness before going to school, and one temporary episode of bed wetting, but never were these crippling in length or severity, similar to the kinds of complaints which bring children to child guidance clinics. Parental anxiety could trigger a child's, but none of these parents ever seemed near collapse in their ability to cope with daily living. Subjected to curses, warnings of attack and death, job loss, these people drew close together, inside their own families, and within the shared assurance of a group. Very important, often overlooked, is the fact that theirs has been lifetime of adaptation to what they know and feel to be unfair treatment.

If the young child can draw and play his sense of himself and the white world outside, draw and play his fears and tensions under desegregation, and how they fit into his problems of growing up, the older child will have a style of living, a manner of thinking and feeling and acting which is similar. In our adolescent children we have heard over the months conscious and unconscious expressions of lessons learned earlier, now organized into patterns of behavior: meekness masking rage, anger expressing shame or anxiety, fearful guilt following expression of frustration or humilitiation, and, in general, a thicket of rationalizations, apprehensive controls, and bland denials, covering sadness, bitterness, and, always with "the Man" around, uncertainty and fear.

We must emphasize, again and again, that for many of these conscious feelings of rage or intimidation to be revealed, for many dreams and fantasies to be spoken, much time and trust were needed. The parallel with any psychotherapy is obvious, but the contrast between these "healthy"

Negroes, the "normal," even "superior" teenagers and their white classmates in ease and freedom of social expression, is startling. White children can and often do say what they feel about Negroes to themselves and often to Negroes, say friendly things, say hurtful things. Negroes not only have no such relatively ready confidence or articulation with a white person, a fact which is obvious to many persons but still denied by many more, but Negroes often are unable to talk with one another about their lot and problems as Negroes. Self-hate, learned from a culture which has bound, ostracized, and proscribed them, becomes a hate for one another. The dreams, associations, comments and written remarks, drawings and paintings, and letters of these children richly illustrate such developments.

Of course, all adolescents have their problems of quick growth and heightened awareness, of torn feelings, of doubts and dreams charged with new hormonal strength; but, again, the Negro adolescent has a special problem with the all too real social and cultural truth of his past exploitation and still curbed apartness. The whole play of the mind will be seen in these youths, their projections and introjections, their distortions and excuses, a host of justifications and remedial responses to the lessons of inferiority, well learned by them, gathered by them from the white ocean of habits, customs, its culture which surrounds them and controls them from their first years to their last moments.

At fifteen, old ideas, old experiences, racially tinged admonitions from parents, racially connected epithets from the outside world, bulk larger and stronger. Negro adolescents in Atlanta met no mobs, but their age and its own burdens, as with the younger children in New Orleans, merged with the stresses of desegregation. Months before nine of them entered four high schools, their names and pictures were in the local papers. While police can deter public throngs, threatening, highly abusive phone calls, similar messages by mail, pressures on employers, pressures by employers, these are less easily contained. Night after night calls came, menacing, foreboding ill, and in the morning special deliveries of the same kind appeared. Atlanta was determined to have law and order, but during that anxious summer before initial desegregation, no one could really know what bomb and dynamite threats, what murderous innuendo might materialize, even with the most elaborate precautions. Nor was the heritage of these families, several poor and recently from farms, conducive to reassurance and thoughtful serenity.

Among the children and their parents too anticipatory anxiety and defenses against it were quite prominent. Especially common were the use of denial and rationalizations, such as "I'm not worried about anything," or "What's there to worry about?" or "They'll bother the boys" from a girl, and "At least we can protect ourselves" from a boy. Frequent projective maneuvers were seen, like "I'll bet they're plenty scared [the whites] of what'll happen," and "It's nothing for us to worry about; it's their problem, not ours." Loss of appetite, migraine headaches, insomnia were seen. One boy read a law book, versing himself on "assault and battery" sections particularly, but he "just happened to be doing" this then. Another boned up on karate, keeping his once slack muscles newly poised. Many, wishing for a foreseeable end to increasing tension, trapped themselves into forecasts of early ease once the first few days of school were over.

Such, sadly, was not to be the case. Long moments of history may disappear, but not necessarily at the speed hoped by an anxious child. During the school year the trials of hard work, frequent loneliness and isolation, and occasional insult had to be met, and each child had his or her own way of dealing with these pressures. One of the original ten selected, a very bright girl of seventeen, dropped out just before school opened, she and her family unable and unwilling to bear the daily rising tension ("We're too far out from the police, and it's too exposed here. They could shoot to kill or dynamite us, and I'm not going to take the chance," the girl admitted, guilty about her decision, but more afraid of what would happen were it not made.) Another slightly younger girl, also quite intelligent, asked to return to her old Negro high school a short while after her first report card showed two A's and three B's, the highest grades of all the Negro students, obtained in a tough academic program. A well-spoken, attractive girl, from a middle-class family, of the three Negroes probably the favorite of teachers and students alike, confused her city a great deal by her withdrawal for what she called "an emotional upset." Front-page publicity, local and national, with editorials and heavy television coverage, remarks of concern by the Governor of Georgia, conflicting explanations of the cause, such attention gives some hint of the interest, hopes, and fears and guilts attached by so many to desegregation in the South.

Having known her for six months, I had seen her respond to apprehension with heavy reliance upon detailed, increasingly burdensome rituals. She was getting little sleep, eating sweets sporadically and excessively, and not simply studying but exhausting herself in tangled, midnight minutiae, in unrealistic and obviously hurtful anticipation of all of the next day's possible recitals. While these classrooms were in fact for these children an awkward, silent, closely watched and often fearful stage, this girl could not handle her role as easily as the eight others who stayed and survived. The reasons seemed more connected with her fragile personal and familial history, frought with divorce and parental strife, than desegregation alone. A crisis had sharpened old angers and sorrows, had upset a delicately balanced relationship with her father and sister, and had found her abilities, even assisted by a relatively friendly school body, newly weakened by rigid and frail defenses.

Depressions, headaches, stomach aches, all these were seen with some frequency. Three of the five girls developed definite increase in premenstrual pain and thereby temporary absence from school. All three boys seemed prone to headaches and dreams of thinly disguised rage:

> I dreamt I was in Little Rock,[2] and they [whites] were coming after me. . . . I started to run, then I found I had a gun, and I turned and it turned out to be a machine gun, so I fired away; I opened up right at them, and I think they stopped. . . . I had another dream later that night. I dreamed I got arrested for driving too fast, and my father said I'd better go to jail because he didn't have the money to pay the fine, and then I woke up. . . . What a night, it was as bad as school!

This dream-pair of anger, of crime and punishment, was followed weeks later, in the same boy, with

> I was running real fast, but then I looked around and no one seemed to be chasing me, so I relaxed. Then I saw a girl walking down the street. She was a white girl, so I looked the other way. You know. Then, a funny thing happened. She told me it was o.k. to look, 'cause I was white, too, and she smiled at me. I think we talked for a while or something but suddenly she turned Negro and ran from me, and I felt bad, like I'd done something wrong. Then I woke up. I couldn't get the dream out of my mind, it was one of those that stick with you real clear all day. . . . Mostly the feeling of doing something wrong . . . I think I woke up before I could get caught or something, maybe the police were around the corner.

Though harassed, these children were also honored, and, just as the New Orleans parents averred, so did they, that here was a chance for purposeful stress in contrast to the arbitrary stresses of the daily rejections and exclusions encountered by the Negro in his living and working. Indeed, eight of ten selected in Atlanta successfully finished these last one or two years of their high school education, though not without signs of tiring and of strain. As wearying as their psychological problems were their academic ones. They come from overcrowded, poorly equipped schools, from years of education in a separate school system which reflects inevitably the position of the Negro, the restricted hopes, the frequent poverty and the precarious affluence (16), the continual fears, the lessened pace, the boredom, apathy, and, too, the reactive rage and its impulsive, expressive violence.

They studied hard, curtailed social activities, graduated. Those who followed them the next year, over forty in number, less carefully picked,

[2] Events like Little Rock stick fast to these children. They were eleven or twelve at the time, awakening to themselves and their world. All refer to memories of Little Rock on television and indentification with the Negro students there. No less than four have used that event in their dreams, to symbolize such feelings as isolation, determined resistance, anger, etc.

struggled even harder with some of these problems of academic adjustment required at so late a time as the last years of high school.

One of the most common difficulties encountered by these colored students was their own ambivalence, their pride of person and new self-respect as Negroes, their identification with changes in Africa, in Asia, in their own country; but also, accompanying such feelings, the results of their ancient exile from American culture, a sense of their own lowly, tainted nature, an aloof, hesitant distrust of whites, a self-hatred appearing as suspicion of them even when they are kindly and affectionate:

> I don't know what to do when they get nice. It's easier when they ignore you, at least then you're used to it, and we know what to do . . . when they get friendly, I feel as if they've got some angle to it, or trying to pull something. . . . I guess I don't really believe we've got some equality now. . . . I finally got up enough courage, and walked over and sat with them. . . . Oh, I knew they were friendly, and they wanted me to come over . . . but you know how it is. . . . But it all went good. I think they were as nervous as I was, but now when I sit with them, no one thinks anything of it.

A less successful outcome, in a school less hospitable to its two Negro children, shows one way of dealing with daily rebuke and insult:

> I don't care what they say, or what they think of me. . . . I'm going to graduate from that school if it's the last thing I do. They can all leave if they want. . . . Man, that's my school as much as theirs, and they don't scare me and if I scare them, they can go . . . go and give me the whole school to myself. . . .

White Children

Since the white children mostly do not leave these schools or when they do, usually return after the crisis of violence or mob action subsides, they too have reactions to desegregation, a new and often surprising reality to them. Born and reared in a region which has, however recently (2), as succession to slavery developed an elaborate array of social customs, strongly bulwarked by law, which regulate from birth to death doings if not feelings between the races, these children are now asked to participate in the beginning of the end of these ways. For the young white child, this ordinarily requires little effort. His age and its momentum appear to overcome even parental injunctions. I have watched them play, children from segregationist homes, urged by their mothers to keep good distance from the one or two colored children among them, yet forgetful of these ideas and pleas in the abandon of play. Teachers say, as this one did:

> They forget what they're told at home when they start playing. You can't

> expect children of this age to worry about skin color when they want to play a game or jump rope. . . . I suppose if their parents scare them to death about it [skin color] they'll remember, but it'll take work to prevent children from being friendly one minute and nasty the next regardless of what color they are.

Two such children and their parents showed us how dark skin can, in fact, become an object of anxious avoidance, a focus for emotions communicated from parents to child by daily example, by daily intimacy of phrase and example. Their parents were not simply voting segregationists, but preoccupied segregationists, fervently involved believers in distance between the races, in Negro inferiority and harmfulness. Their mothers' associations, culled from many interviews, indicate that for them the Negro is dirty, diseased, promiscuous, a source of contamination, and that all these thoughts, held by many others in a casual or quiet fashion, are of strong, constant concern to them. Joining together the words of one mother from several tapes, we must emphasize the urgency and passion of her delivery:

> They're dragging us down, and you know why? Because they're made lower and they're born with less intelligence than we have. Just look at them[3] . . . they're more animal than spiritual, and the Bible tells us not to mix the two if we can help it. . . . They have their place, and it's not anywhere near us. . . . They're dirty, they're drunks, and they slash one another with razor blades and flood Charity Hospital with their V.D. and illegitimate children . . . I don't think they know what family life is, and now they want us to catch it from them, get all our children sick, and start them off with little nigger kids in school. I sent him back there [Frantz School] because I figure there's only one or two, and we can get by if he stays away from them, but what about our grandchildren, when they flood us in twenty years? . . . I get sick near them, because they smell and I almost want to vomit. . . .

She *had* felt sick in the past, with nausea, cold sweats, loss of appetite, in a northern restaurant, seeing them nearby. She and her husband and two children had hastily left. She avoided them, particularly any chance of touching them or being touched, and she had taught this to her young girl and boy. In turn, her son, aged six, could draw a desegregated school under dark clouds, brown colored with black smoke emerging from its chimney. In one of its corners, encircled by a black line, the thinnest skeletal outlines of a Negro child could be seen. He drew another picture of *his* school, the same school, but of orange brick under a blue sky and full sun, scarcely any brown there but the land (dirt) under the flowers and grass. The two separate drawings had come about through the following conversation: (Dr.): "Would you want to draw a picture of your school now?" Nodding,

[3] There is no absence of curiosity in people like her, albeit from a distance.

"Should I draw the whole school?" Hesitating a moment (Dr.): "As much as you want to." He sketched one, then another picture, telling me it was a big school, "except it still isn't filled up again." (The boycott by whites had collapsed, and he, like other white children, had now been allowed to attend.) He was a sturdy and tidy boy, well mannered, rather close to his mother, bright and a good talker, a good artist, too. He wanted to be a doctor, had so for about a year. Doctors, he told me once, had to be clean, and had to scrub their hands very hard before they cut into people. We had embarked on that line of thought after a small discussion of the little Negro girl in his school. "I don't go near her, because I don't like her looks and she's a bad girl. . . . My best friend is Jimmie and he said he wouldn't play with her either because she'd ruin the game, but I don't like her around *anyways*." He did indeed stay away from her, and sought friends who did likewise, or simply were doing likewise not in any exclusiveness, but in the natural order of play. His mother, affectionate toward him and truly confident of his loyalty, felt his assurances to her that he remained clear from the Negro child were unnecessary. "Sometimes he tells me he hasn't gone near her when he comes home, but I know he hasn't regardless." He hasn't, but he is not without a silent interest, a certain preoccupation. He will draw the colored child, then cross her out, he will wipe her out in play and shoot her, but not fail to mention her, bring her into his assertive, frolicsome games.

Other children seem less anxious to draw the Negro into our talk and play. Their parents may have welcomed desegregation, and there is no issue for them, though several of them are very much aware at five or six what it means in our country to be a Negro. One girl in class with a little Negro girl informed me that "most white people don't like them, but I don't mind them." Their parents may not have welcomed desegregation at all, but may place little emotional store in its slow onset. A mother told me she did not want her children at desegregated schools, but she wanted them attending school, and there seemed little alternative but certain private schools, which, she felt, were not the best choice, educationally, for her children.

Attitudes like this, long held and deeply, honestly believed, are less likely to kindle a young child's resolute fear and horror. This woman states her opposition, but is capable of weighing other values and concerns in determining her position. Unlike several of our parents, she will not march out on any street in agitated demonstration. Her children, therefore, duly understand her views, and adopt them as their own, but her daughter occasionally will acknowledge a rope game with the little Negro girl in her class, and one day settled the entire problem for herself by telling me she'd let one Negro girl into her doll house, but no others, "I mean all the others will have to be maids, but she [the one] can have a cup of tea with us." (She loved serving tea in her doll house, and was a most gracious hostess.)

With adolescence, old ideas, stored but forgotten in games, emerge newly influential. In Atlanta, the problems of white boys and girls, now fast growing, struggling with problems of identity, of belief and career, were aptly summarized for me by a high school senior who found a Negro in several of his classes, and felt torn by old traditions and perceptions and new realities:

> It's a shock to us, I'll have to say that. I mean, I'm not against Lawrence. I've grown to respect him; but it's like this . . . they were slaves here and we had them in bondage. We fought a war over the thing and got beat, but even so, we've all grown up with them living one way, and us another. . . . They've worked in our kitchens and taken care of us, and suddenly you wake up and find them in your schools sitting across from you. . . . It's not the way we've ever lived with them before, and you can't expect us not to find it hard for a while.

Attitudes varied from immediate sympathy through indifference to marked hostility. One form of hostility was the direction of overworked attention and conversation to the Negroes with a paternal attitude which they could scarcely not notice and frequently described. Others, less impelled by unconsciously wrought reaction formations, develop slow sympathy with these children, often identifying with them in their loneliness or suffering. One adolescent boy of fifteen, in so many ways reminiscent of Anna Freud's (15) description of the teenaged youth in all his ambivalent turmoil, remarked feelingly:

> I know what it is to be alone. . . . They must be boiling underneath, but I guess they know how to keep up a good front. I think I was against desegregation, and I still am in a way, but I really feel for those two kids, and I've got to admit I like both of them in spite of what I thought I'd feel. . . .

Slowly, as the months go along, many of these white students, less and less afraid of social ostracism, have made overtures to their Negro schoolmates, words, a smile, a question, a nod. Desegregation has then moved from its legal and physical reality to its emotional onset.

Others feel and continue to feel differently: "They're no damn good, and never will be. . . . just look at Africa. . . . All they do is sit and drink. . . . They live like animals and they sure look like them, too. Have you ever smelled them, real close? They make me sick just to go near them!" Strong are the words and feelings these young men and women have and much the notice they are willing to give the Negro: "Most kids ignore us, and a few are really nice, but there are three or four who won't leave us alone. They just stare all day long. . . . I think they've got a kind of 'thing' about us, and it won't let them alone, so how can they let us alone?"

Three of these young persons, two boys and a girl, have never felt unwilling or unable to describe their anger and scorn for the two Negroes at their school. Listening to their words, and feelings underneath those words, one finds, however, sensitivity and guilt as well as rage, revealing them, unconsciously, troubled and conscience-stricken by the very harm they would so freely cause. One would describe with apparent pleasure the isolation and fear experienced by a Negro girl in his class, then wish, always after telling of her painful moments, that she go to Africa, surely an escape as well as an exile. Another talked quietly about his segregationist views, becoming loudly, angrily critical when he described a bullying word or deed toward a Negro in which he shared. He always had to fend off his own sense of right and wrong in these specific situations by such excesses of rationalized pique.

Our Negro and white children, regardless of their differing views and hopes, all agreed on the interesting point that most white children, despite their declared opinions, in practice were indifferent to the Negroes who entered their schools. Over and over from the Negroes we heard, "They couldn't care less." Time and again we heard from white children, "I've got more on my mind than a couple of niggers. . . . My work and basketball are what I think of all day . . . Who cares? I mean, they're here, and we didn't want it, but what difference does it make now? I don't think *any* in your everyday life."

Changes in feelings are reported and observable, too. Several of our high school juniors and seniors have noted changes in sentiments from mild antipathy or disinterest to respect or even affection, often to their own surprise. The unpredictable nature of human behavior, the cumulative effects of conscious and unconscious guilt and atonement, of changed perceptions and newly developing social customs, are all perhaps summarized in the experience of a Negro girl who went through a year of silence and sometimes much worse, only to be herself surprised in her last day of school at the requests of eighteen white classmates to autograph her yearbook. Their messages, from "best wishes" to "I am sorry for my silence, but you will understand why," from "I know you have had an unpleasant time, but we have all had a difficult year, and we have come through it successfully" to "You have proven yourself a fine person, and I have been sorry we could not know you better," speak for themselves, self-evident example of changing emotions, slowly incubating amid the complicated silence of an unusual school year.

Parents and Teachers

We may only briefly here refer to the parents and teachers of these children. It is a commonplace that these children are much affected by the attitudes of both. We have been particularly interested to note, in city after city of the South, that teachers may have felt tension between their roles as teachers and their private feelings. In the two battered schools of New Orleans, in

Central High School in Little Rock, in the unharassed but tensely protected (some police scrutiny all year long) high schools of Atlanta, we have heard teachers reflect what their society obviously faces, the anxieties and mixed feelings of social and cultural change. One teacher put it very well, speaking to me in a group of several teachers discussing their observations of how desegregation was proceeding:

We felt betwixt and between our traditions as Southeners and the law. I wouldn't disobey my school board, because I know I have responsibilities as a teacher, but I had my private opinions about this, of course. You know, we were as nervous as the students were. . . . You could hear a pin drop. I think I've learned as much this year as my students, and maybe a little more. You can't change a whole way of living without a little trouble here and there, but I'm right proud of our children here in Atlanta.

Parents, of course, will vary and teach their young children what they themselves feel, from kindness to fear. Many children in Atlanta obviously reflect their parents' views. Many do not, however. A white boy, son of a militant segregationist father, found friendliness toward Negroes, spoken and acted, a means of teenage rebellion. Several young students differentiated most precisely their views from those of their parents. "We're from another generation," a refrain as old as man himself, was not absent among these youths. Nor do Negro students heed their parents so resolutely. Often their parents, all too aware of past hurts, all too fearful and suspicious, had opposed letting their children transfer to white schools. Family splits — mother approving, father dissenting, or vice versa — were not rare, with younger and older children alike. Once under the crisis of actual participation, however, these splits seemed to be welded together in the common familiar need for mutual support against an outside world potentially harmful.

The Psychiatrist's Roles and Reactions

We can only mention in passing at this time some of the problems encountered by a psychiatrist working in the midst of active social and political crisis, though this subject deserves to be pursued further. For a white psychiatrist to go to Negro homes in the deep South, regularly and over a two-year period, was, in itself, as unusual as desegregation. We met apprehension and fright; their expressions in suspicion, in ingratiation or withdrawal, in awe or in ire, and slowly, in testing and teasing. We found that the more freely these persons felt they could talk with us, the more honestly they could express their feelings, their fears, angers, sorrows, and the more frankly they could acknowledge their initial, conscious deceptions. In several instances, as in individual psychotherapy, long-standing grievances were slowly mentioned, and attitudes of docility, followed by those of resentment and deeply felt hostility, appeared and yielded slowly to a sustained relationship coupled with discussion of these very feelings — for these Negroes the first such relationship with a white person. Well over a year after we had started our visits, we began to hear in many homes remarks like "Now we're telling you what we *really* think," or "We just didn't know at first what to say, and to be honest, we just don't trust any white people to be really truthful to us." Increasing involvement between these children, their parents and me allowed them to forgo many of their stylized maneuvers of dealing with what they most often feel to be alien white lord. They felt free to share their phrases, their humor, their bitterness and sadness at their fate, and they could talk about how they mask these feelings, summoning constant energy in conscious guile, bluff, and falsehood: "Tell them what they want to hear," or "Smile and hope they die on the spot." Such searing wrath demands constant vigilance lest it escape in open violence, sometimes against whites, though this is made hard by the very nature of the society which legally, socially, politically, economically subordinates them, more often against themselves, in quick fury, in long self-hatred and despair.

As with any group of people, different types of relationships developed, depending upon the child, his or her family, and the investigator. This was a study to determine attitudes, reactions, adjustment in a crisis for presumably "normal" individuals, but the therapeutic quality in the work soon appeared. For Negroes under stress to talk with a white psychiatrist about their feelings, their trials, and hopes and tensions over a period long enough for a firm relationship to develop, is obviously of some value for them as well as for the psychiatrist. Briefer contacts in several Southern cities helped us obtain information and make observations less intimately related to the child's life history, but more pertinent to his everyday experiences and adaptations. An effort was made to achieve both, to obtain general information about daily psychological phenomena, and to make the deeper, longer investigation of the individual personality and life history.

With white families, particularly segregationists, we had a different order of problems, not racial but idiosyncratic, our initial reception varying from the friendly to the suspicious and guarded. We had expected difficulty talking with segregationists, but soon learned that those strongly moved and agitated by this issue wanted very much to talk about it, indeed, need to do just that. Changes among white children both did and did not occur, and do and do not occur in those children untouched by studies like this. Several of our children attribute new feelings to talks with us: "I think, just thinking about all these things when you come makes you see more of what's going on. . . . I've wondered how I'd feel about them if I hadn't talked to you this year." We have met children in small rural towns, in larger cities, who have come to new ideas and responses to the Negro children beside them by themselves, through their own experiences and thinking, through discussions with friends and relatives, out of their own life history with its development of conscience, awareness, available intelligence, vitality and quiet of mind.

The psychiatrist, as always, must learn from his own anxieties, his own affections, angers, and depressions, as they arise in his work with other people. In a study such as this, these feelings are sure to be met, telling us about the range and intensity of the emotional forces hugging racial attitudes hard; showing us, surely, how such forces derive much of their powerful and persuasive momentum from sources deep within the individual's emotional life.

Conclusions

From our work, we would venture the following impressions:

1. Negro children, young and old, in most cases endure the stresses of initial desegregation in Southern cities without evidence of significant psychiatric illness. In order to understand why this occurs, the general problems of adjustment required of the Negro in his daily life, from birth to death, must be considered. Many of the strains met by these children and their families are similar to those often encountered elsewhere. In order to cope with such conditions, these children must learn, must be also taught, a variety of psychological maneuvers, and these will serve in desegregated schools as well as on busses, outside theatres, and restaurants. These defenses described in this paper against anxiety, fear, retaliatory anger, shame, and guilt, develop slowly, require constant emotional energy, the continual vigilance of a precarious position in our society. For these children, therefore, desegregation is an opportunity, a kind of trouble which, in contrast to other experiences, makes sense and offers hope of a better life with less hurt in the future. So long as their home life is reasonably satisfactory, so long as they do not have serious personality disorders stemming from their personal psychological development, they seem able to withstand mobs, and the variety of pressures arising from their entry into white schools. We have tried to document some of these pressures and responses to them, and have seen that the trials of desegregation fit into the other problems of the growing child, reflecting them, increasing them, or serving their expression.

2. For the white child, desegregation will have a variety of meanings, depending on his family, his age, and his own nature, needs, and wishes. Compassion, lack of interest, annoyance, strong aversion, all these reactions have been seen, and their roots traced in individual children and their parents. We can clinically distinguish between those families where segregation is a consuming, highly charged interest, a passionate

expression of other problems, and those where it is less integrated in earlier, more basic emotional troubles; thus more changeable as a social habit or custom, strongly and long held, but slowly dispensable when opposed by other wishes or customs. Even among the poorest and most insecure of our families, socially and economically, these differences obtain.

3. Some of the meanings of color, of black and white, to children have been explored. Deeply present in our culture, they indicate the sources of later attitudes and behavior in both whites and Negroes. We have attempted to convey the feelings of these children toward one another, helped by their actual words, their dreams and drawings, their play. We have also tried to indicate some of the ways their parents and teachers choose to behave, how they, too, get along in these tense situations. Finally, we have indicated some of the hurdles along the path of such study of active social and political crises by a psychiatrist.

4. We have seen changes in these Southern children, in Negroes toward less doubt and more trust, in whites toward less hostility and more freedom to see another person as a person, regardless of skin color. We have also seen hardening of suspicion, congealing of hatreds. Since we are dealing with a problem in human development, in stress and response to stress, in early anxieties, rivalries and strains within the family being reflected and expressed in later social and economic realities, there is every reason for both hope and despair. Genes do not present an ineradicable barrier, grounded in biochemistry, against easing conflicts or harsh legal and social inequalities among individuals and groups; but the fatal and reciprocally stimulating engagement of private disorders, personal anxieties, fears, and frustrated hopes, with publicly sanctioned expressions for them, solidified in history, allowed by law, encouraged by communities for their various reasons, is not easily broken. These children have shown us both possibilities, of growth and of continuing affliction.

References

1. Cash, W. J. *The Mind of the South*. New York: Knopf, 1941.
2. Woodward, C. Vann. *The Strange Career of Jim Crow*. New York: Oxford University Press, 1955.
3. Jordan, Winthrop J. *J. South. Hist.*, 28: 1, 1962.
4. Taylor, William R. *Cavalier and Yankee*. New York: Braziller, 1961.
5. Myrdal, Gunnar. *An American Dilemma*. New York: Harper, 1944 and 1962.
6. Allport, Gordon W. *The Nature of Prejudice*. Cambridge: Addison-Wesley, 1954.
7. Clark, Kenneth B. *Prejudice and Your Child*. Boston: Beacon Press, 1955.
8. Davis, Allison, and Dillard, John. *Children of Bondage*. Washington, D.C.: Am. Council on Education, 1940.
9. Kardiner, A., and Ovesey, I. *The Mark of Oppression*. New York: Norton, 1951; and New York: Meridian Books, 1962.
10. Erikson, Erik H. *Childhood and Society*. New York: Norton, 1950.

11. Group for the Advancement of Psychiatry. *Psychiatric Aspects of Desegregation*, report #37, May 1957.
12. Jahoda, Marie. *Race Relations and Mental Health*. Paris: *UNESCO*, 1960.
13. Bernard, Viola W. *J. Am. Psychoanalyt. Ass.*, 1: 256, 1952.
14. ———. *Am. J. Orthopsychiat.*, 26: 459, 1956.
15. Freud, Anna. *The Ego and the Mechanisms of Defense*. New York: International Universities Press, 1946.
16. Frazier, E. Franklin. *Black Bourgeoisie*. New York: Free Press, 1957.

D. SOCIETAL STRUCTURE

CHAPTER 10

The Split-Level American Family

URIE BRONFENBRENNER

Social Structure Affecting Parent-Child Relations

Children used to be brought up by their parents.

It may seem presumptuous to put that statement in the past tense. Yet it belongs to the past. Why? Because *de facto* responsibility for upbringing has shifted away from the family to other settings in the society, where the task is not always recognized or accepted. While the family still has the primary moral and legal responsibility for developing character in children, the power or opportunity to do the job is often lacking in the home, primarily because parents and children no longer spend enough time together in those situations in which such training is possible. This is not because parents don't want to spend time with their children. It is simply that conditions of life have changed.

To begin with, families used to be bigger — not in terms of more children so much as more adults — grandparents, uncles, aunts, cousins. Those relatives who didn't live with you lived nearby. You often went to their houses. They came as often to yours, and stayed for dinner. You knew them all — the old folks, the middle-aged, the older cousins. And they knew you. This had its good side and its bad side.

On the good side, some of these relatives were interesting people, or so you thought at the time. Uncle Charlie had been to China. Aunt Sue made the best penuche fudge on the block. Cousin Bill could read people's minds (according to him). And all these relatives gave you Christmas presents.

But there was the other side. You had to give Christmas presents to all your relatives. And they all minded your business throughout the years. They wanted to know where you had been, where you were going, and why. If they didn't like your answers, they said so (particularly if you had told them the truth).

Not just your relatives minded your business. Everybody in the neighborhood did. Again this had its two sides.

If you walked on the railroad trestle, the phone would ring at your house. Your parents would know what you had done before you got back home. People on the street would tell you to button your jacket, and ask why you weren't in church last Sunday.

But you also had the run of the neighborhood. You were allowed to play in the park. You could go into any store, whether you bought anything or not. They would let you go back of the store to watch them unpack the cartons and to hope that a carton would break. At the lumber yard, they let you pick up good scraps of wood. At the newspaper office, you could punch the linotype and burn your hand on the slugs of hot lead. And at the railroad station (they had railroad stations then), you could press the telegraph key and know that the telegraphers heard your dit-dah-dah all the way to Chicago.

These memories of gone boyhood have been documented systematically in the research of Professor Herbert Wright and his associates at the University of Kansas. The Midwestern investigators have compared the daily life of children growing up in a small town with the lives of children living in a modern city or suburb. The contrast is sobering. Children in a small town get to know well a substantially greater number of adults in different walks of life and, in contrast to their urban and suburban agemates, are more likely to be active participants in the adult settings that they enter.

As the stable world of the small town has become absorbed into an ever-shifting suburbia, children are growing up in a different kind of environment. Urbanization has reduced the extended family to a nuclear one with only two adults, and the functioning neighborhood — where it has not decayed into an urban or rural slum — has withered to a small circle of friends, most of them accessible only by motor car or telephone. Whereas the world in which the child lived before consisted of a diversity of people in a diversity of settings, now for millions of American children the neighborhood is nothing but row upon row of buildings inhabited by strangers. One house or apartment is much like another, and so are the people. They all have about the same income, and the same way of life. And the child doesn't even see much of that, for all the adults in the neighborhood do is come home, have a drink, eat dinner, mow the lawn, watch TV, and sleep. Increasingly often, today's housing projects have no stores, no shops, no services, no adults at work or play. This is the sterile world in which many of our children grow, the "urban renewal" we offer to the families we would rescue from the slums.

Neighborhood experiences available to children are extremely limited nowadays. To do anything at all — go to a movie, get an ice cream cone, go swimming, or play ball — they have to travel by bus or private car. Rarely can a child watch adults working at their trades. Mechanics, tailors, or shopkeepers are either out of sight or unapproachable. A child cannot listen to gossip at the post office as he once did. And there are no abandoned houses, barns, or attics to break into. From a young point of view, it's a dull world.

Hardly any of this really matters, for children aren't home much, anyway. A child leaves the house early in the day, on a schoolbound bus, and it's almost suppertime when he gets back. There may not be anybody home when he gets there. If his mother isn't working, at least part-time (more than a third of all mothers are), she's out a lot — because of social obligations, not just friends — doing things for the community. The child's father leaves home in the morning before the child does. It takes the father an hour and a half to get to work. He's often away weekends, not to mention absences during the week.

If a child is not with his parents or other adults, with whom does he spend his time? With other kids, of course — in school, after school, over weekends, on holidays. In these relationships, he is further restricted to children of his own age and the same socioeconomic background. The pattern was set when the old neighborhood school was abandoned as inefficient. Consolidated schools brought homogeneous grouping by age, and the homogenizing process more recently has been extended to segregate children by levels of ability; consequently, from the preschool years onward the child is dealing principally with replicas of the stamp of his own environment. Whereas social invitations used to be extended to entire families on a neighborhood basis, the cocktail party of nowadays has its segregated equivalent for every age group down to the toddlers.

It doesn't take the children very long to learn the lesson adults teach: Latch onto your peers. But to latch he must contend with a practical problem. He must hitch a ride. Anyone going in the right direction can take him. But if no one is going in that direction just then, the child can't get there.

The child who can't go somewhere else stays home, and does what everybody else does at home. He watches TV. Studies indicate that American youngsters see more TV than children in any other country do. By the late 1950s, the TV-watching figure had risen to two hours a day for the average five-year-old, three hours a day during the watching peak age period of twelve to fourteen years.

In short, whereas American children used to spend much of their time with parents and other grownups, more and more waking hours are now lived in the world of peers and of the television screen.

What do we know about the influence of the peer group, or of television, on the lives of young children? Not much.

The prevailing view in American society (indeed in the West generally) holds that the child's psychological development, to the extent that it is susceptible to environmental influence, is determined almost entirely by the parents and within the first six years of life. Scientific investigators — who are, of course, products of their own culture, imbued with its tacit assumptions about human nature — have acted accordingly. Western studies of influences on personality development in childhood overwhelmingly take the form of research on parent-child relations, with the peer group, or other extraparental influences, scarcely being considered.

In other cultures, this is not always so. A year ago, at the International Congress of Psychology in Moscow, it was my privilege to chair a symposium on "Social Factors in Personality Development." Of a score of papers presented, about half were from the West (mostly American) and half from the Socialist countries (mostly Russian). Virtually without exception, the Western reports dealt with parent-child relationships; those from the Soviet Union and other East European countries focused equally exclusively on the influence of the peer group, or, as they call it, the children's collective.

Some relevant studies have been carried out in our own society. For example, I, with others, have done research on a sample of American adolescents from middle-class families. We have found that children who reported their parents away from home for long periods of time rated significantly lower on such characteristics as responsibility and leadership. Perhaps because it was more pronounced, absence of the father was more critical than that of the mother, particularly in its effect on boys. Similar results have been reported in studies of the effects of father absence among soldiers' families during World War II, in homes of Norwegian sailors and whalers, and in Negro households with missing fathers, both in the West Indies and the United States. In general, father absence contributes to low motivation for achievement, inability to defer immediate for later gratification, low self-esteem, susceptibility to group influence, and juvenile delinquency. All of these effects are much more marked for boys than for girls.

The fact that father-absence increases susceptibility to group influence leads us directly to the question of the impact of the peer group on the child's attitudes and behavior. The first — and as yet the only — comprehensive research on this question was carried out by two University of North Carolina sociologists, Charles Bowerman and John Kinch, in 1959. Working with a sample of several hundred students from the fourth to the tenth grades in the Seattle school system, these investigators studied age trends in the tendency of children to turn to parents versus peers for opinion, advice, or company in various activities. In general, there was a turning point at about the seventh grade. Before that, the majority looked mainly to their parents as models, companions, and guides to behavior; thereafter, the children's peers had equal or greater influence.

Though I can cite no documentation from similar investigations since then, I suspect the shift comes earlier now, and is more pronounced.

In the early 1960s, the power of the peer group was documented even more dramatically by James Coleman in his book *The Adolescent Society*. Coleman investigated the values and behaviors of teenagers in eight large American high schools. He reported that the aspirations and actions of American adolescents were primarily determined by the "leading crowd" in the school society. For boys in this leading crowd, the hallmark of success was glory in athletics; for girls, it was the popular date.

Intellectual achievement was, at best, a secondary value. The most intellectually able students were not those getting the best grades. The classroom wasn't where the action was. The students who did well were "not really those of highest intelligence, but only the ones who were willing to work hard at a relatively unrewarded activity."

The most comprehensive study relevant to the subject of our concern here was completed only a year ago by the same James Coleman. The data were obtained from more than 600,000 children in grades one to twelve in 4,000 schools carefully selected as representative of public education in the United States. An attempt was made to assess the relative contribution to the child's intellectual development (as measured by standardized intelligence and achievement tests) of the following factors: (1) family background (e.g., parents' education, family size, presence in the home of reading materials, records, etc.); (2) school characteristics (e.g., per pupil expenditure, classroom size, laboratory and library facilities, etc.); (3) teacher characteristics (e.g., background, training, years of experience, verbal skills, etc.); and (4) characteristics of other children in the same school (e.g., their background, academic achievement, career plans, etc.).

Of the many findings of the study, two were particularly impressive; the first was entirely expected, the second somewhat surprising. The expected finding was that home background was the most important element in determining how well the child did at school, more important than any or all aspects of the school which the child attended. This generalization, while especially true for Northern whites, applied to a lesser degree to Southern whites and Northern Negroes, and was actually reversed for Southern Negroes, for whom the characteristics of the school were more important than those of the home. The child apparently drew sustenance from wherever sustenance was most available. Where the home had most to offer, the home was the most determining; but where the school could provide more stimulation than the home, the school was the more influential factor.

The second major conclusion concerned the aspects of the school environment which contributed most to the child's intellectual achievement. Surprisingly enough, such items as per pupil expenditure, number of children per class, laboratory space, number of volumes in the school library, and the presence or absence of ability grouping were of negligible significance. Teacher qualifications accounted for some of the child's achievement. But by far the most important factor was the pattern of characteristics of the other children attending the same school. Specifically, if a lower-class child had schoolmates who came from advantaged homes, he did reasonably well; but if all the other children also came from deprived backgrounds, he did poorly.

What about the other side of the story? What happens to a middle-class child in a predominantly lower-class school? Is he pulled down by his classmates? According to Coleman's data, the answer is no; the performance of the advantaged children remains unaffected. It is as though good home background had immunized them against the possibility of contagion.

This is the picture so far as academic achievement is concerned. How about other aspects of psychological development? Specifically, how about social behavior — such qualities as responsibility, consideration for others, or, at the opposite pole, aggressiveness or delinquent behavior? How are these affected by the child's peer group?

Comparison Between USA and USSR

The Coleman study obtained no data on this score. Some light has been shed on the problem, however, by an experiment which my Cornell colleagues and I recently carried out with school children in the United States and in the Soviet Union. Working with a sample of more than 150 sixth-graders (from six classrooms) in each country, we placed the children in situations in which we could test their readiness to engage in morally disapproved behavior such as cheating on a test, denying responsibility for property damage, etc. The results indicated that American children were far more ready to take part in such actions.

The effect of the peer group (friends in school) was quite different in the two societies. When told that their friends would know of their actions, American children were even more willing to engage in misconduct. Soviet youngsters showed just the opposite tendency. In their case, the peer group operated to support the values of the adult society, at least at their age level.

We believe these contrasting results are explained in part by the differing role of the peer group in the two societies. In the Soviet Union, *cospitanie,* or character development, is regarded as an integral part of the process of education, and its principal agent — even more important than the family — is the child's collective in school and out. A major goal of the Soviet educational process, beginning in the nursery, is "to forge a healthy, self-sufficient collective" which, in turn, has the task of developing the child into a responsible, altruistic, and loyal member of a socialist society. In contrast, in the United States, the peer group is often an autonomous agent relatively free from adult control and uncommitted — if not outrightly opposed — to the values and codes of conduct approved by society at large. Witness the new phenomenon of American middle-class vandalism and juvenile delinquency, with crime rates increasing rapidly not only for teenagers but for younger children as well.

How early in life are children susceptible to the effects of contagion? Professor Albert Bandura and his colleagues at Stanford University have conducted some experiments which suggest that the process is well developed at the preschool level. The basic experimental design involves the following elements. The child finds himself in a familiar playroom. As if by chance, in another corner of the room a person is playing with toys. Sometimes this person is an adult (teacher), sometimes another child. This other person behaves very aggressively. He strikes a large Bobo doll (a bouncing inflated figure), throws objects, and mutilates dolls and animal toys, with appropriate language to match. Later on, the experimental subject (i.e., the child who "accidentally" observed the aggressive behavior) is tested by being allowed to play in a room containing a variety of toys, including some similar to those employed by the aggressive model. With no provocation, perfectly normal, well-adjusted preschoolers engage in aggressive acts, not only repeating what they had observed but elaborating on it. Moreover, the words and gestures accompanying the actions leave no doubt that the child is living through an emotional experience of aggressive expression.

It is inconvenient to use a live model every time. Thus it occurred to Bandura to make a film. In fact, he made two, one with a live model and a second film of a cartoon cat that said and did everything the live model had said and done. The films were presented on a TV set left on in a corner of the room, as if by accident. When the children were tested, the TV film turned out to be just as effective as real people. The cat aroused as much aggression as the human model.

As soon as Bandura's work was published, the television industry issued a statement calling his conclusions into question on the interesting ground that the children had been studied "in a highly artificial situation," since no parents were present either when the TV was on or when the aggressive behavior was observed. "What a child will do under normal conditions cannot be projected from his behavior when he is carefully isolated from normal conditions and the influences of society," the statement declared. Bandura was also criticized for using a Bobo doll (which, the TV people said, is "made to be struck") and for failing to follow up his subjects after they left the laboratory. Since then, Bandura has shown that only a ten-minute exposure to an aggressive model still differentiates children in the experimental group from their controls (children not subjected to the experiment) six months later.

Evidence for the relevance of Bandura's laboratory findings to "real life" comes from a subsequent field study by Dr. Leonard Eron, now at the University of Iowa. In a sample of more than 600 third-graders, Dr. Eron found that the children who were rated most aggressive by their classmates were those who watched TV programs involving a high degree of violence.

At what age do people become immune to contagion from violence on the screen? Professor Richard Walters of Waterloo University in Canada, and his associate, Dr. Llewellyn Thomas, showed two movie films to a group of thirty-four-year-old hospital attendants. Half of these adults were shown a knife fight between two teenagers from the picture *Rebel Without a Cause;* the other half saw a film depicting adolescents engaged in artwork. Subsequently, all the attendants were asked to assist in carrying out an experiment on the effects of punishment in learning.

In the experiment, the attendants gave an unseen subject an electric shock every time the subject made an error. The lever for giving shocks had settings from zero to ten. To be sure the assistant understood what the shocks were like, he was given several, not exceeding the level of four, before the experiment. Since nothing was said about the level of shocks to be administered, each assistant was left to make his own choice. The hospital attendants who had seen the knife-fight film gave significantly more severe shocks than those who had seen the artwork film. The same experiment was repeated with a group of twenty-year-old females. This time the sound track was turned off so that only visual cues were present. But neither the silence nor the difference in sex weakened the effect. The young women who had seen the aggressive film administered more painful shocks.

These results led designers of the experiment to wonder what would happen if no film were shown and no other deliberate incitement were introduced in the immediate setting of the experiment. Would the continuing emotional pressures of the everyday environment of adolescents — who see more movies and more TV and are called on to display virility through aggressive acts in teenage gangs — provoke latent brutality comparable to that exhibited by the older people under direct stimulation of the movie of the knife fight?

Fifteen-year-old high school boys were used to test the answer to this question. Without the suggestive power of the aggressive film to step up their feelings, they pulled the shock lever to its highest intensities (levels eight to ten). A few of the boys made such remarks as "I bet I made that fellow jump."

Finally, utilizing a similar technique in a variant of what has come to be known as the "Eichmann experiment," Professor Stanley Milgram, then at Yale University, set up a situation in which the level of shock to be administered was determined by the lowest level proposed by any one of three "assistants," two of whom were confederates of Milgram and were instructed to call for increasingly higher shocks. Even though the true subjects (all adult males) could have kept the intensity to a minimum simply by stipulating mild shocks, they responded to the confederates' needling and increased the degree of pain they administered.

Peer Influence and Socialization

All of these experiments point to one conclusion. At all age levels, pressure from peers to engage in aggressive behavior is extremely difficult to resist, at least in American society.

Now if the peer group can propel its members into antisocial acts, what about the opposite possibility? Can peers also be a force for inducing contructive behavior?

Evidence on this point is not so plentiful, but some relevant data exist. To begin with, experiments on conformity to group pressure have shown that the presence of a single dissenter — for example, one "assistant" who refuses to give a severe shock — can be enough to break the spell so that the subject no longer follows the majority. But the only research explicitly directed at producing moral conduct as a function of group experience is a study conducted by Muzafer Sherif and his colleagues at the University of Oklahoma and known as the "Robber's Cave Experiment." In the words of Elton B. McNeil:

> War was declared at Robber's Cave, Oklahoma, in the summer of 1954 (Sherif et al., 1961). Of course, if you have seen one war you have seen them all, but this was an interesting war, as wars go, because only the observers knew what the fighting was about. How, then, did this war differ from any other war? This one was caused, conducted, and concluded by behavioral scientists. After years of religious, political, and economic wars, this was, perhaps, the first scientific war. It wasn't the kind of war that an adventurer could join just for the thrill of it. To be eligible, ideally, you had to be an eleven-year-old, middle-class, American, Protestant, well-adjusted boy who was willing to go to an experimental camp.

Sherif and his associates wanted to demonstrate that within the space of a few weeks they could produce two contrasting patterns of behavior in this group of normal children. First, they could bring the group to a state of intense hostility, and then completely reverse the process by inducing a spirit of warm friendship and active cooperation. The success of their efforts can be gauged by the following two excerpts describing the behavior of the boys after each stage had been reached. After the first experimental treatment of the situation was introduced,

> good feeling soon evaporated. The members of each group began to call their rivals "stinkers," "sneaks," and "cheaters." They refused to have anything more to do with individuals in the opposing group. The boys . . . turned against buddies whom they had chosen as "best friends" when they first arrived at the camp. A large proportion of the boys in each group gave negative ratings to all the boys in the other. The rival groups made threatening posters and planned raids, collecting secret hoards of green apples for ammunition. To the Robber's Cave came the Eagles, after a defeat in a tournament game, and burned a banner left behind by the Rattlers; the next morning the Rattlers seized the Eagles' flag when they arrived on the athletic field. From that time on name-calling, scuffles, and raids were the rule of the day.
>
> . . . In the dining-hall line they shoved each other aside, and the group that lost the contest for the head of the line shouted "Ladies first!" at the winner. They threw paper, food, and vile names at each other at the tables. An Eagle bumped by a Rattler was admonished by his fellow Eagles to brush "the dirt" off his clothes.

But after the second experimental treatment

> the members of the two groups began to feel more friendly to each other. For example, a Rattler whom the Eagles disliked for his sharp tongue and skill in defeating them became a "good egg." The boys stopped shoving in the meal line. They no longer called each other names, and sat together at the table. New friendships developed between individuals in the two groups.
>
> In the end the groups were actively seeking opportunities to mingle, to entertain and "treat" each other. They decided to hold a joint campfire. They took turns presenting skits and songs. Members of both groups requested that they go home together on the same bus, rather than on the separate buses in which they had come. On the way the bus stopped for refreshments. One group still had $5 which they had won as a prize in a contest. They decided to spend this sum on refreshments. On their own initiative they had invited their former rivals to be their guests for malted milks.

How were each of these effects achieved? Treatment One has a familiar ring:

> . . . To produce friction between the groups of boys we arranged a tournament of games: baseball, touch football, a tug-of-war, a treasure hunt, and so on. The tournament started in a spirit of good sportsmanship. But as the play progressed good feeling soon evaporated.

How does one turn hatred into harmony? Before undertaking this task,

Sherif wanted to demonstrate that, contrary to the views of some students of human conflict, mere interaction — pleasant social contact between antagonists — would not reduce hostility.

> . . . we brought the hostile Rattlers and Eagles together for social events: going to the movies, eating in the same dining room, and so on. But far from reducing conflict, these situations only served as opportunities for the rival groups to berate and attack each other.

How was conflict finally dispelled? By a series of strategems, of which the following is an example:

> . . . Water came to our camp in pipes from a tank about a mile away. We arranged to interrupt it and then called the boys together to inform them of the crisis. Both groups promptly volunteered to search the water line for trouble. They worked together harmoniously, and before the end of the afternoon they had located and corrected the difficulty.

On another occasion, just when everyone was hungry and the camp truck was about to go to town for food, it developed that the engine wouldn't start, and the boys had to pull together to get the vehicle going.

To move from practice to principle, the critical element for achieving harmony in human relations, according to Sherif, is joint activity in behalf of a *superordinate goal.* "Hostility gives way when groups pull together to achieve overriding goals which are real and compelling for all concerned."

Here, then, is the solution for the problems posed by autonomous peer groups and rising rates of juvenile delinquency: Confront the youngsters with some superordinate goals, and everything will turn out fine.

What superordinate goals can we suggest? Washing dishes and emptying wastebaskets? Isn't it true that meaningful opportunities for children no longer exist?

This writer disagrees. Challenging activities for children can still be found; but their discovery requires breaking down the prevailing patterns of segregation identified earlier in this essay — segregation not merely by race (although this is part of the story) but to an almost equal degree by age, class, and ability. I am arguing for greater involvement of adults in the lives of children and, conversely, for greater involvement of children in the problems and tasks of the larger society.

We must begin by desegregating age groups, ability groups, social classes, and once again engaging children and adults in common activities. Here, as in Negro-white relations, integration is not enough. In line with Sherif's findings, contact between children and adults, or between advantaged and disadvantaged, will not of itself reduce hostility and evoke mutual affection and respect. What is needed in addition is involvement in a superordinate goal, common participation in a challenging job to be done.

Where is a job to be found that can involve children and adults across the dividing lines of race, ability, and social class?

Here is one possibility. Urbanization and industrialization have not done away with the need to care for the very young. To be sure, "progress" has brought us to the point where we seem to believe that only a person with a master's degree is truly qualified to care for young children. An exception is made for parents, and for babysitters, but these are concessions to practicality; we all know that professionals could do it better.

It is a strange doctrine. For if present-day knowledge of child development tells us anything at all, it tells us that the child develops psychologically as a function of reciprocal interaction with those who love him. This reciprocal interaction need be only of the most ordinary kind — caresses, looks, sounds, talking, singing, playing, reading stories — the things that parents, and everybody else, have done with children for generation after generation.

Contrary to the impression of many, our task in helping disadvantaged children through such programs as Head Start is not to have a "specialist" working with each child but to enable the child's parents, brothers, sisters, and all those around him to provide the kinds of stimulation which families ordinarily give children but which can fail to develop in the chaotic conditions of life in poverty. It is for this reason that Project Head Start places such heavy emphasis on the involvement of parents, not only in decision-making but in direct interaction with the children themselves, both at the center and (especially) at home. Not only parents but teenagers and older children are viewed as especially significant in work with the very young, for, in certain respects, older siblings can function more effectively than adults. The latter, no matter how warm and helpful they may be, are in an important sense in a world apart; their abilities, skills, and standards are so clearly superior to those of the child as to appear beyond childish grasp.

Here, then, is a context in which adults and children can pursue together a superordinate goal, for there is nothing so "real and compelling to all concerned" as the need of a young child for the care and attention of his elders. The difficulty is that we have not yet provided the opportunities — the institutional settings — which would make possible the recognition and pursuit of this superordinate goal.

The beginnings of such an opportunity structure, however, already exist in our society. As I have indicated, they are to be found in the poverty program, particularly those aspects of it dealing with children: Head Start, which involves parents, older children, and the whole community in the care of the very young; Follow Through, which extends Head Start into the elementary grades, thus breaking down the destructive wall between the school on the one hand and parents in the local community on the other; Parent and Child Centers, which provide a neighborhood center where all generations can meet to engage in common activities in behalf of children, etc.

The need for such programs is not restricted to the nation's poor. So far as alienation of children is concerned, the world of the disadvantaged simply reflects in more severe form a social disease that has infected the entire society. The cure for the society as a whole is the same as that for its sickest segment. Head Start, Follow Through, Parent and Child Centers are all needed by the middle class as much as by the economically less favored. Again, contrary to popular impression, the principal purpose of these programs is not remedial education but the giving to both children and their families of a sense of dignity, purpose, and meaningful activity without which children cannot develop capacities in any sphere of activity, including the intellectual.

Service to the very young is not the only superordinate goal potentially available to children in our society. The very old also need to be saved. In segregating them in their own housing projects and, indeed, in whole communities, we have deprived both them and the younger generations of an essential human experience. We need to find ways in which children once again can assist and comfort old people, and, in return, gain insight to character development that occurs through such experiences.

Participation in constructive activities on behalf of others will also reduce the growing tendency to aggressive and antisocial behavior in the young, if only by diversion from such actions and from the stimuli that instigate them. But so long as these stimuli continue to dominate the TV screen, those exposed to TV can be expected to react to the influence. Nor, as we have seen, is it likely that the TV industry will be responsive to the findings of research or the arguments of concerned parents and professionals. The only measure that is likely to be effective is pressure where it hurts most. The sponsor must be informed that his product will be boycotted until programming is changed.

My proposals for child rearing in the future may appear to some as a pipedream, but they need not be a dream. For just as autonomy and aggression have their roots in the American tradition, so have neighborliness, civic concern, and devotion to the young. By re-exploring these last, we can rediscover our moral identity as a society and as a nation.

CHAPTER 11

Children's Environments and Child Welfare

HENRY S. MAAS

Where children are concerned, the primary aims of the social services are:

> To preserve and strengthen the protection afforded the child by the basic unit of his social environment, the family.
>
> To provide special protection for the abandoned, neglected, or malnourished child.
>
> As part of broader, long-range objectives, to promote a better social environment for the child and adolescent, both in the city and rural areas.[1]

In child welfare, however, the term "social environment" denotes a rather ill-defined area surrounding the child. The purpose of this article is to propose clarifying perspectives on children's environments, with implications for child welfare policy, program, and practice.

The Canadian Conference on Children has echoed UNICEF's statement of child welfare's primary aims — to reinforce, support, and supplement normal family life for children.[2] In effect, though, the secondary aims of compensatory, substitutive, and treatment services preoccupy the field and have become the most prestigious of child welfare services. It is possible that the dominance of clinical orientations in North American child welfare and our inability to examine children's environments in much more than structural terms — bearing, for example, on the size and composition of children's living units — have too narrowly limited child welfare's concerns. With a keener sense of the usefulness of interactional and ecological perspectives on children's environments, together with clinical and structural viewpoints, we might attend more actively to, for example, what Jenkins describes as the separation experience of parents whose children are in foster care.[3] We might learn how to interrupt constructively the cycle of environmental (including medical) nonrecognition of the familial and anomic community situations of battered children.[4] We should be able to act upon the bureaucratization of child welfare agencies and the rapid turnover of staff[5] and to work at a neighborhood level to help integrate otherwise anomic families into mutual self-help services. We should develop economic plans and other social policy proposals for governmental action. For the social environments of a child are far broader and more complex than the influence upon him of his immediate caretaker.

This paper deals with three matters: (1) general ideas about environment, (2) specific ideas about children's environments conceived in ecological and social interactional terms, and (3) implications for child welfare.

General Conceptions of Environment

The environment of children is simplistically conceived when it is seen to be *the* shaper of the child's development and behavior. Much of child welfare's efforts devoted to "matching" is based in this simple assumption. The notion of environment as the shaper of the child ignores the influences of the child's genetic potentials; different children's ways of perceiving, interpreting, and experiencing the same environment; and different children's varying capacities to influence or act upon and control their environments.

We should recall a fundamental biological assumption: "There is a boundary, though not a precise one, between organism and environment."[6] This imprecise boundary varies in permeability from organism to organism. The permeability varies also from time to time or situation to situation involving the same organism. Thus the mutual influences of child upon environment and environment upon child differ when a child is fatigued and hungry or when the same child is rested and well fed. To try to "match" in any way beyond the lowest limits of parental tolerances and the most potentially intolerable behavior of a given child is to give more weight to statics than a dynamic conception of child-environment interaction suggests is necessary — to predict in realms that are too complex and multivariate for such foresight. Children can change their caretakers — and vice versa — unpredictably.

The complexity of organism-environment interaction is further compounded by the constant changes that both organism and environment are naturally and simultaneously engaged in. Commenting on "the psychological implications of social or cultural changes," anthropologist Hughes notes diametrically opposed conclusions about the effects of rapid environmental change. He alludes to behavioral scientists "basing their arguments on the 'function of the similar' in . . . life [who] emphasize . . . the disruptive effects of changes through heightened anxiety and stress, especially if the change is rapid. Others, however, point to the psychological beneficence of change and stress the advantages of rapidity in the turnover of an entire way of life. The evidence, to say the least, is mixed. . . ."[7] Clearly, on the issue of environmental change for the young child, child welfare has been on the side of those who argue for maintenance of the familiar. On the pace of change, we have been for the gradual and accommodative rather than the rapid — but I'm not sure what beyond hunch has convinced us that this is uniformly best, or whether we have thought differently about children for whom rapid change might be better than the gradual, when we have genuine choice.

Finally, environments are effective at different levels or distances in relation to children. For child welfare purposes, it is useful to distinguish between immediate (or "proximal") environments and more remote (or "distal") environments.[8] The latter include, for example, social class and the culture of poverty. These distal environments may be somewhat, but never completely, mediated by a child's family or caretakers. For example, research indicates that children's "imaginative capacity" is most likely to be developed when the child is engaged in "interaction with a variety of physical objects and toys, as well as with meaningful contacts with adults."[9] In describing one *vecinidad* of eighty-three persons composing fourteen families, as part of his long-term study of the culture of poverty, Lewis observes, "Only half of the families in the tenement were able to invest in any toys for their children."[10] Furthermore, in view of the overcrowded living quarters of these families, averaging six persons sleeping and eating in each one-room shack, we note with dismay that "imaginative capacity" flourishes in an environment that provides also "some opportunities for solitary play or privacy. . . . There is evidence that firstborn or only children, or children with relatively few siblings, are likely to have the time and privacy for practice and the greater contact with adults to permit full development of fantasy play."[11] Clearly remote environments expand or limit the opportunities for children's development.

In general, then, environments should be seen as changing and variously responsive to children, who can more or less effectively act upon and control their surroundings. An interactional framework amplifies this point of view. In addition, ecological perspectives elaborate relationships between proximal and distal environments — as they ultimately affect and are affected by such living organisms as the young.

Some Ecological Perspectives

By contrast with structural perspectives, which lead to analyses of relationships among component parts in fragmentary ways, the ecological approach uses a wide-angled lens to encompass the interplay between living organisms and their proximal and distal environments. Ecology, with its biological and social perspectives, provides evidence that every living organism has "in addition to minimum requirements of things such as food, minimal requirements of living space and distance from others of its own species."[12] Cognizant of the great variety of genetic potentials in man, Dubos says our environments must offer diversity or the availability of a "wide range of experiences."[13] Regarding food, he writes:

> Nutritional deprivations or imbalances occurring early in life (prenatal or postnatal) will interfere with the normal development of the brain and of learning ability. Furthermore, bad dietary habits acquired early in life tend to persist throughout the whole life span. . . . People born and raised in an environment where food intake is quantitatively or qualitatively inadequate achieve a certain form of physiological and behavioral adaptation to low food intake. They tend to restrict their physical and mental activity and thereby to reduce their nutritional needs; in other words, they become adjusted to undernutrition by living less intensely. . . . Physical and mental apathy and other forms of indolence . . . [result] especially when nutritional scarcity has occurred during very early life.[14]

Regarding population density or crowding, Dubos generalizes from experiments with various animal species that "as the population pressure increases . . . varieties of abnormal behavior" appear, of a kind that reflects "social unawareness" of others. "Behavior is asocial rather than antisocial." He continues:

> The humanness of man is not innate; it is a product of socialization. Some of the peculiarly "human" traits disappear under conditions of extreme crowding, probably because man achieves his humanness only through contact with human beings *under the proper conditions* [my italics, HSM].[15]

Ecological perspectives direct our attention to more than basic needs for proper diet and for spacial and population arrangements that allow for balances in togetherness and separateness, conducive to optimal human growth and development. One British ecologist complains, "There are too many variables, and it requires very extensive work for long periods of time, at the end of which you have only unraveled one corner of the situation," pursuing such questions as, "What impact did the rabbit have on the environment and vegetation?"[16] Increasingly in child welfare we must ask the analogous question about the child: What impact does this child have on this or that environment? Answers to this question — provided that we have a useful view of relevant environmental parameters — can tell us as much that is practically fruitful as the traditional child welfare question: How does (or, more precariously, how will) this environment influence this child?

Some Interactional Perspectives

By contrast with the fundamentally intrapersonal focus of clinical perspectives, the interactional approach is primarily interpersonal. Although the clinical tends to be concerned with internal constancies, the interactional deals more with the changing flow and continual feedback and patterning of human communication. The clinical effort is highly inferential, involving judgments about such only indirect observables as motives and defenses. The interactional stays closer to observed behavior and its configurations, or patterns and sequences, or stages.

Within this framework, the ecologically oriented biologist and the interactionally oriented psychologist start their observations with complementary premises. Writes Dubos, "Environmental information becomes formative only when it evokes a creative response from the organism."[17] Writes child psychologist J. L. Gewirtz, ". . . our key assumption [is] that the infant's behavior is not simply a function of evoking stimuli in the environment, but a function also of whether and how the environment responds to his behavior."[18] The interactional approach postulates neither merely a shaping environment nor a shaped person. Rather, both environment and person variously, if complexly and at times innovatively and thus unpredictably, respond to one another. And with interactional perspectives, nonresponse is a response too.

Consider Kagan's hypothetical descriptions of the experiences of a lower-middle-class infant and an upper-middle-class infant:

> The upper-middle-class child is lying in her crib in her bedroom on the second floor of a suburban home. She wakes, the room is quiet, her mother is downstairs baking. The infant studies the crib and her fingers. Suddenly the quiet is broken as the mother enters and speaks to the child. The auditory intrusion is maximally distinctive and likely to orient the infant to her mother and to the vocalization. If the child responds vocally, the mother is apt to continue the dialogue. Contrast this set of events with those that occur to an infant girl in Harlem lying on a couch in a two-room apartment with the television going and siblings peering into her face. The child lies in a sea of sound, but like the sea, it is homogeneous. The mother approaches the child and says something to her. The communication, however, is minimally distinctive from background noise and, as such, is not likely to recruit the infant's attention. Many of the infant's vocalizations during the day are not likely to be heard nor are they likely to elicit a special response. The fundamental theme of this argument seeks to minimize the importance of absolute amount of stimulation the child receives and spotlights instead the distinctiveness of that stimulation.[19]

So concludes Kagan, talking like an ecologist about noise levels and related phenomena in two infants' proximal environments. The key differentiating environmental process is, however, the suburban infant's control over her environment. She has a power, in time internalizable as a sense of competence, potentially missing from the slum child's interaction with her environment.

An environment upon which the infant or young child can act in ways which make it do things, under his controlling behavior, is very different from an environment that is nonresponsive. Psychologist John S. Watson has recently been experimenting with colorful mobiles, hung over an infant's crib and rotating as the infant learns to activate them himself. Watson finds learning and social emotional responses in his experimental infants that do not occur among his control infants.[20] The infant enjoys a reciprocal interaction with an environment to which he becomes increasingly attentive and responsive as it seems attentive and responsive to him. He can re-create his environment.

This human-responsiveness component — or psychological aspect of environments — is sometimes overlooked in social interaction research that contrasts caretaking settings. Described in terms of the intentions of caretakers — for example, as monitoring, guidance, support, and integration[21] — social environments take on differential forms, but another set of categories is needed to define the crucial difference such environments make to children and youth. If the boundary between organism and environment is not a precise one, and the critical difference in child welfare environments is the child's "creative response," then interactional configurations[22] should be defined in terms that include the degree of their modifiability by the young.[23] Though Watson cautiously questions the long-term effects of infants' activation of mobiles,[24] for older children, and most clearly for adults, environments which give evidence that one is heard are markedly different psychologically from those in which one is not.

Finally, interactional perspectives upset other ways of "seeing." In his study of social functioning as a social work concept, Alary argues that "interactional determinants" are "in contradistinction [to the view of] personal needs and capacities as internal determinants and environmental demands and opportunities as external determinants."[25] His argument cannot be pursued here, but what he alludes to as a split between "external" and "internal" seems still today to permeate child welfare thinking and practice. The child welfare worker who thinks of a child's needs and capacities and a foster or adoptive home's demands and opportunities may be doing her very best to "match" yesterday's family with a child as of yesterday. Tomorrow, placed together, or after a longer period of interaction, neither child nor environment may look the same as the opportunities/demands and capacities/needs that had been so carefully "matched," unless a pathological stereotypy binds all persons involved.

Implications for Child Welfare

Four sets of proposals for child welfare can be derived from the ideas and observations sketched in the foregoing.

(1) Child welfare should both expand and reallocate its existing resources for a more concerted effort in preventive programs. Prevention means keeping more and more children out of child placement even temporarily or for short-term care. Every relevant study I know indicates that at least some children are placed who need never have been if . . . if . . . and if. We must get to work on these "ifs" — though it means that professional personnel now engaged in adoptive and foster home activities redirect their efforts to the kinds of preventive program to be cited later. We have at present no evidence — in Ripple's followup study of adoptions[26] or elsewhere[27] — that the time invested in "matching" pays off for children or their substitute parents. Why continue it, if our primary concern is the welfare of children?

First, we do know that the vast majority of children cared for by child welfare agencies are poor children in or removed from poor families. Adequate amounts of money for poor families are not a panacea; money alone does not cure all. But money, available at the right time in the family cycle of otherwise poor families, provides opportunities. Money contributes to the prevention of interpersonal and psychic strains and stresses that result in family breakdown and the ostensible need to place children. We need a Family Starting Allowance for the reasons that follow.

An interactional approach to familial environments calls for the conceptualization of stages or phases through which families pass, from one interactional context to another. There are a few such conceptual sequences proposed by students of family life. All give some focus to the new or starting family, on through variously conceived subsequent phases, to the empty nest stage of family life, and finally to the aging parental dyad or widowhood. Schorr describes a four- or five-stage poor family's cycle in relation to income development. Using demographic data, he writes of stage 1, Initial Marriage and Child Rearing; stage 2, Occupational Choice; stage 3, The Family Cycle Squeeze; and stage 4, Family Breakdown. The culmination for many such poor families is breakdown, in part because of the squeeze between aspirations and always drastically limiting amounts of money. His appended stage 5 involves the children in these families, the next generation, who emerge from family breakdown to repeat the same cycle of a family entrapped in poverty.[28]

How does the poor family cycle develop? A combination of factors, beginning with too-early marriage, often before age eighteen, leads to the following, in Schorr's research-based statements:[29]

> . . . low income is likely to be continual experience for those who marry before 18.
>
> . . . earlier marriages tend to be less stable.
>
> . . . young marriage is associated with less education for the husband, [and] a poor first job, a chaotic work history. . . .
>
> When a couple starts out together early, they are not only likely to have their first child earlier than usual; they are likely to have more children.
>
> They are more than ordinarily likely to suffer separation or divorce.
>
> Mothers in broken families are likely to have more children than those in stable families. [There is a high] prevalence of poverty among families headed by women; obviously mothers who start out without a husband are no better off.
>
> The path that opens before [any such] . . . family is a sequence of marriage or liaisons in which the notion of a stable, intact marriage, if it was ever present, becomes fainter.

The poor family cycle Schorr describes is a familiar and well-documented one. Out of such families come a large proportion of the children thrust into placement. The provision of a Family Starting Allowance, granted to such families when the cycle is just beginning, might do much to set the family's interaction sequence on a different course. It should help to keep out of placement many children who travel the long-term foster care path. And most of the children in long-term foster care are the poorest of the poor children in foster care.[30]

Child welfare workers can document from thick case files how poverty contributes directly and indirectly to current child placement caseloads. Child welfare workers can provide evidence of the rising costs of expanding substitute care programs. Child welfare workers can consult with knowledgeable co-professionals on the economics of income supplementation programs. Child welfare workers can assess the pros and cons of various possible alternative plans, for poor families, of some kind of Family Starting Allowance. Child welfare workers can shepherd their proposal through to the appropriate governmental authorities, and stay with their proposal until it becomes reality. Not a simple series of steps at all — but one that has an ultimate outcome of great promise. That is one kind of preventive child welfare program.

(2) We now know a good deal about the psychosocial conditions of familial environments from which children are catapulted into foster care. Long-term foster care involves primarily families who neglect their children or who contribute to the children's development of serious emotional problems. By contrast, children who are brief-care charges

(under three months) come mostly from homes in which mother's illness or confinement precipitates the children's placement. I am primarily concerned here with this latter group, brief-termers, who may so easily become long-termers. A Scottish study that tells us much about these families is useful in preventive work aimed at keeping children from ever entering care. The Schaffers contrasted 100 such families who placed their children in care during mother's confinement with 100 families who made other arrangements.[31] The child-placing families were essentially alienated families, isolated from their extended families and from neighbors. Compared with nonplacing families, the child-placing families were more likely to live in electoral wards with higher rates of population turnover and a higher population density — e.g., more persons per room. They moved more frequently, and more often lived in redevelopment areas or postwar housing estates. Though both groups of families had the same number of relatives and the same number living within the same city, fewer relatives of child-placing families lived within a ten minute walk, and there was less contact with relatives, as well as with church and political activities or with neighborhood mutual self-help services. Only medical facilities were more extensively used by the child-placing family.

Within the family, dissociation prevailed — e.g., far fewer of the parents in these families ever went out together and far fewer of the fathers ever helped in the home. Though the Schaffers nowhere make this explicit in their study report, judging by the age distributions of grandparents (more younger ones in the childplacing families) and of the children (more children under age four), I estimate that these were younger families, much of the kind Schorr describes, very much in need of some kind of Family Starting Allowance, but, in addition, also in need of professional services aimed at family social integration, both intrafamilially and extrafamilially.

I cannot specify in any detail what the intervention strategies and social work practice methods should be, but the goals are definable. They involve increasing the meaningful social interaction of these parents so that they become more integrated and less alienated members of their neighborhoods. For those who would leap in with clinical strategies — such as family therapy programs — I propose that working through the neighborhood environment to engage these families in self-help services, for example, around the development or expansion of parent-staffed child care facilities in their housing units, may be a complementary if not equally effective alternative strategy. Knowledgeable and skillful neighborhood workers, keenly aware of family life and its possible enrichment through mutual neighborhood aid, may induce alienated families to participate in local enterprises. Social workers may support them on a casework basis through their initial efforts to engage in meaningful interaction with neighbors. Social workers may use group work skills and community organization techniques as needed, providing these young families, at an early and critical point in their family cycle, with what may be for them a completely new kind of human experience. For child welfare workers who think first in terms of psychopathology — of possible character disorders, affective deprivation, or the ego impoverishment of these adults — I remind them of therapeutic milieu or therapeutic community approaches. Community mental health programs aim to keep persons from entering — or to bring patients back from — distant governmental hospitals, to live and be cared for in their home communities. As "sick" as these parents may be, they may, like others, rediscover health in a neighborhood milieu that evidences concern for them. More specifically, the professional worker who helps to launch a day care neighbor service, of the kind with which Emlen has been concerned in Portland,[32] may be serving many neighborhood needs, and at the same time be providing a crucial preventive family service.

Most of the child-placing parents in the Scottish study never went out together. But such parents could be informed of and encouraged to participate in a neighborhood child care group, which, long before an emergency or crisis situation arises, might provide a few hours every other week or so for a couple to get away from their children. Relevant here are the previously cited ecological observations on the effects of overcrowding and the need for periods of distance from others, as well as closeness.

(3) The theoretical perspectives outlined here suggest some "should nots" as well as "shoulds" for child welfare workers. I raised questions earlier about the dominance of the clinical approach in present-day child placement. In their review of the literature on foster parenting, Taylor and Starr conclude, "While specific emphases are attached by different authors to various aspects of the home study, the model for the study is the diagnostic assessment process as utilized in a child guidance or family agency setting. This model is based on an untested assumption that such an assessment is necessary for the appropriate placement of foster children."[33] So far as the assumption has been tested, the answer is essentially negative. For example, in his study of 101 foster families, Fanshel observes, "The capacities foster parents display when they first apply to an agency . . . must . . . be seen as potentialities . . . [and] parental capacity must be seen as a variable performance,"[34] since different children evoke different foster parent behavior.

Clinical assessments have questionable payoff not only in foster care but in adoption. In a properly cautious summary following his review of research on adoptions, Kadushin concludes, "Research indicates that the assessments of parents' motivation for adoption are of questionable significance as a predictor of successful adoptive parenting, as are attitudes toward infertility. . . ."[35] Ripple concludes from her follow-up study of adopted children, "None of the factors believed to be important to realistic integration of the adopted child into the family — 'matching' and handling the facts of adoption — showed the expected association with favorable outcome."[36]

In regard to day care services, the picture is somewhat clearer. "It is small wonder that agency family day care programs have remained small in scope, considering the elaborate formal requirements of professionally supervised family day care," writes Emlen. ". . . Family day care is presented to the community as a social agency service based on diagnostic assessment of a family problem. . . . Ruderman and Mayer have pointed to the problem-oriented character of the services offered as unattractive to the general consumer."[37] In short, perspectives developed out of efforts to understand psychopathology may, when applied in child placement situations, communicate to parents an unacceptable and seemingly irrelevant set of questions about their own possible sickness.

Even the clinical approach to family as the context for childhood disturbance has shaky empirical referents. At least, psychological research has been unable to give us any consistent findings about the influence of family dynamics on the development of psychopathology in children. One recent review of the literature on parental deprivation and the etiology of psychiatric illness points only to the contradictory or conflicting nature of the existing evidence.[38] In another, the author concludes, "No factors were found in the parent-child interaction of schizophrenics, neurotics, or those with behavior disorders which could be identified as unique to them or could distinguish one group from another, or any of the groups from the families of the controls."[39] Until we know much more than we do now about such matters, child welfare workers might well consider whether their time is not better invested in other than clinical procedures in their home-finding and placement activities.

I am not putting down clinical orientations and procedures for those parts of the child welfare enterprise that are concerned with the treatment of distrubed children or their troubled parents. I am questioning, however, the relevance of clinical approaches in such activities as home finding and related services that are not typically aimed at psychotherapy.

(4) Finally, I suggest that an ecological approach seems more promising for child welfare than a structural approach to children's environments. We have thus far had only a relatively small yield of useful structural formulations from research. For example, regarding the composition of living groups for children, Wolins' findings on the SOS Kinderdorf[40] and my research on preadolescent peer relations[41] give some support to the placement of school-age children in group-care living units patterned on family-like sibships, mixed by sex and age — rather than sex-divided and strictly age-graded. Benson's review of the research literature on

fatherlessness and the conflicting evidence regarding the effects on children of father absence, concludes, "We could very well give more thought to changing legislation to allow the placement of children in one-parent homes."[42] In a structural framework, child-development research has examined the effects on children's cognitive or social-emotional development of birth order and the sex composition of sibships, but as yet there is little practical guidance for child welfare workers in such research on family structure.

With an ecological framework, however, the world of child and family life widens to larger vistas. For example, there is some ecologically based research on the effects of social isolation. One such inquiry is a study of Norwegian farm families, living miles apart from one another in Norway's mountain country. Drs. Anna von der Lippe and Ernest A. Haggard have been assessing the psychological effects of such isolated living.[43] In briefest terms, they find that these mountain farm families tend to be composed of affectively nonexpressive and cognitively deprived children. If overcrowding in urban neighborhoods makes for a chaotic and overwhelming environment, isolated families apparently experience another kind of sensory deprivation, in an environment that also inhibits optimal human development.

Earlier, using ecological and social interactional frameworks, I alluded to the desirability of helping to integrate anomic families into their neighborhood's services. Richard Titmuss, in his collection of essays entitled *Commitment to Welfare,* remarks that the unifying aim of the social services is social integration and, conversely, the discouragement of alienation.[44] Perhaps child welfare needs to reconsider not only its conceptual views of children's environments, but the extent to which its services promote a constructive bringing together of children, parents, and in UNICEF's terms, "a better social environment . . . both in the city and rural areas." At a distal level, this better social environment excludes extremes of poverty and both social isolation and high density "living," for only in their absence do optimal human growth and development begin.

References

1. UNICEF, *Children of the Developing Countries* (Cleveland and New York: World Publishing Co., 1963), 89.

2. Ray Godfrey and B. Schlesinger, *Child Welfare Services: Winding Paths to Maturity* (Toronto: Canadian Conference on Children, 1969), 179.

3. Shirley Jenkins, "Separation Experience of Parents Whose Children Are in Foster Care," *Child Welfare,* XLVIII, no. 6 (1969), 334-40.

4. Angela E. Skinner and Raymond L. Castle, *Seventy-Eight Battered Children: A Retrospective Study* (London: National Society for the Prevention of Cruelty to Children, 1969).

5. Harry Wasserman, "Early Careers of Professional Social Workers in a Public Child Welfare Agency," *Social Work,* XV, no. 3 (1970), 93-101.

6. C. F. A. Pantin, "Organism and Environment," *Psychological Issues,* IV, no. 2., Monograph 22 (1969), 114.

7. Charles C. Hughes, "Psychocultural Dimensions of Social Change," in Joseph C. Finney, ed., *Cultural Change, Mental Health, and Poverty* (Lexington: University of Kentucky Press, 1969), 179.

8. *Perspectives on Human Deprivation: Biological, Psychological, and Sociological* (Washington, D.C.: National Institute of Child Health and Human Development, U.S. Public Health Service, 1968).

9. Ibid., 10.

10. Oscar Lewis, "The Possessions of the Poor," *Scientific American,* CCXXI (Oct. 1969), 123.

11. *Perspectives on Human Deprivation.*

12. Roman Mykytowycz, "Territorial Marking by Rabbits," *Scientific American,* CCXVIII (May 1968), 126.

13. René Dubos, "Environmental Determinants of Human Life," in D. C. Glass, ed., *Environmental Influences* (New York: Rockefeller University Press and Russell Sage Foundation, 1968), 139.

14. Ibid., 144-45.

15. Ibid., 146.

16. Anne Chisholm, "Nature's Doctors," *Manchester Guardian Weekly,* CII, no. 14 (1970), 16.

17. Dubos, *op. cit.,* 150

18. H. B. Gewirtz and J. L. Gewirtz, "Caretaker Settings, Background, and Events and Behavior Differences in Four Israeli Child-Rearing Environments: Some Preliminary Trends," in B. M. Foss, ed., *Determinants in Infant Behavior* IV (London: Methuen & Co., 1969), 247.

19. Jerome Kagan, "On Cultural Deprivation," in *Environmental Influences,* 238-39.

20. John S. Watson, "Cognitive-Perceptual Developments in Infancy: Setting for the Seventies," paper presented at the Merrill-Palmer Conference on Research and Teaching of Infant Development, Detroit, 1970; also, John S. Watson and Craig T. Ramey, "Reactions to Response-Contingent Stimulation in Early Infancy," paper read, in part, at meeting of the Society for Research in Child Development, California, 1969, and in part at the Institute of Human Development Symposium, University of California, Berkeley, 1969.

21. Howard W. Polsky and Daniel B. Claster, *The Dynamics of Residential Treatment: A Social System Analysis* (Chapel Hill: University of North Carolina Press, 1968), 12-20.

22. Henry L. Lennard and Arnold Bernstein, *Patterns in Human Interaction* (San Francisco: Jossey-Bass, 1969).

23. H. B. and J. L. Gewirtz, *op. cit.*

24. Watson, *op. cit.*

25. J. O. Jacques Alary, "A Meaning Analysis of the Expression 'Social Functioning' as a Social Work Concept," doctoral dissertation, School of Social Work, Tulane University, New Orleans, 1967, 32.

26. Lilian Ripple, "A Follow-up Study of Adopted Children," *Social Service Review,* XLII, no. 4 (1968), 496.

27. Alfred Kadushin, "Child Welfare," in Henry S. Maas, ed., *Research in the Social Services: A Five-Year Review* (New York: National Association of Social Workers, 1970).

28. Alvin L. Schorr, *Poor Kids: A Report on Children in Poverty* (New York: Basic Books, 1966).

29. Ibid. 26-43.

30. Henry S. Maas, "Children in Long-Term Foster Care," *Child Welfare,* XLVIII, no. 6 (1969), 321-33, 347.

31. H. R. Schaffer and Evelyn B. Schaffer, *Child Care and the Family: A Study of Short-Term Admissions to Care* (London: Bell & Sons, 1968). See also findings on rootlessness of families in high-risk child-placement regions in Jean Packman, *Child Care Needs and Numbers* (London: Allen & Unwin, 1968).

32. Arthur C. Emlen, "Realistic Planning for the Day Care Consumer," paper presented at the National Conference of Social Welfare, Chicago, 1970; also, Alice H. Collins and Eunice L. Watson, *The Day Care Neighbor Service: A Handbook for the Organization and Operation of a New Approach to Family Day Care* (Portland, Ore.: School of Social Work, Portland State University, 1969).

33. Delores A. Taylor and Philip Starr, "Foster Parenting: An Integrative Review of the Literature," *Child Welfare,* XLVI, no. 7 (1967), 374.

34. David Fanshel, *Foster Parenthood: A Role Analysis* (Minneapolis: University of Minnesota Press, 1966), 155, 162.

35. Kadushin, *op. cit.*

36. Ripple, *op. cit.*

37. Emlen, *op. cit.,* 8, 9.

38. Alistair Munroe, "The Theoretical Importance of Parental Deprivation in the Aetiology of Psychiatric Illness," *Applied Social Studies,* I (June 1969), 81-92.

39. G. H. Frank, "The Role of the Family in the Development of Psychopathology," *Psychological Bulletin* (1965), 191.

40. Martin Wolins, "Group Care: Friend or Foe," *Social Work,* XIV, no. 1 (1969), 35-53.

41. Henry S. Maas, "Preadolescent Peer Relations and Adult Intimacy," *Psychiatry: Journal for the Study of Interpersonal Processes,* XXXI (May 1968), 161-72.

42. Leonard Benson, *Fatherhood: A Sociological Perspective* (New York: Random House, 1968), 267.

43. Ernest A. Haggard and Anna von der Lippe, "Isolated Families in the Mountains of Norway," in E. James Anthony and Cyrille Koupernik, eds., *The Child in His Family* (New York: Wiley-Interscience, 1970), 465-88.

44. Richard Titmuss, *Commitment to Welfare* (London: Allen & Unwin, 1968), 22.

CHAPTER 12

Social Variables and Their Effect on Psychiatric Emergency Situations Among Children

WILLIAM C. SZE

A Conceptual Model

This study is directed toward the examination of certain variables involved in the process of generating psychiatric emergency needs in children. In order to understand what and how social variables affect children's psychiatric emergency situations, we must first explore the following four questions. (1) What is a psychiatric emergency of a child? (2) What are the variables involved? (3) What are the patterns and relationships among such variables? and (4) What is the "breaking point" of an emergency situation? In fact, these questions constitute the theme of this article. To construct a definition of psychiatric emergency which would be inclusive of all operating factors is not easy. However, if the processes of psychiatric emergency can be conceptualized, the concept of psychiatric emergency of children becomes less obscure.

We may ask how one can systematically study the circumstances which supposedly trigger the child's "psychiatric emergency" needs. Inquiry into this process is a complex task in which many aspects of human behavior and the social system are intricately involved. In the following schema, five variables are used: (1) behavioral adaptation; (2) alleviating processes in a social system; (3) tolerance level in a social system; (4) labeling; and (5) control processes. These five variables are central to our effort to understand the processes which lead to the children's psychiatric emergency situations; on the basis of these variables, we will now formulate a conceptual model of children's psychiatric emergency.

As illustrated in the model, biological, psychological, or social factors, or a combination of these, cause manifest behavior (Boxes 1 and 2). Since

A CONCEPTUAL MODEL

The Conceptual Model in Children's Psychiatric Emergency

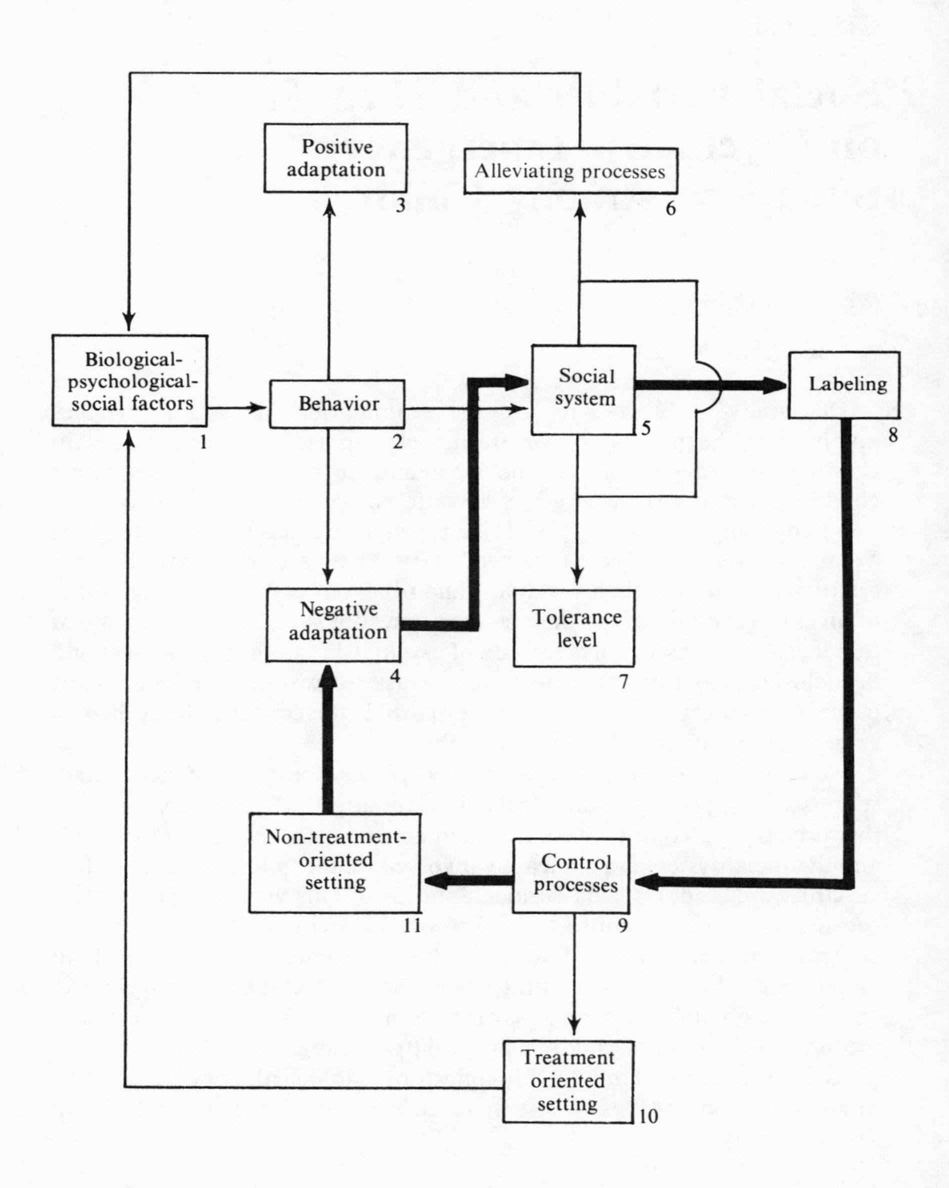

each individual has different resources for coping with his problems, some may have a strong capacity for adaptation and others may have less adaptatability (Boxes 3 and 4). When the maladaptive behavior (negative adaptation) involves a social system (Box 5), it is believed that three types of mechanisms become activated; one is alleviating processes (Box 6), another is tolerance level (Box 7), and a third is labeling (Box 8). The alleviating processes are to be considered as an opportunity mechanism whereby, when maladaptive behavior appears in the system, therapeutic intervention will be initiated for alleviating the maladaptive behavior.

For example, when Johnny manifested behavior problems in school primarily as a result of an unstable family situation and marital difficulties between his parents, he and his family were provided with casework service by the school social worker. While they were in the process of treatment, the classroom teachers and school principal were very cooperative with the treatment process. The teachers and principal exercised patience although Johnny's school performance and behavior were below the standards of the school. The expanding of the tolerance level on the part of the school system helped the child through the turmoil stage. Without having both the social service program and the raised tolerance level of the faculty of the school, Johnny could very easily have been expelled from school before he could have received help. If we assume that the school system did not or was unable to provide these two types of mechanism within the system, then the third mechanism would naturally have come into being; that is, labeling (Box 8). When Johnny is given a label, such as "bad child," "nuts," or "crazy kid," this can greatly affect his social role and self-perception. He is labeled and stigmatized as a deviant which separates him from the main group. The label is usually determined by societal reaction. It also denotes the level of tolerance and general social attitudes toward a certain type of behavior. Once Johnny has been labeled, he is compelled to adapt himself to a "deviant role" or "sick role."

After Johnny becomes stabilized as a career deviant, his long-standing problem further aggravates the social system of which he is a member. Finally, a social control mechanism has to be called in. In the control processes (Box 9) two types of processes can be distinguished: one is to commit Johnny to a treatment-oriented setting (Box 10); the other is to place Johnny in a non treatment-oriented setting (Box 11). Since the latter setting provides only custodial care, such as a penal institution, the results usually do not alleviate Johnny's problem, but only serve to reinforce his stabilized deviant behavior as role-taking processes from other inmates. Therefore, his problem remains the same and, as pointed out by the model (shown by the heavier lines), the vicious circle continues. However, if Johnny were sent to the treatment-oriented setting, he would receive various treatments which would help in alleviating his problem. Hence, the outcome would be different.

Children's Psychiatric Emergency

Sometimes the prima-facie evidence of the emotional symptom of a person is insufficient in determining the need for psychiatric emergency help unless the precipitating circumstance can also be counted. Also, in the development of a psychiatric emergency, the social variables are as crucial as, if not more so than, the psychological makeup of the individual. Therefore, the word "psychiatric" has no direct meaning without a qualified explanation. The reverse is also true — i.e., the social emergency cannot be totally isolated from psychological processes. Often, a psychiatric emergency reflects the importance attached to alternatives of a situation by the value system of the perceiver. In addition, a psychiatric emergency results when the tolerance level in a given system and/or in significant others with respect to the manifested behavior of the actor reaches the saturation point. Therefore, the emergency situation frequently follows a kind of "last straw." Often the perceiver was reluctant to initiate an action to restrain the actor's behavior in the past, but due to the frequency of recurrence or length of the disturbed behavior or inadvertent social consequences which reflect the temper of the significant others or a given social system, the situation comes to be viewed as a psychiatric emergency.

The Assumptions Involved

It is assumed that psychiatric emergency situations among children tend to develop under the following circumstances: (1) when the behavioral adaptation of the child is viewed by significant others as inadequate; (2) when the alleviating processes within the social system are insufficiently developed to deal with the situation; (3) when the tolerance level of significant others within the social system is low; and (4) when societal reaction assigns a deviant-type label to the child's behavior.

Study Procedures

The cases in this study were obtained from two mental health clinics which served the population of the city of Pittsburgh. Sixty children's psychiatric emergency cases were chosen for this study.

The cases are almost equally divided racially, with thirty-one white and twenty-nine Negro. A high percentage (67 percent) come from low socioeconomic school districts. The ages of the forty-one male and nineteen female children range from ten to fifteen years, and they are primarily in grades six through nine.

In each case, the parents and school personnel were interviewed and the school and clinic records also were examined regarding reasons for the emergency call and the circumstances under which the child's problem arose.

The Patterns and Relationships Among Such Factors and the "Breaking Point"

Incidents which led to the crises as viewed by parents, schools, and clinicians respectively were sought in this study. The following three cases represent typical triggering factors in the majority of the cases of our findings.

Case Illustration One

> A thirteen-year-old girl with an I.Q. of 83 attended the sixth grade in a working-class neighborhood elementary school. The principal referred the child to the mental health services for emergency help.
>
> The principal stated that the student became involved in some behavioral situation almost daily with her classmates, other children in the school, teachers, and school safety patrol guards. The school had tried every approach with this child and it was apparent that she needed immediate evaluation by a psychiatrist. The principal further pointed out that she was retarded, but had never been in a special class during her school life. Last year, the school recommended that the mother seek some help from a social agency but she was unable to find an agency which would help her child. At the beginning of this semester, the principal requested the Board of Education to transfer the child to another school. The request was denied. During that period, the child had been given short-term suspensions several times for her misbehavior. The suspensions did not help the child's behavior. After she returned from suspension she was disobedient and displayed misconduct almost daily. The mother, on the other hand, blamed the school for lack of cooperation and inconvenience to her, especially by repeatedly sending the child home for short-term suspensions and calling her for conferences.

This case illustrates the child's maladaptation in behavior and learning situations (behavior adaptation of the conceptual model). The school realized that the child needed special help, but neither the school nor the community had facilities available (alleviating processes of the conceptual model). The school finally turned to the Board of Education for assistance (transfer the child to another school); but the request was denied. The lack of administrative support served to further diminish the tolerance level of the school staff. The only means which the principal had at his disposal was to give short-term suspensions. This really did not get at the root of the problem. The parent resented the inconvenience which occurred during the suspension period because the parent had to provide supervision for the child while she went out to work. After all these processes, finally the so-called "psychiatric emergency" emerged (control process of the conceptual model).

Case Illustration Two

> A fifteen-year-old boy was attending the eighth grade. His parents and teachers were concerned about his school performance because he had been receiving failing grades during the last two years. Teachers who had taught him were all complaining that they could not contain him or settle him down to the point where he could apply himself to his school work. The boy's mother was not too concerned about his academic achievement but was concerned about his effeminate manner and his threats to kill himself. She reacted by rejecting him and calling him "retarded" and "stupid." The child, on the other hand, resented and showed hatred toward his mother's boy friend, thus arousing an intensified hostile reaction from the mother. This situation had been going on for some time, but what made the mother at this particular time ask the school to seek help for the boy is not answered. Clinical examination after referral stated the child was probably a "depressive personality." The difficulty pointed to the necessity of his leaving the home environment because he was being pushed out by the mother, who was under the stress of her own developing relationship with her boy friend. It is interesting to note that the labels given by parent and teachers as their reasons for psychiatric emergency referral were as follows: "effeminate, depressed, suicidal and unable to cope with many situations, poor school work."

Analysis of this case illustration supports every assumption of the conceptual model of child's psychiatric emergency. It is hoped by significant others that, through the labeling, as part of the legitimizing processes, further action against the child will be justified. For example, the mother may want to get rid of the child and place him in a resident home so that she can develop her own affair with her boy friend. The school also wants to have some action taken because the child's poor school performance may cause the teacher's tolerance level to reach the saturation point. The child is getting more disturbed because he is constantly labeled and castigated. This vicious circle is perpetuated until some variable intervenes. As the circular movement continues, the incidents which led to the crises become less understood. For instance, the child's poor academic performance dates back two years. Why does it upset the school at this time? The answer will be a "convenient point." The "convenient point" can also be called a "crisis" or "emergency situation."

Case Illustration Three

> A fifteen-year-old eighth-grade boy was referred to the mental health services for "disturbing classes." After a review of his school records and a talk with the vice-principal, the so-called incident which led to the crisis was revealed. The child acted immaturely in school but was not a "mental case," as the vice-principal explained. He commented that the child was the victim of circumstance. He further pointed out that the school principal and teaching staff had not been working together to resolve some of the problems facing them every day. Each side had become very hostile. Mutual tolerance was very low and the sufferers were frequently the children. In this school, the teachers would send problem children to the principal's office and the principal became upset by this. He then permitted some of the students to get into situations which normally would not have happened if the staff problem had been solved.

This case supports the assumption that poor working relationships among staff members will decrease tolerance levels, which in turn increase the incidence of crisis situations.

Discussion

The complete analysis and findings of this study could not be presented in this article due to space limitations. However, it should be pointed out that one of the most useful variables of the conceptual model is the level of tolerance, especially in understanding why these youngsters were not brought to a psychiatric attention earlier, assuming the degree of the behavior problem has remained unchanged. The decisions of the significant others of these sixty children regarding the need for psychiatric intervention have, in varying degrees, been affected by their level of tolerance.

In light of our findings, we would like to make a few comments concerning the effective way of preventing children's psychiatric emergency occurrences.

It is apparent from the cases studied that school administrators, teachers, and parents have often been utilizing the psychiatric referral as a means for resolving various problems which might face them at any given time. Whether the referred child needs help at that particular time is frequently a moot question, and in some cases, even the stated reasons and the timing of the referral are questionable. It seems that most cases are referred at the convenience of significant others rather than because of the emergency needs of the child. This is not entirely due to the ignorance of the significant others, but reflects many other factors which generate the moment of "urgent need."

There is no doubt that many referred children do have problems of various kinds. The question is whether all these problems can be best met by mental health services; for instance, poor academic achievement may require remedial services instead of mental health services. Indiscriminate utilization of mental health clinics often creates a discrepancy between the expectations of the referring agent and the reality which the clinician is able to deliver.

In order to narrow the gap, a new emphasis on mental health education through the consultative process would be useful. Also, just emphasizing the emergency aspect is not enough; what is needed is a preventive approach which will interrupt the vicious circle, as the conceptual model indicated. For instance, administrative consultation with school administrators may increase understanding of the real needs and problems which confront school personnel every day. If the school has inadequate facilities for those who need extra help, the staff is burdened with an extra load of responsibilities. After a period of time, tension is bound to build up, with the result that a crisis is in the making. Finally, the tolerance of the individual teacher or the school system runs out, and the child is labeled as a problem child who needs immediate psychiatric intervention. The conceptual model of this study has very well demonstrated this point. What can be done is to examine the school situation with reference to "problem children" in order to identify the problem according to the conceptual model and, once the problem area is identified, to find a solution for coping with it. Sometimes it may not be the referred child, but the system, which needs the help. Hence, preventive measures would decrease "psychiatric emergency" incidents to a great extent.

References

Clausen, John A. Paths to the Mental Hospital, *Journal of Social Issues*, 11:25-32, 1955.

Coleman, Donald M. Problems in an Emergency Psychiatric Clinic, *Mental Hospitals*, 11:5:26-27, 1960.

Mechanic, David. Some Factors in Identifying and Defining Mental Illness. *Mental Hygiene*, 46:66-74, 1962.

Miller, Abraham. A Report on Psychiatric Emergencies, *Canadian Hospital*, 36:12:36-37, 82-84, 1959. Scheff, Thomas J. *Being Mentally Ill: A Sociological Theory*. Chicago: Aldine, 1966.

Smith, Kathleen; Pumphrey, Muriel W.; and Hall, Julian C. The Last Straw: The Decisive Incident Resulting in the Request for Hospitalization in 100 Schizophrenic Patients, *American Journal of Psychiatry*, 120:228-233, 1963.

Szasz, Thomas S. *The Myth of Mental Illness: Foundations of a Theory of Personal Conduct*. New York: Harper & Row, 1964.

Waldfogel, Samuel, and George E. Gardner. Intervention in Crisis as a Method of Primary Prevention. In *Prevention of Mental Disorders in Children*, Gerald Caplan, ed. New York: Basic Books, 1961.

Whitmer, Carroll A., and Conover, Glenn C. A Study of Critical Incidents in the Hospitalization of the Mentally Ill. *Social Work*, 9:1:89-94, 1959.

PART II

Adolescence: An Introduction to the Determinants Affecting the Adolescence Stage of the Life Cycle

In current literature on adolescence, three main approaches may be noted. One approach stems from the early writings of C. Stanley Hall (1) more than a half century ago. Hall viewed adolescent difficulty as the result of the biological maturation process, and thus as inevitable. Later writers such as Anna Freud, Peter Blos, and Arthur Jersild (2, 3, 4), influenced by classic Freudian theory, recognized that psychological changes are as important in adolescence as biological changes. They define adolescence as the psychological processes of adaptation to the condition of pubescence — the physical manifestation of sexual maturation. In other words, it is generally thought that the adolescent has little control over his behavior. Typical adolescent behavior patterns, often including irrational, impulsive, and self-centered behavior, are viewed as reactions to the sudden spurt in physical change, such as rapid growth in physique, development of secondary sexual characteristics, and hormonal imbalance, in conjunction with the residue of early unresolved psychological development. According to both these viewpoints adolescence is a stage for all youngsters and is a universal developmental phenomenon.

The third approach to adolescence seems to reject the notion that adolescence as we know it is universal and inevitable. It views behavioral disturbance during adolescence as the result of interplay between personal circumstances and societal factors. In other words, the phenomena of adolescence are seen as by-products of the modern society rather than as the results of biological and psychological maturation processes. Margaret Mead, H. W. Bernard, and Ruth Benedict (5, 6, 7) are leading proponents of this approach.

Despite the influence of the cultural determinant approach, the definition of adolescence has been well institutionalized, particularly in the

Western world, in the sense implied by the bio-psychological determinant approach. In recent years, however, the traditional dichotomy of human behavior into body vs. mind or innate vs. environmentally determined has been rapidly broken down and replaced by a new view of human behavior. The work of Erik Erikson and Kurt Lewin (8, 9) has attested to this. Modern writings have given added weight to the recognition of the importance of an integrated view of human behavior. Erikson, in his writings on adolescence, has identified the central problem of adolescence as the problem of identity and role diffusion. In fact, the so-called identity crisis has eventually become a central theme of many books on adolescence. In Erikson's view, the individual's ego identity is formed through the synthesis of his ego foundations and the values and norms of his social context. Successful resolution of the identity crisis requires both social acceptance and role confirmation. Some writers point out that the lack of well-defined pubertal rites of passage in our modern world makes achievement of social acceptance and role confirmation difficult indeed. In a complex society, eligibility for entry to the adult world requires much more than meeting a simple age criterion. Entry depends on educational achievement and occupational potential, both of which require the adolescent to increase his efforts at a time when psychological and biological demands leave him with a depleted repertory of responses and fewer personal resources to cope and adapt. In addition to these multifaceted social demands, the adolescent must also relinquish his early dependent-parental love-object relationship. It is not surprising that these transitions leave him unsure and vulnerable. This "discontinuity" phenomenon has been advanced and widely discussed in the literature (10).

The combination of psychological affects and social expectations has given rise to the so-called adolescent identity crisis. Robert Merton describes the adolescent as one who is "poised on the edge of several groups but fully accepted by none of them" (11). One aspect of our modern society that enforces the "marginal man" position of the adolescent is the reluctance of our rigid social structure to accommodate those young people who are not especially well equipped to cope with the technological society. An example is the youth who has had difficulty in modeling adult roles. In the modern nuclear family a youth's only opportunity to observe his father in role tasks may be in the brief time the father has at home. The youth sees his father, tired from the day's work, regaling the family with dinner-table complaints about his job. Attempts at effective role modeling under these conditions are likely to be abortive. Furthermore, the youth's opportunities for the meaningful and significant relationships with adult males other than the father of other authority figures are limited by the impersonality of urban living, the nuclear family structure, social mobility, and geographical dispersion. Gone are the days when there was a variety of adult relatives, neighbors, and townspeople to whom a youth could relate and on whom he could model his behavior. The opportunities for what Merton termed "anticipatory socialization" are much reduced for boys today. As a consequence, we have the presently acute youth identity crisis with its youth counterculture and alienation syndrome. Young people have explored various avenues in an attempt to shelter their loneliness and search out their identity.

The importance of the process of identity search and role confirmation in explaining the adolescent developmental crisis has also been recognized by Richard Cloward and Lloyd Ohlin (12), who explain the deviant behavior of adolescence by anchoring their theoretical position in the concept of "anomie," introduced by the renowned European social theorist Emile Durkheim, and further advanced by the American sociologist Robert K. Merton. According to Merton, "anomie develops not because of a breakdown in the regulation of goals alone, but, rather, because of a breakdown in the relationship between goals and legitimate avenues of access to them" (13). Therefore, "aberrant behavior may be regarded sociologically as a symptom of disassociation between culturally prescribed aspirations and socially structured avenues of realizing these aspirations" (14). This hypothesis is illustrated in today's world by the discrepancy between cultural expectation and social accommodation. The increasing imbalance between the two is at the core of today's youth unrest. A college education, for instance, has become a part of the cultural expectation of a majority of youngsters. Despite social class differences, even lower-socioeconomic-class parents aspire to a college education for their children without considering whether the goal is realistic in view of their meager financial resources. The striving for upward occupational mobility is a part of America's dream, and justifiably so in a free and democratic society. Now we have to face the problem this expectation encounters; namely, the limitation in legitimate avenues of access to this high goal.

Inspired by Merton's anomie interpretation, Cloward and Ohlin derived a hypotheses that is invaluable for those interested in adolescent subculture and deviance: "Adolescents who form delinquent subcultures, we suggest, have internalized an emphasis upon conventional goals. Faced with limitations on legitimate avenues of access to these goals, and unable to revise their aspirations downward, they experience intense frustration; the exploration of nonconformist alternatives may be the result" (15). This discussion has widened our understanding of adolescents as a social group and their problems, some of which have deep roots in the culture and societal structure of our society.

Now we turn our attention to the readings that will give us further intellectual challenge and stimulation to aid our understanding of adolescence. In Chapter 13, Blos describes adolescence from a psychoanalytic viewpoint, in terms of the ego — its nature, operation, and function. In this classic work Blos illustrates beautifully the dynamic aspects of the ego mechanism and its function in serving a most difficult period of human growth, the adolescent period. The interplay among the id, ego, and superego is most important in shaping the adolescent's struggle with his past unresolved emotional residues, his overwhelming current instinctual drive energy, and his future perspective of a mature self. As a result of psychic restructuring, the ego eventually becomes the adolescent's savior as he negotiates the tumultuous passage through adolescence.

In Chapter 14, Lea Barinbaum presents two cases that illustrate one of Erikson's developmental tasks of adolescence; that is, the concept of role identification vs. role diffusion. Barinbaum points out that both over- and understructured environments can contribute to identity crisis. The first case illustrates role confusion attributed to cultural change, leaving the adolescent with inadequate guidance in developing his identity. The other case illustrates role confusion attributed to an overprotective environment that affords the adolescent too little autonomy to permit him to develop his own identity.

Anna Freud, in Chapter 15, "Adolescence as a Developmental Disturbance," has noted that during adolescence, many facets of the individual's interpersonal systems undergo extensive change. There are the endocrinological changes that lead to the upsurge of sexual drives, and psychological composition changes that heighten aggression, hostility, and anxiety. Anna Freud considers all these changes as developmental processes that can serve positive functions for growth, preparing the individual to move into adulthood.

The intrapersonal realignment processes of adolescence have been illustrated by Blos and Anna Freud in Chapters 13 and 15. The mechanisms for this realignment rely greatly upon the social and psychological environment outside the intrapersonal system; namely, the assistance of interpersonal relationships. As Irene Josselyn points out in Chapter 16, as the adolescent moves toward adulthood he relinquishes his early love-object relationship, especially with parents, in order to assert himself in the struggle between dependence and independence. As he gives up the old relationship he makes a choice of new love objects from among his peers. They provide him with reassurance and his identity is secured or reaffirmed. This in turn aids the adolescent in his intrapersonal realignment. Josselyn also stresses the importance of interpersonal elements in fulfilling the inherent needs of the individual: "The maturation of the inherent need for relationships with others is expressed in the urge to be an interrelating part of large social groups." This statement implies a broadened concept of interpersonal relationships, which, rather than being limited to a one-to-one relationship. reaches many other societal structures as well.

Each age group has its unique significance and problems, and this is especially true of the group that lies in between childhood and adolescence, the period that is variously called latency, juvenile, school age, and pre-adolescent.Fritz Redl, in Chapter 17, vividly describes some troublesome behavior of the pre-teenage group and pleads for adult understanding. Redl considers this a period of temporary disorganization which paves the way for future growth and is beneficial to future personality makeup. It is a process by which the young person relinquishes his identification with adult society and establishes a strong identification with his peers. This chapter gives additional dimension to Erikson's on various stages of identification.

Nathan Ackerman, in Chapter 18, attributes the formation of adolescent problems to forces beyond those of developmental crises. He suggests that their roots are perhaps in family disorder, ill-conceived social and cultural patterns, and the disordered character of the whole human relations pattern. Ackerman discusses the types of delinquent behavior and their relationship to family and interpersonal patterns in general, and he suggests specifically that we cannot examine behaviors meaningfully without giving consideration to the context (family and interpersonal relationship patterns) in which the adolescent behavior originates.

The effect of cultural factors is much more clearly apparent in adolescence than in childhood. In his early years the youngster is passively subject to cultural influence, but in adolescence the youth is active in creating a new culture. He searches for his identity through cultural means, such as peer culture influence. That is, he learns his behavior and attitudes from his contemporaries or from a subculture with which he is identified. The adolescent in conflict with the dominant culture experiences cultural factors as pressure and stress as he seeks resolution of his conflicts.

Lawrence Schiamberg, in Chapter 19, gives a comprehensive view of cultural factors affecting adolescent-parent conflict based on cross-cultural comparison, and suggests some solutions to the problem Schiamberg indicates that the extended family system broadens the scope of interpersonal relationships open to the adolescent, which in turn alleviates the intensified interpersonal relationships between the adolescent and his parents. The age-graded system of India and the *rites de passage* of some African societies also provide mechanisms for smooth transition of responsibilities from one generation to another. In the traditional Chinese society the generational boundary was unconditionally respected. That is, the father was respected not because he was older or a "good father," but because he was of a different generation. Schiamberg compares the traditional Chinese society with the modern society of Communist China, in which incidence of intergenerational conflict has been greatly increased.

He concludes that traditional behavior norms and role expectations serve to ease adolescent-parent conflict in certain societies; but when the norms and role expectations become less stable in these cultures and more similar to those of industrial societies, the generation gap appears.

In contrast to the majority of writers, who characterize adolescence as a period of storm and stress and of peer culture influence, Frederick Elkin and William Westley, in Chapter 20, report their study of adolescence in a predominantly upper-middle-class community in which they found little resemblance to descriptions of adolescence current in sociological literature. They report few sharp conflicts between adolescents and parents, and no observable signs of a generation gap in the upper-middle-class families studied.

In Chapter 21 Merwyn Garbarino vividly describes the barriers and disadvantages affecting an American Indian girl in her struggle between old and new, majority and minority cultures, and role identity and role diffusion. It is a lively illustration of an adolescent struggling to adulthood in a marginal culture.

When we speak of an individual's physical environment, it is relatively simple for most people to understand it in terms of climate, geographical terrain, and so on. But when one speaks of social environment, many things may be implied. For instance, the term may simply refer to the people around the individual and their interpersonal relations with each other, or it can refer to a very complicated societal structures — complex organizations, governmental processes, or societal systems such as the family or educational institutions.

In Chapter 22, Kenneth Keniston views adolescent problems in the context of the social situation. He sees the post-industrial and technological society of the Western world as the source of a series of new public anxieties. The major contributing factor identified is the prolongation of the educational process and subsequent delay of social role fulfillment. This delay results in "cultural lag," in which the rapid rate of social change becomes frightening. In terms of the adolescence problem, Keniston points out that the central conscious issue during youth is the tension between self and society, which in turn produces in part the phenomenon we currently call alienation. He points out, however, that this phenomenon is not universal among adolescents. He also contends that the youth culture, although somewhat antagonistic to the conventional social order, is in the vanguard of social progress.

Erikson's "Memorandum on Youth," Chapter 23, is an attempt to predict the position and outlook of youth by the year 2000. Erikson assesses the present problems and stance of youth and sees two ideological viewpoints emerging. At one pole is an emphasis on finding identity through technology. A view of the future containing a promise of indefinite progress will replace traditional authority and youth will find its identity through doing and being a part of the technological system. At the opposite pole Erikson sees the adolescent identity search finding expression and resolution through a humanist orientation, insisting on the widest range of human possibilities and opposing mechanization and regimentation. These viewpoints will necessarily have to reach a synthesis, and each has a valuable contribution to make to the other. Their accommodation, Erikson feels, will be forced by the pressures of our common fate, and will result in radically new modes of thought and innovations in both culture and society. As an example, Erikson predicts the waning of age distinction as a basis of authority, and recognition of more life stages and new roles for both sexes in each. Authority will need to be redefined in terms of mutual responsibility for the present, rather than being based on tradition.

Whatever one's prediction for the future of youth, adolescence at the present time seems to be well accepted as a tumultuous period, and there is little question raised as to whether this is a natural course of events or is environmentally precipitated. Most people seem to feel it is a normal phenomenon and that the annoying behavior will diminish spontaneously when the stage is passed through. Thus the adult waits passively for things to get better. In Chapter 24, Roy Menninger points out that the "sit and wait" attitude blindly overlooks some genuine needs of the adolescent. One need, he emphasizes, is that of real participation and involvement as a responsible member of society. As a result of the denial of this need in our present societal structure, we see the social infantilization of the adolescent, which becomes a bombshell for many young people as their feelings of disappointment in themselves, in adults, and in society are expressed as antisocial behavior or retreat from the real social world.

References

1. G. Stanley Hall, *Adolescence*, 2 vols. New York: Appleton, 1904.
2. Anna Freud, *The Ego and the Mechanisms of Defense*, chaps. 11 and 12. New York: International Universities Press, 1946.
3. Peter Blos, *On Adolescence*. New York: Free Press, 1962.
4. Arthur Jersild, *The Psychology of Adolescence*. New York: Macmillan, 1957.
5. Margaret Mead, *Sex and Temperament in Three Primitive Societies*. New York: Morrow, 1935.
6. H. W. Bernard, *Human Development in Western Culture*. Boston: Allyn & Bacon, 1962.
7. Ruth Benedict, *Patterns of Culture*. Boston: Houghton Mifflin, 1964.
8. Erik H. Erikson, *Childhood and Society*. New York: Norton, 1950.
9. Kurt Lewin, "Field Theory and Experiment in Social Psychology Concepts and Methods," *American Journal of Sociology*, vol. 44, 1939, pp. 868-97.
10. Kenneth Keniston, "Social Change and Youth in America," *Daedalus*, Winter, 1962, pp. 145-71; Ruth Benedict, "Continuity and Discontinuity in Cultural Conditioning," *Psychiatry*, vol. 1, 1938, pp. 161-67; and Kingsley Davis, "The Sociology of Parent-Youth Conflict," *American Sociological Review*, vol. 5, August 1940, pp. 523-35.

11. Robert K. Merton, *Social Theory and Social Structure*. New York: Free Press, 1947, p. 265.

12. Richard A. Cloward and Lloyd E. Ohlin, *Delinquency and Opportunity: A Theory of Delinquent Gangs*. New York: Free Press, 1960.

13. Ibid., p. 83.

14. Robert Merton, *Social Theory and Social Structure*. New York: Free Press, 1947, p. 134.

15. Richard A. Cloward and Lloyd E. Ohlin, *Delinquency and Opportunity: A Theory of Delinquent Gangs*. New York: Free Press, 1960, p. 86.

16. Theodore Lidz, *The Person*. New York: Basic Books, 1968, p. 265-66.

A. INTRAPERSONAL SYSTEM

CHAPTER 13

The Ego in Adolescence

PETER BLOS

The topic of the ego in adolescence has been singled out for special consideration because its study will permit a more detailed view of the adolescent process in terms of psychic restructuring as manifested in transient ego activities and permanent ego alterations. Taking the ego as the focus of study will enable us to weave many strands of observation and theory into a coherent texture.

Introductory Remarks

The ego — its nature, operation, and function — can best be studied during periods of maturational dislocations when the balance between drive and ego is upset. Under these conditions the mental apparatus is confronted with the task of accommodating new instinctual drives — new in quantity and quality and also involving new demands from the outer world. The ego is, by definition, the sum total of those mental processes which aim at safeguarding mental functioning; toward this end the ego mediates between drive and outer world. We recognize in the sense of reality the fruits of this mediation process. To these pressures — id and outer world — should be added a third which is, however, a derivative of the environment, namely, the superego.

The methods by which the ego mediates as well as its pattern of operation can be observed during adolescence. Considering the origin of the ego, we have to keep in mind that it is the result of differentiating processes and constitutes a corollary to the young human organism's prolonged dependency on the external world. Throughout its lifetime, the ego preserves the earmarks of this origin; in fact, it continues to be stimulated toward progressive differentiation by the twofold impact of the instinctual drives and the external world. This fact has been previously alluded to by saying that ego development takes its cue from the phase-specific drive organization.

The ego in its operational definition is a relative concept, determined by the degree of pressures exerted on it. Consequently, an ego which has maintained adequate mental functioning during times of relative calm can be overwhelmed by increasing pressures evidenced during puberty; the character of its resources will then prove itself as either sufficient or insufficient for the acute task. This latter condition is precipitated during adolescence by the increased pressures from the three sources mentioned. Anna Freud (1936) referred to the adolescent ego by saying that "a relatively strong id confronts a relatively weak ego." Furthermore, the support which the ego obtained in childhood from education ceases to operate in the accustomed fashion, due to the adolescent's massive rejection of external controls as mere excrescences of infantile dependency.

The patterned reaction to inner dangers is modeled along early experiences of threats emanating from the environment. "Frightening instinctual situations can in the last resort be traced to external situations of danger" (Freud, 1933). In other words, all defensive processes once had an adaptive function in the face of external exigencies.

In both instances — those of internal and external danger — the problem of overexcitation has to be considered: in the infant an inborn apparatus provides a stimulus barrier against the outer world, but the stimulus barrier against instincts has to be developed (Hartmann, 1939, a). In this sense, any psychic mechanism which protects the mental organism against overstimulation serves a positive function; this is equally true for early childhood and for adolescence. The developing psychic apparatus must, so to say, constantly catch up with maturational conditions, which in turn give impetus and direction to ego differentiation and integration. This process reflects the mutual influences of ego and id on their respective development. When this process is not kept fluid, but produces instead a premature crystallization of character or neurotic symptom, the function of the ego has miscarried. What should normally operate as an adaptive or defensive mechanism has taken on a different quality: instead of initiating progression and differentiation, the development of certain ego functions has either miscarried or been arrested.

In any crisis the ego resorts to emergency measures aimed primarily at the protection of its basic function — the maintenance of psychic cohesion and reality contact. For instance, quite apart from its defensive aspects rooted in conflictual anxiety and in the fear of drive intensity the withdrawal of cathexis from the outer world during adolescence operates to preserve and protect basic ego functions. This is to say that both quantitative and qualitative factors have to be considered as separate variables in evaluating ego activities directed at self-protection. Here, as in so many psychic phenomena, the principle of multiple function (Waelder, 1936) has to be kept in mind in order to appreciate the individual variability and combination by which the demands from several sources are effected with least effort. The adolescent situation no doubt calls for extreme measures to avert trauma or disintegration. The huge amount of psychic energy absorbed in this task reduces, if only temporarily and intermittently, adaptive processes to a minimum. When psychic energy is bound in defensive operations or in countercathexis, a depletion of movable energy within the ego will result.

The conflictual aspect of adolescence has dominated the discussion of adolescence for a long time; but it is proper to call attention to the fact that not every balance disturbance in the psychic apparatus is *ipso facto* a conflictual manifestation (Hartmann, 1939, a). This consideration has special importance for adolescence, a time when disturbances in balance, with ensuing regulatory mechanisms often extreme in nature, are the rule rather than the exception.

The Ego at the Onset of Adolescence

The phase-adequate adolescent ego can only develop appropriately if the preparatory phase of the latency period has been traversed more or less successfully. Otherwise, as in the case of an abortive latency period, a prelatency ego must cope with pubertal drives. The result is a mere reintensification of infantile sexuality; nothing new or specifically adolescent makes its appearance. Pseudoadolescent manifestations are simulated attempts at being an adolescent: the ego resorts to the use of tension regulators in direct continuation of early childhood. The ego, in order to be able to cope with puberty and adolescence, requires the achievements of the latency period; only then can it deal with the approaching maturational tasks in terms of new differentiating and integrative processes.

The onset of puberty brings with it a quantitative increase of instinctual drive energy. A recathexis of pregenital instinctual positions occurs which in many ways resembles drive diffusion. Component instincts come blatantly to the fore, the attempt to control them is evidenced in typical adolescent reactions: scoptophilia leads to shyness, embarrassment or blushing; exhibitionism to modesty and self consciousness; sadomasochistic trends to impassivity and indifference; and smelling and tasting are drawn into the sexual realm of conflictual body sensations. Of course, the direct and uncontrolled expression of these instinctual trends during adolescence is well known to every observer of this age.

It is well to remember that "all pregenital impulses, in their aims of incorporation, seem to possess a certain destructive component. Unknown constitutional factors, and above all, experiences of frustration, greatly increase the destructive element" (Fenichel, 1945, b). In the total picture of adolescent instinct control we may have underestimated the ego's effort to tame the aggressive drive by directing our attention almost exclusively to the libidinal conflicts and the attempts to master them.

It might be helpful to define the preconditions which the ego must possess at the onset of adolescence to an appreciable degree in order to develop those qualities and functions that are specifically adolescent and that will bring about those ego transformations which result in the ego of adulthood. The essential ego achievements of the latency period are the following: 1) an increase in cathexis of inner objects (object- and self-representations) with resultant automatization of certain ego functions; 2) an increasing resistivity of ego functions to regression (secondary autonomy) with a consequent expansion of the nonconflictual sphere of the ego; 3) the formation of a self-critical ego which increasingly complements the functions of the superego, so that the regulation of self-esteem has reached a degree of independence from the environment; 4) reduction of the expressive use of the whole body and increase in the capacity for verbal expression in isolation from motor activity (Kris, 1939); 5) mastery of the environment through the learning of skills and the use of secondary process thinking as a means to reduce tension. The reality principle stabilizes the use of postponement and anticipation in the pursuit of pleasure.

Often, what at the onset of adolescence looks like a regressive phenomenon turns out, on closer inspection, to be the result of retarded ego development, or, indeed, of an abortive latency period. Schematically one might say that the psychic achievement of early childhood lies in the mastery of the body, that of the latency period in the mastery of the environment, and that of adolescence in the mastery of the emotions. The completion of one of these tasks and its stabilization can, by and large, be defined in terms of an orderly sequence of ego functions which have to parallel the maturation of the body in order to safeguard normal development.

Stabilizing Mechanisms

The ego is prepared by the achievements of the latency period to cope with increasingly complex conditions, both inner and outer, in differentiated and more economical ways; it is not prepared, however, for the magnitude of the task with which puberty confronts it. The pressures to which the ego is exposed in adolescence change both quantitatively and qualitatively. Not capable of mastering the critical situations with which it is presented, the ego resorts to various stabilizing mechanisms as temporary devices for safeguarding its integrity. Almost exclusive attention has been paid to the defenses against the instincts; and indeed, they do play a major role in the adolescent effort to ward off anxiety which cannot be mastered by integrative processes. However, the anxiety is not necessarily conflictual anxiety; neither is it always due to a repudiation of the sexual drive. Partly, it is the consequence of dammed-up drive energy or of the imperfection of phase-adequate discharge channels. Spiegel (1958), who agrees with some of Bernfeld's (1935) ideas on this subject, comments: "Indeed, some of the symptomatology of adolescence can be viewed as direct sequelae, as actual neurotic symptoms, of the instinctual influx which the still imperfect psychic apparatus is not able to handle at the beginning of adolescence."

Another source of tension is to be found in the parents' retaliatory restrictions meant to deal with the sexual maturation of their offspring. We often witness an exaggerated prohibiting and punitive attitude of parents toward their child as he reaches sexual maturity, gains in physical stature, and acquires a more independent turn of mind. Jones (1913) and Pearson (1958) have shown that the arrival of puberty in the child arouses in the parent fearful and retaliatory reminiscences which hark back to the times when their own smallness foiled their attempts at interfering with the envied and exaggerated privileges of their own parents. The greater sexual permissiveness of the modern parent creates a situation in which the parent accepts the sexual strivings of the adolescent and offers himself as an equal, as a comrade or pal to his pubertal child. This parental attitude aggravates the oedipal involvement through the contrived actuality of an earlier parent-child relationship.

It is obvious that diverse tensions confront the ego of the adolescent and a variety of processes are employed to keep these tensions within manageable limits. What needs emphasis is the fact that the stabilizing mechanisms are not limited to the defenses, in the strict meaning of that term. Hartmann (1939, b) speaks of the two sides of a defense, the unsuccessful pathological one and the adaptive one; Fenichel (1945, b) differentiates between pathogenic and successful defenses, *i.e.*, sublimation; Lampl-de Groot (1957) introduces the distinction between adaptive processes as part of normal development and their distortions, called neurotic defenses; Anna Freud (1936) discusses the normal aspects of defensive reactions as "preliminary stages of defence," — all these authors stress the fact that we are dealing here with a multifunctional concept. To differentiate between the heterogeneous aspects of this concept seems relevant in a discussion of the adolescent ego.

One further distinction must be mentioned which was introduced by Hartmann (1956), who speaks of "defensive maneuvers" and "defense mechanisms." The latter are located in the unconscious; defensive maneuvers operate in the preconscious and are kept from consciousness by censorship. What appears in adolescence as easy access to unconscious determinants in terms of insight is, in fact, more often than not preconscious content. Consequently, the fear that in the treatment of adolescents the weakening of defenses will always mobilize primary processes is in many cases without foundation. The stabilizing mechanisms characteristic for adolescence include defensive, adaptive, restitutive, and compensatory mechanisms. Ego deformation which occurs, for example, in the splitting process takes its place alongside these stabilizing mechanisms; it was described by the author (1954) in a clinical study of prolonged adolescence. Dissociative processes shown in this adolescent syndrome are more often employed to avoid conflict formation than to control conflictual anxiety.

The delineation between the various mechanisms enumerated above is obviously tenuous. This fact, however, should not deter us from an attempt to systematize clinical observations. Several or all of the stabilizing mechanisms can be employed simultaneously; we deal with intertwined psychic mechanisms. In order to single out their specific natures, it is necessary first to establish dynamic characteristics which will permit, by definition, the tracing of discrete unitary processes.

Implicit in the definition of the mechanism of defense is the fact that it is maintained by countercathexis; consequently it results in a permanent deficit in available, i.e., movable, psychic energy.[1] The depletion of available energy due to its being bound in the maintenance of a defense may become critical in adolescence. It must be kept in mind that the employment of each one of the stabilizing mechanisms may reach a point where a pathological state begins to set in. The etiology of specific stabilizing mechanisms takes into account the individual's history, his endowment, and, furthermore, the encouraging or discouraging influence of social institutions such as family, class, caste, school, and church which favor or disfavor certain types of control or mastery.

The stabilizing mechanisms, as stated, should not be thought of as sharply delineated from each other. In fact, one may gradually blend into the other in terms of a slow shift in emphasis until the process has acquired a distinctly new character. This, for example, occurs in the various forms of identification. Besides defensive identification, we also speak of primitive, transient, and adaptive identification. Primitive identification, for example, has annihilated by regression the distinction between object and self; it is, more or less, a merger with the object. In this connection, Geleerd (Solnit, 1959) speaks of a "partial regression to the undifferentiated phase

[1] The question about the source and nature of the energy employed in countercathexis (defense), namely, whether the energy is the same as the one of the warded-off drives, or whether neutralized aggressive energy is to be postulated, cannot be taken up here. Hartmann (1950, a) has discussed this question extensively.

of object relationship"; she claims this is a normal occurrence during adolescence. I am of the opinion it is always of pathognomic significance. Defensive identification and counteridentification leave permanent marks on the character of the ego, while transient identification remains close to experimental psychic maneuvers ranging from fantasy to ego ideal formation and sublimation. Finally, an adaptive identification is to be thought of as a function of the autonomous ego. Transitions and combinations of these various forms of identification are typical of adolescent development. The case of Tom offered a clinical illustration which shows the transition from a defensive (intellectualization) to an adaptive (scholarship) ego function.

Intellectualization was described by Anna Freud (1936) as a typical defense mechanism of adolescence. It represents an attempt to master instinctual dangers by displacement; beyond that, it also has an adaptive aspect which becomes apparent whenever the displacement has acquired the status of an ego interest, i.e., whenever the intellectual activity has become disengaged from instinctual involvement. Hartmann (1939) commented on this by saying that a defense mechanism such as intellectualization has "another reality oriented aspect also, showing that this mechanism of defense against instinctual drives may at the same time be regarded as an adaptive process." In this connection Hartmann makes the point that denial and avoidance as defenses have also an adaptive side — considering the avoidance of the too dangerous and the seeking of the at least possible. Another remark by this same author (1950, b) illuminates the problem further: "Every reactive character formation, originating in defense against the drives, will gradually take over a wealth of other functions in the framework of the ego. Because we know that the results of this development may be rather stable, or even irreversible in most conditions, we may call such functions autonomous, though in a secondary way."

Creativity, especially artistic creation, must be mentioned in this context. Spiegel (1958) has alluded to its special function in ego differentiation and to its stabilizing faculty. "Through artistic creation," he says, "what is self may become object and then externalized, and thus may help to establish a balance of narcissistic and object cathexes." Bernfeld (1924) speaks of the creative products of adolescents as "also-objects." In addition to this function, creativity serves the internal mastery of emotional conflicts. Wolfenstein (1956) has shown the intricacies of this process by her analysis of a poem written by A. E. Housman when he was fifteen years old. The decline in creative activity at the close of adolescence coincides with the emergence of a stable ego organization and the establishment of firm boundaries between self- and object-representations. In the ego organization of the artist this delineation is probably never drawn as sharply as it is in other people.

Adaptive ego functions operate in the nonconflictual sphere of the ego. Consequently "adaptation" is not a value-determined concept, but is defined in terms of an intrasystemic organization. Behavior can be adaptive yet still be in conflict with the environment. The point at issue is whether the behavior is the externalization of an infantile wish, or whether it originates in the nonconflictual sphere of the ego and contains no secondary gain of a sexual nature. In this connection, a mechanism which is akin to the repetition compulsion should be mentioned. Simply by repeating an action, thought, emotion, or affect the adolescent can establish a familiarity with a tolerance of them. This method is especially effective if the dosage of the quantitatively and qualitatively new drive discharge is regulated and kept within tolerable limits. This repetition, then, is another stabilizing mechanism used by the ego in its effort to master instinctual tension.

The restitutive stabilizing mechanism can be observed, for example, in the transient identifications of adolescence proper. As explained earlier, these transient identifications prevent object libido from being totally drained through deflection of the self. The adolescent's need for group belongingness as an expression of social hunger has characteristics of a restitutive process. By gaining access to a full and exciting outer life, the adolescent counteracts his unbearable feelings of emptiness, isolation, and loneliness. Anna Freud (1958, b) has presented a dramatic example of the restitutive mechanism from the study of orphaned children. The children under study "were deprived of the relationship to a stable mother figure in their first years. This lack of a mother fixation, far from making adolescence easier, constitutes a real danger to the whole inner coherence of the personality during that period. In these cases adolescence is preceded frequently by a frantic search for a mother image; the internal possession and cathexis of such an image seems to be essential for the ensuing normal process of detaching libido from it for transfer to new objects, i.e., to sexual partners."

Compensatory mechanisms are a means to maintain the narcissistic balance. Mental or physical shortcomings which are experienced as a narcissistic slight stimulate the — often forced — proliferation of special endowments and thus offset the threatening decline of self-esteem. Observation shows us that the re-establishment of object relations renders the narcissistic balance less precarious and reduces compensatory mechanisms in scope and intensity. The uneven rate of pubertal development which results in the striking maturational differences to be found in the same age group is often modified individually by overplaying or underplaying the respective maturational level. Here again the narcissistic problem lies at the root of the compensatory ego activity. This device for stabilizing the narcissistic balance often initiates an "experimental accident" which brings to the fore latent abilities which may then flourish under the positive acknowledgement of others and self. The transition to an adaptive ego function under such auspices is most favorably promoted.

The vicissitudes of the instinctual drives, in conjunction with the environmental influences on adolescence (such as greater freedom of movement and enforced social responsibilities), both stimulate certain ego functions toward accelerated development, while they stunt and retard other ego functions. In this connection, Hartmann (1950, b) comments: "Influences acting on the ego's development do not always exert a parallel effect on all of its functions in the sense of developing or retarding them. We know that in some cases not only single ego functions but whole sectors of the ego may be retarded." This distinction is particularly relevant for adolescence, when ego development progresses unevenly, at one time advancing the defensive functions, at another the experimental (acting out, imitating, and learning by repetition), at another the adaptive — in brief, giving priority at different times to distinctly disparate ego mechanisms. As Hartmann remarked, "The intellectual or the defensive functions of the ego have prematurely developed, while, for instance, the tolerance for unpleasure is retarded."

Not only is conflictual anxiety responsible for "lopsided" ego development in adolescence, it also underlies the intolerance of tension, and this immaturity of the ego often leads to pathological ego formations. This last consideration is especially pertinent to an understanding of "modern youth," for whom sexual promptings and activities cause little conflictual anxiety. However, taking prolonged refuge in sexual relief through masturbation without advancing to meaningful object relations only perpetuates the state of low tension tolerance characteristic for the immature ego. Sexuality, which in previous generations constituted a source of anxiety based on the more or less conscious, ego-dystonic sexual strivings described as the typical conflictual distress of adolescence, has in recent times become overshadowed — at least for an appreciable section of so-called sophisticated adolescents — by a typical ego-syntonic condition of retarded maturation of certain ego functions. The result is an intrasystemic structural imbalance in the ego. This conflict within the ego leads usually to transitory splitting processes as a means of forestalling a state of disorganization or ego regression.

A typical adolescent trait, the adolescent's proclivity to action, must be mentioned here because it touches on a fundamental antithesis of this period — that between passivity and activity. The fear of passivity in terms of infantile receptivity and submission is equally strong in both sexes. The merger of passivity with aspects of femininity, of course, makes it an anathema to the boy, in whom action and self-assertion often serve as negations of passivity. By projection one experiences an inner threat as if it existed in the outside world; hence the adolescent predilection for "acting out." Closely related to this phenomenon is Anna Freud's (1951) adolescent "negativism" as a defense against emotional surrender and the loss of the sense of identity.

I shall close this discussion of stabilizing mechanisms with a quotation from Freud (1938, b) in which he views this problem in dualistic terms, and precludes any comforts of a simple alternative. "Whatever defensive efforts the ego makes in warding off dangers, whether it is repudiating a position of the external world or whether it seeks to reject an instinctual demand from the internal world, its success is never complete or unqualified; there results always two opposing attitudes, of which the defeated, weaker one, no less than the other, leads to psychological complications."

The Stage of Consolidation

The decisive test of the ego, at least as far as its integrative and synthetic capacity is concerned, comes with late adolescence, the phase of personality consolidation in terms of preferential, fixed ego interests as well as highly personalized love needs. These integrative efforts of the ego are carried over into postadolescence, with the specific aim at activating the inner gains on the environment. Several observers of adolescence have been struck by the fact that the period of emotional reorganization is followed by a period during which ego integrative and adaptive processes absorb a major part of psychic energy. Wittels (1948) spoke of a "second latency"; Braatoy (1934) of an "interregnum"; Erikson (1956), who emphasized the time span required by ego-integrative processes and their activation on the environment, has called this period the adolescent "moratorium."

It was stated earlier than conflicts are only partially resolved at the close of adolescence; but nonetheless, a synthesis is achieved which proves to be highly individualistic and stable. One might say that certain conflictual complexes attain the rank of a leitmotiv by being rendered ego-syntonic. At any rate, it is my contention that this definitive ego synthesis at the close of adolescence incorporates unresolved (traumatic) remnants of early childhood, and that these dynamically active remnants in turn furnish an urgent and determined driving force (repetition compulsion) which becomes apparent in the conduct of life. These ego processes are subjectively felt as an awareness of a purposeful and meaningful existence. The intrasystemic organization of the ego is affected by differentiative and stratificatory processes; that is to say, ego interests become more narrowly defined with the result that a halt is called to the unlimited scope of "possible lives." This organization marks the end of a childhood state which is typical up to and inclusive of adolescence proper.

The consolidation at the close of adolescence is accompanied by repressions, producing a state of amnesia reminiscent of the beginnings of the latency period. However, there is an essential difference between the two: at the end of early childhood, memories are closer to the emotions experienced, and the facts are deeply repressed. In contrast, at the end of adolescence, memories contain precise factual details, but the emotions experienced are repressed. This was illustrated by a male patient of twenty-one who said, "I remember that wonderful feeling from the time I was five or six — I can feel it now — that I could fly up and down the staircase." The factual details of this period were not remembered. He continued, with reference to his memories of adolescence, "I remember clearly my friend and I masturbating and trying to observe the spermatozoa under the microscope. But the only feeling I can remember is the embarrassment at having made a spot on the rug when I ejaculated." It has often been pointed out that the reconstruction of adolescent emotional life deserves greater attention than is usually accorded to it in the analysis of adults.

The above-mentioned ego alterations, which are essential for the attainment of adulthood, considerably tax the synthetic and integrative capacity of the ego; therefore the major psychic mortality of adolescence falls in this phase. Descriptive psychiatry of the past denoted dementia praecox as the psychotic condition which typically has its onset in adolescence. Braatoy (1934) in a study of men between the ages of fifteen and twenty-five commented on the frequency of psychosis (schizophrenia) in males occurring "exactly in those years when the individual should make a start of putting into practice what up till then had been training, dream and school work."

Whenever the ego is victorious in the struggle of this phase, a legitimate narcissistic gratification — pride, self-reliance and self-regard — lends durability and stability to the achievement. The study of the ego at the close of adolescence has increasingly supported the opinion that rectifications and reparative changes can be instituted spontaneously at a developmental phase as advanced as late adolescence. Studies (Beres and Obers, 1950) of children who suffered extreme deprivation in infancy have indicated that the "distortion of psychic structure" which they had experienced was "not immutably fixed." As late as adolescence reparative processes counteracted, at least partially, early deficits, and considerable growth in ego functions took place. We still are in a quandary as to the factors that enable children who were severely damaged emotionally in early childhood to effect, nonetheless, significant reparative changes in their adolescence.

The ego at adolescence has the task of counteracting the disruptive influence of infantile trauma by pathological solutions; this is achieved by the employment of stabilizing mechanisms, and finally by processes of differentiation, stratification, and integration which are the psychological hallmarks of a cohesive personality. Individuality is determined by the specific set of conflictual themes (traumata) which have become permanent, integrated aspects of the ego. Their resolution is destined to remain the task of a lifetime.

References

Beres, D., and Obers, S. J. (1950) "The Effects of Extreme Deprivation in Infancy on Psychic Structure in Adolescence: A Study in Ego Development" *Psychoanalytic Study of the Child*, vol. 5 International Universities Press, New York.

Bernfeld, S. (1924) Vom dichterischen Schaffen der Jugend *Internationaler Psychoanalytischer Verlag*, Vienna.

Bernfeld, S. (1935) "Uber die einfache mannliche Pubertat" *Zeitschrift fur Psychoanalytische Padagogik*, vol. 9.

Blos, P. (1954) "Prolonged Adolescence: The Formulation of a Syndrome and Its Therapeutic Implications" *American Journal of Orthopsychiatry*, vol. 24.

Braatoy, T. (1934) *Manner Zwischen 15 and 25 Jahren.* Fabritius & Sonner, Oslo.

Deutsch, H. (1954) "The Imposter" *Psychoanalytic Quarterly*, vol. 23.

Erikson, E. (1956) "The Problem of Ego Identity" *Journal of the American Psychoanalytic Association,* vol. 4.

Fenichel, O. (1945, b) *The Psychoanalytic Theory of Neurosis.* Norton, New York.

Freud, A. (1936) *The Ego and the Mechanisms of Defense* International Universities Press, New York, 1946.

Freud, A. (1951) "A Connection Between the States of Negativism and of Emotional Surrender" In *International Journal of Psycho-analysis*, vol. 33, 1952.

Freud, A. (1958, b) "Child Observation and Prediction" *Psychoanalytic Study of the Child*, vol. 13. International Universities Press, New York.

Freud, S. (1933) *New Introductory Lectures on Psychoanalysis* Norton, New York, 1933.

Freud, S. (1938, b) *An Outline of Psychoanalysis* Norton, New York, 1949.

Hartmann, H. (1939, a) *Ego Psychology and the Problem of Adaptation* International Universities Press, New York, 1959.

Hartmann, H. (1939, b) "Psychoanalysis and the Concept of Health" *International Journal of Psycho-analysis*, vol. 20.

Hartmann, H. (1950, b) "Psychoanalytic Theory of the Ego" *Psychoanalytic Study of the Child*, vol. 5, International Universities Press, New York.

Hartmann, H. (1956) "Notes on the Reality Principle" *Psychoanalytic Study of the Child*, vol. 11. International Universities Press, New York.

Jones, E. (1913) "The Phantasy of the Reversal of Generations" *Papers on Psychoanalysis.* Williams & Wilkins, Baltimore, 1948.

Kris, E. (1939) "Laughter as an Expressive Process" *Psychoanalytic Explorations in Art* International Universities Press, New York, 1952.

Lampl-de Groot, J. (1957) "On Defense and Development: Normal and Pathological" *Psychoanalytic Study of the Child*, vol. 12 International Universities Press, New York.

Pearson, G. H. J. (1958) *Adolescence and the Conflict of Generations* Norton, New York.

Solnit, A. J. (1959) "The Vicissitudes of Ego Development in Adolescence" Panel Report, *Journal of the American Psychoanalytic Association*, vol. 7.

Spiegel, L. A. (1958) "Comments on the Psychoanalytic Psychology of Adolescence" *Psychoanalytic Study of the Child*, vol. 13. International Universities Press, New York.

Waelder, R. (1936) "The Principle of Multiple Function" *Psychoanalytic Quarterly*, vol. 5.

Wittels, F. (1948) "The Ego of the Adolescent" In *Searchlights on Delinquency*, ed. K. R. Eissler International Universities Press, New York.

Wolfenstein, M. (1956) "Analysis of a Juvenile Poem," *Psychoanalytic Study of the Child*, vol. 11. International Universities Press, New York.

CHAPTER 14

Role Confusion in Adolescence

LEA BARINBAUM

The concept of "role confusion" (as opposed to "identity") in adolescence is one of the most adequate modes of describing what happens to some adolescents who undergo what Erikson calls "delinquent and outright psychotic episodes" (Erikson, 1950, p. 262).

In countries like Israel and the U.S., with large immigrant populations, where different cultures are represented often in the same neighborhoods, where ego-ideals derived from "old country" models are at war with the ideals of the new country, the danger of the onset of "role confusion" is even more immanent than in countries with stable populations. In Israel, as a counselor, I have found two kinds of adolescents who are sometimes especially near the psychotic and/or delinquent fringe. The first are children from oriental countries, brought up in a world of "magical thinking" and fatalistic belief in Fate, and the second, strangely enough, are some kibutz children.

Before actually presenting the two cases which form the core of this paper, let me explain why this is so. The oriental child coming to Israel sees the rewards of higher education — a higher level of living materially (cars, comfortable apartments, foreign travel and, most important, honor and recognition) — without having any idea of how much this cost the owner in "sweat and tears." So when he comes to the school counselor's office, the first thing he says is, "I want to become an engineer." "Engineer" is the magic word used to describe any male adult who is prosperous-looking and is driving his own car. When some of these adolescents find that in order to reach a certain educational level, they have to work hard and persevere, even when things become more and more complicated, they sometimes drop out in one of the ways described by Erikson. They either seek a short cut to prosperity through delinquency or indulge in daydreaming. While they indulge in dreams of grandeur, they suffer from a complete or partial paralysis of will and do not function either as pupils or as apprentices.

The kibutz (communal settlement) child faces a different problem. Every kibutz child receives twelve years of education within the kibutz school system. Classes are small, and methods, in most cases, being very progressive, children's mental abilities can be developed fully. No expense is too great for the children, who are provided with every possible assistance — from teaching aids to special help for backward students and from psychological treatment to high-level science courses. But on leaving school, the pupil has not been prepared for matriculation, which is the gate to higher education in Israel. After three years of Army service (two for girls), the young kibutznik may matriculate and wait his turn until he is sent to a university or technical college, depending on the needs of his kibutz. The idea of identification with a role, as Erikson sees it, or that of "self-actualization" as used by Maslow, is suppressed by the young kibutznik, for he knows that he will work wherever there is a job to be done. "Each according to his ability" is the maxim, and the kibutznik, in turn, receives "each according to his needs." This precludes role identification on the part of a population of adolescents who previously had enjoyed a much more intellectually stimulating education than their city counterparts. (This statement is a nonjudgmental description of the actual circumstances.)

The following cases illustrate the state of role confusion in which two adolescents react similarly to the description given by Erikson:

Understructured Environment

MAZAL: "My problem is not the choice of a profession, as that of others who come to your office," says M., age 16. "I am on my way to becoming a very famous actress. You'll hear about me when the time comes. I happen to be terribly lonely, since most of the students of Actors' College are much older than I. As a matter of fact, I may look older to you, but I am only thirteen, so . . ." When asked about former friends and classmates who might be joined in their activities, M. answers that she is a new immigrant, doesn't know anybody. I asked, "How about the excellent Hebrew you are speaking?" I received an astonished look and a statement to the effect that I should have noticed by then that I was in the presence of a child prodigy to whom everything comes easily, including Hebrew. After further talks in which M. supplied more details of her glamorous life, she is advised to join a certain youth club where there is, along with other courses, a dramatic society. As a word of friendly advice, Mazal is asked not to make too much of her outstanding intellect in front of the other youngsters, since this might embarrass them. After her first visit to the club, M. called me. She was furious. How, for heaven's sake, did I dare send her to such a bunch of liars and cheats, who claim that she isn't a newcomer at all, but has attended kindergarten with one of them, elementary school with another? Told that this was not a matter for a phone call, M. came to the office, armed with a popular journal of parapsychology, and explained to me that there are cases in which two souls inhabit one body at different times. She claimed to be such a case: her one "self" a new immigrant, a child prodigy, etc., the other, an ordinary Sabra, an Israel-born child. (Sabra means cactus — thus something with a prickly surface which is soft inside.) At this point, a consultation with the regional "prevention worker"[1] took place in which the following facts came to light: Mazel, the fourth of eight children, had been born in Israel sixteen years ago. A bright pupil in elementary school, she was recommended by the school counselor to a local high school. Since the demands of high school are much greater than those of elementary school, Mazal's grades, like those of many others, fell. From that time on, Mazal had started to impersonate different characters — the elder sister of a kindergarten child who brought her little sister to the kindergarten and fetched her at noon — until one day she left with money taken from the teacher's pocketbook; another time she posed as a girl soldier, hitchhiking through the countryside, asking her "hosts" for pocket money, since she had lost her purse. Having been discharged by the first secondary school, the prevention worker persuaded another school to accept her while she was on probation in order to "give her another chance," but this school could not keep her either, so she found herself again without a school. She then started to attend all kinds of evening courses and lectures in various community centers. Her favorite subject became parapsychology. Her parents, too simple to be able to find out the truth, were told by her that she had switched from morning to evening classes at the same school. During the last six months she had stuck with the image of the thirteen-year-old child prodigy, destined to become a great actress. She held this image even while being hypnotized at the occasion of one of her participations as a medium of a hypnotist. Mazal's obstinacy in claiming that she was studying to be an actress made her completely unreachable by all counseling personnel. Attempts are being made by the prevention worker to persuade her to report to the mental health center of the Workers' Sick Fund, but Mazal sees no reason to comply with this wish.

Overstructured Environment

ADY (a name used by both sexes): "I am a blind boy, age eighteen. When disaster struck when I was eight years old, I found that I was living in a grave. Without sight I felt like a walking corpse. I shall never be able to say to a girl, 'How lovely you look today,' for I shall not know how she looks. I feel an urge to kill and destroy all living things around me since I can't see

[1] A specially trained social worker devoted to the prevention of delinquency.

them. I can't see my own face, but I'm sure it is the face of a murderer." This "reader's letter" was published in one of the daily papers, ending with a plea for help and advice as to what he could do with his wretched life and signed with his full name, Ady X., Kibutz Y in the Jordan Valley — a place under constant fire at the time this is being written — where all children sleep in bunkers every night.

Although I have some experience with blind children, I had never met such a case of depression and desperation before. The assumption of "looking like a murderer" seemed especially terrible and completely unlike the spirit of other blind people I had met. After some hesitation I wrote a letter to Ady in which I simply invited him to Haifa, where we have a club for the blind, attended by seeing young people as well, but equipped with a Braille library and special chess sets, etc. With the returning post a letter arrived, signed by Ady. "Since I wrote that letter, I've got lots of answers from all parts of the country, even from Eilat [the most southern town]. I wrote the letter as a kind of an exercise. I am neither blind nor a boy, I just tried to "creep inside" a blind boy's skin and feel how it looks like. I'm very happy to learn how much compassion and sincere feelings people have for the blind. In the nearest future I shall be going to the U.S. as an exchange student to spend six weeks with a family whose daughter has been our guest here at the kibutz, and then I shall join the army." Ady didn't display any awareness of having led people astray. On the contrary, she was pleased by all the attention she received. When asked whether she had told all those who had answered what she had disclosed to me, she said, "Only those who came up with concrete proposals, like you, had to be told, before they actually came to see me! Your letter showed me that you meant business. All the others must not necessarily be told the truth; I'm letting them believe that a friend is writing for me."

Discussion: Im both cases we have an adolescent girl playing a role (or roles in the case of Mazal). In both we may speak of "role confusion" in the Eriksonian sense. Mazal, a girl from a lower class oriental home, finds herself in a community where a girl can actually have a career — unlike the generation of her own mother. Being a bright student in elementary school, she succeeds in entering secondary school. But at the first signs of failure she is unable to face facts and starts what Erikson calls a "psychotic episode" coupled with delinquency. When all rational sources of education fail, parapsychology is her obvious choice. Her interest in this subject probably has deeper sources, belief in spirits and ghosts being inherent in the oriental folklore in which she was brought up. At the same time there is the matter of a career. Since higher education is open only to those who matriculate and Mazal is evidently not able to do so, she resorts to inventing a career already started. At the same time she is troubled by her social isolation; hence her emphasis on "not being in need of professional guidance," and being aware only of the social problem. Mazal insists that she is three years younger than her actual age. Her regression to age thirteen is symbolic, for at that age she was still successful and hoping for further academic advancement. In Mazal's case, "role confusion" is due to social as well as psychological factors — the social role of oriental women, including M.'s mother, is an inferior one. Mazal is the first girl to be educated with a profession as the objective, at least in her family. Psychologically, she finds herself unable to cope with the requirements of this role and, being unable to return to the old feminine identity, namely the traditional role of the obedient daughter expecting to be married off by her father, she resorts to daydreaming, role-playing, and the acting out of her subconscious wish for power — in this case, money, glamour, and possessions.

Ady, in contrast to Mazal, has grown up in an atmosphere of social security and educational fulfillment (to a certain extent). Her role has been mapped out for her by the life style fostered in the kibutz: twelve years of education, two years of army service, followed by re-entry into the kibutz as a full-fledged member. But on the other hand, she has learned all about physical insecurity: soldiers being killed near the border, houses hit, civilians and kibutz members killed or wounded by land mines or artillery fire from across the border. Her role confusion may stem from her not having to search for an identity. The idea of blindness may have occurred to her in connection with her "blindly following" the path mapped out for her (the expression exists in Hebrew, too), paired with secret fears of actually being hit. The "living grave" she described could be the bunker in which she is obliged to spend her nights. Ady claims that she (he, in her letter) can't see her own face — does she feel a lack of identity, a facelessness among the masses? But she speaks about "a murderer's face," an aggressive image which conjures up an association of the almost complete suppression of any signs of aggression within the children's house in the kibutz. The most obvious sign of role confusion in her letter is the plea for guidance and help. Here she departs not only from her personal "self" but also from the communal "kibutz self" which includes help of all members for all members, especially in an emergency. In her letter to me in which she discloses the truth, she asks me to read her letter to a blind boy in order to find out his reactions. This lack of empathy stands in complete contrast to her self-professed wish to enter "into the shoes" of a blind boy and "see the world through his eyes." Ady shows the following signs of role confusion, at least in her letters. Not knowing her own identity, the wish to be (or fear of being) somebody else — a boy, not a girl, a blind boy who attracts a lot of attention and compassion, someone who doesn't know what to do with "his" life. . . .

Reference

Erikson, Erik H. *Childhood and Society*, New York: Norton, 1950.

CHAPTER 15

Adolescence as a Developmental Disturbance

ANNA FREUD

The Psychoanalytic View of Mental Health and Illness

Our psychoanalytic investigations of individuals have convinced us that the line of demarcation between mental health and illness cannot be drawn as sharply as had been thought before. Especially so far as the neuroses are concerned, neurotic nuclei are found in the minds of normal people as regularly as large areas of normal functioning are part of the makeup of every neurotic. Also, people cross and recross the border between mental health and illness many times during their lives.

There is the further point that the concept of health as it is derived from the physical field cannot be taken over to the mental side without alteration. Physically, we are healthy so long as the various organs of the body function normally and, via their specific action, contribute to an over-all state of well-being. Mentally, more than this is needed. It is not enough if each part of the mind, as such, is intact, since the various parts of our personality pursue different aims and since these aims are only too often at cross purposes with each other. Thus, we may be healthy so far as our instinctual drives are concerned; or our sense of reality plus adaptation to the environment may be well up to the mark; or our ideals may be considered admirable by other people. Nevertheless, these single items do not yet add up to the result of mental health. To achieve this, all the agencies in our mind — drives, reasonable ego, and ideals — have to coincide sensibly and, while adapting to the external world, resolve the conflicts inherent in the total situation. To say it in other words: mental health depends on workable compromises and on the resulting balance of forces between the different internal agencies and different external and internal demands.

The Concept of Developmental Disturbances

It is implied in the view above that this balance and these compromises are precarious and easily upset by any alteration in the internal or external circumstances. It is obvious also that such changes are as inevitable as they are continuous and that they occur especially frequently on the basis of development. Every step forward in growth and maturation brings with it not only new gains but also new problems. To the psychoanalyst this means that change in any part of mental life upsets the balance as it had been established earlier and that new compromises have to be devised. Such change may affect the instinctual drives, as happens in adolescence; or it may occur in the ego, that is, in the agency whose function it is to manage and control the drives; or what undergoes change may be the individual's demands on himself, his aims and ideals or his love objects in the external world or other influences in his environment. Changes may be quantitative or qualitative. Whatever they are, they affect the internal equilibrium.

Developmental disturbances of this type are frequent occurrences, for example, in the area of sleep and food intake in early childhood. Infants may be perfect sleepers in the first half-year of life, that is, drop off to sleep whenever they are tired and when no stimuli from inside or outside their bodies are strong enough to disrupt their peace. This will alter with normal further growth when the child's clinging to the people and happenings in his environment make it difficult for him to withdraw into himself and when falling asleep thereby is turned into a conflictful process. Likewise, the disturbing food fads of childhood are no more than the impact on eating of various infantile fantasies, of dirt, of impregnation through the mouth, of poisoning, of killing. These fantasies are tied to specific developmental phases and are transitory accordingly, as are the feeding disorders based on them. In fact, in clinical practice with children, the concept of transitory developmental disturbances has become indispensable to us as a diagnostic category.

It is worth noting here that developmental change not only causes upset but can also effect what is called spontaneous cures. A case in point here is the temper tantrums that serve young children as affective-motor outlets at a time when no other discharge is available to them. This is altered by the mere fact of speech development that opens up new pathways and by which the earlier turbulent and chaotic behavioral manifestation is rendered redundant.

The Adolescent Reactions as Prototypes of Developmental Disturbances

Let us return to the problems of adolescence that, in my view, are the prototypes of such developmental upsets.

Although in the childhood disorders of this nature we are confronted usually with alterations in one or the other area of the child's personality, in adolescence we deal with changes along the whole line. There are, as a basis on the physical side, the changes in size, strength, and appearance. There are the endocrinological changes that aim at a complete revolution in sexual life. There are changes in the aggressive expressions, advances in intellectual performance, reorientations as to object attachments and to social relations. In short, the upheavals in character and personality are often so sweeping that the picture of the former child becomes wholly submerged in the newly emerging image of the adolescent.

Alterations in the Drives

So far as the sexual drive in adolescence is concerned, I have found it useful to differentiate between quantitative and qualitative changes. What we see first, in the period of preadolescence, is an indiscriminate increase in drive activity that affects all the facets which have characterized infantile sexuality, that is, the pregenital, sexual-aggressive responses of the first five years of life. At this juncture, the preadolescent individual becomes, as a first step, hungrier, greedier, more cruel, more dirty, more inquisitive, more boastful, more egocentric, more inconsiderate than he has been before. This exacerbation of the pregenital elements is followed then, shortly after, by a change in the quality of the drive, namely the changeover from pregenital to genital sexual impulses. This new element involves the adolescent in dangers which did not exist before and with which he is not accustomed to deal. Since, at this stage, he lives and functions still as a member of his family unit, he runs the risk of allowing the new genital urges to connect with his old love objects, that is, with his parents, brothers, or sisters.

Alterations in the Ego Organization

It is these temptations of giving way, first to sexual-aggressive pregenital behavior and, next, to incestuous fantasies or even actions that cause all those ego changes which impress the observer as the adolescent's personal upheaval and also as his unpredictability. Serious attempts are made by the preadolescent to keep the quantitative drive increase under control as drive activity has been controlled in earlier periods. This is done by means of major efforts on the side of the defenses. It means bringing into play more repressions, more reaction formations, more identifications and projections, sometimes also more determined attempts at intellectualizations and sublimations. It means also that the entire defensive system of the ego is overstrained and breaks down repeatedly and that therefore the frantic warding off of impulses alternates with unrestrained upsurges of drive activity. When we approach a young adolescent at this stage, we never know which of these two aspects we are going to meet: his overstrict, highly defended personality or his openly aggressive, openly sexual, uninhibited primitive self.

Alterations in Object Relations

What serves the preadolescent as some protection against the quantitative pressure of the drives proves wholly inadequate against the qualitative change to a primacy of the genital urges, that is, adult sexuality proper. Nothing helps here except a complete discarding of the people who were the important love objects of the child, that is, the parents. This battle against the parents is fought out in a variety of ways: by openly displayed indifference toward them — by denying that they are important — by disparagement of them since it is easier to do without them if they are denounced as stupid, useless, ineffective; by open insolence and revolt against their persons and the beliefs and conventions for which they stand. That these reactions alternate also with returns to helplessness and dependence on the part of the young persons does not make it any easier for the parents. Obviously, the task imposed on them is a double one: to be thick-skinned, self-effacing, and reserved, but also to change over at a moment's notice to being as sympathetic, concerned, alert, and helpful as in former times.

The closer the tie between child and parent has been before, the more bitter and violent will be the struggle for independence from them in adolescence.

Alterations in Ideals and Social Relations

The adolescent's change in social relationships follows as the direct consequence of his stepping out of his family. He is not only left without his earlier object ties. Together with the attachment to his parents, he has thrown out also the ideals that he shared with them formerly, and he needs to find substitutes for both.

There is a parting of the ways here which, I believe, produces two different types of adolescent culture. Some adolescents put into the empty place of the parents a self-chosen leader who himself is a member of the parent generation. This person may be a university teacher, a poet, a philsopher, a politician. Whoever he may be, he is considered infallible, Godlike, and is followed gladly and blindly. At present, though, this solution is comparatively rare. More frequent is the second course where the peer group as such or a member of it is exalted to the role of leadership and becomes the unquestioned arbiter in all matters of moral and aesthetic value.

The hallmark of the new ideals as well as of the new emotionally important people is always the same: that they should be as didfferent as possible from the former ones. In the remote past, when I myself was adolescent, there had come into being in central Europe the so-called Youth Movement, a first attempt at an independent adolescent culture. This was directed against the bourgeios complacency and capitalistic outlook of the parent generation of the period, and the ideals upheld by it were those of socialism, intellectual freedom, aestheticism, and so on. Poetry, art, and classical music were what parents did not believe in, although adolescents did. We know how far the tide has turned in the last two generations. At present, adolescents are hard put to set up new ideals — constructive or disastrous — which can serve to mark the dividing line between their own and their parents' lives.

Concluding Remarks

To the abbreviated summary of the main theme given above, I add a few concluding remarks that concern more general issues.

First, it has struck me always as unfortunate that the period of adolescent upheaval and inner rearrangement of forces coincides with such major demands on the individual as those for academic achievements in school and college, for a choice of career, for increased social and financial responsibility in general. Many failures, often with tragic consequences in these respects, are due not to the individual's incapacity as such but merely to the fact that such demands are made on him at a time of life when all his energies are engaged otherwise, namely, in trying to solve the major problems created for him by normal sexual growth and development.

Second, I feel that the obvious preponderance of sexual problems in adolescence is in danger of obscuring the concomitant role of aggression that, possibly, might be of great significance. It is worth noting that countries which are engaged in a struggle for existence, such as, for example, Israel, do not report the same difficulties with their adolescents as we do in the Western world. The main difference in their situation is that the aggression of the young people is not lived out within the family or community but directed against the enemy forces that threaten the state and therefore usefully employed in socially approved warlike activities. Since this is a factor outside the sphere of sexual growth, this should extend our thinking into new directions.

Third and last, it seems to me an error not to consider the details of the adolescent revolt in the light of side issues, disturbing as they may be. If we wish to maintain the developmental point of view, it is of less significance how the adolescent behaves at home, in school, at the university, or in the community at large. What is of major importance is to know which type of adolescent upheaval is more apt than others to lead to a satisfactory form of adult life.

B. INTERPERSONAL RELATIONS

CHAPTER 16

The Adolescent Today

IRENE M. JOSSELYN

The Adolescent's Confusion

Adolescence is a psychological step in development that has always had a dual impact on any society that is not rigidly and statically structured. On one hand, adults have been frightened by the confusion rooted in the adolescent's internal struggles that typically find expression in behavior that, to a greater or lesser extent, violates the mores that have been established by the previous generation. On the other hand, as the adolescent becomes adult, his confusion resolves itself into a pattern of behavior and into attitudes that — as the young person gradually assumes the responsibilities of adulthood and participates in the social and political life of his environment — become expressed in a modification of the culture.In a culture that is still in a fluid state political, social, and ethical philosophies undergo change; the change is either instigated by, or endorsed by the younger adults who believe that the change is desirable. From this develops a new, dominant point of view which then becomes the groundwork upon which it is hoped a new and better society will be built. Many of these shifts are in essence the result of the resolution of the confusion of the adolescents of the past who have now become leaders of the present.

In a nation whose political and social philosophy is dictated by a few it is apparently only through revolution that social trends can be shifted to new goals. In contrast, the basic tenet of democracy is that cultural, political, and social shifts occur through evolution, not through revolution. Since, in a democracy, the change is the result of the dominant group rather than of a rebellious few who bind others to the philosophy, young adults invest themselves in the change and therefore make a dynamic adjustment to it and tend to fulfill it, on the whole constructively, in spite of manifest utilization by a few of the new point of view for destructive purposes. The latter then become a minority, and the trend of the majority is to counteract the negative effects of that minority.

There is another aspect to this evolutionary process in which the adolescent's experiences play a significant role. The adolescent is, due to his confusion, looking for a leader who will offer some solution to the pain of his conflicts. Thus he readily responds to those who are cognizant of his fluid state and who will capitalize upon it for good or bad. It is because cultural evolution in a democracy is feasible, because the resolution of adolescent conflict fosters leadership in that evolution, and because the adolescent himself is seeking an answer to his problems through leadership of others that work with the adolescent is of such major importance. It is also why leaders with goals that appear to be dangerous to our culture may arouse in adolescents enthusiasm and emotionally charged dedication to a "Cause," unless the work of those leaders can be counteracted by the effective leadership of those adults who would preserve the good of our culture and who are willing to consider modifications that might militate against the bad.

Consider, for example, the burning of draft cards. The young man of draft age finds not only many of his peer group but also many adults opposed to the war in Vietnam. If he himself is ambivalent or resistant to becoming a soldier, he finds support in this opposition. If he burns his draft card, he is confronted by others with two contradictory points of view. The majority condemn him, reject him and/or punish him. But to another group he is a brave hero, a martyr, a person who will risk all for an ideal. Is it surprising that he turns to the latter group for homage? Admiration is important to all of us, but particularly important to a frightened, insecure adolescent. The support of others, and particularly those who formalize and eulogize a social philosophy that is alien to the majority of the parent generation of the culture, gives the particular adolescent a "Cause," the implications of which appear to him to go far beyond his own self-interest, representing, as he comprehends it, the interest of a downtrodden part of the social structure.

Those who criticize the adolescent for burning his draft card failed him prior to his incendiary act. They did not offer him the meaningful leadership that would have helped him conceptualize himself as a person who could not be a draft-dodger. It is a tantalizing, but at present unanswerable question, as to whether in the American culture of tomorrow it will be the right of every individual to decide whether a battle that is being fought is valid or not, participation in it to be determined by each individual's own judgment.

The Adolescent's Conflicts

I will not review in detail in this paper the typical conflicts of the adolescent who is reared in a culture that allows him freedom of development (Josselyn, 1948, pp. 93-118). In summary, however, I would state that the characteristic of the growth period to which we refer as adolescence is in itself a revolution resolved by an evolution. In the typical adolescent it is possible to see every stage of psychosocial development of childhood repeated in adolescence. The defenses and adaptive patterns that gradually evolved to handle infantile conflicts during childhood, to a greater or lesser extent, disintegrate during adolescence. This partial disintegration is in itself important because it frees the adolescent to mature with patterns of adaptation and defenses that are compatible with adulthood instead of those effective in childhood.

However, when the adolescent of today struggled as an infant and young child to find solutions to his conflicts, he did so in a somewhat orderly fashion. While typically he did not completely solve each level of development before he undertook the next, he did attain a partial solution of one level of development before he dealt with the next one; therefore, his major effort was devoted to the new task, with the previous task making secondary demands on his psychological energy. The adolescent is confronted with all conflicts of early childhood and struggles to answer all at the same time. For example, he must deal with the intensification of sexual impulses, and at the same time he must free himself from his childhood type of dependency upon others, particularly upon his parents. This confusion is frequently manifested in his choice of love objects among his peer group. He makes what is often referred to as a "narcissistic object choice." This choice may serve multiple functions; it may provide a re-experiencing of the infantile symbiotic relationship in which the object and the self are not truly differentiated. He loves his friend because his friend is one with him; his friend thinks the same, desires the same results from activities, has the same interests, has the same attitudes. In spite of protests from his parents and others, he wears his hair the same way, wears the same style of clothes, walks in a similar fashion, and talks like his friend. At the same time this object choice, or the multiple object choices that he makes, serve to help him structure his own adult ego ideal and superego. In discussions and mutual participation in activities with another peer, or a group of peers, the adolescent gradually conceptualizes what he wants to be, how he wants to attain that self-image, and how he wants to function in that role. He and his companions gradually formulate a hypothetically ideal person whom they try to emulate.

The World of the Adolescent

The adolescent also uses his peer group to create a world separate from a childhood world. Inhabitants of that world have their own language, they have their own customs, and they have their own attitudes toward the adults of their childhood. His conformity to the peer group is not necessarily idiosyncratic to the individual adolescent but rather may be evidence of his adaptation to a new world separate from his past. The populace of the world of his adolescent peers is more or less effectively united against a common enemy, the adults of the reality world. He confides in his single or multiple object choices, thus seeking the reassurance which in childhood he attained by dependence on his parents.

Another aspect of the narcissistic object choice is a striving to find an adult who represents an ideal or a guide to adulthood. The adolescent, as a result, often chooses as an object choice a person in the culture who is himself in rebellion against the culture, but who is — at least chronologically — adult. This person, because he is older and has placed his confusion into what looks like an orderly form, appeals to the adolescent; he seems to have found an answer to the maze in which the adolescent is struggling.

At the same time both his culture and inherent patterns of maturation stimulate the adolescent to find what appears on the surface to be a heterosexual object choice. He becomes interested in the opposite sex, tests out his sexual competence of which he is unsure, but also uses that object as a parent figure while choosing one that is as disimilar as possible to the parent figure.

As far as a sexual object choice is concered, it is important to bear in mind that with the breakdown of childhood defenses the adolescent is confronted again with the multiple aspects of the early oedipal conflict. There are many reasons why this is untenable to him. Whatever the reasons, he must deny the oedipal wishes and seek a love object of the opposite sex who, for the boy for example, is either like his mother (but recognized as not being his mother) or as unlike her as possible in order to deny the unconscious gratification of the oedipal desires. He also seeks to find some outlet for his sexual longings through love for a masculine figure. His primary love for a person of his own sex is his father; this, again for many reasons, is transiently untenable, and he seeks a substitute either in an older man or someone in his own peer group, with his sexual impulses expressed in pseudo-homosexuality.

At this point of discussion of the object choice of the adolescent (which in fact covers only a part of the subject), I have probably sufficiently confused you so that you realize that the adolescent himself is more confused than you are. Furthermore, the multiple facets of the object choice in adolescence constitute only one of the problems with which he struggles. All of his problems are encompassed by Erikson's concept that the adolescent is seeking his self-identity as an adult (Erikson, 1956, pp. 56-121).

Freedom and Limits

Although adults may seem to be the enemy — even to the courageous and confident adolescent — they are highly valued. Parents especially are of great importance to the adolescent. For example, while the adolescent seeks freedom, he is fearful of it, often with justification because of his inexperience. He therefore wishes an external force to place limits upon his expression of his freedom. He protests but values the parent or other adults who impose reasonable limits. When these reasonable limits are imposed, the adolescent feels more confident in exploring the areas within those limits; he is much less frightened than he is when adults, in their eagerness not to mold the adolescent, refrain from imposing limits.

The adolescent's choice of a person to impose limits is determined in part by his childhood experiences; often he turns to his parents. To accept their limits, however, is to feel like a child again, so he must protest. His protest is not serious if it is against the backdrop of a valued relationship in childhood and if the parents are willing to let him, in safe areas, experiment with his strong impulse to be independent. The valued relationship of the past remains intact, superficial responses to the contrary notwithstanding.

However, we see many adolescents who will not accept, for one reason or another, the limits imposed by parents. It is extremely unrealistic, for example, to place the burden of guilt solely on parents for the disturbed behavior of the adolescent who refuses to go to school or who is delinquent. Such behavior may be due to parental neglect or failure to establish a satisfactory child-parent relationship in the earlier years. As one parent said to me, "Should I put bars on his bedroom windows? I tell him that he can't go out on a certain night; I recently discovered that after we retire he goes out his bedroom window. If I tell him he can't have the car, he rides in his friend's car; if I tell him he cannot do that, he walks a block away — his friend picks him up there. While it is my lack of control that undoubtedly is the root of his delinquency, no one has been able to tell me how to get control unless I put him in chains."

It is this group of adolescents who are referred to us. Be they delinquent or seriously emotionally disturbed, they have, for whatever reason, been unable to accept parental limits and/or have been unable to find, through their parents, answer to their psychological problems. What answer do we have for them? Before that question can be answered, a broader one must be explored.

Morality and the Freudian Approach

There has been a great deal of discussion lately as to the effect of Freudian concepts upon our culture. Many commentators have stressed that present theories of therapy for emotionally disturbed individuals have no moral or ethical goals. I question this assumption. Freud originally indicated that the repressed material of childhood caused psychological ills; and he believed that bringing the repressed to consciousness would cure these ills. The next broad step in the evolvement of his concepts of psychological development was the recognition of the role of the ego as mediator between id impulses, reality demands, and the superego. The ego concept thus not only described why impulses in their primitive form were repressed or expressed in a distorted form, but also why man lived by certain defined standards of behavior in the framework of a reality world. The disturbed individual was seen as a victim of the struggle between id impulses, the superego and/or reality — a struggle the ego moderated by mechanisms that crippled the individual rather than offering adaptation to and modification of the forces caught in the maelstrom.

The challenge of this conceptualization to the therapist was to investigate the problems of the patient to determine whether it was possible for the patient to find modes of discharge for internal impulses through the resources of and within the demands of the external reality, and whether the superego would modify so as to foster positive expression of those impulses and cooperate with the ego in forbidding destructive expression. While Freud at times bemoaned the fate of man who could not remain a primitive, impulsive psychic organism, he accepted the fact that, in man, the struggle between the three aspects of the psychic gestalt was inherent and that some non-crippling answer should be sought. Theoretically, the answer lies in freeing the patient so that those aspects of his early life that crippled him or prevented maturation could be re-evaluated, distorted developmental patterns corrected, and maturation fostered.

This is not a goal devoid of ethical or moral criteria. A theoretically "totally" mature individual would be an individual who finds expression of internal drives in a form that provides optimum fulfillment of them in the world in which he lives. Therapists following Freud, who accepted the concept of the role of the ego, have developed a very strong morality even though they may not express it in moralistic terms. This morality is based upon the assumption that an individual should be an adult and should be freed to become an adult. This freedom is the moral and ethical goal of therapy.

There is a further implied assumption; the human being does not find fulfillment except in a transactional relationship with others. He is not basically an isolate and does not get optimum gratification except in relatedness to others. Relatedness to others is not transient, or solely on a one-to-one basis. The maturation of the inherent need for relationships with others is expressed in the urge to be an inter-relating part of larger social groups. Cultures have not been inflicted upon people; they are the varied attempts of individuals functioning as a part of a group to fulfill inherent needs. The degree of maturity of the dominant individuals in the group will determine the efficacy of the structure in fulfilling the social needs of the individual.

An ethical and moral goal of therapy is to help the individual break through the crippling defenses that make this true relatedness to others difficult, absent or distorted. The question for the therapist is, "What is preventing the maturation of the personailty?" This certainly does not mean that the therapist who is a Democrat should try to convince his patient who is a Republican of the errors of his political conviction. The therapist's responsibility is to free the individual of neurotic aspects of his political thought so that he will be a socially constructive, self-expressed Republican.

The significance of this morality of therapy is particularly important during adolescence. The adolescent, in an attempt to find an answer to his internal problems, may in some cases be easily convinced to mold himself in the image of his therapist, or he may be equally determined to be the opposite, in order to feel that he is a separate person. Our goal in therapy is not to mold the adolescent to our ideas or to provoke his rebellion against our domination, but rather to help him to struggle with his problems so he can become an integrated and mature individual. Thus, we must be tolerant of many points of view, but our aim must be to assist the individual so that *his* particular point of view is ultimately integrated into a mature functioning adult.

Cultural Values

In many ways the adolescent of today is the adolescent of yesterday and, we hope, the adolescent of tomorrow. He faces many difficulties which in their specific form are characteristic of this era. Many of today's adolescents are the children of parents who in their own childhood experienced directly, or through the attitude of their parents, the insecurity imposed by the depression. That insecurity undoubtedly led many to believe that security lies only in attaining money; it was the lack of money that undermined their parents' security when they were children. During the life of the parents of the present generation of adolescents, there has never been real peace. War, increasingly terrifying in nature, has always been possible the next day; the parents and the adolescent do not know what tomorrow will bring.

Adolescents are living in an era in which empathy for those who suffer because of lack of basic needs of life has led to a program in which it sometimes appears that those who are competent must take care of the incompetent, even sacrificing their own rewards from their labors.

Adolescents now live in a culture in which much of the productivity is in the hands of impersonal corporations and, at least in the business world, only a few can aspire to be at the top. The rest must be satisfied at lower echelons, always somewhat subservient to the man above.

Because we are living in this period, we tend to feel that it is both better and more painful than the past. As life has become better to live, it is assumed that it is more painful to think of dying on the battlefield. At one time the individual was directly at the mercy of droughts, floods, hailstorm, or uncontrollable disease either in himself or in the food he grew. Now that food has become more available and its supply more predictable, starvation, lack of the basic necessities, disease, and the inability to have the frills that make life exciting *all* seem threatening. I doubt, however, that the soldier of the past anticipated dying any more than the soldier of today. If he did, he wouldn't have been a good soldier. I doubt that a farmer of the past, seeing his family on the borderline of starvation and himself feeling the pangs of hunger, suffered any less than the laborer of today who is out of work. Human problems take different forms, but basically they have an awesome similarity throughout history.

However, I believe a change has gradually evolved in our culture that makes the task of the adolescent striving for maturation more difficult than perhaps it was in the past. In the past the average adolescent was reared in a culture whose philosophy transcended the practicalities of living and provided a way to fit the deficiencies of the day into an overall plan. If the adolescent of today has any problem different from those of preceding generations, it is in terms of the philosophy of the culture that gives little voice or audience to the basic idealism and sensitivity that has always been characteristic of adolescents.

It is important to bear in mind that when a therapist discusses any problem or any personality group, he does so primarily through a distorted glass. We see, in some depth, the disturbed person. Our social contacts provide us experiences with less confused people, but we rarely know them as deeply as we know the disturbed person with whom we are working. Thus, the remarks I will make are related to what in essence is a caricature of the impact of our present culture upon the adolescent. My observation of adolescents who are not patients to a certain extent militates against the following generalizations. This makes me feel hopeful for the future, because I believe there is a good chance the adolescent of today will, in his adulthood, create a different world and a different philosophy of life from that which we have created.

The disturbed adolescent whom I see in my office today frequently presents a picture different from the majority of my adolescent patients in years past. There is now a much larger number whose attitudes toward life are crassly egocentric, with their basic philosophy being that one of the inalienable rights of the individual is license rather than freedom.

Philip Rieff has indicated that our culture has aims that contrast with the cultures of the past (Rieff, 1966, pp. 1-27). The present philosophy is that the worst possible sin is boredom and individual gratifications of the moment the highest goal, in contrast to the philosophy of life that governed people in the past when Judean-Christian concepts determined ideals.

Attitudes Toward Education

Consider the average parent's attitude toward the reason for becoming educated. The adolescent of today seems to be encouraged to have as his goal the acquisition of money — achieved with minimal effort — in order to have self-centered pleasure. I have done an informal study of parents who are struggling with their child's poor school work; I have asked them why it concerns them. Almost always the answer is, "If he doesn't get good grades in high school, he won't get into a good college, and you can't get a good job unless you have a college education. I keep telling him that." When I ask for a definition of a good job, it is one in which the individual will advance to increasingly higher income. It is a rare parent who reports what a parent said to me the other day, "What bothers me is that John is not experiencing the joy of learning and of mastery so that he has the personal satisfaction of doing something well. Also, he will be deprived of the richness of life that education could give him." Perhaps this parent, a scholar himself, could be so atypical because inherited wealth minimized his need to earn a livelihood.

The adolescent does not have many opportunities in our culture to observe people who enjoy doing a job well. He does not see the garage mechanic who really cares about having the car that has been entrusted to him for repairs become miraculously functional again as a result of that mechanic's unique investment in fulfilling a responsibility. Whatever pleasure his lawyer father may have had in successfully trying a case, the enjoyment is expressed too often in the father's excessive drinking to celebrate the event and/or in discussion of the fees to be paid by the client. A surgeon who saves a woman's life is more likely to complain to his family about the hours of sleep he missed to take care of a patient who would never pay than to reveal the excitement of having fulfilled his function as a saver of life. Even those parents who find their own careers the source of inner fulfillment, as well as financial remuneration, stress to their offspring that they must go to college in order to get into a good graduate school and to be assured of *financial* success in the future. They shyly hide the non-financial remunerations that success in their work has provided.

Because education to the average child and adolescent of today is presented as such a sterile experience, with only the ephemeral future goal of making money (which they may not achieve), it is not surprising to me that those who find educational standards hard to meet wish to drop out. Nor is it surprising that the bright individual finds the mechanics of school work dull and wishes to seek fulfillment outside of the drudgery of school. Both groups find the world outside of the classroom as promising and much more exciting, as of the moment, than their routine studies. Few people work to attain a final goal unless in the process of attaining that goal there are intermediate gratifications. If the final goal is to have enough money so that one can have the luxuries of life in the future — and be able to complain of taxes that support poor people and wars — with no real pleasure in the process of attaining that goal because there is no financial reward present, it does not seem surprising that education and the demands of the classroom become the source of boredom and something from which young people wish to escape.

This attitude toward education is particularly tragic if one considers the small child's response to learing. The small child is driven by curiosity to find out about his world, and each new segment that he discovers and understands is a source of pleasure. The small child strives to master any new situation that offers itself if it is within the scope of his capacity. Mastery of such a segment is a source of gratification to him. For example, the infant has a truly exciting experience, judging by his response, when he takes his first steps. He doesn't feel bored with achieving that mastery. His goal of the moment is not to become a World Sprint Champion; it is to take a few more steps. Our contemporary educational system for the most part seems inadvertently to minimize and ultimately cause to atrophy those very inherent characteristics that make for a richer life. When the individual reaches adolescence, he is more sensitive, more responsive and more eager than is typical of the pre-adolescent of any preceding generation, but at the same time, during his past, he has failed to find in education the source of gratification of intermediate goals, except the intermediate goal of passing from one grade to another.

Attitudes Toward Individual Responsibility

Interwoven with this sterility of educational goals is, in my opinion, an atrophy of the inherent gratification in depth relationships. One of the common characteristics of adolescents, no matter what the culture may be, is egocentricity. This is not surprising in view of the multiple problems they are trying to solve. They don't have time to think to much of others, except those who are serving an immediate purpose. But the egocentricity of the disturbed adolescent in our present culture has perhaps more malignant implications. The implications are expressed in a guilt-free, ego-syntonic conviction that they have rights but no obligations. This has always been a common complaint against adolescents of all generations, but I feel that in certain cases we are seeing a more malignant form. For example, there have always been adolescents who have said, "I don't owe anything to my parents, they brought me into the world without asking me, and therefore they have to take care of me and tolerate my peculiarities."

Now there seem to be more adolescents who find this attitude completely ego-syntonic. Adolescents often say to me, "My parents should give me a car; my parents should give me a larger allowance; they owe it to me." When I point out that maybe they can't afford to do this, the answer is, "That is their problem." Recently an adolescent was referred to me who was making a good adjustment outside of the home but was making her mother quite miserable by her refusal to follow any rules of the household and by her attack on her mother for any unwelcome comment. Actually her mother was critically ill, both physically and psychologically. There was no question that she was a difficult person with whom to live. Her illness, however, was being seriously intensified by the difficulties with her daughter. I empathized with the girl, acknowledging that it was difficult for her at home, but that there was the reality situation of ther mother's illness about which many people were justifiably concerned. Her answer was, "That's her problem, not mine." One could put this down as a character disorder were it not that it is not as atypical as one would wish among those adolescents who come to our attention.

If there is an increase of malignant egocentricity in the adolescent, its roots may be in the child-rearing philosophy of the recent past. We went through a period in which we were proud that ours was a child-centered world. Adults existed primarily to give to children that which they wished. The rights of the child erased the rights of the adult. Children did not experience the struggle to become a part of the family, and then a part of the world. The inherent need to be really a part of something atrophied from disuse. The idea that the individual was important was translated in child-rearing into the assumption that the individual was not important in terms of his fulfillment as a part of a social structure, but rather the social structure was considered the enemy of the individual; and therefore the child should be protected from its attack. The child was deprived of the gratification that comes from creative self-amusement; his amusement was provided for him by activities planned by adults. He was encouraged to believe that he had a right to be happy; and adults, his parents, teachers, and others failed him if they did not provide him with that happiness. Those children, now that they are adolescents, show the effects of their childhood experience; they are convinced that they have the right to be happy and other people fail if they do not provide that happiness.

One young man in his early 20's was discussing his high school period. He was a bright boy but had done only barely passing work in high school.

He had spent a great deal of his time drag racing, usually under the influence of alcohol, on the main street of the suburban town in which he lived. At the time of our interview he had settled down and was not in any serious difficulty. I asked him why he thought that a group of the high school students took recourse to alcohol. He said the answer was simple; they were bored, and alcohol dulled the boredom. When I expressed my idea that it was unfortunate that at adolescence life was so jaded that it was boring, he answered, "That isn't really the point; we didn't have any idea how to have a good time. During childhood a supposedly good time was always provided for us. I had to learn the hard way how to really enjoy my life. I am only beginning to feel that it is fun to study on my own rather than with a tutor, to enjoy taking a long walk with a friend, or to play records that my friends and I enjoy. I have finally found that when you have been drinking you don't really enjoy what you are doing as much as you do when you haven't had alcohol." This boy was the product of a child-centered world; a world that was the epitome of what I feel is the source of some of our malignant adolescent disturbances. His childhood was so full of planned activities he never learned to have fun on his own. He was forced to attend school so that he could get into high school, so he could get into a good college, so he could become a professional person, so that he could rise above the economic level of his family and would marry a desirable girl, desirable because she had a college education.

One notes an interesting development as some adolescents take on adult responsibilities. There have always been parents of small children who have complained to their therapists that their rights should also be considered, rather than the emphasis being on the rights of the child. However, again it seems to me that there are more young parents who take the attitude — when they are having difficulty with their child — that the child should adjust to the parents, as if in young adulthood the individuals are *still* protesting that it should be an egocentric world for them.

From Adolescence to Adulthood

The adolescent who accepts the educational goals defined by parents and other adults, and who is dedicated to the philosophy of privilege without concomitant responsibility, is not the adolescent who arouses the greatest mass anxiety — unless his attitudes are expressed in delinquency. It is rather the adolescent who endorses a cause which appears to invite iconoclasm instead of constructive paths for the future that shifts the public interest from the individual to a defensive mobilization against certain dissident groups. But it is the members of these groups who are most dramatically seeking answers that transend their egocentric motivations. They are striving to find expression for an idealism whose goals are a part of our democratic heritage. Their idealism is youthful, untempered by experience that would amalgamate theoretical idealism and practical reality into the building stones for a better world.

Are we listening to them or not hearing them? Are we talking to them, or at them? Are we offering them an opportunity to test their ideals realistically, or are we trying to force an adaptation to our definition of reality? Are we entrenching ourselves behind the status quo, thereby necessitating, if they seek a better world than ours, that they turn to leaders who may prove destructive? If so, as the young people become more experienced in the role of assuming responsibility for the cultural development of the future, they will face disillusion with the destructive leadership they have followed, and will have to retrace their steps rather than be free to advance. We cannot help them if our sole aim is to prevent our own apple-cart — which undoubtedly has among its shiny apples some that are rotten — from being upset; we can help them sort the good apples from the bad, with a recognition on our part that perhaps sometimes they have a more sensitive nose for the odor of an apple that is beginning to spoil.

In spite of the negative pictures that I have presented of malignancy in adolescence, and of some of our own shortcomings, I am convinced that most adolescents of today will resolve their childhood conflicts and make an adaptation to an adult role surprisingly well in light of what has happened to them in the past. Adolescents today, like the adolescents of the past, have a remarkable ability to reactivate inherent drives and to find finally a constructive outlet for them. If they need our help, however, it is important to bear in mind that there is a morality underlying all therapeutic approaches. The morality is one that conceptualized adulthood as the attainment of a psychological state in which inherent drives find expression in the framework of a sound superego and a dynamic, not static, adaptation to reality. Our goal in working with the adolescent is to help him resolve his adolescent conflicts so that his impulses can finally find expression constructively for him and for others in the adult world. His adult world, if our culture continues to become richer, will be different from ours. From his standpoint it will be compatible with his desires. The fact that the therapeutic goal may not be ever attained at all for some, or may be only relatively reached for others, indicates the magnitude of our therapeutic task and the present limits in our skills.

References

Erikson, Erik H. "The Problem of Ego Identity." *Journal of the American Psychoanalytic Association*, vol. 4, 1956, pp. 56-121.

Josselyn, Irene M. *Psychosocial Development of Children.* New York: Family Service Association of America, 1948.

Rieff, Philip. *The Truimph of the Therapeutic*. New York: Harper & Row, 1966.

CHAPTER 17

Pre-adolescents— What Makes Them Tick?

FRITZ REDL

The period of pre-adolescence is a stretch of no-man's-land in child study work. By pre-adolescence I mean the phase when the nicest children begin to behave in a most awful way. This definition cannot exactly be called scientific, but those who have to live with children of that age will immediately recognize whom I am talking about. This also happens to be the age about which we know least. Most of our books are written either about Children or about Adolescents. The phase I am talking about lies somewhere between the two — crudely speaking, between about nine and thirteen, in terms of chronological age, or between the fifth and eighth grade in terms of school classification.

It is surprising that we know so little about this age, but there certainly is no doubt that it is one of the most baffling phases of all. Most referrals to child guidance clinics occur around this age, and if you look for volunteers to work on programs in recreation or child care, you will make this peculiar discovery: you will have no trouble finding people who just love to bathe little babies until they smell good and shine. You will have a little more, but not too much, trouble finding people who are just waiting for a chance to "understand" adolescents who "have problems" and long for a shoulder on which to cry. But the pre-adolescent youngster offers neither of these satisfactions. You won't find many people who will be very happy working with him.

Are they children? No. Of course, they still look like young children. Practically no visible change has as yet taken place in their sex development. The voice is about as shrill and penetrating as it ever was, the personal picture which they represent is still highly reminiscent of a child — of about the worst child you have met, however, definitely not of the child they themselves were just a short time ago.

Are they adolescent? No. While filled with a collector's curiosity for odd elements of information about human sex life at its worst, they are not as yet really maturing sexually. While they occasionally like to brag about precocity in their sex attitude, "the boy" or "the girl" of the other sex is still something they really don't know what to do with if left alone with it for any length of time. While impertinent in their wish to penetrate the secrets of adult life, they have no concepts about the future, little worry about what is going to happen to them, nothing they would like to "talk over with you."

The reason why we know so little about this phase of development is simple but significant: it is a phase which is especially disappointing for the adult, and especially so for the adult who loves youth and is interested in it. These youngsters are hard to live with even where there is the most ideal child-parent relationship. They are not as much fun to love as when they were younger, for they don't seem to appreciate what they get at all. And they certainly aren't much to brag about, academically or otherwise. You can't play the "friendly helper" toward them either — they think you are plain dumb if you try it; nor can you play the role of the proud shaper of youthful wax — they stick to your fingers like putty and things become messier and messier the more you try to "shape" that age. Nor can you play the role of the proud and sacerdotal warden of the values of society to be pointed out to eager youth. They think you are plain funny in that role.

So the parent is at a loss and ready for a desperate escape into either of two bad mistakes — defeatism or tough-guy stubbornness. The teacher shrugs her shoulders and blames most of this pre-adolescent spook on the teacher the youngster had before her, or on lack of parental cooperation, and hopes that somehow or other these children will "snap out of it." Even the psychiatrist, otherwise so triumphantly cynical about other people's trouble with children, is in a fix. For with these children you can't use "play techniques" any longer. They giggle themselves to death about the mere idea of sitting in one room with an adult and playing a table game while that adult desperately pretends that this is all there is to it. And one can't use the usual "interview technique" either. They find it funny that they should talk about themselves and their life, that they should consider as a "problem" what has "just happened," that they should try to remember how they felt about things, and that they are constantly expected to have "worries" or "fears" — two emotions which they are most skillful at hiding from their own self-perception, even if they do occur. Most of these youngsters seriously think the adult himself is crazy if he introduces such talk, and they naively enjoy the troubles they make, rather than those they have, and would much rather bear the consequences of their troubles than talk about them, even though those consequences include frustration or a beating or two.

Research, too, with very few exceptions, has skipped this period. If you study adolescence, you certainly can have graphs and charts on the rate at which the growth of pubic hair increases, the timing between that and the change of voice, and the irrelevance of both in terms of psycho-sexual development. Unfortunately, at the age we are talking about little of all this seems to take place. No drastic body changes occur, and whatever may happen within the glands is certainly not dramatic enough to explain the undoubtedly dramatic behavior of that phase. For a while some Yale biologists tried to discover an increase in hormone production around the age of eight, long before there is any visible sex maturation. However, they had trouble in making their research results useful for practical purposes. It took them weeks for one specimen of urine to be boiled in the right way so as to show up the existence or nonexistence of these hormones, and in the meantime Johnny would probably have been kicked out of five more schools anyway. In short, research has discreetly left this phase alone, and has retired from it, as it always does from things which are either too hard to demonstrate by statistical methods, or too hot to talk about after they have been discovered.

Thus, the practitioner — the parent, the teacher, counsellor, or group worker — is left to his own devices. Fortunately, most of the things which are characteristic symptoms of this phase are known to us all.

Pre-adolescent Behavior—Bad and Improper

Here are some of the most frequent complaints adults raise in connection with their attempts to handle pre-adolescents: Outwardly, the most striking thing about them is their extreme physical restlessness. They can hardly stand still, running is more natural to them than walking, the word sitting is a euphemism if applied to what they do with a table and a chair. Their hands seem to need constant occupational therapy — they will turn up the edges of any book they handle, will have to manipulate pencils, any objects near them, or any one of the dozen-odd things they carry in their pockets, or even parts of their own body, whether it be nose and ears, scratching their hair, or parts of the anatomy usually taboo in terms of infantile upbringing. The return to other infantile habits is surprisingly intensive in many areas: even otherwise well-drilled and very housebroken youngsters may again show symptoms like bed-wetting, soiling, nail-biting, or its substitutes, like skin-chewing, finger-drumming, etc. Funny gestures and antics seem to turn up overnight with little or no reason — such things as facial tics, odd gestures and jerky movements, long-outgrown speech disorders, and the like.

In other areas these youngsters do not return to exactly the same habits of their infancy, but they go back to typical problem areas of younger childhood and start again where they had left off. Thus their search for the facts of life which had temporarily subsided under the impact of partial parental explanations will be resumed with vehemence, and with the impudence and insistence of a news correspondent rather than with the credulity of an obedient young child. It is the oddity, the wild fantastic story and the gory detail which fascinate them more than parental attempts at well-organized explanations of propagation, which they find rather boring.

Their old interpretation of the difference of sexes is revived too. Girls seem obviously inferior to boys, who again interpret the difference in sex as that of a minus versus plus rather than of a difference in anatomical function. Thus girls are no good unless they are nearly like boys; and where the direct pride in masculine sexuality is subdued, indirect bragging about the size and strength of the biceps takes its place and becomes and sets the evaluation of anybody's worth. The girls go through somewhat the same phase, accept the interpretation of the boys all too eagerly and often wander through a period of frantic imitation of boyish behavior and negation of their female role. What sex manipulation does occur at this age usually happens in terms of experimentation and is on a highly organic level and very different from the masturbation of later adolescent years.

The fantasy life of youngsters of this age is something to look into, too. Wild daydreams of the comic-strip type of adventure, on the one hand, long stages of staring into empty space with nothing going on in their conscious mind on the other, are the two poles between which their fantasy life moves rapidly back and forth. Often manipulative play with a piece of string or the appearance of listening to the radio cover long stretches of quickly changing flights of ideas, and youngsters who reply "nothing," when you ask them what they were thinking about, do not necessarily lie. This description really fits the content as far as it could possibly be stated in any acceptable logical order and grammatical form.

The most peculiar phenomena, though, are found in the area of adult-child relationships. Even youngsters who obviously love their parents and have reason to do so will develop stretches of surprising irritability, distrust, and suspicion. Easily offended and constantly ready with accusations that adults don't understand them and treat them wrongly, they are yet very reckless and inconsiderate of other people's feelings and are quite surprised if people get hurt because of the way they behave. The concept of gratitude seems to be something stricken from the inventory of their emotions. The worst meal at the neighbors', at which they weren't even welcome, may be described more glowingly in its glory than the best-planned feast that you arranged for their birthday. The silliest antics, the most irrelevant possessions or skills of neighbors will be admired 'way beyond any well-rooted qualities and superior achievements of father and mother.

Daily life with Junior becomes a chain of little irritations about little things. The fight against the demands of obeying the rules of time and space is staged as vehemently as if the children were one or two years old again. Keeping appointed meal times, coming home, going to bed at a prearranged hour, starting work, stopping play when promised — all these demands seem to be as hard to get across and as badly resented, no matter how reasonable the parents try to be about them, as if they were the cruel and senseless torments of tyranny.

Lack of submission to parent-accepted manners becomes another source of conflict. If these youngsters would listen as attentively to what Webster has to say as they do to the language of the worst ragamuffin on the street corner, their grades in English would be tops. Dressing properly, washing, keeping clean are demands which meet with obvious indignation or distrust. In a way, they seem to have lost all sense of shame and decency. Previously clean-minded youngsters will not mind telling the dirtiest jokes if they can get hold of them, and the most charming angels of last year can spend an hour giggling over the acrobatics which a youngster performs with his stomach gas and consider it the greatest joke.

And yet, while unashamed in so many ways, there are other areas of life where they become more sensitive rather than more crude: the idea of being undressed or bathed by their own parent may all of a sudden release vehement feelings of shame hitherto unknown to their elders, and the open display of affection before others makes them blush as though they had committed a crime. The idea of being called a sissy by somebody one's own age is the top of shamefulness and nearly intolerable because of the pain it involves.

One of the most interesting attitude changes during this period is that in boy-girl relationships. The boy has not only theoretical contempt for the girl but he has no place for her socially. Social parties which adults push so often because they find the clumsiness of their youngsters so cute, and because it is so safe to have boys and girls together at that age, such social dances are a pain in the neck to youngsters, who would obviously much rather have a good free-for-all or chase each other all over the place. The girls have little place in their lives for the same-age boys either. It is true that with them the transition through this pre-adolescent period usually is shorter than with the boys. But for a time their actual need for boy company is nil. The picture is different, though, if you watch the children within their own sex gangs. Then, all of a sudden, in talks under safe seclusion with their buddies, the boys or girls will display a trumped-up interest in the other sex, will brag about their sexual knowledge or precocity or about their success in dating. All this bragging, however, though it is about sex, is on an entirely unerotic level; the partner of the other sex only figures in it as the fish in the fisherman's story. The opposite sex, like the fish, serves only as an indirect means for self-glorification.

What Makes Them Tick?

The explanation of this peculiar phenomenon of human growth must, I think, move along two lines. One is of an *individualistic* nature, the second is a chapter in *group psychology.*

Explanation No. I: During pre-adolescence the well-knit pattern of a child's personality is broken up or loosened, so that adolescent changes can be built into it and so that it can be modified into the personality of an adult.

Thus, the purpose of this developmental phase is not *improvement* but *disorganization*; not a permanent disorganization, of course, but a disorganization for future growth. This disorganization must occur, or else the higher organization cannot be achieved. In short, a child does not become an adult by becoming bigger and better. Simple "improvement" of a child's personality into that of an adult would only produce an oversized child, an infantile adult. "Growing" into an adult means leaving behind, or destroying some of what the child has been, and becoming something else in many ways.

The real growth occurs during adolescence: pre-adolescence is the period of preliminary loosening up of the personality pattern in order that the change may take place. It is comparable to soaking the beans before you cook them. If this explanation is true, then we can understand the following manifestations:

1. During this "breaking-up-of-child-personality" period, old, *long-forgotten or repressed* impulses of earlier childhood will come afloat again, for a while, before they are discarded for good. This would explain all that we have described about the return to infantile habits, silly antics, irritating behavior, recurring naughty habits, etc.

2. During this period of the breaking up of an established pattern, we also find that *already developed standards and values* lose their power and become ineffectual. Therefore the surprising lack of self-control, the high degree of disorganization, the great trouble those youngsters have in keeping themselves in shape and continuing to live up to at least some of the expectations they had no difficulty living up to a short time ago. The individual conscience of the child seems to lose its power, and even the force of his intelligence and insight into his own impulses is obviously weakened. This would explain all we have said about their unreliability, the lowering of these standards of behavior, the disappearance of some of the barriers of shame and disgust they had established and their surprising immunity to guilt feelings in many areas of life.

3. During a period of loosening up of personality texture, we would

expect that the whole individual will be full of conflict, and that the *natural accompaniments* of conflict will appear again, namely, *anxieties and fears,* on the one hand, and *compulsive mechanisms of symbolic reassurance*, on the other. This is why so many of these youngsters really show fears or compulsive tics which otherwise only neurotic children would show. Yet this behavior is perfectly normal and will be only temporary. This would explain the frequent occurrence of fantastic fears in the dark, of ghosts and burglars, and it would also explain the intensity with which some of these youngsters cling to protective mechanisms, like the possession of a flashlight or gun as a symbol of protection, or the display of nervous tics and peculiar antics which usually include magic tricks to fool destiny and assure protection from danger or guilt.

For a long time I thought that all this about finishes the picture of pre-adolescent development, until a closer observation of the group life of pre-adolescents showed me that such a theory leaves much unexplained. It seems to me that there is still another explanation for a host of pre-adolescent symptoms.

Explanation No. II: During pre-adolescence it is normal for youngsters to drop their identification with adult society and establish a strong identification with a group of their peers.

This part of pre-adolescent development is of a group psychological nature, and is as important to the child's later functioning as a citizen in society as the first principle is for his personal mental and emotional health. This group phenomenon is surprisingly universal and explains much of the trouble we adults have with children of pre-adolescent age. To be sure that I am rightly understood, I want to emphasize that what happens during this age goes 'way beyond the personal relationship between Johnny and his father. Johnny's father now becomes for him more than his father: he becomes, all of a sudden, a representative of the value system of adult society versus the child, at least in certain moments of his life. The same is true the other way around. Johnny becomes more to his father than his child: at certain moments he isn't Johnny any more but the typical representative of Youth versus Adult. A great many of the "educational" things adults do to children, as well as many of the rebellious acts of children toward adults, are not meant toward the other fellow at all; they are meant toward the general group of "Adults" or "Youth" which this other person represents. To disentangle personal involvement from this group psychological meaning of behavior is perhaps the most vital and so far least attempted problem of education in adolescence.

If this explanation is true, then it seems to me that the following phenomena of pre-adolescent behavior will be well understood:

1. In no other age do youngsters show such a deep need for *clique and gang formation* among themselves as in this one. From the adult angle this

is usually met with much suspicion. Of course, it is true that youngsters will tend to choose their companions from among those who are rejected, rather than approved of, by their parents. Perhaps we can understand why the more unacceptable a youngster is on the basis of our adult behavior code, the more highly acceptable he will be in the society of his own peers. The clique formation of youngsters among themselves usually has some form of definitely "gang" character: that means it is more thoroughly enjoyed by being somewhat "subversive" in terms of adult standards. Remember how youngsters often are magically fascinated by certain types of ringleaders, even though this ringleadership may involve rather harmless though irritating activities — such as smoking, special clothes, late hours, gang language, etc.

From the angle of the adult and his anxieties, much of this seems highly objectionable. From the angle of the youngster and his normal development, most of it is highly important. For it is vital that he satisfy the wish for identification with his pals, even though or just because such identification is sometimes frowned upon by the powers that be. The courage to stick to his pal against you, no matter how much he loves you and otherwise admires your advice, is an important step forward in the youngster's social growth.

2. In all groups something like an *unspoken behavior code* develops, and it is this unwritten code on which the difference between "good" and "bad" depends. Up to now the youngster has lived within the psychological confines of the adult's own value system. Good and bad were defined entirely on the basis of adult tastes. Now he enters the magical ring of peer codes. And the code of his friends differs essentially from that of adult society. In some items the two are diametrically opposed. In terms of the adult code, for instance, it is good if children bring home high grades, take pride in being much better than the neighbor's children, in being better liked by the teacher, and more submissive to the whims of the teaching adult than are other people's children. In terms of peer standards things are directly reversed. Studying too much exposes you to the suspicion of being a sissy, aggressive pride against other children is suspiciously close to teacher-pet roles, and obedience to the adult in power often comes close to being a fifth columnist in terms of "the gang."

Some of the typically adult-fashioned values are clearly rejected by peer standards; others are potentially compatible at times, while conflicting at other times; some of them can be shared in common. Thus, a not too delinquent "gang" to which your youngster is proud to belong may be characterized by the following code range: It is all right in this gang to study and work reasonably well in school. It is essential, though, that you dare to smoke, lie, even against your own father, if it means the protection of a pal in your gang, that you bear the brunt of scenes at home if really important gang activities are in question. At the same time this gang would not want you to steal, would be horrified if your sex activity went beyond the telling of dirty stories, and would oust you tacitly because they would think you too sophisticated for them. The actual group life of pre-adolescents moves between hundreds of different shades of such gang codes, and the degree to which we adults have omitted opening our eyes to this vital phase of child development is astounding.

3. The change from *adult code* to *peer code* is not an easy process for a youngster but full of conflict and often painful. For, while he would like to be admired by his pals on a peer-code basis, he still loves his parents personally and hates to see them misunderstand him or have them get unhappy about what he does. And, while he would love to please his family and be again accepted by them and have them proud of him, he simply couldn't face being called a sissy or be suspected of being a coward or a teacher's pet by his friends. In most of those cases where we find a serious conflict between the two sets of standards, we will find the phenomenon of *social hysteria.* This applies to youngsters who so overdo their loyalty to either one of the two behavior standards that they then have to go far beyond a reasonable limit. Thus, you find youngsters so scared of being thought bad by their parents that they don't dare to mix happily with children of their own age; and you find others so keen to achieve peer status with friends of their own age that they begin to reject all parental advice, every finer feeling of loyalty to the home, and accept all and any lure of gang prestige even if it involves delinquent and criminal activity. It is obvious that a clear analysis by the adult and the avoidance of counter-hysteria can do much to improve things.

How to Survive Life with Junior?

If any of the above is true, then it should have an enormous impact on education. For then most of this pre-adolescent spook isn't merely a problem of things that shouldn't happen and ought to be squelched, but of things that should happen, but need to be regulated and channelized. Of course, you can't possibly just let Junior be as pre-adolescent as he would like without going crazy yourself, and you definitely shouldn't think of self-defense only and thus squelch the emotional and social development of your offspring. How to do both — survive and also channelize normal but tough growth periods without damage to later development — is too long a story to complete in a short article. But here are a few general hints:

1. Avoid Counter-Hysterics

It seems to me that 90 percent of the more serious problems between children and parents or teachers on which I have ever been consulted could have been easily avoided. They were not inherent in the actual problems of growth. They were produced by the hysterical way in which the adults reacted to them. Most growth problems, even the more serious ones, can be outgrown eventually — though this may be a painful process — provided the adults don't use them as a springboard for their own overemotional reactions. This does not mean that I advocate that you give up and let everything take its course. I do suggest you study the situation and decide where to allow things and where to interfere. The problem is: whichever you decide to do — the *way* you do it should be realistic, free from hysteric overemotionalism. With this policy in mind you can enjoy all the fun of having problems with your child without producing a problem child.

2. Don't Fight Windmills

Let's not forget that pre-adolescents are much more expert in handling us than we ever can be in handling them. Their skill in sizing us up and using our emotions and weaknesses for their own ends has reached a peak at this age. It took them eight to ten years to learn, but they have learned by thorough observation. While we were worrying about them, they were not worried about us, and they had ample time and leisure to study our psychology. This means that if they now go out on the venture of proving to themselves how emancipated they are, they will choose exactly the trick which will irritate us most. They will develop pre-adolescent symptoms in accordance with their understanding of our psychology. Thus, some of them will smoke, curse, talk about sex, or stay out late. Some will stop being interested in their grades, get kicked out of school, or threaten to become the type of person who will never be acceptable in good society. Others again will develop vocational interests which we look down on, will choose the company we dread, talk language which makes us jump, or may even run away at intervals.

But whatever surface behavior they display — don't fall for it. Don't fight the behavior. Interpret the cause of it first, then judge how much and in what way you should interfere. Thus, Johnny's smoking may really mean he is sore that his father never takes him to a football match, or it may mean he thinks you don't appreciate how adult he already is, or it may mean he has become dependent on the class clown. Mary's insistence upon late hours may mean she doesn't know how to control herself, or it may mean she is sore because her school pals think you are social snobs who live a life different from theirs, or it may mean she is so scared that her sex ignorance will be discovered that she has to run around with a crowd more sophisticated about staying up late, so as to hide her lack of sophistication in another respect.

In any case, all these things are not so hard to figure out. Instead of getting excited and disapproving of the strange behavior, just open your eyes for a while and keep them open without blinking.

3. Provide a Frame of Life Adequate for Growth

No matter how much you dislike it, every pre-adolescent youngster needs the chance to have some of his wild behavior come out in some place or other. It will make a lot of difference whether or not he has a frame of life adequate for such growth. For example: Johnny needs the experience of running up against some kind of adventurous situation where he can prove he is a regular guy and not just mother's boy. Cut him off from all life situations containing elements of unpredictability and he may have to go stealing from the grocery store to prove his point. Give him a free and experimental camp setting to be adventurous in and he will be happily pre-adolescent without getting himself or anybody else in trouble. All youngsters need some place where pre-adolescent traits can be exercised and even tolerated. It is your duty to plan for such places in their life as skillfully as you select their school or vocational opportunities.

4. Watch Out for Pre-adolescent Corns

Most people don't mind their toes being stepped on occasionally. But if there is a corn there, that is a different matter. Well, all pre-adolescents have certain corns, places where they are hypersensitive. Avoid these as much as possible. One of the most important to avoid is harking back to their early childhood years. The one thing they don't want to be reminded of is of themselves as small children, yourself as the mother or father of the younger child. If you punish, don't repeat ways you used when they were little. If you praise, don't use arguments that would please a three-year-old but make a thirteen-year-old red with shame or fury. Whether you promise or reward, threaten or blackmail, appeal to their sense, morals, or anything else, always avoid doing it in the same way you used to do when they were little.

I have seen many pre-adolescents reject what their parents wanted, not because they felt that it was unreasonable or unjustified, but on account of the way in which the parents put the issue. There is something like a developmental level of parental control as well as developmental levels of child behavior. The two have to be matched or there will be fireworks.

5. If in Doubt, Make a Diagnostic Check-up

Not all the behavior forms we described above are always merely "pre-adolescent." Some of them are more than that. After all, there are such things as juvenile delinquents and psychoneurotics, and we shouldn't pretend that everything is bound to come out in the wash.

Usually you can get a good hunch about dangerous areas if you check on these points: How deep is the pre-adolescent trait a youngster shows? If it is too vehement and impulsive, too unapproachable by even the most reasonable techniques, then the chances are that Johnny's antics are symptoms not only of growth but also of something being wrong somewhere and needing repair. Often this may be the case: Five of Johnny's antics are just pre-adolescent, pure and simple, and should not be interfered with too much. However, these five are tied up with five others which are definitely serious, hangovers from old, never really solved problems, results of wrong handling, wrong environmental situations, or other causes. It will do no good to brush off the whole matter by calling it pre-adolescent. In that case the first five items need your protection and the other five need a repair job done. Whenever you are very much in doubt, it is wise to consult expert help for the checkup — just as you would in order to decide whether a heart murmur is due to too fast growth or to an organic disturbance.

CHAPTER 18

Adolescent Problems: A Symptom of Family Disorder

NATHAN W. ACKERMAN

The adolescents of our time are hoisting distress signals. In many ways, both direct and indirect, they let the rest of us know that they are in trouble. Their disordered behavior today is an almost universal phenomenon. We have in the United States of America the teenage gangs and beatniks; in England, the "angry young men"; in Germany, the "Bear-Shirts"; in Russia, the "Hoodlums"; in Japan, the split of the teenagers into "wet" and "dry." These are but a few examples of semi-organized group expressions of widespread adolescent conflict. Conspicuously in evidence are signs of disorientation, confusion, panic, outbursts of destructiveness and moral deterioration. The disordered behavior of the adolescent needs to be understood not only as an expression of a particular stage of growth, but beyond that, as a symptom of parallel disorder in the patterns of family, society, and culture.

In a setting of world crisis the distress of the adolescent may be viewed as a functional manifestation of the broader pattern of imbalance and turbulence in human relations. The family, as a behavior system, stands intermediate between the individual and culture. It transmits through its adolescent members the disorders that characterize the social system. In our native community we confront the special challenge of the anarchy of youth. The recurrent bursts of bizarre teenage violence are emblazoned for us in the daily papers and other mass media. This is dramatic and frightening. But the problem embraces far more than juvenile delinquency. While some adolescents explode crudely in extremes of destructive antisocial action, others manifest their distress in a more subtle, indirect and concealed way, no less serious for its inconspicuousness. Fundamentally, what underlies the entire range of disorders is the adolescent's fierce, often failing struggle to find himself in this chaotic world. He is searching for a sense of identity, for a sense of wholeness and continuity, in a society that is itself anything but whole and anything but steady in its movement through time.

But let us not imagine that it is the adolescent only or exclusively who experiences this painful struggle. It is all of us, at all stages of life, who echo in our personal lives the disorder of the social system. The agitation of the adolescent surely does not exist in isolation. It is matched and paralleled by the emotional insecurity of his parents, the imbalance of the relations between them, and the turbulence and instability of the family life as a whole. The family, as family, does not know clearly what it stands for; its resources for solving present-day problems and conflicts are deficient. Not only are families confused, disoriented, fragmented, and alienated; whole communities sometimes exhibit these same trends.

Let us glance for a moment at the community response when there is an eruption of teenage destructiveness. Generally, there is an immediate outcry, a show of fright, shock, worry, righteous indignation; then talk and more talk. Soon the excitement simmers down, until the next shocking eruption, and the process repeats itself. The recurrent bursts of savage, inhuman violence among juveniles strike a note of alarm in the community. They stir a deep-rooted anxiety among parents, teachers, and community leaders. The community turns desperate. There is a loud call for action, for a program; something must be done. But the demand for action arises not only out of a sense of desperation; it expresses also a profound helplessness, a feeling of sheer impotence to do anything about it. Why? Because the finger of accusation points responsibility not to the delinquent adolescent alone, but to the whole disordered character of the human relations pattern of the present-day family and community as well. Yes, there is something deeply wrong, and it is not just with our adolescents; it is with our whole way of life. Our social health is failing; and the effects of the failure cast a long shadow on our mental health.

With each outbreak of juvenile destructiveness comes a spate of suggested remedies — vocational training, work camps, recreational facilities, group activity and guidance programs, and finally, freer use of the big stick, physical punishment at home and in the schools; ultimately, stricter policing and even penalization of the parents, who are presumed to be negligent. At the peak of such community agitation, there is much talk and a strong resolve to act, but as is the case with New Year's resolutions, nothing much comes of it. The conviction grows that these measures are mere sops, a feeble attempt to plug the hole in the dam. Then comes widespread disillusionment and finally, perhaps, the grudging admission that the real problem is not the adolescent alone, but rather the sources of dinintegrative influence in a sick or broken family. And exactly at this point the loud talk subsides because no one yet has a program to offer for what ails the family. It is this that explains the adults' sense of helpless resignation and their temptation, shame-facedly, to turn away from the problem. In effect, then, these waves of panic and agitation are a kind of shadow boxing. People rant and rave, they lash out at the shadow, but are impotent with the real thing — the disorder of family life itself.

Trends in Juvenile Problems

Let us examine now the range of adolescent symptoms. The outstanding features are: (*a*) a tendency to antisocial behavior, specifically expressed in acts of violence; in close association, a vulnerability toward mob action with a propensity toward organized prejudice, bullying, and scapegoating of innocent victims; (*b*) a revolution in sexual mores, shown in a tendency to promiscuity; (*c*) a wave of contagion that makes an obsession of everything "hot" — hot jazz, hot dancing, hot rods; closely associated with a compulsive quest for an ever new kind of kick, as in the use of such drugs as benzedrine, marijuana, and even cocaine; (*d*) a leaning toward overconformity with the peer group; (*e*) a quest for a safe niche, a sinecure, a steady job with a pension in industry, a "couple of little mothers," as they say in teenage jargon; a split-level home with an electrified kitchen in suburbia; (*f*) a tendency to withdrawal; a closely associated tendency toward a loss of hope and faith, disillusionment and despair with a gradual destruction of ideals; a conspicuous product of this trend is the static-mindedness of many adolescents, the lack of adventuresomeness and loss of creative spark; (*g*) disorientation in the relations of the adolescent with family and community; an inability to harmonize his personal life with the goals of family and society; in consequence, a trend toward confusion or loss of personal identity; (*h*) as the outcome of this magnified social disorder, the intensified vulnerability of adolescents toward mental breakdown.

The anarchy of youth in contemporary society is a thorny problem. Since 1948 there has been an overall increase of juvenile delinquency of over 70%, although the child population has increased only 16%. Since 1950, this represents a rise four times as fast as the population. F.B.I. data show that a sizable percentage of the major aggressive felonies (murder, manslaughter, rape, assault, robbery, larceny, and automobile theft) are committed by young people under the age of eighteen. In 1956, the youth in this age group accounted for 45% of all arrests for such crimes, and two-fifths of these offenders were under fifteen. According to the F.B.I., the number of killings committed by juveniles continues to rise. About half of these young killers are under eighteen, and one-third are children below sixteen. It is estimated that by 1965, more than a million children will appear before the courts. As J. Edgar Hoover put it, "Gang style ferocity, once the domain of hardened adult criminals, now centers chiefly on cliques of teenage brigands."

A psychiatrist, Dr. Wertham, expressed his alarm a different way: "This is the age of violence and it began with the dropping of the first atom bomb on Hiroshima. . . . Younger and younger children commit more and more serious and violent acts. . . . The current spate of child murders is like the difference between a disease and an epidemic. One such killing would be a crime, but ten or a hundred or a thousand, that is a social phenomenon."

Three main features can be discerned in this trend: the gross numerical increase of these acts of violence, the deepening severity of these acts, and the universality of the trend, as many nations throughout the world affirm the findings of the F.B.I.

A few illustrations: Not long ago, the public was stunned by the strange tragedy of Cheryl Turner Crane, who stabbed to death the lover of her mother, actress Lana Turner. The pathos of this story is still sharply etched in our memory. Shortly after this event, I saw in consultation a friend of Cheryl, a young girl of thirteen. She had just tried to poison her father's whiskey.

In *The New York Times* there appeared the following item: A young army man, assigned to security work, was found dead in the burned wreckage of his car. He had sprayed the interior of his car with gasoline, poured gasoline all over himself and then set fire. What possibly could have been this young man's state of mind to induce this fiendishly violent form of suicide?

Again, not long ago, we were struck with horror by the photographs of a fanatic Japanese boy who stabbed to death a leading political figure.

Another aspect of the anarchy of youth is the tendency to indiscriminate sexual behavior. Mostly the grownups in the community are loath to look squarely at this problem, not because it is sex, but rather because it is sex in such a shocking form. It is the subhuman or even inhuman quality of adolescent sexual conduct that compels us to turn our faces away. There are the recurrent reports of organized sex clubs for teenagers, group orgies that combine violence and indiscriminate sexual indulgence. Such behavior is not in any sense confined to members of any one social class, certainly not confined to the working class or the uneducated. It also appears in our institutions of higher learning, the universities. In occasional instances a single disturbed girl in a college dormitory can wreak havoc with the student population, not only by multiple seduction of college boys, but also by disrupting the sexual standards of the female students. In such situations, the college authorities are hard put to know what to do. The dean and faculty have no desire to inflict harsh punishment, but neither can they permit a contagion of disorganized, irresponsible sex orgies. The community demands action and somebody's head must roll. Frequently enough, the reputation of innocent persons is hurt.

Closely related, emotionally speaking, is the almost hypnotic trance into which adolescents fall as they become absorbed in hot music, hot dancing, hot rods, and the crazy craving for ever new kinds of thrill and excitation, even the use of narcotic drugs. Observers have been forcibly struck by the madness that seems to take hold of these young people. It is as if they were caught in a spell. At the very peak, they become transformed into elemental beings and are oblivious of all else but the excitement of their senses. They seem to be drawn off to another world. Most of them recover, to be sure; for some, however, there is a point of no return.

Of course, it is not one-sidedly implied here that such behavior is purely regressive and emotionally sick. It is only to be expected that the adolescent should have hot blood and venture forth in the search for the rhythm of life. It is normal for them to try to capture the rhythm of life for themselves; in fact, to become one with it. In many instances, it is surely true that an active quest for nature's rhythm brings an enormous relief of tension and actually offsets the danger of destructive outbursts. If adolescents can feel the rhythm of life down to their bone marrow, they can more easily reorient their bodily surgings to the existing world and do so in more constructive ways.

At the opposite extreme there is a tendency toward a rigid conformity with the standards and expectations of the peer group. Such behavior is mostly undramatic and inconspicuous. It is nevertheless important insofar as it reflects the severity of the pressure to conformity, even to the extent of the adolescent's experiencing a submergence or loss of his individual self. In the extreme, this is exemplified by the fashions of the zoot-suiters, the black stocking fad, the contests involving the swallowing of goldfish and the squeezing of college boys like sardines into a phone booth. Also, the habit of going steady and getting pinned, just because everyone else is doing it. This leaning toward submission to the peer group consitiutes a danger precisely because of its implied renunciation of qualities of individual uniqueness and difference.

A great segment of the adolescent population is turning painfully cautious and conservative. In a topsy-turvy world with a radical change proceeding at a galloping pace, adolescents paradoxically turn cautious. They want to hold the world still long enough to catch up with it. So we see them seeking routine, dependable safe jobs in industry. They become scared of adventure, withdraw from the center of strife and turmoil, pull away from politics, move to the suburbs and raise babies. They become good organization men, but they lose their sense of adventure and surrender their creative dreams.

The adolescent experiences the greatest hardship in building a satisfying sense of personal identity. He is pained by his failing effort to harmonize his personal life goals with the goals of family and community. This induces in him a terrible dilemma. It is almost a commonplace that adolescents in our day cannot say who they are, where they belong, where they are going in life. They complain of a sense of emptiness, futility, and boredom — so why try!

Then there is a tendency of the adolescent to flee from conflict into a state of isolation and withdrawal. Often some depression is linked to this withdrawal. Both of these tendencies serve the purpose of self-preservation in the face of excessive conflict and anxiety. However, these efforts boomerang insofar as they can be indulged only up to a point and at the cost of progressive weakening of the personality. One manifestation of this withdrawal is the inability to face up to the urgent life questions such as boy-girl relations and choice of career. On the part of many adolescents, there is an urge to escape from these necessary decisions, or at least to stall them.

From still another angle, there is the frequent trend toward loss of faith or resignation to the defeats of life, a disintegration of ideals, and a shift of mood toward cynicism. This is especially evident in relation to the future rewards of work and family life. It is conspicuously manifested in attitudes of disillusionment toward authority figures — parents, community and political leaders. Sometimes there is a cynical rejection of interest in the larger affairs of community, nation and world.

From all this there are serious consequences; the social disorders of the adolescents tend ultimately to sexual deviation and mental breakdown. In our time there is a tendency not only toward a rising incidence of breakdown, but also toward an increasing severity of breakdown.

So far I have had to indulge in some sweeping generalizations, which inevitably carry with them unavoidable risks of misunderstanding. There is the whole question of establishing the distinction of normal and abnormal adolescent behavior. Let us therefore introduce a few needed qualifications. A healthy, vibrant, sparkling adolescent is a pleasure to behold. There are still enough of them, even in our troubled world. Adolescent behavior is characterized by enormous variability. Within this range, it is entirely possible for a healthy teenager for a time to exhibit some of the above trends and yet recover his emotional balance, relatively untouched and unharmed. This is simply a phase of growth. In most instances the tendency toward such disturbance is mild, transitory, benign and reversible. It is influenced by a multiplicity of factors — social, economic, cultural, geographic, familial and personal. Of central importance, of course, is the emotional conditioning of the adolescent within a particular type of family. Nonetheless, while we recognize explicitly that most youngsters survive, the fact remains that these trends represent a distinct vulnerability for the adolescent group as a whole.

What I have described cannot be viewed simply as a phase of the normal growth, the inherent instability of adolescent personality. It is much more than this. The significant source of such disturbance is something outer rather than inner. Basically these adolescent trends are a sign and symptom of a sick kind of family living in a sick community. Significant anthropological studies have demonstrated that adolescents can mature in certain cultures without undergoing critical emotional storms and without serious explosions of destructive conduct. The trend in our particular culture is a dangerous one. It is a disposition to a type of social and emotional disorder that can have profound consequences for mental health.

Illustration of Disturbed Family and Adolescent

Let us view the above adolescent disorders as symptoms of the social psychopathology of the adolescents of our time. We must then make the further step of relating this to the social psychopathology of the family of our time. The misbehavior of the adolescent may be regarded as a symptom of chronic pathology in the whole family. The adolescent acts as a kind of carrier of the germs of conflict in his family. In many instances, he seems to act out in an unrational manner among his extrafamilial relationships the conflicts and anxieties of his family, particularly disturbances existing in the relations of his two parents.

The following illustration is typical:

A man of middle years makes an appointment for psychiatric consulatation for his eighteen-year-old son. John has flunked out of three schools, cannot concentrate, steals money from his mother, gambles, and while playing truant, escapes to the low-class movie houses on 42nd Street. He is frightened of encounters with homosexual men, and yet paradoxically goes exactly to those places haunted by them. He has been accosted and propositioned by these characters many times.

The psychiatrist requests that the whole family come for an office interview. Instead, John appears only with his mother. When asked, "Where are your father and older brother?" John and his mother in a single voice instantly exclaim, "Oh, you'll never get Pop here. He's against psychiatrists; he doesn't believe in them."

John is a boy of meduim height, with attractive, delicate, girlish features. He has almost a baby face. He is nattily dressed and his skin-tight trousers stand out conspicuously. Despite the obvious care invested in clothing, his posture is slouched over, his head is bowed; he observes the interview proceeding in a detached, blank way. Now and then, he bursts into a fit of childlike giggling. He is amused at the silly quarrels between his parents. He thinks his father is cute, puts on quite an act. He handles his body awkwardly. He is self-conscious about his body. He is ashamed of his well-developed breasts and his "big can." He admits his fear of looking like a girl. Therefore, he won't appear in a bathing suit. He is fearful of homosexuals, but also very curious. He is apathetic toward girls. He has only a single interest — gambling; he wants to play the Wall Street game as his father does. His father brags of his talent in gambling. He swears he can beat the Wall Street game. With a false, exhibitionistic modesty, he declares he is a "gambling degenerate."

In subsequent sessions, Pop did come and treatment was begun with the whole family on the assumption that John's disturbance echoed the family disturbance. Pop put on a show. He was a real ham. He insisted on talking for everyone else in the family. He barely gave the others a chance to open their mouths; he intruded on them time and time again. He is a bright, quick, but tense, agitated, overaggressive man. He is plagued by diffuse hypochondriacal fears, cancer, cardiac collapse, bleeding from the rectum, etc. From the word go, the father launched a critical, reproachful, intimidating tirade against his family. In the most dramatic manner, he depicted himself as a hero and and martyr to the family cause. He described vividly how he came up the hard way on the streets of the East Side. He is a devoted father who "gives and gives and gives" to his family out of his pocketbook. He talked of nothing but the dollar, how hard it is to "make a buck," and how his family robs him. They demand more and more; they're never satisfied. He has to make sacrifices to keep "doling out the dough."

Severely provocative, the father stirred his family to counterattack. Mother kept quiet, but her face was depressed, embittered, and a sly contemptuous grin played about the corners of her mouth. She was intensely hostile. It was the older son, Henry, who assumed for the mother the role of attacking the father. He did so viciously. On the other hand, the younger son, John, merely grinned. He felt unable really to talk with his father. He was frightened of him. The only way he could reach him was to love him up physically. Each evening, when his father comes home, John hugs him and kisses the top of his head, he caresses him; sometimes he playfully pokes him in the belly with his fist, or as he says lovingly, "I wring the neck my father hasn't got." Father has a very short neck.

Here is a bird's-eye glimpse of a disturbed family in action, containing one predelinquent adolescent boy; a family in which John's failure in school, his stealing, his tendency toward homosexuality, and his insatiable craving to gamble are symptoms of the disturbance of his entire family group. This disturbance has its origin in certain basic unsolved problems in the relations between mother and father.

From the word go in this marriage, there was trouble. According to father, mother shunned his love, she rejected him sexually, she was cold, ungiving. Mother, on the other hand, declared that he was excessively demanding. For the first week, he did not leave her bedroom. By his own admission, father was afraid to let mother out of his sight because he was jealous of other men. He imagined that she might be unfaithful. On the other hand, mother was filled with bitter feelings against father. He left her alone night after night; he never stayed home, he went out with the boys, drinking and gambling, and looking at other women. He amused himself with dirty pictures and dirty stories. Throughout the entire twenty years of marriage, mother and father could not agree about anything. If mother said white, father said black. Father terrorized mother with his explosive tantrums. She in turn took vengeance in sly, covert ways. She emasculated him. She declared openly for years that he was undersexed. She allied both boys with her against father; she made it three against one. The father felt rejected and exiled. He then turned about to become seductive with his younger son, John. He carried on an open flirtation with him. John became the second woman for his father, substitution for mother. Mother in turn trained her older son, Henry, to be her fighting arm against father.

Disturbances in Social Norms

Now, besides the familial conflicts one must consider the wider disturbances in social norms that reverberate in the behavior of adolescents. Today, the goals of society and family life are unclear. Between the two there is little harmony. In general, human relations are agitated, turbulent, out of balance. Families as families are confused and disoriented; it is difficult to be clear as to aims, ideals and standards. People marry earlier, separate and divorce more frequently. The small nuclear family gets separated from the extended family representations. Grandparents and other relatives fade out of the picture. Mother has no built-in mother's helper, no dependable sustained support for her duties with the children. The family moves about from place to place; it fails to grow roots in the community; it is weakly buttressed by church, school, social services, etc. In effect, there is a long gap between the old and new way of life — a discontinuity from one generation to the next. The connection of the family with its past is severed; its horizontal supports in the wider community are thin and unpredictable. It cannot see what lies ahead.

The family of our day therefore does not succeed in planning ahead. It deals with its problems in hit-or-miss fashion; it improvises, it often reacts in extremes. It fails frequently to achieve a real unity; parents are confused as to the kind of world for which they are training their children. The binding power of love and loyalty in family living is fickle and undependable. The contemporary family cannot somehow hold itself together very well. Instead, each member tends to go his own way.

The emotional climate of the family is often pervaded by mistrust, doubt and fear; there is less feeling of closeness, less sharing, less intimacy and affection. The symbols of authority, the patterns of cooperation, the division of labor, are confused. No one knows clearly what to expect of anyone else in the family. There is considerable fuzziness concerning appropriate behavior for father, mother, and child. Because of this, it is extremely difficult to maintain the needed balance and harmony of essential family functions and thus the stability of family life is thrown into jeopardy.

Since father and mother do not know what to expect of one another, neither is sure what to expect of the child. Emotional splits develop within the group; one part of the family sets itself against another. There is a battle between the generations, or a battle between the sexes. Under the stress and tension of daily living, the family sacrifices one essential family function in order to maintain another. Family life gets rigid, it gets stereotyped; it also grows sterile and static. No longer is the kitchen the place of cozy exchange of family feeling; instead, it is the shiny, cold, impersonal, machine-like kitchen. Family relations get cold and almost dehumanized.

The two parents compete; neither can take the allegiance of the other for granted. The father tries to be strong, but fails. Mom takes the dominant position; Pop recedes to the edge of the family. Out in the world, Pop is a somebody; he counts for something. At home, he is not sure he has any role to play. According to Loomis, the modern father is something of a cross between an absentee landlord, a vagrant and a sap. He pursues what is called the suicidal cult of masculinity. It is not enough to be man, he must be a superman. The sexual relation ceases to be love-making; it becomes a proving-ground for technique, the struggle for competitive dominance. It becomes impersonal, dull, a hollow ritual; it dies a slow withering death. Within the family, often it is unclear who is the man and who is the woman. In fact, all the family roles tend to be confused; sometimes the children act like parents, they usurp control. The parents lean on their children for a sense of guidance which they do not derive from other sources. In this way, the natural difference between the two generations becomes obscured.

Parents seem to be afraid to love their children. It is as if loving is losing something they need for themselves. They function with an image of profit and loss in family relations. They fear to sacrifice that which they may need to spend on themselves. They feel that the demands of their children are exorbitant, they project to their children their fears and hates and unwanted qualities of their inner selves; they intimidate and scapegoat their children, and in turn are scapegoated by them. They do not give their children an appropriate sense of responsibility; often they are overprotective. They become self-conscious, stilted professional parents. They compete with their own children and sometimes try to play the role of big brothers and sisters rather than parents. Out of a sense of weakness and guilt they take recourse to manipulation and coercion. Their efforts to discipline the children are feeble and ineffective, and they resort weakly and artificially to devices of deprivation, or so-called reasoning with the child, but have no confidence that the discipline will work.

So, in our adolescents we see anarchy and violence, sexual revolt, excessive conformity, a quest for a safe, secure, unadventurous, uncreative niche in industry, or we see withdrawal, a sense of defeat and cynicism and a loss of faith, hope, and idealism; certainly a profound and bitter disillusionment in the symbols of parental authority. In this setting, the adolescent lives out unrationally the elements of conflict, imbalance and destructive competition in the relations of parents and grandparents and also the conflicts between family group and the wider community.

From these considerations, one learns one simple lesson: the disturbed behavior of adolescents is not only a reflection of a personality problem, but also a symptom of a disordered family. The two components must be correlated, the adolescent disturbance and the parallel imbalance in family relationships. One needs to trace the fluid, shifting processes of identity relations of adolescent with family and family with the wider community. More precise diagnosis of these multiple interdependent, interpenetrating sets of influence would enable us to develop a more effective remedy, an appropriate merging of family life, education, social service and psychotherapy.

C. CULTURAL VARIATIONS

CHAPTER 19

Some Sociocultural Factors in Adolescent-Parent Conflict: A Cross-Cultural Comparison of Selected Cultures

LAWRENCE SCHIAMBERG

Some level of adolescent-parent conflict has been virtually a constant factor in human societies. The problem of intergenerational relations has been so widespread as to have required some societal response, whether in the form of initiation rites or rules governing intergenerational behavior (1). It is generally during the period of adolescence that youth-parent conflicts are intensified because it is during this period that the youth must begin to make progress toward becoming an adult.

Several reasons have been suggested for the so-called conflict of generations in Western societies: (*a*) the different content of experience for youth of the present and for their parents when they were young; (*b*) the lack of clearly defined steps marking the recession of parental authority over children; and (*c*) the resulting differences between parents and youth on the psychological and sociological levels (youthful imagination versus adult experience, on the psychological level, and parental role as supervisor of child development versus child's need for independent experience on the sociological level (2).

The brevity of these statements should not disguise the ultimate complexity of adolescent-parent conflict. For example, the idea of "different experiences" of youth and adults involves a large number of possible combinations, such as the particular style of family relationships, and lack of a sense of historical relatedness due to continual social change and particular traumatic events — to name two of the more general categories of experience. The main point of this paper is that adolescent-parent conflicts, or the so-called conflict of generations in the West, are not arbitrary and inscrutable but are directly related to the sociocultural background in which they occur. This point will become somewhat clearer upon examination of several different cultures — their values, norms, and their handling of adolescent-parent conflicts.

However, before examining the "generation gap" in its cross-cultural perspective, an examination of the various explanations of the adolescent-parent conflict will serve to further clarify that point. Explanations of the conflict of adolescent and parent (father) have followed the pattern of initially treating the problem as an intrapsychic personality problem (Freud) and more recently emphasizing the equally important influence of the sociocultural milieu in helping to shape the personality.

Some Selected Theories

Perhaps the most famous explanation of the "conflict of generations" was Freud's notion of the "Oedipus complex." Freud thought that the male youth's relationship with his father, and ultimately with his culture, was detrmined by how well the son resolved the problem of identifying with his father versus his desire for sexual relations with his mother. The dilemma would hopefully be resolved when the adolescent son observed the father's dominance over the desired sexual object, the mother, and then would identify with the father as a source of power and control. These intrapsychic conflicts of the Oedipal period reappeared during the adolescent stage and then had to be permanently resolved (3).

The weakness of the Oedipus complex theory is that it seems to underplay the fact that father-adolescent son relationships are conditioned largely by their social and cultural background. Freud's theory treats the problem of the "conflict of generations" from an intrapsychic point of view rather than as a problem of interpersonal relationships influenced by sociocultural norms and values. This is not to deny the existence of intrapsychic aspects of the problem but rather to suggest that Freud perhaps laid an undue amount of stress upon somewhat impressionistic and unverifiable constructs, such as the Oedipus complex.

Erik Erikson's notions of adolescent-parent conflict differ from Freud in that the latter places much more emphasis on the role of the Oedipus complex whereas Erikson emphasizes the social nature of the conflict (4). Erikson's notion of adolescence involved the problem of establishing one's "ego identity." This concept primarily involved the individual's relationship to his parents and other individuals in his world, and the establishment of a general "stance" toward the world. Erikson was to some extent influenced by the work of cultural anthropology because he recognized that the method and content of adolescent ego identity would differ from culture to culture. Erikson did suggest that the achievement of adolescent ego identity had one element in common for all cultures: the adolescent must receive meaningful recognition of his achievements from his parents and from his society (5).

Erikson recognized two major sources of conflict between adolescents and parents: (1) the failure of parents to accord recognition of adolescent achievement; and (2) adolescent revolt against the values and dominance of the parents. Erikson indicated that youths rarely identified with their parents during adolescence and often rebelled against their parents in their quest for ego identity. Ultimately, Erikson thought the adolescent must establish his ego identity by adopting and formulating a stance toward the world. Erikson seems to extend the Freudian concept of same-sex identification to include the social implications of a failure to achieve ego identity. Thus, in Erikson there is an emphasis upon the social nature of adolescent-parent conflicts, although he does not divorce himself from strict Freudian interpretation such as the Oedipus complex theory.

Perhaps the work of Kurt Lewin applies most specifically to the nature of adolescent-parent conflicts, since Lewin was concerned with both the stage of adolescence and the resolution of social conflicts. Lewin approached the problem of intergenerational conflicts from the point of view of his general theory of behavior which stated that behavior is a function of the person and the environment [B=F(P.E.)]. The sum of all environmental and personal factors (motivation, needs, perception, etc.) in interaction is called the life space, and behavior is also a function of this construct [B=F(L.Sp.)] (6). Lewinian field theory recognizes that life spaces may differ between individuals and cultures. Field theory views adolescence as a transition from the life space of the child to the life space of the adult. Because of the rapid and somewhat abrupt shifts in the life space of the adolescent, he often becomes a "marginal man" with one foot in the world of childhood and the other in the world of the adult. The adolescent often experiences emotional tension due to the ambiguity of his social position. The resulting tension may affect adolescent-parent relationships and often produces conflicts over role expectations. Perhaps the most important contribution of Lewin to the study of intergenerational conflicts is his notion of the general cultural atmosphere or background for social situations: "In sociology, as in psychology, the state and event in any region depend upon the whole of the situation of which this region is a part" (7).

Some recent ideas on the relationship between personality and culture — and by implication the relationship between the generations — seem to have followed in the footsteps of Kurt Lewin. The background of parent-adolescent conflicts is seen to be interaction of the personality system and the sociocultural systems, which in turn are mediated by the family system (8). The social system has certain functional prerequisites such as role differentiation, shared goals and values, and communication, while the personality system has certain requirements such as the satisfaction of needs and recognition of achievement. The family system mediates between the personality and social systems by ensuring the presence in individuals of societal goals and values and by providing an atmosphere in which the achievement of these goals is recognized. The socialization processes of the society — of which the family system is a prominent force — serve to create a congruence between the functional prerequisites of the society and the motivational patterns of the personality system (9). Parent-adolescent crisis, if it does develop, occurs within this complex arrangement of personality, cultural, social, and familial variables. The problem of the conflict of the father with the male adolescent — the main concern of this paper — cannot then be divorced from this complex interrelationship of variables. Whether one approaches the problem from the viewpoint of a learning theory or a cognitive theory or any other point of view, the parent-adolescent relationship is ultimately limited by the nature of the particular sociocultural values in each particular culture.

The conflict of generations would seem to be best explained from a cultural anthropological view which recognizes that the parent-adolescent relationship is related not only to particular personalities and temperaments, and particular circumstances, but is also related to cultural and societal values and norms which influence the parent-adolescent relationship through the medium of the family. In light of this stance, it makes no sense to condemn industrialized Western societies for what appears to be a greater prevalence of such conflicts and rivalries, since many Western societies are so much more complex and more rapidly changing than some Eastern societies which apparently have less parent-adolescent conflict. Further, as Margaret Mead has indicated, the *Sturm und Drang* characteristic of American adolescence is perhaps the price we pay for our high level of technology and material welfare (10).

In many of the cultures and societies to be discussed in this paper, Western technological innovations are beginning to reshape the society, and therefore the parent-adolescent relationship. With increased industrialization in Eastern countries such as India and China, adolescent-parent problems begin to appear which bear a marked resemblance to those of the West. For example, in China, as more job opportunities have become available to adolescents, the traditional dependence of the young upon their elders has begun to disappear and with it has gone the centuries-old tradition of unqualified respect for one's elders and ancestors (11). A "generation gap" has begun to appear as the experiential worlds of Chinese youth have become different from the experiential world of their elders and ancestors. These experiential differences have formed a basis for more conflict between adolescents and their parents. When this situation in China (a situation which has not, as yet, permeated all of Chinese society) is compared with the United States — in which the so-called computer revolution is further extending the effects of industrialization — then perhaps it is not so surprising that differences of values and intergenerational conflicts are perhaps more prevalent in the West.

The question of adolescent-parent conflict has important ramifications for society at large and for the schools in particular. In Western societies, as is the case in many non-Western societies, the family is the primary influence on the developing child. However, in static (unchanging) societies parental influence is often continued into latter childhood and often into adolescence and adulthood, as children learn virtually all they need to know from their parents. However, in American society the schools have developed as a means of training youth for adult responsibilities (especially those skills necessary in a technological era). The schools (especially those devoted to higher education) in advanced technological societies serve to provide youth with the kind of information and skills which are not the specific function of the family or other societal institutions to transmit. If the schools serve only to transmit and teach technological and social skills without confronting the problem of adolescent-parent conflict and, more generally, the conflict of generations, then the problem can only become worse. The purpose of this paper is not to dispute whether the apparent "price" paid for social change is too high in terms of individual discontent, but rather to place the problem of adolescent-parent conflict in perspective: (1) by showing the intimate relationship between sociocultural values and methods of preventing adolescent-parent conflict in both nontechnological societies and in societies which are only beginning to be influenced by Western technology; and (2) to present some of the educational implications of adolescents and parents who are out of step with one another.

Some Selected Cultures

Indian society reflects certain characteristics of Eastern societies which have led to much lower levels of parent-adolescent differences and conflicts. Indian society has traditionally placed a great deal of emphasis on the quality of interpersonal relationships. Relatively little value is placed on the quality of material existence. Indian society has been relatively static for centuries, and as such the traditional values and norms of the society have remained unchanged and unchallenged. Generational gaps have not developed because life styles have remained constant for centuries.

A characteristic of Indian society which alleviates the strain of potential adolescent-parent conflicts and is consistent with the societal emphasis upon quality of interpersonal relationships is the Indian "extended family."

This is the characteristic type of household in India (12). The extended family allows for a number of interpersonal relationships whereas the nuclear family tends to produce more developed and intense relationships with fewer people. The extended family allows for more distribution of emotions and feelings over a greater number of family members, as compared with the nuclear family with fewer members, a potentially higher concentration of emotions per relationship, and therefore a higher likelihood of potentially explosive intergenerational relationships (13).

In India the institution of *asrama* ensures the smooth transfer of authority from generation to generation. According to this tradition, sons owe complete obedience to their fathers, while the fathers are required eventually to relinquish their authority so as to avoid conflicts between adolescent males and their fathers. The Indian male is supposed to go through four asramas, or "age grades" in Western terminology. At the age of eight the boy enters the first asrama or the celibate stage. During this period the child is guided by one or more instructors who teach him about sacred love, the arts, the use of weapons, and the profession which he will eventually take up (14).

After his educatory period, the adolescent (age of twenty) is admitted into the next age grade — that of "householder." In this stage, he gets married, starts a home, and establishes himself in the profession for which he has been trained. This stage lasts until his own son reaches the householder stage or, if he has no children, until he is middle-aged (determined when his hair turns grey). At this point, he enters the third age grade, or *vanaprasthasrama.* During this stage the man is relieved of his household duties and is now free to devote his time to meditation or to worship. The man may live at home if he so desires, although an orthodox Hindu usually lives in a nearby house. The vanaprasthasrama age grade ensures that by the time a man's son is able to enter the householder stage and accept the responsibilities of manhood, the man of the house is leaving the household, thus allowing a smooth transfer of authority (15).

Underlying the social structure of the asrama is the basic Hindu ideal of *dharma* (ideal duty), which consists of the right behavior appropriate to one's particular stage and station in life. According to the Hindu religion a man is born with three debts: (1) a debt to the gods; (2) a debt to the sages; and (3) a debt to his ancestors. The individual pays these debts by worship and ritual, learning and teaching, and by raising children (16). These activities comprise the various age grades or asramas. Indian life is carefully organized so that each individual is constantly "paying" one of his three debts, while respecting his elders at all times.

Another underlying current in Indian life which reduces intergenerational tension is the emphasis on the ascetic life (as manifested in worship and meditation). Respect for the aged, retired Indian who devotes his last years to religious meditation is encouraged because of the widespread Hindu ideal of respect for the ascetic life. The Hindu belief that self-denial is superior to self-indulgence unites all Hindus in their respect for the aged and further contributes to the lessening of conflicts between adolescents and parents (17). Respect for one's elders is an absolute requirement of behavior for adolescents in India.

The above description of the nature of the parent-adolescent conflict in the Indian family is perhaps most applicable to the more orthodox Hindus and higher caste families and somewhat less descriptive of lower caste families, Muslim communities in India, and families influenced by Western ideals of family relationships. Even for the orthodox Hindu families and the upper caste families, the pattern has been changing very slowly. With increasing trends toward urbanization and the greater availability of factory jobs, the adolescent has begun to liberate himself from complete dependence on the extended or joint family. For thousands of years the extended family has been the main economic force in Indian society. All members of the joint family operate as a single unit, contributing all their earnings to the entire extended family. One might expect greater resistance to this tradition, once individuals are able to establish themselves independently of the family. Although India is still far from being an industrialized nation, increasing industrialization and concomitant social change would appear to be factors in the breakdown of the traditional structure of adolescent-parent relationships.

Unlike the family structure of the United States or that of India, the traditional Chinese society (before the Communist takeover) placed great emphasis on differences of generation as the basis of role differentiation. In the traditional Chinese society the father was respected not because he was older, or a "good father," but because he was of a different generation. The younger generation always had the burden of responsibility toward the older generation. Traditional Chinese society was a patriarchal society in which the men dominated the women and the older generation dominated the younger generation by control of the economic roles of the family. This control was often exercised by the older generation to restrain adolescents from leaving the household or otherwise causing problems, since these adolescents often had no alternative employment except that offered by their parents (18). (This situation began to change in the twentieth century as Chinese cities became industrialized and more job opportunities were available to Chinese adolescents.) The fact that the traditional Chinese family was the primary economic unit in the society placed a great deal of power and authority in the hands of the older generation who controlled the extended family. Respect for one's elders by the Chinese adolescent was not only a traditional value of the society but also a socioeconomic necessity.

Traditional Chinese society also had a hierarchal chain of power. Older brothers were more important than younger brothers and were responsible to their parents for ensuring the proper behavior of their siblings. Through these chores and responsibilities, adolescents usually gained experience in controlling young children which would be of enormous assistance when raising their own children. In the determination of this power hierarchy, traditional Chinese society emphasized "particularistic" factors such as one's generation and the sequence of siblings. Certain Western societies (e.g., United States) place more emphasis on "universalistic" factors such as the ability to run a family (19).

In comparison with the United States, the dominance of the older generation in the traditional Chinese society holds for both the early and later years of life. In American society the older generation dominates for the early years of growth and not necessarily for the later years (20). This dominance of the Chinese youth and adolescents was the result of traditional Chinese inheritance customs. The land and property were not supposed to be divided among the children until the father died. No matter how ineffectively an old man carried out the household chores or the work in the fields, he was still recognized as the head of household with authority over all those of a younger age. This remained the case even though younger family members might, in fact, be doing more efficient work than the titular household head. The inheritance customs buttressed the tradition of ancestor worship and respect for one's elders. There was no problem dealing with the rate of recession of parental authority as in the United States, since parental authority was virtually a lifelong fact. Thus, three main factors established the locus of power in traditional Chinese society: (1) sex (males dominate); (2) generation; and (3) relative age (oldest siblings having preference).

From childhood on, the youth and adolescent was virtually immersed in a culture which stressed filial piety. There simply were no exceptions to the rule — that is, no successful exceptions. The child and the adolescent were confronted with examples of filial piety in their daily lives, in novels, textbooks, and nursery schools. Veneration of the older generation and of one's ancestors was possible in traditional China because it was essentially a static culture in which the life experiences of one generation were almost identical with those of any other generation (21). Perhaps the respect for age and experience rather than youth and imagination is best expressed by Confucius. When asked how to farm, Confucius replied: "I do not know as much as an old farmer" (22).

Respect for the aged was in part derived from the Confucian ideal that the good life consisted of the proper behavior between individuals. According to Confucius there were five types of interpersonal relationships: (1) the parent-child relationship; (2) the king-minister relationship; (3) the husband-wife relationship; (4) the older brother-younger brother relationship; and finally (5) the friend-friend relationship. Confucius felt that the highest form of respect that could be shown was between father and son. More specifically, that respect which the son showed for his father. Confucius felt that the father-son relationship was the archetypal pattern for the other four basic types of relationships. This Confucian ideal was in fact quite practical, in light of the traditional Chinese inheritance customs (23).

The traditional precedence of the older generation has broken down in modern China. With the coming of industrialization and its emphasis on youth and adolescents as those best qualified to learn the new technological and factory jobs, the unqualified respect for the older generation began to deteriorate. The role of the traditional Chinese family as the educator of the young for occupational work is being usurped.

The gradual transition from an emphasis on age to an emphasis on youth was one of the primary trends which the Communist movement seized upon to gain support among Chinese youth. When the Communists finally gained control of China after World War II, they encouraged progressive young people to disregard existing kinship ties and to ignore the concept and practice of ancestor worship (24). Soon progressive adolescents and young people became feared throughout China because any word from them to the Communist leadership of the practice of age prestige rituals could lead to stiff reprisals against the "guilty." Youth in "transitional" China have also rebelled against the traditional parental control of marriages and divorces. Marriage soon became focused upon the husband-wife relationship, the nuclear family (husband and wife rather than the clan), and on the free choice of one's mate. Thus, with the collapse of the family as the main economic unit in Chinese society and with the subsequent decline in age prestige, traditional adolescent respect for parents has begun to wane.

Indian and Chinese societies have traditionally had considerably less adolescent-parent conflicts primarily because the traditional concept of adulthood has been coherent and meaningful to the adolescents of these societies. There has been little difficulty in becoming an adult because the prerequisites for adult status were within the reach of virtually all adolescents. The only requirement was that adolescents become reasonably proficient in such tasks as working in the fields and caring for the young — tasks for which they had been trained all their lives. There were no frustrations involved in the choice of vocation, level of education, or style of life since there were virtually no alternatives to the existing social framework, nor any notion that society could be any different than it always had been. In short, where there were no choices and virtually everyone accepted the existing life styles, there were fewer individual frustrations — that is, fewer frustrations caused by having to choose between two or more alternative styles of life and accepting the consequences of that choice, but probably more frustrations in getting enough food — and fewer frustrations in interpersonal relationships and interpersonal role expectations.

The traditional simplicity and integrated social organization of China and India have permitted an emphasis on the quality of interpersonal relationships. In Erikson's terms, the attainment of "ego identity" has been so much easier in India and China that much smoother interpersonal relationships have been more likely to occur than in societies where ego identity is more difficult. Where there is greater agreement on the means and ends of life, and where adulthood is both possible and meaningful, adolescent-parent respect is greatly facilitated. Respect for one's elders and one's ancestors is more likely when one can be sure that Confucius' old farmer does, in fact, know more about farming than Confucius.

In some African societies the parent-adolescent relationship is linked primarily to an individual sense of reciprocity with the groups of which one is a member (25). Principles of mutual rights and mutual duties — especially between the father and his adolescent son — run strongly through interpersonal relationships in many of the African societies such as the Tallensi (26). Religious beliefs often further reinforce the influence of social groups ("spirits" — especially the spirits of the wrongdoer's ancestors — are frequently seen as forces of retribution) (27). Further, many African societies have clearly delineated systems of courts and methods of hearing disputes between the various clan members and between the various clans or joint families. Many African societies handle the problem of parent-adolescent conflict with clearly spelled-out age grades or initiation ceremonies or some combination of the two (e.g., the pastoral Massai tribe). Other societies such as the Tallensi or the Mossi, lacking in highly specific age grades or initiation ceremonies, rely on other societal institutions or practices such as the principle of reciprocity (the societal norm of mutual duties and rights in interpersonal relationships, the practice of parent-adolescent avoidance, or refined methods of parent-adolescent dispute arbitration), (28).

In the social system of the pastoral Massai, the male sex is divided into boys, warriors, elders, and old men (29). The first age grade ("boys") lasts until circumcision, somewhere between the ages of thirteen and seventeen. Those who are circumcised automatically become members of a peer group of circumcised youth called by some distinctive name such as "white swords" (30). The newly circumcised youths do not become adults until they reach a senior age grade and are then allowed to marry. The newly circumcised live in a separate age-set village and are entirely under the authority of the elder age grades. Circumcision is thus only a first step to becoming an adult man. It might be thought of as the prerequisite for passage from childhood to adolescence, whereas marriage is the *rite de passage* from adolescence to adulthood (31). Once the youth acquires a wife he becomes a "junior elder." Junior elders are mostly concerned with family matters (their interests are mostly of a private nature) and they play a rather small role in tribal politics. Their major concern is to increase their status by having more wives and by having their own children circumcised and then initiated. Once junior elders have had one of their children initiated, they become "senior elders" with a full share in the tribal political life and assume the prime responsibility of initiating new members into Massai manhood (32).

The major importance of the Massai system of age grades and initiation rites is that they regulate an individual's conduct in relation to those in similar or different age grades. The basic prerequisite for effective initiation ceremonies and age grades — the solidarity of the initiators in their concept of the role of an adult — is fulfilled in Massai society (33). Unless this solidarity exists, initiation rites do not have a very powerful effect. Solidarity exists not only among the elders but also among the lower age grades, all of whom subscribe to the personality type preferred for each of the various age grades. These strong forces of cohesion among the various age grades direct attention away from inter-age grade struggles and channel it toward fulfillment of the specified goals of each age grade.

Not all societies require such rigid maintenance of age grades and *rites de passage* as do the pastoral Massai. For example, in the Tallensi society the transition from childhood to adulthood is a gradual process. The Tallensi do not have rigidly defined age grades but two rather loose criteria on which they determine the developmental level of an individual: the physiological criterion of pubescence and full physical maturity, and the economic criterion of development of skills requisite to doing a man's work in the fields. The attainment of adulthood in Tallensi society is based on a rather variable and somewhat flexible schedule. There is no social break to mark the transition from childhood to adulthood (34).

For the Tallensi, the cornerstone of filial piety is the basic notion that the bearing or begetting of a child is a difficult matter commensurate with a great deal of respect. Because childbearing is thought to be so difficult, children are taught to have respect for the fact that their parents brought them into the world, even though they may or may not have respect for their parents as persons. In this respect, the Tallensi believe that the bonds between the youth and the parents can never be obliterated or repudiated. The Tallensi concept of filial piety is a diffused norm rather than a specifically elaborated doctrine as in the case of the Chinese (35).

The principle of reciprocity, or the mutual rights and duties of fathers and their adolescent sons, operates to prevent conflict between them (36).

In the parent-adolescent relationship, reciprocity operates in two notable instances to prevent conflict: (1) in the case of the use of the son's property by the father; and (2) in the arrangement of a marriage by the father. In the former case, if a man uses one of his son's cattle or goats he must eventually make an equivalent return. In the second instance, while the son is always under the authority of the father, it is the latter who arranges the son's marriage. Thus, the priciple of reciprocity helps to reduce parent-adolescent conflicts by ensuring that, in the long run, mutual services and favors must balance.

Other means of reducing parent-adolescent conflict are held in common with the Mossi tribe of the Voltaic Republic, just north of Ghana. The basic economic activities of the tribe include the production of livestock and cotton, and caravan trading. If a wife gives birth to a male child as her firstborn, the father sends the child away to live with the boy's maternal relatives until he reaches puberty. The father then has little contact or conflict with his future successor. This practice of initial removal of the newborn child is a rather drastic means of avoiding parent-adolescent conflict.

Since the Mossi firstborn males eventually inherit the wives of their fathers, it is often those wives who produce friction between fathers and their adolescent sons. In order to avoid conflict, Mossi firstborn sons are usually allowed to visit the family compound for important purposes only. It is an unfortunate circumstance when a father encounters his oldest adolescent son at the doorway of the household compound. In order to avoid this situation, the father or son usually shouts something loud upon entering or leaving the household (37).

Mossi fathers usually do not want to procure wives for their sons because the Mossi father sees the possession of a wife as representing the advantages of majority status. Since the son will gain this status only when the father dies, the antagonism between father and son is increased. It is not suprising that fathers often become extremely jealous of the development of their adolescent sons since this usually signals — at least in their minds — their eventual demise and decline in power. Mossi fathers are careful to avoid situations in which they are or could be directly compared with their adolescent sons. Often a Mossi father will not be seen walking with his eldest son in the village, lest someone accidentally fail to acknowledge their difference in age (38).

Thus, in Mossi society the major indication of majority status and the major basis for adolescent-parent conflict is the possession of a wife by the adolescent. This state of affairs is further complicated by the fact that the eldest adolescent son usually inherits his father's wife upon the death of the father, and also because the adolescent youth remains completely dependent upon the father for obtaining a wife. (Adolescent sons do not possess the property to trade for a wife and must rely upon their fathers to get them one.) Interpersonal relationships between fathers and their adolescent sons are strained because of the social structure of Mossi society, in which a confrontation between father and son is openly acknowledged, although adequate precautionary measures are taken. Perhaps the adherents of the Oedipal theory of generational conflict would point out that the very fact that the Mossi mothers are the subject of conflict between fathers and sons is evidence that the Oedipus complex produces adolescent-parent conflict. The problem with such a position is that there is little or no objective evidence that what appears to be a manifestation of the Oedipus complex in Mossi society does in fact occur in other cultures with any kind of regularity.

Besides the Mossi practice of avoidance, their system of intergenerational conflict resolution helps to alleviate adolescent-parent conflicts. When a serious decision or problem arises between parents and their adolescent youth, the head of the extended family is consulted. No decision regarding such intergenerational problems is ever made without consulting the head of the extended family. This type of decision-making process helps to improve parent-adolescent relations (1) by ensuring that adequate reflections and thought precede each decision, and (2) by preventing direct confrontations between parents and adolescents (39).

Further supplementing the practices of avoidance and dispute arbitration in maintaining smooth relationships between parents and adolescents is the practice of shared adolescent discipline. Disciplining of Mossi adolescents is a shared responsibility between the father — the family head — and the head of the adolescent work group. After Mossi adolescents are circumcised at about the age of thirteen or fourteen, they join a "work group" composed of other youths recently circumcised. In the work group the youths perform work in the fields and are given their food and lodging for the seven-month period that they work there. During this work period the adolescents are responsible to the work group leader — usually an adult — for proper behavior and may be disciplined by the leader for improper actions. The division of disciplinary responsibility in this fashion reduces the possibility that conflict regarding disciplinary practices will develop between father and adolescent (40).

Should the situation ever arise that the eldest adolescent has been given a wife, the son is required to leave the family household and build his own *soukala* or hut, even though he will eventually return to his father's house to inherit his property when the latter dies. The Mossi do this because of a strong belief that young married adolescents have a strong need for independence and that if the father were to keep his married son at his side, the latter might wish his father's death so that he could achieve independence (41). (Mossi fathers maintain the right to control the lives of their children until the death of the children or the father. Mossi children rarely achieve independence until the death of the father, at which time the property is divided and the father's wives are inherited.) (42)

Thus, the problem of adolescent-parent relationships is kept within reasonable bounds by means of several rules of behavior which govern the relationship of the adolescent and the parent. Potential intergenerational rivalry is fostered primarily by the social custom of inheritance of wives. In order to prevent such conflict the following rules of behavior have been established: (1) the enforced removal of the eldest male child from the household; (2) the dispute arbitration process; (3) the divided disciplinary responsibility; and (4) the departure of the eldest married son from the household. The parent-adolescent conflict is rooted deeply in the structure and functions of the Mossi culture. The primary economic and social force in Mossi culture is the joint family. In order to maintain the solidarity of the family unit, the act of acquiring wives is reserved for the elders of the extended family — the father, and the clan chief or head. All marriages are arranged to enhance the extended family and to ensure that all new family members and new wives will remain loyal to the extended family. The inheritance of wives not only indicates the relatively low position of women in Mossi society but also serves to maintain the continuity and effective operation of the Mossi extended family by retaining effective women workers in the family unit. The code of behavior governing adolescent-parent relationships is the Mossi method of successfully integrating personal needs of both fathers and their adolescent sons with the functional prerequisites of a strong and efficient extended family.

In the Arab world the general guidelines for parent-adolescent relationships are laid down in the religious mores and teachings which pervade all aspects of Arab life. Muhammad is reported to have said: "Whoever has a son born to him, let him give him a good name, teach him good manners, and when he reaches puberty get him married . . . if he reaches puberty and has not been married and falls into sin, it is the father who is responsible . . . " (43). The Qur'an (holy book of the Muslim faith) advocates proper respect toward one's parents and states that filial piety is the highest form of good works. In practice these ideals are usually translated into a rather stern father who exercises his absolute authority over his children in arranging their marriages and ensuring that they do not "fall into sin," lest he be responsible. Adolescents are required to show absolute respect and obedience to their father.

Among Arab families there usually exists a close relationship between mother and youth and a rather loose one between father and adolescent boy. The Arab father has complete authority over both sons and daughters. Generally the father is the disciplinarian while the mother helps to keep the family together as an integrated unit and often acts as a buffer between the father and the adolescent. If the adolescent gets into any kind of trouble, help is usually sought from the mother rather than from the father. The parent-child relationship is usually a short one for girls, who generally marry at a rather early age and leave the household, whereas sons have an extended relationship with their parents. The son eventually marries and brings his wife with him to live with his family (44).

The rights of Arab youth and adolescents are usually fixed by custom and/or law. The child has the right to food, care, and upbringing. The child has three types of guardianship: (1) guardianship of upbringing (*tarbiya*), which is accomplished by the mother and usually ends when the child is seven or nine years old; (2) guardianship of education (spiritual guardianship), which involves proper training in the values and rules of the society; and (3) guardianship of property, which involves the maintenance of the adolescent's property until he reaches majority status (45).

In the villages of Turkey, the male population is divided into four groups: (1) children, (2) unmarried youths (*delikani*), (3) young men (*genc*), (4) old people (*ihtiyan*). Marriage is the only necessary requirement to move from the delikani age grade to the genc stage. Passage from childhood to delikani is accomplished by growing a mustache, while passage from genc to ihtiyan is accomplished by growing a beard to accompany the mustache. Circumcision is a necessary requirement before one can marry (46).

Villages in Turkey are made up of households containing extended families. Status in this society is designated by difference in age. Old people are held in greater esteem than the young. Respect for age difference is demonstrated dramatically at household dinners and community feasts. Old men and guests always sit in places of honor and are served first (47). Adolescents must always address their elders with specific titles or kinship terms indicating their age grade, and are rarely permitted to address them with their actual names. Competition between adolescent and parent is further diminished by the social concept of status in Turkish life. One always outranks those in lower age grades. Status is determined by one's position in his age grade (a man is judged by his achievements in comparison with other individuals of his age) (48). In this manner, interage grade comparisons are usually avoided and are relatively insignificant if they occur at all. Such comparisons between adolescents and adults are relatively unimportant since both are virtually agreed on the basic goals of adolescence — to get the adolescent married and to prevent him from "falling into sin." Thus,there is less intergenerational conflict than in the West because life experience in Turkey has been relatively static for hundreds of years. There is no conflict because there are no alternative styles of life.

Transmission of knowledge to the adolescent presents no real problem because this knowledge is virtually the same from generation to generation and specific social institutions have long satisfied this need. The most important semiformal situation for learning the values of the village culture and the role of man in village society are the evening meetings at the Muhtar's home (the Muhtar is the village head — usually an older experienced man) (49). The meetings are usually attended by all the males in the village, although they are primarily directed at the adolescent males.

The relatively homogeneous values of the Turkish village, which promoted a smooth transition from adolescence to adulthood and which greatly lessened the incidence of parent-adolescent conflict, have nonetheless been threatened by social change. Problems have begun to arise due to the increasing number of youths going on to higher education after grade school, and therefore not getting married at the customary age of about nineteen (50). Most of the towns and villages now send their children to a "grade school" from which qualified graduates continue on to a junior high school and to a high school (*lise*). As of 1961, very few of the villagers had finished grade school (51). However, with the increasing number of village children and adolescents who complete high school and continue on to some form of higher education, youths are delaying marriage. This failure to marry at the traditional age has perpetuated the status of Turkish adolescence such that, in the future, there may no longer be a clear and distinct difference between adolescents and adults. The implications of this social change for the adolescent-parent relationship are that experiential differences between adults and adolescents and the special opportunities afforded to the educated youth may, in fact, lead to a kind of "generation gap."

Summary, Conclusions and Observations

In conclusion, the behavior norms and role expectations provide the basis for the smooth adolescent-parent relationships surveyed in the several cultures presented in this paper. Nontechnological societies which have relatively clear-cut and broadly accepted societal goals and values have less adolescent-parent conflict than societies in which individuals are confronted with the choice of many different occupations and life styles. This is not to suggest that the so-called generation gap has reached the critical stage in industrialized nations or that choice of life styles is a bad thing, but that where conditions exist which could create experiential gaps between parents and adolescents, the stage is set for more numerous intergenerational conflicts.

Surely, not all adolescent-parent conflicts are caused by social change or are avoided merely by the absence of social change. Personality variables undoubtedly play a role. Perhaps a solution to the problem of the generation gap lies in the development of certain cognitive styles which will promote understanding between the generations in the face of ever increasing social change. Perhaps a feasible method of partially alleviating the problem is for the school to assume some leadership responsibility in teaching individuals to accept change and to adapt to a world of ever increasing change and complexity in which the establishment of "ego identity" (a stance toward the world) in adolescence becomes ever more difficult. Perhaps what is needed — and is most difficult to develop in individuals — is a tolerance for uncertainty. As the rate of social change increases, the shape of the future becomes somewhat fuzzy and the schools become less certain of how to prepare people for that future. To the extent that this kind of situation exists, it would seem logical that the more successful individuals will be those who can tolerate and live with this uncertainty. David E. Hunt of the Ontario Institute of Educational Studies has developed a model of cognitive development which has as its goal a conceptual style which permits numerous combinations of information processing, as opposed to monolithic or stereotyped categorization of information (52). Higher conceptual levels are associated with more advanced information processing and with greater tolerance of frustration and uncertainty.

Perhaps also the school could exercise some leadership in reexamining some basic ideas of human development such as Maslow's notion of self-actualization. Maslow postulates two basic kinds of human needs: (1) deficit needs, or those shared by all members of the human species (e.g., safety, love), and (2) idiosyncratic or self-actualization needs which are peculiar to each person.

> Just as all trees need sun, water, and foods from the environment, so do all people need safety, love and status from their environment. However, in both cases this is just where real development of individuality can begin, for once satiated with these elementary, species-wide necessities, each tree and each person proceeds to develop in his own style, uniquely, using these necessities for his own private purposes. In a very meaningful sense, development then becomes more determined from within rather than from without [53].

As the distribution of wealth and the standard of living have increased, notions of self-actualization such as the acquisition of material goods have become somewhat less meaningful to a growing minority of so-called "alienated" youth. Although it may have been very meaningful for a man to work to provide food and shelter for his family during the depression of the 1930s, when food and jobs were hard to get, it is perhaps somewhat less meaningful to work to pay off a mortgage on a split-level suburban home. David Riesman has raised the poignant question: "Abundance, for what?"

(54). Perhaps adolescent-parent conflict and the attendant generation gap might be eased if there were some reinterpretation of the notion of self-actualization to encompass the broad range of human possibilities.

References

1. S. N. Eisenstadt. *From Generation to Generation: Age Groups and Social Structure.* Glencoe, Ill.: Free Press, 1956.

2. Kingsley Davis. "The Sociology of Parent-Youth Conflict." *American Sociological Review,* 5, August 1940, pp. 523-35.

3. Sigmund Freud. *An Outline of Psychoanalysis.* New York: Norton, 1940, pp. 25-33.

4. Rolf E. Muuss. *Theories of Adolescence.* New York: Random House, 1962, pp. 34-39.

5. Muuss, ibid.

6. Muuss, pp. 84-93.

7. Kurt Lewin. *Resolving Social Conflicts.* New York: Harper & Row, 1948, p. 4.

8. David Aberle et al. "The Functional Prerequisites of a Society." In Albert D. Vilman, *Sociocultural Foundations of Personality.* Boston: Houghton Mifflin, 1965, p. 396.

9. Bert Kaplan, ed. *Studying Personality Cross-Culturally.* Evanston, Ill.: Row, Peterson, 1961, p. 3.

10. Margaret Mead. *Coming of Age in Samoa.* New York: Morrow, 1928.

11. Marion J. Levy. *The Family Revolution in Modern China.* Cambridge: Harvard University Press, 1949, p. 155. (Human Relations Area Files: China, Source 8.)

12. The "extended" or joint family is to be distinguished from the "nuclear" family which is composed of a husband, a wife, and their children. An extended family is a group of nuclear families living together in the same household. The above definitions are taken from Stuart A. Queen, Robert W. Halsenstein, and John B. Adams, *The Family in Various Cultures.* Philadelphia: Lippincott, 1961, p.12.

13. Francis L. K. Hsu. *Americans and Chinese: Two Ways of Life.* New York: Henry Schuman, 1953, p. 28.

14. Irawati Karve. *Kinship Organization in India.* Deccan College Monograph Series, Poona, 1953, pp. 60-62.

15. Ibid., p. 62.

16. Ibid.

17. L. S. O'Malley, ed. *Modern India and the West.* London: Oxford University Press, 1941, p. 59. (Human Relations Area Files: India, Source 1.)

18. Levy, p. 155

19. Ibid., p. 161.

20. Ibid.

21. Ch'in-kun Yang. *The Chinese Family in the Communist Revolution.* Cambridge: Massachusetts Institute of Technology Center for International Studies, 1953, p. 150. (Human Relations Area Files: China, Source 70.)

22. Ibid., p. 150.

23. Cornelius Osgood. *The Koreans and Their Culture.* New York: Ronald Press, 1951, pp. 38-39. (Human Relations Area Files: Korea, Source 22.)

24. Ch'in-kun Yang, p. 151. Also see the following for an analysis of the results of this policy: T. H. Chen, C. Wen-hui, and Chen, "Changing Attitudes Towards Parents in Communist China," *Sociology and Social Research,* 43, pp. 175-182.

25. Simon Ottenberg. *Cultures and Societies of Africa.* New York: Random House, 1960, p. 57.

26. Meyer Fortes. *The Web of Kinship Among the Tallensi.* London: Oxford University Press, 1949, p. 209.

27. Ottenberg, p. 58.
28. Fortes, p. 209.
29. A. C. Hollis. *The Massai: Their Language and Folklore.* Oxford: Clarendon Press, 1905, p. 298. (Human Relations Area Files: Mossi, Source 1.)
30. Hollis, p. 298.
31. Ibid.
32. B. Bernardi. ("The Age System of the Massai.") *Annali Lateranensi,* 18. Cittá de Vaticano: Pontificio Museo Missionanio Ethnologico, 1955, pp. 257-318. (Human Relations Area Files: Massai.)
33. Frank W. Young. *Initiation Ceremonies — A Cross-Cultural Study of Status Dramatization.* Indianapolis: Bobbs-Merrill, 1965, p. 141.
34. Fortes, p. 198.
35. Ibid., p. 171.
36. Ibid., p. 207.
37. Elliot P. Skinner. "Intergenerational Conflict Among the Mossi: Father and Son." *Journal of Conflict Resolution,* 5, March 1961, pp. 55-60.
38. Ibid., pp. 55-60.
39. Ibid.
40. Eugene Mongin. *Essay on the Manners and Customs of the Mossi People in the Western Sudan.* Paris: Augustin Challamel, 1921, p. 92. (Human Relations Area Files: Mossi, Source 2.)
41. Louis Tauxier. *The Black Population of the Sudan, Mossi and Gourounsi Country, Documents and Analyses.* Paris: Emile Larose, Librairie-Editeur, 1912, p. 49.
42. Ibid., p. 49.
43. Raphael Patai. "Relationship Patterns Among the Arabs." *Middle Eastern Affairs,* 5, New York: Council for Middle Eastern Affairs, Inc., 1951, pp. 180-85. (Human Relations Area Files: Middle East, Source 41.)
44. Ibid., p. 184.
45. Majid Khadduri and Herbert J. Liebesny, eds. *Law in the Middle East: Origin and Development of Islamic Law,* 1, Washington: The Middle East Institute, 1955, p. 155. (Human Relations Area Files: Middle East, Source 56.)
46. Paul Stirling. *Turkish Village.* London: Weidenfeld & Nicholson, 1965, p. 223.
47. Joe E. Pierce. *Life in a Turkish Village.* New York: Holt, Rinehart, & Winston, 1964, p. 83.
48. Ibid., p. 84.
49. Ibid., p. 91.
50. Ibid.
51. Ibid.
52. O. J. Harvey, David E. Hunt, and Harold M. Schroder. *Conceptual Systems and Personality Organization.* New York: Wiley, 1961
53. Abraham H. Maslow. *Toward a Psychology of Being.* New York: Van Nostrand, 1962, p. 31. Maslow presents a possible approach to the question of "self-actualization" by defining motivation in terms of two concepts: "self-actualizing" motivation and "deficiency" motivation.
54. David Riesman. *Abundance, for What? and Other Essays.* Garden City: Doubleday, 1964.

CHAPTER 20

The Myth of Adolescent Culture

FREDERICK ELKIN and WILLIAM A. WESTLEY

In the current sociological literature, adolescence is often described as a unique period in development, distinct from childhood and adulthood. It is an age-grade period characterized by "storm and stress" and participation in a "youth culture," both of which arouse the serious concern of parents. In this paper, we propose to examine this characterization of adolescence, to explore its relationship to certain empirical evidence, and to discuss the pattern of development discovered in an upper-middle-class suburban setting.

Nature of Adolescent Culture

Storm and Stress

Noteworthy in the characterization of adolescence by current writers is the emphasis on "storm and stress" which is said to be inherent in the position of the adolescent in the contemporary American social structure. Sexual frustrations arising out of physical maturation and social restrictions, problems of occupational choice, difficulties in emancipation from small family groups, inconsistencies in authority relationships, conflicts between generations in a changing society, and discontinuities in socialization patterns — all are said to contribute to the general conflicts, insecurities, and uncertainties of the adolescent. The following statements by Kingsley Davis, Robin Williams, and Norman D. Humphrey respectively are illustrative of this emphasis in theoretical analyses and textbooks.

> If . . . the society is complex and changing, adolescence tends to become a time of difficult choosing. . . . If the element of competition is introduced, it

acts as an individualizing force that makes of adolescence a period of strain . . . [1].

In our society . . . the adolescent finds an absence of definitely recognized, consistent patterns of authority. . . . [He] is subjected to a confusing array of competing authorities . . . [2].

The relative lack of *rites de passage* for adolescence both reflects and contributes to the uncertainty and instability faced by the adolescent. . . . Boys and girls . . . often find it difficult to tell when and how "adult" behavior is expected. This indeterminacy in social expectations is a phenomenon almost unknown to the youth of simpler and more stable societies [3].

Adolescence is a period of crisis. Frequently, therefore, it is also a time of revolt. . . . The cultural patterns and sanctions of one generation rarely, in an era of accelerated change, obtain for another [4].

Peer Group Culture

A social counterpart to this individual tension appears in the distinctive youth culture of the peer group, which functions to ease "the transition from the security of childhood in the family of orientation to that of full adulthood in marriage and occupational status" (5). This youth culture is distinguished in both the sociological literature and the mass media by its affirmation of independence, its rejection of adult standards of judgment, its compulsive conformity to peer group patterns, its romanticism, and a participation in "irresponsible" pleasurable activities. Specific characteristics may include a distinctive dress and argot, rambling telephone conversations, and interests in popular music, movie stars, sports, and dancing. In application to lower socioeconomic levels, the emphasis is likely to be on criminal and delinquent behavior. Illustrative of such a characterization are the following statements of Parsons, Williams, and Green, respectively:

Its principal characteristics may be summarized.

1. Compulsive independence of and antagonism to adult expectations and authority. This involves recalcitrance to adult standards of responsibility. . . .
2. Compulsive conformity within the peer group of age mates. It is intolerable to be "different." . . .
3. Romanticism: an unrealistic idealization of emotionally significant objects [6].

So extreme is this gap between generations in some instances that parents and their adolescent children literally represent subcultures [7].

The so-called youth subculture, for example, is a world of irresponsibility, specialized lingo, dating, athleticism, and the like, which rather sharply cuts off adolescent experience from that of the child and from that of the adult [8].

Assumptions and Implications

Three major assumptions are implicit in this characterization of adolescence. (1) One assumption is that adolescence is a unique period to which the phrase "storm and stress" is distinctively appropriate, and which results from the adolescents' peculiar age-grade position in the American social structure. The discontinuity in socialization, the conflict between generations in a changing society, the problems of occupational and marital choice — all are said to contribute to the tensions of the adolescent. The particular application of these characteristics to the period of adolescence thus implies that the age-grade periods before and after adolescence do not have such tensions and do not merit the distinctive characterization of "storm and stress."

(2) A second assumption is that a youth culture exists in fact and is a widespread and dominant pattern among adolescents in American life. This characterization, apparently derived from life in the metropolis, implies that no alternative patterns are so significant for the youth under discussion; and that those who participate in the culture deeply experience its demands, accept it as a dominant element in their lives, and do not judge their own behavior from an "adult" point of view.

(3) A third assumption is that the youth culture of the adolescent is etiologically and functionally linked to the "storm and stress" of the individual. The adolescent, in becoming emancipated from his family, participates in a conformity-demanding peer group, and this participation serves to balance his needs for independence and security. Parsons expresses this relationship as follows:

> Thus the compulsive independence of youth culture may . . . be interpreted as involving a reaction-formation against dependence needs . . . the compulsive conformity, in turn, would seem to serve as an outlet for these dependency needs, but displaced from parental figures onto the peer group so that it does not interfere with the independence [9].

This assumption of a link between the needs of the adolescent and the youth culture leads to a biased selection of illustrations. The illustrative material cites resistances to adult standards of judgment, behavior oriented towards the pleasure principle, concerns with romanticism and athletics, and conforming "other directed" behavior in peer groups. The writers tend to ignore deferred gratification patterns, the internalization of adult modes of thought and behavior, positive relationships with authority figures, instances of family solidarity, or "inner directed" interests which may set an adolescent apart from his age-mates.

The statements in the literature about adolescence ordinarily purport to describe a modal adolescent pattern. As such, the descriptions should not be inapplicable to upper-middle-class urban groups. With this problem in mind, the authors have been engaged in a study of adolescent socialization in a suburban community of Montreal. The community is a well-to-do suburb in which, according to the 1951 census, 69 percent of the male labor force were executive, managerial, or professional, and the median income per male wage earner was approximately $5,500. A detailed report will be presented in a subsequent article; here we merely summarize some of the relevant findings. The findings are based on an investigation, through interview, of a sample of twenty adolescents and their parents, and of twenty others through life history material (10). The life histories were obtained from college students who lived in the community. The adolescents interviewed were fourteen or fifteen years of age, Protestant, Anglo-Saxon in origin, and from professional and business families. They were selected from lists provided by various residents of the community and community organizations, and comprise approximately 15 percent of the adolescents in the community who met the criteria of selection (11).

The data were drawn from four or more interviews with members of each family. Following a preliminary interview with parents and child together, each adolescent was interviewed two or more times. In these partially directed interviews, the adolescents were asked to discuss certain activities, interests, and beliefs, and to prepare a detailed diary of day-to-day activities over a period of one week. Finally, each mother and child was given an interview covering the same general subjects. These final interviews were held on the same day, in succession, in order to prevent collusion.

Suburban Town

One dominant pattern in Suburban Town is that of adult-directed and -approved activity. The activities of the adolescent take place almost completely within the suburban community and in view of adult figures. The adolescent, in effect, has little unstructured time. Typically, on school days, he spends his time out of school doing two hours of homework, helping in household activities, and participating in school organizations, directed sports, or church and "Y" activities. On weekends, with more free time, he participates in some family projects, has certain allotted household tasks, and often attends gatherings at which adults are present. In summers, he either works, attends camp, or vacations with his family at a summer cottage.

Family ties are close and the degree of basic family consensus is high. The parents are interested in all the activities of their children, and the adolescents, except for the area of sex, frankly discuss their own behavior and problems with them. In many areas of life, there is joint participation between parents and children — the boys may help their fathers build patios, the members of the family may curl together, and the parents may attend the school's athletic competitions. In independent discussions by parents and adolescents of the latter's marriage and occupational goals, there was a remarkable level of agreement. The adolescents also acknowledged the right of their parents to guide them, for example, accepting, at least manifestly, the prerogative of the parents to set rules for the number of dates, hours of return from dates, and types of parties. The parents express relatively little concern about the socialization problems or peer group activities of their children.

In many respects, for this given sample of adolescents, the continuity of socialization is far more striking than the discontinuity. With future education and career possibilities in mind the adolescents discuss their choice of school courses with their parents. The parents encourage their children to play host to their friends. Also, since the parents themselves engage in entertaining, dancing, and community sports, the adolescents observe that in many respects their own pattern of social life is not very different from that of the older generation. The continuity of socialization is especially well exemplified in the organization of parties and the concern with social proprieties. Typically, when a girl gives a party, she sends out formal invitations, and her guest list includes those she *should* invite as well as those she wants to invite. The girls take great pains to play their hostess roles properly, and the boys so strongly recognize their escort responsibilities that they may privately draw straws to decide who walks home with the less popular girls.

The adolescents themselves demonstrate a high level of sophistication about their own activities, in many respects having internalized "responsible" and "adult" perspectives. For example, they took their homework very seriously; they tended to view their household tasks as their contribution to family maintenance; some suggested that the clubs to which they belonged gave them valuable social or organizational experience; and the boys, after telling of their practical jokes in school, spoke of this behavior as "silly" or "kid stuff." Furthermore, the youth culture elements in which they participated were recognized as transitory and "appropriate for their age." Thus the "steady date" was likely to be viewed as a pleasant temporary association, not directly related to marriage, which gave a certain immediate security and taught them about heterosexual feelings and relationships. Some spoke of the dating pattern as "the kind of stuff kids do at our age."

This description of adolescence in a suburban town is not an isolated portrait; it is supported in various aspects by other discussions of upper middle-class and upwardly mobile groups. Most noteworthy is the study of

Elmtown (or Jonesville) by Hollingshead, Warner, et al. Although Elmtown is an isolated community of 10,000 population, we find many similar patterns of behavior. In Elmtown, likewise, the upper-middle-class adolescent spends much of his time in supervised extracurricular activities; the parents know the families of their children's assoriates and bring pressure to bear on their children to drop "undesirable" friends; the children are taught to be polite and refined in their speech and behavior and to repress aggressive tendencies. The children of the middle and upper classes, concludes Hollingshead, "are guided by their parents along lines approved by the class cultures with remarkable success (12).

Similar descriptions are given in other studies which discuss the deferred gratification pattern among upper middle and mobile lower-class adolescents. The child learns to forgo immediate indulgences for the sake of future gains and thus inhibits his aggressive and sexual impulses, strives for success in school, and selects his associates with care (13). This pattern is in direct contradiction to the implications of a strong and pervasive youth culture. The individual who internalizes a deferred gratification pattern does not act solely in terms of irresponsible pleasure seeking and conforming peer group pressures and, much as he may apparently be absorbed in dances, gang activity, and sports, he does not lose sight of his long-run aspirations.

It is to be stressed here that we have been focusing throughout on the overt and behavioral, and not the psychiatric, aspects of adolescent development. No implications of any kind are intended about the psychological health or ill health of the adolescents concerned.

Conclusion

The data from Suburban Town and other empirical studies suggest that the characterization of adolescent culture advanced in the sociological literature needs to be questioned. The empirical data do not deny that there are psychological tensions and distinctive interests among adolescents; however, the data do suggest — at least among these middle-class groups studied — that the current model of adolescent culture represents an erroneous conception. And if so, the theories which employ such a culture to analyze the social structure are without adequate foundation.

It has been an assumption of the writers on adolescent culture that adolescents generally suffer from a storm and stress which results from their peculiar age-grade position in the American social structure. However, in the empirical data of upper-middle-class groups, we do not find the structural etiological elements which allegedly produce this storm and stress. There is more continuity than discontinuity in socialization;

there are few sharp conflicts between parents and children; and there are no serious overt problems of occupational choice or emancipation from authority figures. Considering the tenor of these data, along with the knowledge that the child, the adult, and the aged likewise have problems of adjustment, it seems logical to propose that the emphasis in discussing storm and stress among adolescents should be more on their participation in modern urban society than on the distinctive characteristics of their age-grade period.

It has been a second assumption that a youth culture exists and is a widespread and dominant pattern among adolescents in American life. It should follow that we could rather easily find identifiable patterns of youth culture generic to adolescence among middle-class groups. While our data and those of supporting studies are derived from suburban and small-town settings — rather than from metropolitan areas from which the "youth culture" concept has probably derived — they do not support this assumption. In Suburban Town and other communities studied the youth culture elements exist, but they are less dominant than are accepted family and authority guidance patterns. The adolescents in their peer groups are not compulsively independent and rejecting of adult values; they are not concerned solely with immediate pleasurable gratifications. Furthermore, in regard to those aspects of their lives which might be regarded as youth culture, they are remarkably sophisticated, they themselves pointing out that their dating patterns and their "kidding around" are passing temporary phenomena.

It was likewise assumed, in this characterization of adolescence, that the youth culture was both etiologically and functionally linked to the psychological needs of the individual. The psychological needs underlie participation in youth culture; and the youth culture, in turn, with its independence and conformity demands, serves to balance the dual needs of dependence and family emancipation. There is reason, we have noted, to question this characterization of youth culture; but even to the extent that it is valid, the theory leaves no place for empirical data which evidence continuity of socialization. Such a theory neither explains the correlation between adolescent class position, choice of school courses, and subsequent occupational goals; nor the acceptance by adolescents of adult guidance of many of their activities; nor does it make allowance for deferred gratification patterns, the internalization of adult values, solidary family relationships, or positive relationships with authority figures; all of which are found in studies of middle-class groups.

This contradiction between the current sociological characterization of adolescence and the reported data for middle-class groups suggests that "adolescent culture" has a somewhat mythical character.

References

1. K. Davis, "Adolescence and the Social Structure," *Annals of the American Academy of Political and Social Sciences*, November 1944, p. 11.

2. Ibid., p.13.

3. R. Williams, *American Society*. New York: Knopf, 1952, p. 71.

4. N. D. Humphrey, "Social Problems," in A. M. Lee, ed., *New Outline of the Principles of Sociology*. (New York: Barnes & Noble, 1946), p. 24. Also see M. Mead, *Coming of Age in Samoa* (New York: Morrow, 1928); R. Benedict, "Continuities and Discontinuities in Cultural Conditioning," *Psychiatry*, vol. 1, 1938, pp. 161-67; and T. Parsons, "Age and Sex in the Social Structure of the United States," *American Sociological Review*, vol. 7, October 1942, pp. 604-616.

5. Parsons, *Age and Sex*, p. 614.

6. Parsons, "Psychoanalysis and the Social Structure." *Psychoanalytic Quarterly,* vol. 19, 1950, pp. 371-84.

7. Williams, *American Society*, p. 73.

8. A. Green, *Sociology*. New York: McGraw-Hill, 1952, p. 95.

9. Parsons, "Psychoanalysis and the Social Structure," p. 381.

10. We are indebted to Maurice Leznoff for field research in this community and for helpful suggestions.

11. We cannot be certain that the sample is representative of adolescents in the community. However, the relative size of the sample and the consistency of the findings to some degree temper the significance of this methodological deficiency.

12. A. B. Holingshead, *Elmtown's Youth*. New York: Wiley, 1949, p. 443. For a perceptive semipopular report of another closely knit suburban community, see W. H. Whyte, "The Transients," *Fortune*, May, June, July, August 1953.

13. See A. Davis and R. J. Havighurst, "Social Class and Color Differences in Child Rearing," *American Sociological Review*, vol. 11, December 1946, pp. 698-710; A. Davis and J. Dollard, *Children of Bondage* (Washington, D.C.: American Council on Education, 1940); W. L. Warner and Associates, *Democracy in Jonesville* (New York: Harper, 1949); W. F. Whyte, *Street Corner Society* (Chicago: University of Chicago Press, 1943); A. C. Kinsey, W. B. Pomeroy, and C. E. Martin, *Sexual Behavior in the Human Male* (Philadelphia: Saunders, 1948); St. Clair Drake and H. R. Cayton, *Black Metropolis* (New York: Harcourt, Brace, 1945); L. Schneider and S. Lysgaard, "The Deferred Gratification Pattern: A Preliminary Study," *American Sociological Review*, vol. 18, April 1953, pp. 142-49; and B. M. Spinley, *The Deprived and the Privileged* (London: Routledge & Kegan Paul, 1953).

CHAPTER 21

Seminole Girl

MERWYN S. GARBARINO

One hundred and thirty miles of circuitous road and 250 years of history separate the city of Miami from the four federal reservations that lock the Seminole Indians into the Florida swamplands known as Big Cypress Swamp. A new road is under construction that will trim in half the traveling distance between the city and the reservations scattered along the present winding U.S. Highway 41, or Tamiami Trail, as it is called by the Indians. But the new road will only draw the two communities closer on the speedometer; it will not alter the vastly different lifeways it links; it may only make more apparent the historical inequities that brought the two areas into existence.

The Seminoles are harsh examples of what happened to American Indians caught in the expansion process that saw the United States swell from a federation of thirteen colonies to a nation blanketing more than half a continent. The peoples who came to be known as "Seminole," which means "wild" or "undomesticated," were Indians who fled south from the guns and plows of the whites. Some were Yamassee who were driven from the Carolinas in 1715. Others were Hitchiti-speaking Oconee who moved down the Apalachicola River to settle in Spanish-held Florida. These two groups were joined by others escaping soldiers or settlers or other Indians demanding their lands.

The loose confederation of Seminoles was tripled by a large influx of Creeks after the Creek War of 1813-14. Although the Creeks were linguistically related to the Hitchiti, the primary factor uniting the diverse groups was the hatred and fear they felt toward their common foe, the young United States. But this common bond was enough to regroup the broken political units into a single body that absorbed not only Indians, but renegade whites and Negroes escaping slavery.

In 1817-18, the United States sent Andrew Jackson to Florida, ostensibly to recover runaway slaves. This resulted in the First Seminole

War, one of the three Seminole wars that were among the bloodiest ever fought by American forces against Indians. The war also led to the annexation of Florida by the U.S. in 1821 because Spain was in no position to fight for it.

At the time of annexation, the Indians held extensive farm and pasture lands that the Spaniards had not wanted for themselves. American settlers, however, wanted them very much. Insatiabale, they forced the Seminoles ever southward, until finally they demanded that the Indians relocate to the area of the Louisana Purchase which is now Oklahoma. Some Indians went westward, but a number under the leadership of Osceola fought bitterly. When Osceola was captured under a flag of truce, some of his warriors fled into the Everglades where they could not be flushed out. To this day they haven't recognized the treaty that drove their fellow tribesmen to the West.

In 1911, Florida reservation land was set aside by an executive order. The Seminoles, however, were not pressured at that time into moving on to the federal territory. South Florida was a real wilderness; Miami was little more than a town, and lavish coastal resorts were unforeseen. Literally no one but the alligators, snakes, birds, and Indians wanted the land they lived on. The climate is one of wet summers and dry winters, and the area is often struck by hurricanes from the Caribbean. Annual rainfall is in excess of sixty inches and without drainage, the prairie is almost always under water. Brush fires in the dry season destroy valuable hardwood trees on the hummocks which are the low-lying hills undulating through the swampland. Fire also destroys the highly flammable drained peat, and for the same reason there is an absence of pines in many areas suitable for their growth. Elsewhere, however, there are moderately to heavily wooded places, and sometimes great flocks of white egrets alight in the branches of the trees, looking like puffs of white blossoms. Except for the hummocks, the horizon is flat, a vista of sky and water, broken only by the occasional wooded clumps.

Most inhabitants of this waste and water live on elevated platforms under thatched roofs held in place by poles. Unemployment is an ever present problem helped somewhat by seasonal agricultural work or by crafts such as the gaily colored garments, dolls, basketry, and carvings made for the tourists who are now visiting their homeland on a year-round basis. Lands are leased to commercial vegetable farmers and deer can still be hunted. So subsistence living is still possible for the "wild" Seminoles.

Somewhere along the Tamiami Trail, Nellie Greene — a pseudonym, of course — was born a Seminole, raised in a chickee, and learned the ways of her people. Her father was a frog hunter and could neither read nor write; her mother was a good Seminole mother who later had troubles with tuberculosis and drinking. No one Nellie knew had much more education than her father and mother. Yet, despite the ignorance and illiteracy on the reservation, Nellie Green wanted and was encouraged to get a good education. As it does for most Indians in the United States, this meant leaving her "backward" people, mixing with whites who at best patronized her.

I first met Nellie Greene when she had graduated from college and was living in an apartment in Miami, where she worked as a bank clerk. I knew her background from having spent three summers in the middle sixties, thanks to the National Science Foundation, on the Seminole reservations of Florida. In September of 1966, Nellie wrote me that she had been offered a job as manager in the grocery store back on the reservation. If she didn't take it, a white person would, for she was the only native with the necessary knowledge of bookkeeping. She had accepted the job, she said, but since she had once told me (in Miami) that she could never give up the kind of life she had grown accustomed to there, I was curious to find out why she had returned to the reservation.

She herself said that she took the job to help her people, but she added that it had not been an easy decision; in fact, it had been quite a struggle. I could have guessed that this was so; many Indian tribes that offer educational grants to their younger members do so only with the stipulation that the recipients later return to the reservation. The stipulation is a measure of the difficulty in getting their educated members to come back to the tribe. In any event, I wanted to hear Nellie Greene tell her own story. I went to see her, and this is what she told me.

Nellie Greene's Story

I was born in a Miami hospital on February 6, 1943. At that time my parents were living on the (Tiamiami) Trail, and my daddy was making his living frog hunting. He owned an air boat and everything that goes along with frog hunting. It was during the war, and at that time I guess it was hard to get gas. When it was time for me to be born, my father had to borrow gas from a farmer to get to Miami. But the taillight was broken, so my father took a flashlight and put a red cloth over it and tied it on to the truck and went to the hospital. My daddy often told me about that.

I had an older sister and an older brother. We lived in a chickee until 1961, when my daddy bought a CBS (a cement block structure, "hurricane proof" according to state standards) at Big Cypress, and we moved into it. When I was little, my daddy had to be out in the Everglades a lot, so he would take all of us out to a hummock, and we would make camp there and stay there while he went off to hunt for frogs. When he got back, he'd take the frog legs in to the hotels and sell them. Then he would bring back something for each of us. When he would ask us what we wanted, I always asked for chocolate candy.

About all I remember of the Everglades is that it was a custom when you got up to take a bath or go swimming early in the morning. My mother says they always had to chase me because I didn't like to get wet in winter when it was cold. We were there four or five years, and then we moved near the Agency at Dania (renamed Hollywood in 1966). I had never been to school until then. We were taught at home, the traditional things: to share with each other and with children of other families, to eat after the others — father and grandfather first, then mothers and kids. But lots of times us kids would climb up on our fathers' knees while they were eating. They didn't say anything, and they'd give us something. It just wasn't the custom for families to eat the way we do today, everybody sitting together around the table.

Folktales, too, we learned; they were like education for us, you know. The stories told about someone doing something bad, and then something bad happened to him. That was the way of teaching right and wrong.

When we were growing up we broke away from some family customs. My parents spanked us, for instance, not my mother's brother, who would have been the right person to punish his sister's children — one of the old ways. But they were not close to my mother's family because my daddy was a frog hunter, and we wandered around with him. My parents were chosen for each other by their families. I guess they learned to love each other in some ways, but I have heard my mother say that it is not the same kind of love she would have had if she had chosen her own husband. It was respect, and that was the custom of the Indians.

Most parents here show so little affection. Even if they love their kids, maybe they don't think they should show love. I know a lot of parents who really care, but they don't tell their kids how they feel. We always knew how our parents felt about us. They showed us affection. Sometimes I hear kids say, "My mother doesn't care whether I go to school or not." These kids have seen how others get care from their parents, like the white children at school. And that kind of concern doesn't show up here. A lot of parents don't even think of telling their children that they want them to succeed. They don't communicate with their children. You never see an Indian mother here kiss and hug children going to school. But white parents do that, and when Indian children see this in town or on TV, it makes them think that Indian parents just don't care. Kids are just left to go to school or not as they wish. Often the mothers have already left to work in the fields before the school bus comes. So no one sees whether children even to go school.

I felt loved. My parents never neglected us. We have never gone without food or clothes or a home. I have always adored my mother. She has made her mistakes, but I still feel the same about her as when I was a child.

We moved to Big Cypress around 1951 or 1952. I had been in first grade at Dania. I remember I didn't understand English at all when I started first grade. I learned it then. We moved around between Big Cypress and Dania, visiting, or because my father was doing odd jobs here and there.

Both my parents wanted me to go to school because they had wanted to go to school when they were kids. I can remember my mother telling me that she and her sister wanted to go to school. But the clan elders — their uncles — wouldn't let them. The uncles said they would whip the two girls if they went.

One of my father's greatest desires was to go to school when he was a boy. He said that he used to sneak papers and pencils into the camp so that he could write the things he saw on the cardboard boxes that the groceries came in, and figures and words on canned goods. He thought he would learn to read and write by copying these things. My daddy adds columns of figures from left to right, and he subtracts the same way. His answers will be correct, but I don't know how. Almost everything he knows he learned on his own. He can understand English, but he stutters when he talks. He has a difficult time finding the right word when he speaks English, but he understands it.

When my parents said no, they meant no. That was important to me. They could be counted on. The other thing that was important in my childhood schooling was that my daddy always looked at my report card when I brought it home from school. He didn't really know what it meant, and he couldn't read, but he always looked at my report card and made me feel that he cared how I did in school. Other parents didn't do this. In fact, most of the kids never showed their parents their report cards. But my daddy made me feel that it was important to him. I told him what the marks stood for. It was rewarding for me because he took the time.

"Nothing for Me to Do"

Public school was hard compared to what I'd had before, day school on the reservation and a year at Sequoyah Government School. I almost flunked eighth grade at the public school, and it was a miracle that I passed. I just didn't know a lot of things, mathematics and stuff. I survived it somehow. I don't know how, but I did. The man who was head of the department of education at the Agency was the only person outside of my family who helped me and encouraged me to get an education. He understood and really helped me with many things I didn't know about. For a long time the white public school for the Big Cypress area would not let Indian children attend. A boy and I were the first Big Cypress Indians to graduate from that school. He is now in the armed forces.

After I graduated from high school I went to business college, because in high school I didn't take courses that would prepare me for the university. I realized that there was nothing for me to do. I had no training. All I could do was go back to the reservation. I thought maybe I'd go to Haskell Institute, but my mother was in a TB hospital, and I didn't want to go too far away. I did want to go on to school and find some job and work. So the director of education said maybe he could work something out for me so I could go to school down here. I thought bookkeeping would be good because I had had that in high school and loved it. So I enrolled in the business college, but my English was so bad that I had an awful time. I had to take three extra months of English courses. But that helped me. I never did understand why my English was so bad — whether it was my fault or the English I had in high school. I thought I got by in high school; they never told me that my English was so inferior, but it was not good enough for college. It was *terrible* having to attend special classes.

"I Learned How to Dress"

At college the hardest thing was not loneliness but schoolwork itself. I had a roommate from Brighton (one of the three reservations), so I had someone to talk to. The landlady was awfully suspicious at first. We were Indians, you know. She would go through our apartment, and if we hadn't done the dishes, she washed them. We didn't like that. But then she learned to trust us.

College was so fast for me. Everyone knew so much more. It was as though I had never been to school before. As soon as I got home, I started studying. I read assignments both before and after the lectures. I read them before so I could understand what the professor was saying, and I read them again afterwards because he talked so fast. I was never sure I understood.

In college they dressed differently from high school, and I didn't know anything about that. I learned how to dress. For the first six weeks, though, I never went anywhere. I stayed home and studied. It was hard — real hard. (I can imagine what a real university would be like.) And it was so different. If you didn't turn in your work, that was just your tough luck. No one kept at me the way they did in high school. They didn't say, "OK, I'll give you another week."

Gradually I started making friends. I guess some of them thought I was different. One boy asked me what part of India I was from. He didn't even know there were Indians in Florida. I said, "I'm an American." Things like that are kind of hard. I couldn't see my family often, but in a way that was helpful because I had to learn to adjust to my new environment. Nobody could help me but myself.

Well, I graduated and went down to the bank. The president of the bank had called the agency and said he would like to employ a qualified Indian girl. So I went down there and they gave me a test, I was interviewed. And then they told me to come in the following Monday. That's how I went to work. I finished college May 29, and I went to work June 1. I worked there for three years.

In the fall of 1966, my father and the president of the Tribal Board asked me to come back to Big Cypress to manage a new economic enterprise there. It seemed like a dream come true, because I could not go back to live at Big Cypress without a job there. But it was not an easy decision. I liked my bank work. You might say I had fallen in love with banking. But all my life I had wanted to do something to help my people, and I could do that only by leaving my bank job in Miami. Being the person I am, I had to go back. I would have felt guilty if I had a chance to help and I didn't. But I told my daddy that I couldn't give him an answer right away, and I knew he was upset because he had expected me to jump at the chance to come back. He did understand, though, that I had to think about it. He knew when I went to live off the reservation that I had had a pretty hard time, getting used to a job, getting used to people. He knew I had accomplished a lot, and it wasn't easy for me to give it up. But that's how I felt. I had to think. At one time it seemed to me that I could never go back to reservation life.

But then really, through it all, I always wished there was something, even the smallest thing, that I could do for my people. Maybe I'm helping now. But I can see that I may get tired of it in a year, or even less. But right now I'm glad to help build up the store. If it didn't work out, if the store failed, and I thought I hadn't even tried, I would really feel bad. The basic thing about my feeling is that my brothers and sisters and nieces and nephews can build later on in the future only through the foundation their parents and I build. Maybe Indian parents don't always show their affection, but they have taught us that, even though we have a problem, we are still supposed to help one another. And that is what I am trying to do. Even when we were kids, if we had something and other kids didn't, we must share what we had with the others. Kids grow up the way their parents train them.

By the age of nine, girls were expected to take complete care of younger children. I too had to take care of my little brother and sister. I grew up fast. That's just what parents expected. Now teenagers don't want to do that, so they get angry and take off. Headstart and nurseries help the working mothers because older children don't tend the little ones any more. The old ways are changing, and I hope to help some of the people, particularly girls about my age, change to something good.

There are people on the reservation who don't seem to like me. Maybe they are jealous, but I don't know why. I know they resent me somehow. When I used to come in from school or from work back to the reservation, I could tell some people felt like this. I don't think that I have ever, ever, even in the smallest way, tried to prove myself better or more knowing than other people. I have two close friends here, so I don't feel too lonely; but other people my age do not make friends with me. I miss my sister, and I miss my roommate from Miami. My two friends here are good friends. I can tell them anything I want. I can talk to them. That's important, that I can talk to them That's what I look for in a friend, not their education, but for enjoyment of the same things, and understanding. But there are only two of them. I have not been able to find other friends.

The old people think I know everything because I've been to school. They think it is a good thing for us to go to school. But the old people don't have the kind of experience which allows them to understand our problems. They think that it is easy somehow to come back here. They think there is nothing else. They do not understand that there are things I miss on the outside. They do not understand enough to be friends. They are kind, and they are glad that I am educated, but they do not understand my problems. They do not understand loneliness.

It was hard for me to get used again to the way people talk. They have nothing interesting to talk about. They are satisfied to have a TV or radio, but they don't know anything about good books or good movies or the news. There is almost no one I have to talk to about things like that. Here people don't know what discussion is. That's something I found really hard. They gossip: they talk about people, not ideas.

And it was hard getting used to what people think about time. You know, when you live in the city and work, everything is according to time. You race yourself to death, really. But I got used to that and put myself on a schedule. But here, when you want something done, people take their time. They don't come to work when they should, and I just don't want to push them. I would expect it of the older people, but the younger generation should realize how important time is. When you go to school, you just eat and study and go to school, and not worry too much about time; but on a job, you must keep pace. You are being paid for a certain performance. If you do not do what you are supposed to, you do not get paid. But how do I get that across to my people?

"I Don't Know Why"

I was lonely when I first came back here. I was ready to pack up and go back to Miami. People hardly talked to me — just a few words. I don't know why. I've know these people all my life. I don't know why they didn't know me after just three years. I couldn't carry on a conversation with anyone except my own family. I was working all day at the store, and then I had nothing to do but clean the house, or go fishing alone, or with someone from my family.

Coming back to the reservation to live did not seem to be physically hard. At first I lived in a house with a girl friend because I did not want to stay with my family. I wanted to be sure of my independence. I think this hurt my father. But later, when more of my friend's family moved back to the reservation, I decided it was too crowded with her and went back to live in my old home with my father and family. My father's CBS is clean and comfortable. It is as nice as an apartment in Miami.

My idea was that, being raised on the reservation and knowing the problems here, I could hope that the Indian girls would come to me and ask about what they could do after they finished high school: what they could do on the reservation, what jobs they could get off the reservation. I hoped they would discuss their problems with me, what their goals should be. I'd be more than happy to talk with them. But I can't go to them and tell them what to do. Just because I've worked outside for three years doesn't give me the right to plan for other people. But I thought I had something to offer the girls here, if only they would come for advice.

"They Say I'm Mean"

I would like to see the financial records at the store so well kept that an accountant could come in at any time and check the books, and they would be in perfect order. It is difficult because only Louise and I can run the store, and if either of us gets sick, the other one has to be at the store from 7 A.M to 9 P.M., or else close the store. At first I had to be very patient with Louise and explain everything to her. She had no training at all. Sometimes I started to get mad when I explained and explained, but then I'd remember that she can't help it. People do not know some of the things I know, and I must not get irritated. But if things go wrong, I am responsible, and it is a big responsibility. The younger people are not exactly lazy; they just don't know how to work. I want them to work and be on time. If they need time off, they should tell me, not just go away or not appear on some days.

So some of them start calling me bossy. But that is my responsibility. I tried to talk to them and tell them why I wanted them to come to work on time, but still they didn't. I want them to realize that they have to work to earn their money. It is not a gift. They were supposed to do something in return for their wages. They are interested in boys at their age, and that's why they aren't good workers. But still, the National Youth Corps, operating in Big Cypress, gives kids some idea of how it is to work, to have a job. If I don't make them do the job, they're really not earning their money. That is one thing I had to face. I know that they are going to say I'm mean and bossy. I expect that. But if I'm in charge, they're going to do what they're supposed to do. That's the way I look at it. Everybody talks here. I know that, but I've been away, and I can take it.

I think people my own age are jealous. It is not shyness. Before I left, they were all friendly to me. I came back, and they all look at me, but when I go to talk to them, they just turn around, and it is so hard for me. They answer me, but they don't answer like they used to, and talk to me. That has been my main problem. It is hard for someone to come back, but if he is strong enough, you know, he can go ahead and take that. Maybe someday people will understand. There is no reason to come back if you really think you are better than the people. They are wrong it that is what they believe about me. There is not enough money here, and if I didn't really care about the people, then I would have no reason to return.

I am worried about my mother, and I want to stay where I can help her (my parents are now divorced). It is best to come back and act like the other people, dress like they dress, try to be a part of them again. So even if a person didn't have kinfolk here, it he wanted to help, he could. But he must not show off or try to appear better.

If I didn't have a family here, it would be almost like going to live with strangers. I have to work now. It has become a part of my life. People here just don't understand that. I can't just sit around or visit and do nothing. If there were no work here, I could not live here. It would be so hard for me to live the way the women here do, sewing all the time or working in the fields, but if I had to take care of my family and there was nothing else to do, I guess I would stay here for that. My aunt has taught me to do the traditional sewing, and how to make the dolls, so I could earn money doing that; but I wouldn't do it unless I had to stay here to take care of my family.

I think the reason almost all the educated Indians are girls is because a woman's life here on the reservation is harder than the man's. The women have to take all the responsibility for everything. To go to school and get a job is really easier for a woman than staying on the reservation. The men on the reservation can just lay around all day and go hunting. They can work for a little while on the plantations if they need a little money. But the women have to worry about the children. If the women go away and get jobs, then the men have to take responsibility.

A woman and a man should have about the same amount of education when they marry. That means there is no one at Big Cypress I can marry. The boys my age here do not have anything in common with me. If a girl marries an outsider, she has to move away, because the Tribal Council has voted that no white man can live on the reservation. A woman probably would miss the closeness of her family on the reservation. I would want to come back and visit, but I think I could marry out and make it successful. I would expect to meet and know his family. I would like to live near our families, if possible. I will always feel close to my family.

"I Think About the City"

Sometimes I think about the city and all the things to do there. Then I remember my mother and how she is weak and needs someone who will watch over her and help her. You know my mother drinks a lot. She is sick, and the doctors want her to stop; but she herself cannot control her drinking. Well, I guess us kids have shut our eyes, hoping things will get better by themselves. I know you have not heard this before, and I wish I was not the one to tell you this sad story, but my move back to the reservation was partly brought on because of this. She has been to a sanitarium where they help people like her. It has helped her already to know that I want to see her get help and be a better person. I am having a chickee built for her, and I must stay here until she is well enough to manage alone.

Economic opportunity has been severely limited on the reservation until recently. Employment for field hands or driving farm machinery has been available on ranches in the area, but the income is seasonal. Both men and women work at crafts. The products are sold either privately or through the Arts and Crafts Store at the tribal agency on the coast, but the income is inadequate by itself. Some of the men and one or two Indian women own cattle, but none of these sources of income would appeal to a person with higher education. Until the opening of the grocery store, there was no job on the reservation which really required literacy, let alone a diploma.

Examining these possibilities and the words of Nellie Greene, what would entice an educated Indian to come back to work and live on the reservation? A good paying job; a high status as an educated or skilled person; to be back in a familiar, friendly community; a desire to be with his family and to help them. Perhaps for the rare individual, an earnest wish to try to help his own people. But income from a job on the reservation must allow a standard of living not too much lower than that previously enjoyed as a member of outer society. Nellie never gave any consideration to returning to the reservation until there was the possibility of a job that challenged her skills and promised a comparable income. The salary she receives from managing the store is close to what she had made at the bank in Miami. In Miami, however, she worked forty hours a week, while on the reservation she works nearly sixty hours a week for approximately the same pay, because there are no trained personnel to share the responsibility. Given the isolation of Big Cypress, there is not enough time, after she has put in her hours at the store, to go anywhere off the reservation. It is not merely a question of total pay; it is a problem of access to a way of life unattainable on the reservation. Economic opportunity alone is not sufficient.

It is quite apparent from Nellie's interviews and from observation of the interaction between Nellie and other Indians, both in the store and elsewhere on the reservation, that her status is very low. Her position appears to vary: from some slight recognition that her training places her in a category by herself, to distinct jealousy, to apparent puzzlement on the part of some of the old folks as to just what her place in the society is. Through the whole gamut of reaction to Nellie, only her proud family considers her status a high one.

The primary reason Nellie gave for returning to the reservation was to help her people, but the reservatation inhabitants did not indicate that they viewed her activities or presence as beneficial to them. Older Indians, both male and female, stated that it was "right" that she returned because Indians should stay together, not because she might help her people or set an example to inspire young Indians who might otherwise be tempted to drop out of school. Younger people regard her as bossy and trying to act "white." She does not even have the status of a marriageable female. There is no Indian man on the reservation with the sort of background that would make him a desirable marriage partner, from her standpoint; in their traditional view of an ideal wife, she does not display the qualities preferred by the men. At the same time, there is a council ordinance which prohibits white men from living on the reservation, and therefore marriage to a white man would mean that she would have to leave the reservation to live. There is no recognized status of "career woman," educated Indian, or marriageable girl, or any traditional status for her.

Obviously, with an inferior status, it is unlikely that a person would perceive the community as a friendly, familiar environment. From the point of view of the reservation people, who have had contacts with her, she is no longer truly "Indian," but rather someone who has taken over so much of the Anglo-American ways as to have lost her identity as an Indian woman. Nearly all of Nellie's close acquaintances are living off the reservation. The only two girls she considers friends on the reservation are, like herself, young women with more than average contact with outside society, although with less formal education.

Nellie may have rationalized her decision to return by stressing her determination to help the people, but her personal concern for her mother probably influenced her decision to return more than she herself realized. Nellie was the only person in the family who had the ability, knowledge, and willingness to see that her mother received the proper supervision and help.

The Bureau of Indian Affairs is attempting to increase the economic opportunites on the reservations, but I believe their efforts at holding back the "brain drain" of educated Indians will not be effective. Retraining the reservation people who do not have an education is certainly desirable. But, as the story of Nellie Greene points out, it takes more than good pay and rewarding work to keep the educated Indians down on the reservations. If the educated Indian expects to find status with his people, he is going to be disappointed. White people outside are apt to pay more attention to an educated Seminole than his own Indian society will. If the Indian returns from college and expects to find warm personal relationships with persons of his own or opposite sex, he is going to find little empathy, some distrust and jealousy because of his training and experiences outside the reservation. For Nellie Greene there was a personal goal, helping her sick mother. She was lucky to find a job that required her skills as an educated person, and which paid her as well as the bank at Miami. Her other goal, to help her own people, was thwarted rather than helped by her college education.

D. SOCIETAL STRUCTURE

CHAPTER 22

Youth as a Stage of Life

KENNETH KENISTON

Before the twentieth century, adolescence was rarely included as a stage in the life cycle. Early life began with infancy and was followed by a period of childhood that lasted until around puberty, which occurred several years later than it does today. After puberty, most young men and women simply entered some form of apprenticeship for the adult world. Not until 1904, when G. Stanley Hall published his monumental work, *Adolescence: Its Psychology and Its Relations to Physiology, Anthropology, Sociology, Sex, Crime, Religion, and Education,* was this further preadult stage widely recognized. Hall's work went through many editions and was much popularized; "adolescence" became a household word. Hall's classic description of the *Sturm und Drang*, turbulence, ambivalence, dangers and possiblities of adolescence has since been echoed in almost every discussion of this stage of life.

But it would be incorrect to say that Hall "discovered" adolescence. On the contrary, from the start of the nineteenth century, there was increasing discussion of the "problem" of those past puberty but not yet adult. They were the street gang members and delinquents who made up what one nineteenth-century writer termed the new "dangerous classes"; they were also the recruits to the new public secondary schools being opened by the thousands in the late nineteenth century. And once Hall had clearly defined adolescence, it was possible to look back in history to discover men and women who had shown the hallmarks of this stage long before it was identified and named.

Nonetheless, Hall was clearly reflecting a gradual change in the nature of human development, brought about by the massive transformations of American society in the decades after the Civil War. During these decades, the "working family," where children labored alongside parents in field and factories, began to disappear; rising industrial productivity created new economic surpluses that allowed millions of teenagers to remain outside the labor force. America changed from a rural agrarian society to an urban industrial society, and this new industrial society demanded on a mass scale not only the rudimentary literacy taught in elementary schools, but higher skills that could only be guaranteed through secondary education. What Hall's concept of adolescence reflected, then, was a real change in the human experience, a change intimately tied to the new kind of industrial society that was emerging in America and Europe.

Today, Hall's concept of adolescence is unshakably enshrined in our view of human life. To be sure, the precise nature of adolescence still remains controversial. Some observers believe that Hall, like most psychoanalytic observers, vastly overestimated the inevitability of turbulence, rebellion and upheaval in this stage of life. But whatever the exact definition of adolescence, no one today doubts its existence. A stage of life that barely existed a century ago is now universally accepted as an inherent part of the human condition.

In the seven decades since Hall made adolescence a household word, American society has once again transformed itself. From the industrial era of the turn of the century, we have moved into a new era without an agreed-upon name — it has been called postindustrial, technological, postmodern, the age of mass consumption, the technetronic age. And a new generation, the first born in this new era of postwar affluence, television and the Bomb, raised in the cities and suburbs of America, socially and economically secure, is now coming to maturity. Since 1900, the average amount of education received by children has increased by more than six years. In 1900, only 6.4 percent of young Americans completed high school, while today almost 80 percent do, and more than half of them begin college. In 1900, there were only 238,000 college students: in 1970, there are more than seven million, with ten million projected for 1980.

These social transformations are reflected in new public anxieties. The "problem of youth," "the now generation," "troubled youth," "student dissent" and "the youth revolt" are topics of extraordinary concern to most Americans. No longer is our anxiety focused primarily upon the teenager, upon the adolescent of Hall's day. Today we are nervous about the new "dangerous classes" — those young men and women of college and graduate school age who can't seem to "settle down" the way their parents did, who refuse to consider themselves adult, and who often vehemently challenge the existing social order. "Campus unrest," according to a June, 1970, Gallup Poll, was considered the nation's main problem.

The factors that have brought this new group into existence parallel in many ways the factors that produced adolescence: rising prosperity, the further prolongation of education, the enormously high educational demands of a postindustrial society. And behind these measurable changes lie other trends less quantitative but even more important: a rate of social change so rapid that it threatens to make obsolete all institutions, values, methodologies and technologies with the lifetime of each generation; a technology that has created not only prosperity and longevity, but power to destroy the planet, whether through warfare or violation of nature's balance; a world of extraordinarily complex social organization, instantaneous communication and constant revolution. The "new" young men and young women emerging today both reflect and react against these trends.

But if we search among the concepts of psychology for a word to describe these young men and women, we find none that is adequate. Characteristically, they are referred to as "late-adolescents-and-young-adults" — a phrase whose very mouth-filling awkwardness attests to its inadequacy. Those who see in youthful behavior the remnants of childhood immaturity naturally incline toward the concept of "adolescence" in describing the unsettled twenty-four-year-old, for this word makes it easier to interpret his objections to war, racism, pollution or imperialism as "nothing but" delayed adolescent rebellion. To those who are more hopeful about today's youth, "young adulthood" seems a more flattering phrase, for it suggests that maturity, responsibility and rationality lie behind the unease and unrest of many contemporary youths.

But in the end, neither label seems fully adequate. The twenty-four-year-old seeker, political activist or graduate student often turns out to have been *through* a period of adolescent rebellion ten years before, to be all too formed in his views, to have a stable sense of himself, and to be much farther along in his psychological development than his fourteen-year-old high school brother. Yet he differs just as sharply from "young adults" of age twenty-four whose place in society is settled, who are married and perhaps parents, and who are fully committed to an occupation. What characterizes a growing minority of postadolescents today is that they have not settled the questions whose answers once defined adulthood: questions of relationship to the existing society, questions of vocation, questions of social role and life style.

Faced with this dilemma, some writers have fallen back on the concept of "protracted" or "stretched" adolescence — a concept with psychoanalytic origins that suggests that those who find it hard to "settle down" have "failed" the adolescent developmental task of abandoning narcissistic fantasies and juvenile dreams of glory. Thus, one remedy for "protracted adolescence" might be some form of therapy that would enable the young to reconcile themselves to abilities and a world that are rather less than they had hoped. Another interpretation of youthful unease blames society, not the individual, for the "prolongation of adolescence." It argues that youthful unrest springs from the unwillingness of contemporary society to allow young men and women, especially students, to exercise the adult powers of which they are biologically and intellectually capable. According to this view, the solution would be to allow young people to "enter adulthood" and do "real work in the real world" at an earlier age.

Yet neither of these interpretations seems quite to the point. For while some young men and women are indeed victims of the psychological malady of "stretched adolescence," many others are less impelled by juvenile grandiosity than by a rather accurate analysis of the perils and injustices of the world in which they live. And plunging youth into the "adult world" at an earlier age would run directly counter to the wishes of most youths, who view adulthood with all of the enthusiasm of a condemned man for the guillotine. Far from seeking the adult prerogatives of their parents, they vehemently demand a virtually indefinite prolongation of their nonadult state.

If neither "adolescence" nor "early adulthood" quite describes the young men and women who so disturb American society today, what can we call them? My answer is to propose that *we are witnessing today the emergence on a mass scale of a previously unrecognized stage of life,* a stage that intervenes between adolescence and adulthood. I propose to call this stage of life the stage of *youth,* assigning to this venerable but vague term a new and specific meaning. Like Hall's "adolescence," "youth" is in no absolute sense new: indeed, once having defined this stage of life, we can study its historical emergence, locating individuals and groups who have had a "youth" in the past. But what is "new" is that this stage of life is today being entered not by tiny minorities of unusually creative or unusually disturbed young men and women, but by millions of young people in the advanced nations of the world.

To explain how it is possible for "new" stages of life to emerge under changed historical conditions would require a lengthy excursion into the theory of psychological development. It should suffice here to emphasize that the direction and extent of human development — indeed, the entire nature of the human life cycle — is by no means predetermined by man's biological constitution. Instead, psychological development results from a complex interplay of constitutional givens (including the rates and phases of biological maturation) and the changing familial, social, educational, economic and political conditions that constitute the matrix in which children develop. Human development can be obstructed by the absence of the necessary matrix, just as it can be stimulated by other kinds of environments. Some social and historical conditions demonstrably slow, retard or block development, while others stimulate, speed and encourage it. A prolongation and extension of development, then, including the emergence of "new" stages of life, can result from altered social, economic and historical conditions.

Like all stages, youth is a stage of transition rather than of completion or accomplishment. To begin to define youth involves three related tasks. First, we need to describe the major *themes* or issues that dominate consciousness, development and behavior during this stage. But human development rarely if ever proceeds on all fronts simultaneously: instead, we must think of development as consisting of a series of sectors or "developmental lines," each of which may be in or out of phase with the others. Thus we must also describe the more specific *transformations* or changes in thought and behavior that can be observed in each of several "lines" of development (moral, sexual, intellectual, interpersonal, and so on) during youth. Finally, we can try to make clear what youth is *not*. What follows is a preliminary sketch of some of the themes and transformations that seem crucial to defining youth as a stage of life.

Major Themes in Youth

Perhaps the central conscious issue during youth is the *tension between self and society*. In adolescence, young men and women tend to accept their society's definitions of them as rebels, truants, conformists, athletes or achievers. But in youth, the relationship between socially assigned labels and the "real self" becomes more problematic, and constitutes a focus of central concern. The awareness of actual or potential conflict, disparity, lack of congruence between what one is (one's identity, values, integrity) and the resources and demands of the existing society increased. The adolescent is struggling to define who he is; the youth begins to sense who he is and thus to recognize the possibility of conflict and disparity between his emerging selfhood and his social order.

In youth, *pervasive ambivalence* toward both self and society is the rule: the question of how the two can be made more congruent is often experienced as a central problem of youth. This ambivalence is not the same as definitive rejection of society, nor does it necessarily lead to political activism. For ambivalence may also entail intense self-rejection, including major efforts at self-transformation employing the methodologies of personal transformation that are culturally available in any historical era: monasticism, meditation, psychoanalysis, prayer, hallucinogenic drugs, hard work, religious conversion, introspection, and so forth. In youth, then, the potential and ambivalent conflicts between autonomous selfhood and social involvement — between the maintenance of personal integrity and the achivement of effectiveness in society — are fully experienced for the first time.

The effort to reconcile and accommodate these two poles involves a characteristic stance vis-a-vis both self and world, perhaps best described by the concept of the *wary probe*. For the youthful relationship to the social order consists not merely in the experimentation more characteristic of adolescence, but now in more serious forays into the adult world, through which its vulnerability, strength, integrity and possibilities are assayed. Adolescent experimentation is more concerned with self-definition than are the probes of youth, which may lead to more lasting commitments. This testing, exacting, challenging attitude may be applied to all representatives and aspects of the existing social order, sometimes in anger and expectation of disappointment, sometimes in the urgent hope of finding honor, fidelity and decency in society, and often in both anger and hope. With regard to the self, too, there is constant self-probing in search of strength, weakness, vulnerability and resiliency, constant self-scrutiny designed to test the individual's capacity to withstand or use what his society would make of him, ask of him, and allow him.

Phenomenologically, youth is a time of alternating *estrangement and omnipotentiality*. The estrangement of youth entails feelings of isolation, unreality, absurdity, and disconnectedness from the interpersonal, social and phenomenological world. Such feelings are probably more intense during youth than in any other period of life. In part they spring from the actual disengagement of youth from society; in part they grow out of the psychological sense of incongruence between self and world. Much of the psychopathology of youth involves such feelings, experienced as the depersonalization of the self or the derealization of the world.

Ominipotentiality is the opposite but secretly related pole of estrangement. It is the feeling of absolute freedom, of living in a world of pure possibilities, of being able to change or achieve anything. There may be times when complete self-transformation seems possible, when the self is experienced as putty in one's own hands. At other times, or for other youths, it is the nonself that becomes totally malleable; then one feels capable of totally transforming another's life, or creating a new society with no roots whatsoever in the mire of the past. Omnipotentiality and estrangement are obviously related: the same sense of freedom and possibility that may come from casting off old inhibitions, values and constraints may also lead directly to a feeling of absurdity, disconnectedness and estrangement.

Another characteristic of youth is the *refusal of socialization* and acculturation. In keeping with the intense and wary probing of youth, the individual characteristically begins to become aware of the deep effects upon his personality of his society and his culture. At times he may attempt to break out of his prescribed roles, out of his culture, out of history, and even out of his own skin. Youth is a time, then, when earlier socialization and acculturation is self-critically analyzed, and massive efforts may be made to uproot the now alien traces of historicity, social membership and culture. Needless to say, these efforts are invariably accomplished within a social, cultural and historical context, using historically available methods. Youth's relationship to history is therefore paradoxical. Although it may try to reject history altogether, youth does so in a way defined by its historical era, and these rejections may even come to define that era.

In youth we also observe the emergence of *youth-specific identities* and roles. These contrast both with the more ephemeral enthusiams of the adolescent and with the more established commitments of the adult. They may last for months, years, or a decade, and they inspire deep commitment in those who adopt them. Yet they are inherently temporary and specific to youth: today's youthful hippies, radicals and seekers recognize full well that, however reluctantly, they will eventually become older; and that aging itself will change their status. Some such youth-specific identities may provide the foundation for later commitments; but others must be viewed in retrospect as experiments that failed or as probes of the existing society that achieved their purpose, which was to permit the individual to move on in other directions.

Another special issue during youth is the enormous value placed upon change, transformation and *movement*, and the consequent abhorrence of *stasis*. To change, to stay on the road, to retain a sense of inner development and / or outer momentum is essential to many youths' sense of active vitality. The psychological problems of youth are experienced as most overwhelming when they seem to block change: thus, youth grows panicky when confronted with the feeling of "getting nowhere," of "being stuck in a rut," or of "not moving."

At times the focus of change may be upon the self, and the goal is then to *be moved*. Thus, during youth we see the most strenuous, self-conscious and even frenzied efforts at self-transformation, using whatever religious, cultural, therapeutic or chemical means are available. At other times, the goal may be to create movement in the outer world, to *move others:* then we may see efforts at social and political change that in other stages of life rarely possess the same single-minded determination. And on other occasions, the goal is to *move through* the world, and we witness a frantic geographic restlessness, wild swings of upward or downward social mobility, or a compelling psychological need to identify with the highest and the lowest, the most distant and apparently alien.

The need for movement and terror of stasis often are a part of a heightened *valuation of development* itself, however development may be defined by the individual and his culture. In all stages of life, of course, all individuals often wish to change in specific ways: to become more witty, more attractive, more sociable or wealthier. But in youth, specific changes are often subsumed in the devotion to change itself — to "keep putting myself through the changes," "not to bail out," "to keep moving." This valuation of change need not be fully conscious. Indeed it often surfaces only in its inverse form, as the panic or depression that accompanies a sense of "being caught in a rut," "getting nowhere," "not being able to change." But for other youths, change becomes a conscious goal in itself, and elaborate ideologies of the techniques of transformation and the *telos* of human life may be developed.

In youth, as in all other stages of life, *the fear of death* takes a special form. For the infant, to be deprived of maternal support, responsiveness and care is not to exist; for the four-year-old, nonbeing means loss of body intactness (dismemberment, mutilation, castration); for the adolescent, to cease to be is to fall apart, to fragment, splinter, or diffuse into nothingness. For the youth, however, to lose one's essential vitality is merely *to stop*. For some, even self-inflicted death or psychosis may seem preferable to loss of movement; and suicidal attempts in youth often spring from the failure of efforts to change and the resulting sense of being forever trapped in an unmoving present.

The youthful *view of adulthood* is strongly affected by these feelings. Compared to youth, adulthood has traditionally been a stage of slower transformation, when, as Erik H. Erikson has noted, the relative developmental stability of parents enables them to nurture the rapid growth of their children. This adult deceleration of personal change is often seen from a youthful vantage point as concretely embodied in apparently unchanging parents. It leads frequently to the conscious identification of adulthood with stasis, and to its unconscious equation with death or nonbeing. Although greatly magnified today by the specific political disillusionments of many youths with the "older generation," the adulthood = stasis (= death) equation is inherent in the youthful situation itself. The desire to prolong youth indefinitely springs not only from an accurate perception of the real disadvantages of adult status in any historical era, but from the less conscious and less accurate assumption that to "grow up" is in some ultimate sense to cease to be really alive.

Finally, youths tend to band together with other youths in *youthful counter-cultures*, characterized by their deliberate cultural distance from the existing social order, but *not* always by active political or other opposition to it. It is a mistake to identify youth as a developmental stage with any one social group, role or organization. But youth *is* a time when solidarity with other youths is especially important, whether the solidarity be achieved in pairs, small groups, or formal organizations. And the groups dominated by those in this stage of life reflect not only the special configurations of each historical era, but also the shared developmental positions and problems of youth. Much of what has traditionally been referred to as "youth culture" is, in the terms here used, adolescent culture; but there are also groups, societies and associations that are truly youthful. In our own time, with the enormous increase in the number of those who are entering youth as a stage of life, the variety and importance of these youthful counter-cultures is steadily growing.

This compressed summary of themes in youth is schematic and interpretive. It omits many of the qualifications necessary to a fuller discussion, and it neglects the enormous complexity of development in any one person in favor of a highly schematic account. Specifically, for example, I do not discuss the ways the infantile, the childish, the adolescent and the truly youthful interact in all real lives. And perhaps most important, my account is highly interpretive, in that it points to themes that underlie diverse acts and feelings, to issues and tensions that unite the often scattered experiences of real individuals. The themes, issues and conflicts here discussed are rarely conscious as such; indeed, if they all were fully conscious, there would probably be something seriously awry. Different youths experience each of the issues here considered with different intensity. What is a central conflict for one may be peripheral or unimportant for another. These remarks, then, should be taken as a first effort to summarize some of the underlying issues that characterize youth as an ideal type.

Transformations of Youth

A second way of describing youth is by attempting to trace out the various psychological and interpersonal transformations that may occur during this stage. Once again, only the most preliminary sketch of youthful development can be attempted here. Somewhat arbitrarily, I will distinguish between development in several sectors or areas of life, here noting only that, in fact, changes in one sector invariably interact with those in other sectors.

In pointing to the *self-society relationship* as a central issue in youth, I also mean to suggest its importance as an area of potential change. The late adolescent is only beginning to challenge his society's definition of him, only starting to compare his emerging sense of himself with his culture's possibilities and with the temptations and opportunites offered by his environment. Adolescent struggles for emancipation from external familial control and internal dependency on the family take a variety of forms, including displacement of the conflict onto other "authority figures." But in adolescence itself, the "real" focus of conflict is on the family and all of its internal psychic residues. In youth, however, the "real" focus begins to shift: increasingly, the family becomes more paradigmatic of society than vice versa. As relatively greater emancipation from the family is achieved, the tension between self and society, with ambivalent probing of both, comes to constitute a major area of developmental "work" and change. Through this work, young people can sometimes arrive at a synthesis whereby both self and society are affirmed, in the sense that the autonomous reality, relatedness yet separateness of both, is firmly established.

There is no adequate term to describe this "resolution" of the tension between self and society, but C. G. Jung's concept of "*individuation*" comes close. For Jung, the individuated man is a man who acknowledges and can cope with social reality, whether accepting it or opposing it with revolutionary fervor. But he can do this without feeling his central selfhood overwhelmed. Even when most fully engaged in social role and societal action, he can preserve a sense of himself as intact, whole, and distinct from society. Thus the "resolution" of the self-society tension in no way necessarily entails "adjusting" to the society, much less "selling out" — although many youths see it this way. On the contrary, individuation refers partly to a psychological process whereby self and society are differentiated internally. But the actual conflicts between men and women and their societies remain, and indeed may become even more intense.

The meaning of individuation may be clarified by considering the special dangers of youth, which can be defined as extremes of *alienation, whether from self or from society*. At one extreme is that total alienation from self that involves abject submission to society, "joining the rat race," "selling out." Here, society is affirmed but selfhood denied. The other extreme is a total alienation from society that leads not so much to the rejection of society, as to its existence being ignored, denied and blocked out. The result is a kind of self-absorption, an enforced interiority and subjectivity, in which only the self and its extensions are granted live reality, while all the rest is relegated to a limbo of insignificance. Here the integrity of the self is purchased at the price of a determined denial of social reality, and the loss of social effectiveness. In youth both forms of alienation are often assayed, sometimes for lengthy periods. And for some whose further development is blocked, they become the basis for life-long adaptations — the self-alienation of the marketing personality, the social alienation of the perpetual drop-out. In terms of the polarities of Erikson, we can define the central developmental possibilities of youth as individuation vs. alienation.

Sexual development continues in important ways during youth. In modern Western societies, as in many others, the commencement of actual sexual relationships is generally deferred by middle-class adolescents until their late teens or early twenties: the modal age of first intercourse for American college males today is around twenty, for females about twenty-one. Thus, despite the enormous importance of adolescent sexuality and sexual development, actual sexual intercourse often awaits youth. In youth, there may occur a major shift from masturbation and sexual fantasy to interpersonal sexual behavior, including the gradual integration of sexual feelings with intimacy with a real person. And as sexual behavior with real people commences, one sees a further working-through, now in behavior, of vestigial fears and prohibitions whose origin lies in earlier childhood — specifically, of Oedipal feelings of sexual inferiority and of Oedipal prohibitions against sex with one's closest intimates. During youth, when these fears and prohibitions can be gradually worked through, they yield a capacity for genitality, that is, for mutually satisfying sexual relationships with another whom one loves.

The transition to genitality is closely related to a more general pattern of *interpersonal development.* I will term this the shift from *identicality* to mutuality. This development begins with adolescence[1] and continues through youth: it involves a progressive expansion of the early-adolescent assumption that the interpersonal world is divided into only two categories: first, me-and-those-who-are-identical-to-me (potential soulmates, doubles and hypothetical people who "automatically understand everything"), and second, all others. This conceptualization gradually yields to a capacity for close relationships with those on an approximate level of *parity* or similarity with the individual.

The phase of parity in turn gives way to a phase of *complementarity,* in which the individual can relate warmly to others who are different from him, valuing them for their dissimilarities from himself. Finally, the phase of complementarity may yield in youth to a phase of *mutuality,* in which issues of identicality, parity and complementarity are subsumed in an overriding concern with the other *as other.* Mutuality entails a simultaneous awareness of the ways in which others are identical to oneself, the ways in which they are similar and dissimilar, and the ways in which they are absolutely unique. Only in the stage of mutuality can the individual begin to conceive of others as separate and unique selves, and relate to them as such. And only with this stage can the concept of mankind assume a concrete significance as pointing to a human universe of unique and irreplaceable selves.

Relationships with elders may also undergo characteristic youthful changes. By the end of adolescence, the hero worship or demonology of the middle adolescent has generally given way to an attitude of more selective emulation and rejection of admired or disliked older persons. In youth, new kinds of relationships with elders become possible: psychological

[1] Obviously, interpersonal development, and specifically the development of relationships with peers, begins long before adolescence, starting with the "parallel play" observed at ages two to four and continuing through many stages to the preadolescent same-sex "chumship" described by Harry Stack Sullivan. But puberty in middle-class Western societies is accompanied by major cognitive changes that permit the early adolescent for the first time to develop hypothetical ideals of the possibilities of friendship and intimacy. The "search for a soulmate" of early adolescence is the first interpersonal stage built upon these new cognitive abilities.

apprenticeships, then a more complex relationship of mentorship, then sponsorship, and eventually peership. Without attempting to describe each of these substages in detail, the overall transition can be described as one in which the older person becomes progressively more real and three-dimensional to the younger one, whose individuality is appreciated, validated and confirmed by the elder. The sponsor, for example, is one who supports and confirms in the youth that which is best in the youth, without exacting an excessive price in terms of submission, imitation, emulation or even gratitude.

Comparable changes continue to occur during youth with regard to *parents*. Adolescents commonly discover that their parents have feet of clay, and recognize their flaws with great acuity. Childish hero worship of parents gives way to a more complex and often negative view of them. But it is generally not until youth that the individual discovers his parents as themselves complex, three-dimensional historical personages whose destinies are partly formed by their own wishes, conscious and unconscious, and by their historical situations. Similarly, it is only during youth that the questions of family tradition, family destiny, family fate, family culture and family curse arise with full force. In youth, the question of whether to live one's parents' life, or to what extent to do so, becomes a real and active question. In youth, one often sees what Ernst Prelinger has called a "telescoped re-enactment" of the life of a parent — a compulsive need to live out for oneself the destiny of a parent, as if to test its possibilities and limits, experience it from the inside, and (perhaps) free oneself of it. In the end, the youth may learn to see himself and his parents as multidimensional persons, to view them with compassion and understanding, to feel less threatened by their fate and failings, and to be able, if he chooses, to move beyond them.

In beginning by discussing affective and interpersonal changes in youth, I begin where our accounts of development are least precise and most tentative. Turning to more cognitive matters, we stand on somewhat firmer ground. Lawrence Kohlberg's work on *moral development,* especially on the attainment of the highest levels of moral reasoning, provides a paradigmatic description of developments that occur only in youth, if they occur at all.

Summarized over-simply, Kohlberg's theory distinguishes three general stages in the development of moral reasoning. The earliest or *pre-moral* stage involves relatively egocentric concepts of right and wrong as that which one can do without getting caught, or as that which leads to the greatest personal gratification. This stage is followed, usually during later childhood, by a stage of *conventional* morality, during which good and evil are identified with the concept of a "good boy" or "good girl" or with standards of the community and the concept of law and order. In this stage, morality is perceived as objective, as existing "out there."

The third and final major stage of moral development is *post-conventional.* It involves more abstract moral reasoning that may lead the individual into conflict with conventional morality. The first of two levels within the postconventional stage basically involves the assumption that concepts of right and wrong result from a *social contract* — an implicit agreement entered into by the members of the society for their own welfare, and therefore subject to amendment, change or revocation. The highest postconventional level is that in which the individual becomes devoted to *personal principles* that may transcend not only conventional morality but even the social contract. In this stage, certain general principles are now seen as personally binding although not necessarily "objectively" true. Such principles are apt to be stated at a very high level of generality: for example, the Golden Rule, the sanctity of life, the categorical imperative, the concept of justice, the promotion of human development. The individual at this stage may find himself in conflict with existing concepts of law and order, or even with the notion of an amendable social contract. He may, for example, consider even democratically-arrived-at laws unacceptable because they lead to consequences or enjoin behaviors that violate his own personal principles.

Kohlberg's research suggests that most contemporary Americans, young or old, do not pass beyond the conventional stage of moral reasoning. But some do, and they are most likely to be found today among those who are young and educated. Such young men and women may develop moral principles that can lead them to challenge the existing moral order and the existing society. And Kohlberg finds that the achievement of his highest level, the stage of personal principles, occurs in the twenties, if it occurs at all. Moral development of this type can thus be identified with youth, as can the special moral "regressions" that Kohlberg finds a frequent concomitant of moral development. Here the arbitrariness of distinguishing between sectors of development becomes clear, for the individual can begin to experience the tension between self and society only as he begins to question the absolutism of conventional moral judgments. Unless he has begun such questioning, it is doubtful whether we can correctly term him "a youth."

In no other sector of development do we have so complete, accurate and convincing a description of a "development line" that demonstrably characterizes youth. But in the area of *intellectual development*, William Perry has provided an invaluable description of the stages through which college students may pass. Perry's work emphasizes the complex transition from epistemological dualism to an awareness of multiplicity and to the realization of relativism. Relativism in turn gives way to a more "existential" sense of truth, culminating in what Perry terms "commitment within relativism." Thus, in youth we expect to see a passage beyond simple views of Right and Wrong, Truth and Falsehood, Good and Evil to a more complex and relativistic view; and as youth proceeds, we look for the development of commitments within a universe that remains epistemologically relativistic. Once again, intellectual development is only analytically separable from a variety of other sectors — moral, self-society and interpersonal, to mention only three.

In his work on *cognitive development*, Jean Piaget has emphasized the importance of the transition from concrete to formal operations, which in middle-class Western children usualy occurs at about the age of puberty. For Piaget the attainment of formal operations (whereby the concrete world of the real becomes a subset of the hypothetical world of the possible) is the highest cognitive stage possible. But in some youths, there seem to occur further stages of cognitive development that are not understandable with the concept of formal operations. Jerome Bruner has suggested that beyond the formal stage of thought there lies a further stage of "thinking about thinking." This ability to think about thinking involves a new level of consciousness — consciousness of consciousness, awareness of awareness, and a breaking-away of the phenomenological "I" from the contents of consciousness. This breaking-away of the phenomenological ego during youth permits phenomenological games, intellectual tricks, and kinds of creativity that are rarely possible in adolescence itself. It provides the cognitive underpinning for many of the characteristics and special disturbances of youth, for example, youth's hyperawareness of inner processes, the focus upon states of consciousness as objects to be controlled and altered, and the frightening disappearance of the phenomenological ego in an endless regress of awarenesses of awarenesses.

Having emphasized that these analytically separated "lines" of development are in fact linked in the individual's experience, it is equally important to add that they are never linked in perfect synchronicity. If we could precisely label one specific level within each developmental line as distinctively youthful, we would find that few people were "youthful" in all lines at the same time. In general, human development proceeds unevenly, with lags in some areas and precocities in others. One young woman may be at a truly adolescent level in her relationship with her parents, but at a much later level in moral development; a young man may be capable of extraordinary mutuality with his peers, but still be struggling intellectually with the dim awareness of relativism. Analysis of any one person in terms of specific sectors of development will generally show a simultaneous mixture of adolescent, youthful and adult features. The point, once again, is that the concept of youth here proposed is an ideal type, a model that may help understand real experience but can never fully describe or capture it.

What Youth Is Not

A final way to clarify the meaning of youth as a stage of life is to make clear what it is not. For one thing, youth is not the end of development. I have described the belief that it is — the conviction that beyond youth lie only stasis, decline, foreclosure and death — as a characteristically youthful way of viewing development, consistent with the observation that it is impossible truly to understand stages of development beyond one's own. On the contrary, youth is but a preface for further transformations that may (or may not) occur in later life. Many of these center around such issues as the relationship to work and to the next generation. In youth, the question of vocation is crucial, but the issue of work — of productivity, creativity, and the more general sense of fruitfulness that Erikson calls generativity — awaits adulthood. The youthful attainment of mutuality with peers and of peerhood with elders can lead on to further adult interpersonal developments by which one comes to be able to accept the dependency of others, as in parenthood. In later life, too, the relations between the generations are reversed, with the younger now assuming responsibility for the elder. Like all stages of life, youth is transitional. And although some lines of development, such as moral development, may be "completed" during youth, many others continue throughout adulthood.

It is also a mistake to identify youth with any one social group, role, class, organization, or position in society. Youth is a *psychological* stage; and those who are in this stage do not necessarily join together in identifiable groups, nor do they share a common social position. Not all college students, for example, are in this stage of life: some students are psychological adolescents, while others are young adults — essentially apprentices to the existing society. Nor can the experience of youth as a stage of life be identified with any one class, nation or other social grouping. Affluence and education can provide a freedom from economic need and an intellectual stimulation that may underlie and promote the transformations of youth. But there are poor and uneducated young men and women, from Abraham Lincoln to Malcolm X, who have had a youth, and rich, educated ones who have moved straightaway from adolescence to adulthood. And although the experience of youth is probably more likely to occur in the economically advanced nations, some of the factors that facilitate youth also exist in the less advanced nations, where comparable youthful issues and transformations are expressed in different cultural idioms.

Nor should youth be identified with the rejection of the status quo, or specifically with student radicalism. Indeed, anyone who has more or less definitively defined himself as a misanthrope or a revolutionary has moved beyond youthful probing into an "adult" commitment to a position vis-a-vis society. To repeat: what characterizes youth is not a definitive rejection of the existing "system," but an ambivalent tension over the relationship between self and society. This tension may take the form of avid efforts at self-reform that spring from acceptance of the status quo, coupled with a sense of one's own inadequacy vis-a-vis it. In youth the relationship between self and society is indeed problematical, but rejection of the existing society is not a necessary characteristic of youth.

Youth obviously cannot be equated with any particular age-range. In practice, most young Americans who enter this stage of life tend to be between the ages of eighteen and thirty. But they constitute a minority of the whole age-grade. Youth as a developmental stage is emergent; it is an "optional" stage, not a universal one. If we take Kohlberg's studies of the development of postconventional moral reasoning as a rough index of the "incidence" of youth, less than forty percent of middle-class (college-educated) men, and a smaller proportion of working-class men have developed beyond the conventional level by the age of twenty-four. Thus, "youths" constitute but a minority of their age group. But those who are in this stage of life today largely determine the public image of their generation.

Admirers and romanticizers of youth tend to identify youth with virtue, morality and mental health. But to do so is to overlook the special youthful possibilities for viciousness, immorality and psychopathology. Every time of human life, each level of development, has its characteristic vices and weaknesses, and youth is no exception. Youth is a stage, for example, when the potentials for zealotry and fanaticism, for reckless action in the name of the highest principles, for self-absorption, and for special arrogance are all at a peak. Furthermore, the fact that youth is a time of psychological change also inevitably means that it is a stage of constant recapitulation, reenactment and reworking of the past. This reworking can rarely occur without real regression, whereby the buried past is reexperienced as present and, one hopes, incorporated into it. Most youthful transformation occurs *through* brief or prolonged regression, which, however benignly it may eventually be resolved, constitutes part of the psychopathology of youth. And the special compulsions and inner states of youth — the euphoria of omnipotentiality and the dysphoria of estrangement, the hyper-consciousness of consciousness, the need for constant motion and the terror of stasis — may generate youthful pathologies with a special virulence and obstinacy. In one sense those who have the luxury of a youth may be said to be "more developed" than those who do not have (or do not take) this opportunity. But no level of development and no stage of life should be identified either with virtue or with health.

Finally, youth is not the same as the adoption of youthful causes, fashions, rhetoric or postures. Especially in a time like our own, when youthful behavior is watched with ambivalent fascination by adults, the positions of youth become part of the cultural stock-in-trade. There thus develops the phenomenon of *pseudo-youth* — preadolescents, adolescents and frustrated adults masquerade as youths, adopt youthful manners and disguise (even to themselves) their real concerns by the use of youthful rhetoric. Many a contemporary adolescent, whether of college or high school age, finds it convenient to displace and express his battles with his parents in a pseudo-youthful railing at the injustices, oppression and hypocrisy of the Establishment. And many an adult, unable to accept his years, may adopt pseudo-youthful postures to express the despairs of his adulthood.

To differentiate between "real" and pseudo youth is a tricky, subtle and unrewarding enterprise. For, as I have earlier emphasized, the concept of youth as here defined is an ideal type, an abstraction from the concrete experience of many different individuals. Furthermore, given the unevenness of human development and the persistence throughout life of active remnants of earlier developmental levels, conflicts and stages, no one can ever be said to be completely "in" one stage of life in all areas of behavior and at all times. No issue can ever be said to be finally "resolved"; no earlier conflict is completely "overcome." Any real person, even though on balance we may consider him a "youth," will also contain some persistent childishness, some not-outgrown adolescence, and some precocious adulthood in his makeup. All we can say is that, for some, adolescent themes and levels of development are *relatively* outgrown, while adult concerns have not yet assumed full prominence. It is such people whom one might term "youths."

The Implications of Youth

I have sketched with broad and careless strokes the rough outlines of a stage of life I believe to characterize a growing, although still small, set of young men and women. This sketch, although presented dogmatically, is clearly preliminary; it will doubtless require revision and correction after further study. Yet let us for the moment assume that, whatever the limitations of this outline, the concept of a postadolescent stage of life has some merit. What might be the implications of the emergence of youth?

To most Americans, the chief anxieties raised by youth are over social stability and historical continuity. In every past and present society, including our own, the great majority of men and women seem to be, in Kohlberg's terms, "conventional" in moral judgment, and, in Perry's terms, "dualistic" in their intellectual outlook. Such men and women accept with little question the existing moral codes of the community, just as they endorse their culture's traditional view of the world. It is arguable that both cultural continuity and social stability have traditionaly rested on the moral and epistemological conventionality of most men and women, and on the secure transmission of these conventional views to the next generation.

What, then, would it mean if our particular era were producing millions of postconventional, nondualistic, postrelativistic youth? What would happen if millions of young men and women developed to the point that they "made up their own minds" about most value, ideological, social and philosophical questions, often rejecting the conventional and traditional answers? Would they not threaten the stability of their societies?

Today it seems clear that most youths are considered nuisances or worse by the established order, to which they have not finally pledged their allegiance. Indeed, many of the major stresses in contemporary American society spring from or are aggravated by those in this stage of life. One aspect of the deep polarization in our society may be characterized psychologically as a struggle between conventionals and postconventionals, between those who have not had a youth and those who have. The answer of the majority of the public seems clear: we already have too many "youths" in our society; youth as a developmental stage should be stamped out.

A more moderate answer to the questions I am raising is also possible. We might recognize the importance of havine a *few* postconventional individuals (an occasional Socrates, Christ, Luther or Gandhi to provide society with new ideas and moral inspiration), but notheless establish a firm top limit on the proportion of postconventional, youth-scarred adults our society could tolerate. If social stability requires human inertia — that is, unreflective acceptance of most social, cultural and political norms — perhaps we should discourage "youth as a stage of life" in any but a select minority.

A third response, toward which I incline, seems to me more radical. To the argument from social stability and cultural continuity, one might reply by pointing to the enormous *in*stabilities and gross cultural *dis*continuities that characterize the modern world. Older forms of stability and continuity have *already* been lost in the postindustrial era. Today, it is simply impossible to return to a bygone age when massive inertia guaranteed social stability (if there really was such an age). The cake of custom crumbled long ago. The only hope is to learn to live without it.

In searching for a way to do this, we might harken back to certain strands in socialist thought that see new forms of social organization possible for men and women who are more "evolved." I do not wish to equate my views on development with revolutionary socialism or anarchism, much less with a Rousseauistic faith in the goodness of the essential man. But if there is anything to the hypothesis that different historical conditions alter the nature of the life cycle, then men with different kinds of development may require or be capable of living in different kinds of social institutions. On the one hand, this means that merely throwing off institutional shackles, as envisioned by some socialist and anarchist thinkers, would not automatically change the nature of men, although it may be desirable on other grounds. "New men" cannot be created by institutional transformations alone, although institutional changes may, over the very long run, affect the possibilities for continuing development by changing the matrix in which development occurs.

But on the other hand, men and women who have attained higher developmental levels may be capable of different kinds of association and cooperation from those at lower levels. Relativism, for example, brings not only skepticism but also tolerance of the viewpoints of others, and a probable reduction in moralistic self-righteousness. Attaining the stage of personal principles in moral development in no way prevents the individual from conforming to a just social order, or even for that matter from obeying unreasonable traffic laws. Men and women who are capable of interpersonal mutuality are not for that reason worse citizens; on the contrary, their capacity to be concerned with others as unique individuals might even make them better citizens. Examples could be multiplied, but the general point is obvious: high levels of development including the emergence on a mass scale of "new" stages of life may permit new forms of human cooperation and social organization.

It may be true that all past societies have been built upon the unquestioning inertia of the vast majority of their citizens. And this inertia may have provided the psychological ballast that prevented most revolutions from doing more than reinstating the *ancien régime* in new guise. But it does not follow that this need always continue to be true. If new developmental stages are emerging that lead growing minorities to more autonomous positions vis-a-vis their societies, the result need not be anarchy or social chaos. The result might instead be the possibility of new forms of social organization based less upon unreflective acceptance of the status quo than upon thoughtful and self-conscious loyalty and cooperation. But whether or not these new forms can emerge depends not only upon the psychological factors I have discussed here, but even more upon political, social, economic and international conditions.

CHAPTER 23

Memorandum on Youth

ERIK H. ERIKSON

Search for New Modes of Conduct

In responding to the inquiry of the Commission on the Year 2000, I will take the liberty of quoting the statements put to me in order to reflect on some of the stereotyped thinking about youth that has become representative of us, the older generation. This, it seems to me, is prognostically as important as the behavior of the young people themselves; for youth is, after all, a *generational phenomenon*, even though its problems are now treated as those of an outlandish tribe descended on us from Mars. The actions of young people are always in part and by necessity reactions to the stereotypes held up to them by their elders. To understand this becomes especially important in our time when the so-called communications media, far from merely mediating, interpose themselves between the generations as manufacturers of stereotypes, often forcing youth to live out the caricatures of the images that at first they had only "projected" in experimental fashion. Much will depend on what we do about this. In spite of our pretensions of being able to study the youth of today with the eyes of detached naturalists, we are helping to make youth in the year 2000 what it will be by the kinds of questions we now ask. So I will point out the ideological beams in our eyes as I attempt to put into words what I see ahead. I will begin with questions that are diagnostic and then proceed to those that are more prognostic in character.

I would assume that adolescents today and tomorrow are struggling to define new modes of conduct which are relevant to their lives.

Young people of a questioning bent have always done this. But more than any young generation before and with less reliance on a meaningful choice of traditional world images, the youth of today is forced to ask what is *universally relevant* in human life in this technological age at this junction of history. Even some of the most faddish, neurotic, delinquent preoccupation with "their" lives is a symptom of this fact.

Yet, this is within the context of two culture factors which seem to be extraordinary in the history of moral temper. One is the scepticism of all authority, the refusal to define natural authority (perhaps even that of paternal authority) and a cast of mind which is essentially anti-institutional and even antinomian.

I do not believe that even in the minority of youths to whom this statement is at all applicable there is a scepticism of *all* authority. There is an abiding mistrust of people who act authoritatively without authentic authority or refuse to assume the authority that is theirs by right and necessity. Paternal authority? Oh, yes — pompous fathers have been exposed everywhere by the world wars and the revolutions. It is interesting, though, that the word *paternal* is used rather than *parental*, for authority, while less paternal, may not slip altogether from the parent generation, insofar as a better balance of maternal and paternal authority may evolve from a changing position of women. As a teacher, I am more impressed with our varying incapacity to own up to the almost oppressive authority we really do have in the minds of the young than in the alleged scepticism of *all* authority in the young. Their scepticism, even in its most cynical and violent forms, often seems to express a good sense for what true authority is, or should be, or yet could be. If they "refuse to define natural authority" — are they not right if they indicate by all the overt, mocking, and challenging kinds of "alienation" that it is up to *us* to help them define it, or rather redefine it, since we have undermined it — and feel mighty guilty?

As to the essentially anti-institutional cast of mind, one must ask what alternative is here rejected. It appears that the majority of young people are, in fact, all too needy for, trusting in, and conforming to present institutions, organizations, parties, industrial complexes, super-machineries — and this because true personal authority is waning. Even the anti-institutional minority (whom we know better and who are apt to know our writings) seem to me to plead with existing institutions for permission to rebel — just as in private they often seem to plead with their parents to love them doubly for rejecting them. And are they not remarkably eager for old and new uniforms (a kind of uniformity of nonconformity), for public rituals, and for a collective style of individual isolation? Within this minority, however, as well as in the majority, there are great numbers who are deeply interested in and responsive to a more concerted critique of institutions from a newer and more adequate ethical point of view than we can offer them.

The second factor is an extraordinary hedonism — using the word in the broadest sense — in that there is a desacralization of life and an attitude that all experience is permissible and even desirable.

Again, the word *hedonism* illustrates the way in which we use outdated terms for entirely new phenomena. Although many young people entertain a greater variety of sensual and sexual experiences than their parents did, I see in their pleasure seeking relatively little relaxed joy and often compulsive and addictive search for *relevant* experience. And here we should admit that our generation and our heritage made "all" experience relative by opening it to ruthless inquiry and by assuming that one could pursue radical enlightenment without changing radically or, indeed, changing the coming generations radically. The young have no choice but to experiment with what is left of the "enlightened," "analyzed," and standardized world that we have bequeathed to them. Yet their search is not for all-permissibility, but for new logical and ethical boundaries. Now only direct experience can offer correctives that our traditional mixture of radical enlightenment and middle-class moralism has failed to provide. I suspect that "hedonistic" perversity will soon lose much of its attractiveness in deed and in print when the available inventory has been experimented with and found only moderately satisfying, once it is permitted. New boundaries will then emerge from new ways of finding out what really counts, for there is much latent affirmation and much overt solidarity in all this search. All you have to do is to see some of these nihilists with babies, and you are less sure of what one of the statements as yet to be quoted terms the "Hegelian certainty" that the next generation will be even more alienated.

As for the desacralization of life by the young, it must be obvious that our generation desacralized their lives by (to mention only the intellectual side) naive scientism, thoughtless scepticism, dilettante political opposition, and irresponsible technical expansion. I find, in fact, more of a search for resacralization in the younger than in the older generation.

At the same time society imposes new forms of specialization, of extended training, of new hierarchies and organizations. Thus, one finds an unprecedented divorce between the culture and the society. And, from all indications, such a separation will increase.

Here, much depends on what one means by the word *imposes*. As I have already indicated, in much of youth new hierarchies and organizations are accepted and welcome. We are apt to forget that young people (if not burdened with their parents' conflicts) have no reason to feel that radical change as such is an imposition. The unprecedented divorce we perceive is between *our* traditional culture (or shall I spell it *Kultur?*) and the tasks of *their* society. A new generation growing up with technological and scientific progress may well experience technology and its new modes of thought as the link between a new culture and new forms of society.

In this respect, assuming this hypothesis is true, the greatest strains will be on the youth. This particular generation, like its predecessors, may come back to some form of accommodation with the society as it grows older and accepts positions within the society. But the experiences also leave a "cultural deposit" which is cumulative consciousness and — to this extent I am a Hegelian — is irreversible, and the next generation therefore starts from a more advanced position of alienation and detachment.

Does it make sense that a generation involved in such unprecedented change should "come back to some form of accommodation with the society"? This was the fate of certain rebels and romantics in the past; but there may soon be no predictable society to "come back to," even if coming back were a viable term or image in the minds of youth. Rather, I would expect the majority to be only too willing to overaccommodate to the exploiters of change, and the minority we speak of to feel cast off until their function becomes clearer — with whatever help we can give.

Sources of Identity Strength

Having somewhat summarily disavowed the statements formulated by others, I would now like to ask a question more in line with my own thinking, and thereby not necessarily more free from stereotypy: Where *are* some of the principal contemporary sources of identity strength? This question leads us from diagnosis to prognosis, for to me a sense of identity (and here the widest connotation of the term will do) includes a sense of anticipated future. The traditional sources of identity strength — economic, racial, national, religious, occupational — are all in the process of allying themselves with a new world-image in which the vision of an anticipated future and, in fact, of a future in a permanent state of planning will take over much of the power of tradition. If I call such sources of identity strength *ideological*, I am using the word again most generally to denote a system of ideas providing a convincing world-image. Such a system each new generation needs — so much so that it cannot wait for it to be tested in advance. I will call the two principal ideological orientations basic to future identities the *technological* and the *humanist* orientations, and I will assume that even the great politico-economic alternatives will be subordinated to them.

I will assume, then, that especially in this country, but increasingly also abroad, masses of young people feel attuned, both by giftedness and by opportunity, to the technological and scientific promises of indefinite progress; and that these promises, if sustained by schooling, imply a new ideological world-image and a new kind of identity for many. As in every past technology and each historical period, there are vast numbers of individuals who can combine the dominant techniques of mastery and domination with their identity development, and *become* what they *do.* They can settle on that *cultural consolidation* that follows shifts in technology and secures what mutual verification and what transitory familiarity lie in doing things together and in doing them right — a rightness proved by the bountiful response of "nature," whether in the form of the prey bagged, the food harvested, the goods produced, the money made, the ideas substantiated, or the technological problems solved.

Each such consolidation, of course, also makes for new kinds of entrenched privileges, enforced sacrifices, institutionalized inequalities, and built-in contradictions that become glaringly obvious to outsiders — those who lack the appropriate gifts and opportunities or have a surplus of not quite appropriate talents. Yet it would be intellectual vindictiveness to overlook the sense of embeddedness and natural flux that each age provides in the midst of the artifacts of organization; how it helps to bring to ascendance some particular type of man and style of perfection; how it permits those thus consolidated to limit their horizon effectively so as *not* to see what might destroy their newly won unity with time and space or expose them to the fear of death — and of killing. Such a consolidation along technological and scientific lines is, I submit, now taking place. Those young people who feel at home in it can, in fact, go along with their parents and teachers — not too respectfully, to be sure — in a kind of *fraternal identification*, because parents and children can jointly leave it to technology and science to provide a self-perpetuating and self-accelerating way of life. No need is felt to limit expansionist ideals so long as certain old-fashioned rationalizations continue to provide the hope (a hope that has long been an intrinsic part of an American ideology) that in regard to any possible built-in evil in the very nature of super-organizations, appropriate brakes, corrections, and amendments will be invented in the nick of time and without any undue investment of strenuously new principles. While they "work," these super-machineries, organizations, and associations provide a sufficiently adjustable identity for all those who feel actively engaged in and by them.

All of us sense the danger of overaccommodation in this, as in any other consolidation of a new world-image, and maybe the danger *is* greater today. It is the danger that a willful and playful testing of the now limitless range of the technically possible will replace the search for the criteria for the optimal and the ethically permissible, which includes what can be given on from generation to generation. This can only cause subliminal panic, especially where the old decencies will prove glaringly inadequate, and where the threat or the mere possibility of overkill can be denied only with increasing mental strain — a strain, incidentally, which will match the sexual repression of the passing era in unconscious pathogenic power.

It is against this danger, I think, that the nonaccommodators put their very existence "on the line," often in a thoroughly confounding way because the manifestations of alienation and commitment are sometimes indistinguishable. The insistence on the question "to be or not to be" always looks gratuitously strange to the consolidated. If the question of being oneself and of dying one's own death in a world of overkill seems to appear in a more confused and confusing form, it is the ruthless heritage of radical enlightenment that forces some intelligent young people into a seemingly cynical pride, demanding that they be human without illusion, naked without narcissism, loving without idealization, ethical without moral passion, restless without being classifiably neurotic, and political without lying: truly a utopia to end all utopias. What should we call this youth? *Humanist* would seem right if by this we mean a recovery, with new implications, of man as the measure, a man far grimmer and with much less temptation to congratulate himself on his exalted position in the universe, a self-congratulation that has in the past always encouraged more cruel and more thoughtless consolidations. The new humanism ranges from an *existential* insistence that every man *is* an island unto himself to a new kind of humaneness that is more than compassion for stray animals and savages, and a decidedly *humanitarian* activism ready to meet concrete dangers and hardships in the service of assisting the underprivileged anywhere. Maybe *universalist* would cover all this better, if we mean by it an insistence on the widest range of human possibilities — beyond the technological.

But whatever you call it, the universalist orientation, no less than the technological one, is a *cluster* of ideas, images, and aspirations, of hopes, fears, and hates; otherwise, neither could lay claim to the identity development of the young. *Somewhat* like the "hawks" and the "doves," the technologists and the universalists seem almost to belong to different species, living in separate ecologies. "Technological" youth, for example, expects the dominant forces in foreign as well as in domestic matters to work themselves out into some new form of balance of power (or is it an old-fashioned balance of entirely new powers?). It is willing, for the sake of such an expectation, to do a reasonable amount of killing — and of dying. "Humanist" youth, on the other hand, not only opposes unlimited mechanization and regimentation, but also cultivates a sensitive awareness of the humanness of any individual in gun-sight range. The two orientations must obviously oppose and repel each other totally; the acceptance of even a part of one could cause an ideological slide in the whole configuration of images and, it follows, in the kind of courage to be — and to die. These two views, therefore, face each other as if the other were *the* enemy, although he may be brother or friend — and, indeed, oneself at a different stage of one's own life, or even in a different mood of the same stage.

Each side, of course, is overly aware of the dangers inherent in the other. In fact, it makes out of the other, in my jargon, a negative identity. I have sketched the danger felt to exist in the technological orientation. On the "humanist" side, there is the danger of a starry-eyed faith in the certainty that if you "mean it," you can move quite monolithic mountains, and of a subsequent total inertia when the mountain moves only a bit at a time or slides right back. This segment of youth lacks as yet the leadership that would replace the loss of revolutionary tradition, or any other tradition of discipline. Then there is the danger of a retreat into all kinds of Beat snobbishness or into parallel private worlds, each with its own artifically expanded consciousness.

New Form of Accommodation

As one is apt to do in arguing over diagnosis, I have now overdrawn two "ideal" syndromes so as to consider the prognosis suggested in a further question presented to me:

Is it possible that the fabric of traditional authority has been torn so severely in the last decades that the re-establishment of certain earlier forms of convention is all but unlikely?

I have already indicated that I would answer this question in the affirmative; I would not expect a future accommodation to be characterized by a "coming back" either to conventions or to old-fashioned movements. Has not every major era in history been characterized by a division into a new class of *power-specialists* (who "know what they are doing") and an intense new group of *universalists* (who "mean what they are saying")? And do not these two poles determine an era's character? The specialists ruthlessly test the limits of power, while the universalists always in remembering man's soul also remember the "poor" — those cut off from the resources of power. What is as yet dormant in that third group, the truly under-privileged, is hard to say, especially if an all-colored anticolonial solidarity that would include our Negro youth should emerge. But it would seem probable that all new revolutionary identities will be drawn into the struggle of the two ideological orientations sketched here, and that nothing could preclude a fruitful polarity between these two orientations — provided we survive.

But is not the fact that we are still here already a result of the polarization I have spoken of? If our super-technicians had not been able to put warning signals and brakes into the very machinery of armament, certainly our universalists would not have known how to save or how to govern the world. It also seems reasonable to assume that without the apocalyptic warnings of the universalists, the new technocrats might not have been shocked into restraining the power they wield.

What speaks for a fruitful polarization is the probability that a new generation growing up with and in technological and scientific progress as a matter of course will be forced by the daily confrontation with unheard-of practical and theoretical possibilities to entertain radically new modes of thought that may suggest daring innovations in both culture and society.

"Humanist" youth, in turn, will find some accommodation with the machine age in which they, of course, already participate in their daily needs and habits. Thus, each group may reach in the other what imagination, sensitivity, or commitment may be ready for activation. I do not mean, however, even to wish that the clarity of opposition of the technological and the humanist identity be blurred, for dynamic interplay needs clear poles.

What, finally, is apt to bring youth of different persuasions together is a change in the generational pr cess itself — an awareness that they share a common fate. Already today the mere division into an older — parent — generation and a younger — adolescing — one is becoming superannuated. Technological change makes it impossible for any traditional way of being older (an age difference suggested by the questions quoted) ever to become again so institutionalized that the younger generation could "accommodate" to it or, indeed, resist it in good-old revolutionary fashion. Aging, it is already widely noted, will be (or already is) a quite different experience for those who find themselves rather early occupationally outdated and for those who may have something more lasting to offer. By the same token, young adulthood will be divided into older and younger young adults. The not-too-young and not-too-old specialist will probably move into the position of principal arbiter, that is, for the limited period of the ascendance of his speciality. His power, in many ways, will replace the sanction of tradition or, indeed, of parents. But the "younger generation," too, will be (or already is) divided more clearly into the older- and the younger-young generation, where the older young will have to take over (and are eager to take over) much of the direction of the conduct of the younger young. Thus, the relative waning of the parents and the emergence of the young adult specialist as the permanent and permanently changing authority are bringing about a shift by which older youth will have to take increasing responsibility for the conduct of younger youth — and older people for the orientation of the specialists and of older youth. By the same token, future religious ethics would be grounded less in the emotions and the imagery of infantile guilt, than in that of mutual responsibility in the fleeting present.

In such change we on our part can orient ourselves and offer orientation only by recognizing and cultivating an age-specific *ethical* capacity in *older* youth, for there are age-specific factors that speak for a differentiation between morality and ethics. The child's conscience tends to be impressed with a moralism which says "no" without giving reasons; in this sense, the infantile super-ego has become a danger to human survival, for suppression in childhood leads to the exploitation of others in adulthood, and moralistic self-denial ends up in the wish to annihilate others. There is also an age-specific ethical capacity in older youth that we should learn to foster. That we, instead, consistently neglect this ethical potential and, in fact, deny it with the moralistic reaction that we traditionally employ toward and against youth (*anti-institutional, hedonistic, desacralizing*) is probably resented much more by young people than our dutiful attempts to keep them in order by prohibition. At any rate, the ethical questions of the future will be less determined by the influence of the older generation on the younger one than by the interplay of subdivisions in a life scheme in which the whole life-span is extended; in which the life stages will be further subdivided; in which new roles for both sexes will emerge in all life stages; and in which a certain margin of free choice and individualized identity will come to be considered the reward for technical inventiveness. In the next decade, youth will force us to help them to develop ethical, affirmative, resacralizing rules of conduct that remain flexibly adjustable to the promises and the dangers of world-wide technology and communication. These developments, of course, include two "things" — one gigantic, one tiny — the irreversible presence of which will have to find acknowledgment in daily life: the Bomb and the Loop. They together will call for everyday decisions involving the sanctity of life and death. Once man has decided not to kill needlessly and not to give birth carelessly, he must try to establish what capacity for living, and for letting live, each generation owes to every child planned to be born —anywhere.

One can, I guess, undertake to predict only on the basis of one of two premises: Either one expects that things will be as bad as they always have been, only worse; or one visualizes what one is willing to take a chance on at the risk of being irrelevant. As I implied at the beginning, a committee that wants to foretell the future may have to take a chance with itself by asking what its combined wisdom and talent would wish might be done with what seems to be given.

CHAPTER 24

What Troubles Our Troubled Youth?

ROY MENNINGER

Youth Behaviors and Adult Reactions

One cannot read the newspapers or the popular weekly magazines, watch television, or even travel about our larger cities without being made aware of our youth. Whether it is in their numbers, their outlandish or provocative behavior, their fads, or their economic influence, they make us conscious of their presence all about us. To be aware of their presence in all its forms is to become aware of something more: much that they do is somehow troubling to members of other generations, even to persons but a few years older than they.

Those who have occasion to see these youths professionally, as do we psychiatrists, become conscious of the fact that the youths who trouble so many of us are themselves troubled people. It is no trick, of course, to decide that their troubles are their own — particular difficulties peculiar to the individual who seeks our help, youth who are troubled because of having come from troubled families. It is not a much bigger step to decide that these special cases of trouble should be referred to a physician, a psychiatrist, a counselor or the like, or perhaps just treated with pills and otherwise disregarded.

To be sure, many of these troubled youths do come from troubled families, and do need psychiatric help. But what are often not considered by adults, or intentionally ignored if perceived, are some of the concerns these troubled young people have about themselves and their place in the world. This they have in common not only with each other, but also with many of their peers who have not gone the route of becoming patients. These concerns are not to be dismissed with a wave of an older and wiser hand and a disdainful comment on "the modern generation." These concerns that trouble our troubled youth require a hearing, particularly a hearing from those of us who say, loudly and publicly, that we are concerned about our youth, that we are working to help our youth, that we are humanitarian in our interests. I think we have failed our youth by having failed to listen — or, having listened, failed to hear.

The evidences of their trouble are manifold. Statistics have a way of sounding cold and harsh, of often failing to reveal the human tragedy they imply; but let me share a few of them with you. One out of every six teenagers becomes pregnant out of wedlock; one-third to one-half of all teenagers' marriages are prefaced by illegitimate pregnancy; the number of unwed mothers under eighteen years of age has doubled since 1940; one teenage marriage in every *two* ends in divorce within five years; 40 percent of all the women who walk down the aisle today are between the ages of fifteen and eighteen.

But it does not stop there. Three youngsters in every hundred between the ages of ten and seventeen will be adjudged delinquent this year. There are nearly half a million children haled into juvenile court every year. There is a tremendous increase the use of drugs — amphetamines, barbiturates, LSD. It is estimated that in Nassau County, New York —there is no reason to think things are different here than they are there — one youngster in every six has taken marijuana or LSD. Some estimate that up to 50 percent of the youth on college campuses are experimenting with these drugs. The statistics go on, and they do not get better.

To me, these figures are dismaying, they are troubling; they certainly are a sign of troubled youth. One of the first reactions of most adults to these statistics and the tragedies that lie behind them is fear. So much evidence of disrupted living evokes apprehension within most of us.

Will any of these things happen to my children? If they do, am I, the parent, to blame? These are questions we are likely to ask. These chilling statistics, coupled with our own impressions of adolescence as a stormy and turbulent time, contribute to a sense of apprehension about adolescence in general. "Clearly, they are unpredictable, stormy, and potentially violent people," we think. The sudden sound of screeching tires on a nearby street in a quiet neighborhood brings an immediate reaction: "There goes a teenager" — when, of course, we cannot know whether we are right or not. We walk down a city street and see a clustered group on the corner. For all we know, they are a bunch of happy, contented kids on the way home from a movie. But what do we feel? Fear. What do we think? They might attack us. So often is there conveyed by the word "teenager" an image of turbulence, conflict, explosiveness, unpleasantness, uncontrollability.

Not all of us are so consciously aware of this fear; but its workings are nonetheless evident in the reactions of contempt, disdain, disgust, or distaste so frequently expressed in the wake of some teenage act. This reaction or rejection is born perhaps of some conviction that adolescents are volatile combinations of sex and aggression barely under control. For most of us, it is a short and easy step to a reaction of indignant anger. Made anxious by the visible struggles of our teenagers, we are quick to defend ourselves by righteous proclamations, usually emphasizing our adult wisdom, our greater experience. Out of these anxious and angry feelings come unreasonable constraints on our adolescents, vitriolic attacks on their behavior, ready capitulation to their demands, or, perhaps, what is worst of all, turning our backs on them, their concerns, and their needs.

These adult reactions are problems for all sorts of reasons. They enable the adolescent to feel misunderstood (which he is); they allow us to think we have done something constructive, when we have done nothing of the kind; and, even worse, they lead us to miss the whole point of this troubled behavior. In my view, so much of it speaks of the failure of society to deal with the real issues that adolescence poses for the adolescent, and for the society in which he lives. By their very provocativeness, these behaviors draw our attention to the symptoms, obscuring completely the existence of a more serious problem that may underlie them.

Youth Behaviors and Society

For so many adolescents, their challenging behavior is a reaction of frustration to the failure of society to make a reasonable and sensible and appropriate place for them. To put it bluntly, our adult society tends to regard the adolescent as an unfortunate inconvenience; a sort of bad moment that we half wish would go away; a distraction or maybe a disruption that gets in the way of the real business of living for the rest of us; a kind of incidental way station in life that will surely pass if we wait long enough or hold our breath or look the other way. It is as if adult society regards adolescence as an unattractive extension of childhood that we must somehow put up with, until the magic of time has somehow transmuted that cute little baby of yesteryear into the adult of tomorrow. Most of us feel put upon by the very existence of the adolescent, annoyed with his presence, his unpredictability, his demands, his parasitic nature, and the like, as if we were somehow the victim and he the aggressor. And, as with any victim, the roads of appeasement and bribery are natural resources. So we give him a car when he asks, or a new electric guitar, or an increase in his allowance — anything, just to get him "off our backs" and out of our way.

More than this, we couple our anxious responses with words of moral uplift, sermonizing about how things will have to change when they get out into that cold, cruel world, how they must carry their end of the load, learn to be responsible, put their shoulder to the wheel, and so forth. Often in that vein we tax them with busy work that is meaningless to them and little more than our exploitation of their cheap and available labor.

So it is logical to suggest that our adolescents' provocative behavior may be their way of saying to us, "I object." They may be trying to tell us how they feel about our systematically segregating them from adult society. They may be trying to make us understand how grave is our failure to perceive their legitimate needs for participation, their legitimate needs for genuine challenge and engagement in the real tasks of living.

How is it that adolescents are not greater participants in society? Partly, perhaps, because we look upon the job of the child and the adolescent as a single, narrowly focused task: completing their schooling. No matter how we define it, attending school is their task, and all else is secondary and generally classified under the rubric of play. By virtue of this commitment to schooling imposed by society, the child through late adolescence has no other significant social contribution to make. But beyond this we ask, "How can he make a contribution?" He is too immature, too irresponsible, or too inexperienced, or a drug on the labor market, or without enough social merit in the aggregate to permit anything more than the most token participation in any of the social processes characteristic of adult living. He is not ready for the privileges and responsibilities of this participation until some magical point has been reached — a particular age or an official change in status.

Without regard to their individual talents, their interests, their perceptiveness, their energy, their idealism, or their enthusiasm, we deny them a significant role in society at large.

Nowhere are the starkness and meagerness of this social isolation more apparent than in the lot of the fifteen-year-old. Except for going to school, virtually nothing that he can do is legal. He can't quit school, he can't work, he can't drink, he can't smoke, he can't drive in most states, he can't marry, he can't vote, he can't enlist, he can't gamble. He cannot, in fact, participate in *any* of the adult virtues, vices, or activities.

But consider the consequences of this enforced sidelining of the adolescent. There he waits, champing at the bit, full of energy, drive, and curiosity, intrigued and tempted by the publicly advertised advantages of adulthood. Yet, he is asked to forgo the pleasures that he sees the adults all around him engaging in freely and often to excess. Is it any wonder that he samples these experiences secretly, or in defiance, or inappropriately? And how does it prepare the adolescent for the world of adult responsibilities when he is given no opportunities to test, to try, to experience, to learn by doing? By what magic do we expect this growing adolescent, denied opportunities for the participation that teaches, suddenly to emerge on the stage of adulthood, full-fledged, capable, mature, and responsible? Small wonder that so few are ready for these responsibilities when the time comes, when their predominant experience has been the frustration of waiting, forgoing, postponing, and standing apart from the society flowing all around them.

How does this come to be? How is it that we view our adolescents as overgrown children, treat them as such, and then are perturbed by their acting that way? How is it that we are face to face with a social phenomenon of discontented adolescence that we can neither understand nor manage?

In a paper called "Socio-cultural Dilemmas in the World of Adolescence and Youth," prepared for the pre-congress book for the International Association of Child Psychiatry and Allied Professions, Soskin, Duhl, and Leopold[1] presented several interesting factors about the social phenomenon of discontented adolescence.

One of these factors is the dramatic change in the economic status of the adolescent. The affluence of our society means, in effect, that the adolescent does not have to work, because the money he might earn is not as necessary for support of the family as it once was. Affluence provides him with the means for fantastic self-indulgence. It is estimated that last year the aggregate total for allowances and money earned by adolescents came to the staggering sum of $14 billion. This amount of money (plus a large amount of leisure time, plus a lack of significant involvement in the social fabric) inevitably makes for a pattern of living with the character of endless play. This is a sharp swing of the pendulum to the opposite extreme from the days of child labor of fifty years ago.

Moreover, the affluence in our society provides a devastating contrast between the luxuries available to middle-class youth and the continuing deprivations of the lower class, and particularly to the Negro youth. Undoubtedly this contrast produced part of the inflammatory pressures that led to the riots in the summer of 1967.

This tremendous affluence means, of course, a tremendous consumer market, which develops a self-sustaining and expanding dynamic of its own. And, from the point of view of the adolescent, it could be argued that this self-sustaining market tends to introduce more superficial, materialistic, spurious, shifting, status-centered values that push out the more solid virtues.

A second factor is the upward extension of schooling itself. Compulsory public education for all, initially limited to the elementary grades, was gradually extended to secondary school education, as there was an expansion of knowledge that needed to be mastered and an increasing need for more and better training of people. But out of this virtue of compulsory public education have come a few unexpected disadvantages. Among other things, it has meant an extended period in which the growing adolescent is dependent upon, and controlled by, adults. This spells further delay in permitting him to engage in some of the activities that will teach him how to deal with such ultimate life functions as work and the assumption of citizen and social responsibility. We have watched the age of legal responsibility creep upward from seven, where it used to be, to sixteen, eighteen, twenty-one — surely, for very good reasons, but with not so happy consequences. Many years ago, a boy of sixteen might well have been head of the household, or a soldier in the king's army. Even in our Civil War there were drummer boys and buglers serving at the age of fifteen.

What are some of these unhappy consequences? Perhaps the most serious is the extent to which the adolescent is infantilized — "childized," as Soskin and associates[1] have called it. The adolescent becomes more childish, with the room and the permission to stay that way. This state is a deterrent to healthy growth; it provokes and sustains our perceptions of him as immature. It is a magnificent example of a self-fulfilling prophecy. We deny the adolescent some of the responsibilities of maturity; and, when he responds with childish behavior, we say, "See, I told you all along you weren't ready." As we react by giving him still less responsibility and penning him up more, he reacts with still greater evidence of immaturity, which then justifies another round of adult control and demands for conformity.

I think this infantilizing of the adolescent does something more. I think it probably provokes adventure-seeking, thrill-seeking, serious risk-taking behavior, such as taking drugs, playing "chicken" on the highway, speeding at ninety miles an hour through the city, and so forth. I would suggest that this behavior not only expresses the sense of helplessness and frustration the adolescent feels at being so irrelevant to the adult society all around him, but conveys as well his anger and his resentment for being disregarded and shoved aside by adults.

Adolescents are action-oriented people; they are people seeking a cause and a reason for being. If we fail to supply tasks that are adequate to absorb their energies and relevant to their psychosocial needs, they will do the only thing they can: seek their own outlets, and adults be damned. The fatal combination of their needs plus our indifference necessarily and inevitably leads to behavior that will either embarrass or trouble us and risks being a danger to all.

Our expectations and our presumptions about the adolescent as generally too immature to assume much responsibility embarrass us when he shows unexpected evidence of political or social maturity. Witness our astonishment at the success of the Peace Corps and our amazement at the conviction and effectiveness of the adolescent civil rights worker. We may not always agree with the sentiments adolescents express and work for, but we cannot deny the strength and the effectiveness of their commitment when they are finally given the opportunity to make it.

In our systematic social infantilization of the adolescent, we hang him between the horns of a serious dilemma. By its nature, adolescence forces gradual estrangement of the youth from the support and the nurture that the family gave him as a child, yet does not provide the benefits and supports of adulthood. And there he hangs, able neither to retreat to the warmth and support of the family nor to advance into the companionship of adult society. This limbo in which the adolescent now finds himself was filled in earlier times by the opportunity to serve as an apprentice and by the availability of real work. With these no longer open to him, the adolescent's world is an empty one, populated by church and youth groups and some commercial interests. As others have observed, the former are too selective and exclusive, failing to reach the very youth who may need them most; and the latter only exploit the chaos of adolescence for their own interests, with service neither to youth nor to society.

Even more tragic, enforced schooling combined with enforced infantilization cemented by a systematic absence of real work and real participation in the social process yields an unfortunate fruit. In spite of twelve years of education, the average high school graduate emerges from his educational cocoon with no place to go and nothing to be. He has no occupational identity, no skills worth selling, no systematic practice in the arts of living in a complex society, and not much of a clue about where to go to find what he does not yet have.

The exceptions are the college-bound youths; but they, by that very token, are not average. Even here, though they may continue their schooling through various kinds of higher education, these older adolescents continue to feel isolated from society and are, in fact, excluded from much significant participation in social processes. Without voice or responsibility for the society that they are physically a part of, and still aware of a pervasive sense of irrelevance to the larger adult community, they give vent to their distress and their resentment through overt external action — through acts of social protest, or town-gown riots, or vociferous support of unpopular causes, or internal retreat through LSD or becoming a hippie.

But these differences between the college adolescent and his drifting buddy who barely made it through high school are differences in degree, not kind. Each in his own way is struggling to come to terms with the failure of society to have prepared him better for the adult life it now expects him to lead. The symptoms of this failure, already expressed by the distressing statistics referred to earlier, are portrayed by unhappy premature marriages or excessive drinking or pursuit of crime. This is a message we cannot afford to miss. This is not simply a school problem; this is not a problem of better law enforcement; this is not a problem of stricter laws to prevent illegitimate pregnancies or teenage marriages; this is not a problem to be solved by crying alarm or singling out some group for blame.

It is a problem for which all of us, all of us adults, are to blame. But our response to the problems of our adolescents cannot stop with the expression of a bit of guilt followed by some reassuring pabulum. In fact, it must proceed to a new and more honest look at the adolescent and his relationship to the world around him. We must set aside this moralistic, holier-than-thou attitude of complacent superiority that we adults so often assume in front of our youngsters, and be willing instead to take a closer look at what it is the adolescent needs for his tasks of growth as he moves from childhood to adulthood, and what it is that we as the guardians of society are, or are not, providing him to make that possible. I think we need to take a harder look at the extent to which our society has denied the adolescent the room he needs to experiment, to participate, to engage, and to involve himself in the fabric of real living. We need to consider how we can enable our youth to participate legitimately in the social issues of our time, struggle with the real problems of racial prejudice, social and economic deprivation, self-government, and the development of conceptions of service to others. This means a recognition of our failure to give adolescents a chance to participate in meaningful, active, and effective ways in the social processes of our communities.

We deprive the adolescent of these opportunities at the very moment in his development when he needs the challenge of real situations and real problems to test himself, to define his capabilities and his interests, and to find out what he can do and who he is and of what he is made. We have sidelined him from engagement in the vital concerns of our society at a time when he is motivated more powerfully by idealism and a sense of justice than he ever will be again. The adolescent is task oriented, he is eager, he has enormous energy and a willingness to invest this in useful and meaningful activities — but only if he has the opportunity and the permission to do so. If we fail, as I think we have failed, to provide these opportunities and permissions, we deprive an enormous proportion of our youth of engagement in the social and community activities that will supply the laboratory of learning they need. We also deprive the community of an enormous contribution of vigor, spirit, energy, enthusiasm, and capacity for change that could literally remake society.

What makes this failure of adult society so ironic is the fact that there are so many evidences of acute and immediate need in our communities for service, for assistance, for support, for rehabilitation. To permit these needs, on the one hand, and the terrible waste in our eager, energetic, but uninvolved youth, on the other, is a tragedy of enormous proportions. The barest hint of what alchemy of change is possible when these psychologically needful youth are engaged in the tasks of community living is evident in the work of the Peace Corps and VISTA. It was impressive to note that perhaps the single most effective method of controlling the rioters in the summer of 1967 in Florida and, belatedly, in Newark was the use of adolescents in the responsible role of roving carriers of the word, quieting the restless and resentful citizens and undoubtedly forestalling further riots. No one has yet recorded what this experience must have done for the youth involved, but there is no question about the value to the community of their unique participation.

There are surely ways in which a marriage of these sociopsychological needs of our adolescents and the human needs of our communities can be made, with inestimable profit for both. Perhaps this process can be begun by the many voluntary service organizations devoted to youth. Yet, this cannot occur without a simultaneous look — a hard look — at the artificiality and irrelevance, the busywork and triviality that all too many of our youth organizations put forth under the rubric of "character building." To be sure, some learning does take place, even from the fun-type avocational pursuits that most middle-class-oriented youth organizations offer their constituents. But, too often, this learning is limited to furthering such selfish concerns as one's own advancement, indulgence, or gratification.

Confrontation with the vastly greater needs of the segregated, ignored, deprived kids is rare. Engagement of our middle-class youth in actually working with these less fortunate kids — the ones our youth organizations practically never reach — is even rarer, for it seems that it is the pattern of all too many youth groups to rest in the comfortable complacence of promoting good, solid, middle-class values of achievement, progress, education, competition, and the like, and to ignore or silently avoid confrontation with, let alone engagement in, some of the critical concerns of our current society — racial and social justice, poverty, deprivation, delinquency, and the absence of individual dignity for so many.

Yet it is in work in these areas where the needs are so great that our hungry, identity-searching adolescents can stand to give so much and learn so much. By grappling with real problems and engaging in real situations they develop a truer sense of values than can ever be acquired from simply sitting passively and being told. They learn what so many of us have yet to learn well: the satisfactions and usefulness of service to others. They learn how their needs can be effectively meshed with the needs of others to produce a greater sense of community between both.

It is to be hoped that youth organizations will rise to this tremendous challenge and offer some of the opportunities that both our adolescents and our community so badly need. But even this task, vital as it is, cannot begin until we have first begun to recognize the extent to which we have disenfranchised our adolescent youth from the social processes that engage so much of the attention of the rest of us. Without a doubt, voluntary organizations dedicated to youth have a crucial and significant role to play, so that our adolescents may again become a part of the mainstream that flows on to adult maturity. This is a task well worth all the attention we can possibly give it.

Reference

[1] Soskin, W. F.; Duhl, L. J.; and Leopold, R. L. Socio-cultural Dilemmas in the World of Adolescence and Youth. Paper prepared as a chapter for inclusion in Pre-Congress book for the International Association of Child Psychiatry and Allied Professions.

PART III

Early and Middle Adulthood: An Introduction to the Determinants Affecting the Early and Middle Adulthood Stage of the Life Cycle

Early and middle adulthood covers a long span of life, a period much longer than the two preceding periods, childhood and adolescence, combined. It extends approximately from ages twenty-one to sixty-five. Since many things change in the course of forty-some years, life circumstances and social expectations are not the same in the first twenty years as they are in the second twenty years. Thus some writers prefer to separate this period into two stages — young adult and middle age. Here the discussion will be focused on the young adult stage, in view of its importance as the foundation upon which one's remaining years are built.

It is a complex period indeed, a period in which an individual's past endowments and deficits, present opportunities and restraints all come into play, with consequences that determine the adult's future life prospects. Irving Sarnoff, in his discussion of personality development in adulthood, agrees that "as the adult lives out his existence, his behavior will reflect (*a*) old elements of personality, derivatives of childhood and adolescence; and (*b*) new elements — new motives, attitudes, interests — that he acquires in the course of his adult years" (1). There is a paucity of work on the development of a comprehensive adult personality theory in the professional literature. Erikson has ambitiously conceptualized man's complete life cycle into eight stages, but when he reaches the seventh stage, corresponding to adulthood, his exposition slackens. He briefly sums up the essence of the adult life stage as "the ability to lose oneself in the meeting of bodies and minds leading to a gradual expansion of ego interests and of libidinal cathexis over that which has been thus generated and accepted as a responsibility" (2). He terms the crux of this stage as the resolution of generativity vs. stagnation. Erikson sees generativity as "primarily the interest in establishing and guiding the next generation or whatever in a given case may become the absorbing object of a parental kind of responsibility. Where this enrichment fails, a regression from generativity to an obsessive need for pseudo intimacy, punctuated by moments of mutual repulsion, takes place, often with a pervading sense (and objective evidence) in individual stagnation and interpersonal impoverishment" (3). What this means is that one has to integrate one's self with contemporary vision without too much interference from his past hangups, and be comfortable enough to pursue his goals responsibly. In this context, Freudian analysts would say that one's instinctual drives have to be channeled, or sublimated, into socially acceptable activities. Karl Menninger, in Chapter 27, deals specifically with this issue. The enduring socially acceptable activities in modern society for an adult might be his vocational activities, marriage, and parenthood. Each of the three basic activities has its snags and mutually influencing effects on other activities. Let us briefly examine each of these activities as a normal process of adult life.

Economic security has been one of the most pressing problems for adults in modern society. The Protestant ethic, sometimes called the "work ethic," is deeply rooted in American thought. It is one of the most pervasive and long-lived ideologies in the Western world. In the late twentieth century, simply working or just producing may be sufficient for basic material needs, but it will not necessarily produce psychological and social satisfaction. In a competitive society like the United States, work has been stratified into various social status levels, such as manual, white collar, professional. Social status has a vital symbolic meaning to a person's functioning, his self-worth being reflected by the social symbol on the one hand and by his sense of competence and self-satisfaction on the other. The difficulty is that in a highly industrialized society, most people have little or no opportunity to be creative and are unable to develop a sense of satisfaction from daily routines other than mere job security. Sarnoff observes that "over the years, the grinding implacability of this sort of routine work may well produce first incipient rebellion, then impotent resentment, and finally, dullness, apathy and unthinking resignation — as the impossibility of escape becomes ever more indisputable" (4). Even for the chosen few, high-level business executives, job pressures are no less pressing, although they may differ in content. Pressures to stay on the top may create invisible inner fears in people of every walk of life — doctors, lawyers, and professors alike. Business executives and professionals are expected to devote most of their time, whether on or off the job, to their professional roles and to the success of their respective organizations. Such pressures are part of the modern working world and are a great source of stress for adults.

On the other hand, vocational activities may be prime sources of self-esteem, satisfaction, and fulfillment. One of the key concepts helpful in maintaining the level of motivation for the challenge and sometimes the drudgery of work is a sense of efficacy, described by Robert White in Chapter 26. The lack of such a feeling of efficacy can make our vocational activities taxing and burdensome as we sometimes ask ourselves, "What am I working for?

The aspects of the family most important to young adults are love, sex, marriage, and parenthood. The needs for love and sex can be a prelude to the formation of a family, or they can be dealt with independently. All these aspects necessarily highlight the basic relationships between the family members.

Today man has more choices than ever before, although he may have doubts about his choice in the areas of love, sex, marriage, and parenthood. Typical questions relating to choice are: Whom should I love? With whom and under what circumstances should I make love? Whom should I marry? Do I need a child? How shall I raise my child? We cannot estimate how much energy each adult spends in pondering these questions and how many people are frustrated and disappointed because they cannot arrive at the kind of answers they desire.

Love is indeed a many-splendored thing. We need to give love to others, and we also need to be loved by others. This basic human characteristic gives rise to our comfort and sometimes our despair. Certainly, the concept of love is very illusive, and often is deceptively employed for a self-serving purpose. One interesting illustration is given by William Lederer and Don Jackson in their book *Mirages of Marriage* (5), in which seven spurious assumptions of marriage are listed. One of these assumptions is that people marry because they are in love with each other. "They like to think of themselves as being in love; but by and large the emotion they interpret as love is in reality some other emotion — often a strong sex drive, fear, or a hunger for approval" (6). This provocative view has caused many people to reexamine their marital relationships and the true meaning of love.

Sex is another illusive subject in the minds of adults. Human sexuality is still undergoing limited basic research, and most people know very little about it. Old sexual norms are much relaxed in recent years, but sexual compatibility has obviously not been increased. It is perhaps safe to say that greater sexual freedom has not produced greater sexual satisfaction. The irony seems to be that sexual satisfaction does not arise solely from the sex act, but is a result of a combination of sexual attitudes and intra-interpersonal behavior. In essence, we may consider that human sexuality involves an interplay of social, behavioral, emotional, and physical factors.

There is increasing interest in unconventional marriage in recent years. It is interesting to note that in the past unconventional marriage occurred chiefly among the lower social class, whereas today middle-class people are involved as well. The unconventional marriage occurs in various forms. Some consist of relationships that merely bypass the formal legal marriage procedures. Others go beyond sexual bonds between the conjugal couple in a communal type of family, so that sexual activity is shared with other couples. The recent women's liberation movement has also exerted an impact on our family system. The traditional family relationship has been challenged, and new role models for the various members of the family have to be reinstitutionalized.

The arrival of a child, especially the first child, can be a joyful experience as well as a crisis for the parents; a joy because the child is the product of their common bond and a shared identification, a crisis usually because they feel as though a burden had suddenly fallen upon their shoulders. Adjustments have to be made in the daily routine, life style, and perception of the self. For instance, the working wife may have to quit her job entirely for several years at least in order to care for her child. The financial pinch is due not only to the loss of the wife's earnings but also to the additional cost of rearing a child. The life style must be modified; the couple has to be home more often and sacrifice their weekend social and recreational activities. As the parents begin to learn their new roles, they experience more changes in self-perception. They may see the union as more permanent and important than before and become less self-centered and more family-centered in their attitudes, and become interested in community and neighborhood activities. In some instances, the parents become more aware of the youth culture and sometimes emulate it. It is a sort of inverted socialization process for the parents which may give rise to appreciation of the youth culture and serve to bridge the generation gap.

The discussion so far has been geared to the normal processes of parenthood. I would be remiss if I did not mention briefly some maladjustments of parenthood. There are many couples who simply cannot adjust to their new responsibilities and learn new roles. Frequently their intensive emotional turmoil results in separation or divorce. Personality characteristics and problems of interpersonal relationship are primary sources of difficulties in adjustment to parenthood. In addition, the contemporary culture and societal structure also make adjustment to parenthood difficult. For instance, the nuclear family system and the geographic mobility in our society result in the wide dispersal of close relatives and extended family. Family members are less available to help in times of need or stress, and are not able to serve as role models. The young couple frequently is forced into difficult situations with no adequate preparation or assistance.

Bernice Neugarten, in Chapter 25, reminds us that personality development is not automatically arrested once a person reaches adulthood. On the contrary, the adult is just as busy as the child or adolescent in coping with the many tasks and issues that are crucial to functioning. Most important to the adult is having a consonant ego functioning within himself, including such qualities as "competence," "self," and "effectance." Neugarten further explains how these qualities are related to adult personality development and specifies conditions by which experience is translated to adult personality. The conditions are: "his structuring of the social world in which he lives; his perspectives of time; the ways in which he deals with the major themes of work, love, time and death; the changes in self-concept and changes in identity as he faces the successive contingencies of marriage, parenthood, career advancement and decline, retirement, widowhood, illness and personal death." The biological changes in middle adulthood are not as salient a factor as in childhood, adolescence, and old age. The age-status system, on the other hand, has great impact on the behavior of the adult in view of the internalization of age norms, which have great effect on the adult personality development processes.

Previous discussion has alluded to the importance of "competence" in adult life. In Chapter 26, Robert White, a renowned psychologist, provides a unique conceptual foundation for the concept of "competence" or "effectance." The concept of competence has great value to adult ego maintenance, since adulthood is a period of independence and autonomy. The adult preserves his independence by demonstrating his capacity through successful environmental manipulation and satisfactory emotional intimacy with the people around him. The feeling of efficacy then becomes a prime motivational force for human action.

Work, in common terminology, is equated with earnings. Most people think work is necessary for survival because we are living in a monetary society in which human labor is converted to a monetary unit, which then determines one's way of life. Karl Menninger, in Chapter 27, speaks of work as a psychological term that has inherent value to psychological well-being. Menninger views work as a sublimation, a channeling of aggressive impulses into productive use. He quickly adds, however, that not all work, or work under all circumstances, will result in psychological satisfaction. In other words, there is no indication that work is in itself pleasurable. The pleasure has to be derived from certain external and internal conditions of the individual. This chapter not only illustrates the analytical concept of sublimation in relation to work, but illustrates as well the value of work in a controlled situation as a useful therapeutic measure.

The application of general systems theory has contributed to the better understanding and conceptualization of a complex sociocultural system, family functioning, and interpersonal interactions. In Chapter 28 David Speer reviews family interpersonal relationships based on an open system framework and challenges the inadequacy inherent in the homeostasis concept, which defines equilibrium as the optimum condition of a family unit. Speer eloquently discusses the critical distinction between the two vital concepts of morphastases (no room for change) and morphogenesis (room for growth and development) and their implications for the definition of a "healthy family." He suggests a careful examination of the deficiencies of narrow usage of the systems concept, particularly the concept of equilibrium or homeostasis.

The concept of role is another useful tool for analyzing interpersonal relationships and family interaction, as illustrated in Chapter 29. John Spiegel defines role "as a goal-directed pattern or sequence of acts tailored by the cultural process for the transactions a person may carry out in a social group or situation." Interpersonal relations, especially in a family, can be maintained only by the process of "equilibrium-disequilibrium balance." That is, when role reciprocity is broken down, as a result of such causes as cognitive discrepancy and discrepancy of goals, attempts at restoration will be initiated by the role participants. Spiegel discusses eleven sequential steps for the purpose of achieving complementarity of role system, such as role induction, coaxing, and so on. Spiegel's analysis of the causes of role conflict and its resolution provide a new dimension to the process of maintaining interpersonal relationships in a family unit or a small group.

Interpersonal difficulty between marriage partners or other persons of opposite sex does not necessarily result from a sexual problem. But sexual incompatibility can hinder interpersonal relationships. How much do we know about human sexuality? Little progress has been made in this area since the Kinsey report on human sexuality more than two decades ago. The lack of real understanding of sexuality is great even among mature adults. Many are unwilling to explore sexuality for fear of suffering interpersonal embarrassment and severing marital relationships. Among the leading clinical researchers to deal with this subject in recent years are William Masters and Virginia Johnson. Their unique contribution to one of the significant sectors of human interaction and human sexuality has thrown light on this hitherto taboo subject, which was rarely discussed openly. In Chapter 30, Masters and Johnson explain the fundamental problems of sexual incompatibility from a clinical standpoint. They clearly spell out the attitudinal misconceptions of sex partners and the curative maneuvers for both sexes.

An additional dimension affecting marital sexuality, cultural variation, is explored in Chapter 31. Lee Rainwater reports on a study that examined sexual relation patterning, especially among the lower social classes, in four areas — the United States, England, Puerto Rico, and Mexico. Rainwater shows how both the man and the woman experience their sexual relationship within the structure of a given cultural prescription, which provides them with various rationales to explain their adult sexual behavior. He suggests that often the lack of real sexual satisfaction, particularly for wives, is due to the high degree of segregation in the role relationships of the married couple. This inadequate socialization pattern makes genuine sexual pleasure more difficult to attain.

Another interesting point regarding cultural influence on man's behavior toward woman is brought forward by Jack Balswick and Charles Peek. In Chapter 32 they discuss American males in relation to the ways in which their emotional feelings are expressed. They contend it is culturally undesirable for the American male adult to show great emotion or affection toward the female partner, as such expression is regarded as a sign of weakness and unmanliness. This creates internal inconsistency in view of the nature of modern marital relationships. In the old days the family was viewed as a task-oriented unit; the husband was expected to provide food, clothing, and shelter for the family and the wife was expected to take care of housework and child rearing. Today, affection and companionship are viewed as more important, and the marriage relationship is expected to function as a more romantic, affectively oriented unit. There is increased emphasis on what Balswick and Peek term "the quality of expressiveness," which is somewhat inconsistent with the cultural image of the American male. They differentiate two types of inexpressive male, designating one as the "cowboy" type and the other as the "playboy" type. To resolve the cultural inconsistency, Balswick and Peek point out, men learn "situational expressiveness." That is, the male learns to make an exception with his wife, being expressive toward her, but maintaining inexpressiveness toward other women. If the male were not able to resolve the cultural inconsistency in this manner, it would be disruptive to the marital relationship, perhaps severing it altogether.

The advent of parenthood requires the adult to assume a new set of social roles. Transitional problems common to the role shift are the primary concern of Chapter 33. Like Erikson with his life cycle approach, Alice Rossi contends that each stage of the life cycle has a set of social roles to be accomplished. In some cases adjustment to the new set of roles is necessary because of the lack of adequate cultural alternatives for rejecting or terminating the new role demands. For instance, once a person has become a parent he cannot stop being one, but of course many people terminate the fulfillment of the parental role prematurely. Rossi also explains the various roles, each set of which has four broad stages: anticipatory stage, honeymoon stage, plateau stage, and disengagement-termination stage. Each of these four stages is accompanied by culturally prescribed norms. In view of the current social changes in attitudes toward the marital relationship, child-rearing responsibility, and womanhood in general, the individual finds many incongruencies in role expectations and commitments.

The traditional social role of women has been challenged by the women's liberation movement. To what extent this affects family structure and man-woman relationships is not clear at this time. However, it is certain that reverberations from the current movement will be with us for many years to come. In Chapter 34, Sylvia Clavan discusses the objectives of the women's liberation movement and their impact on the existing family organization. She further suggests some possible changes in the conjugal family system itself and child care responsibility.

In adulthood, nearly everyone has the potential to be a parent. Unless child-care responsibility is totally relinquished from parental to public auspices, there will be increasing discussion of parent-child relationships. There is an abundance of readings on child-rearing in the literature, most of which deal with the subject from psychological or physical perspectives. Unlike most writers, Melvin Kohn, in Chapter 35, approaches the parent-child relationship from the viewpoint of social class. He suggests that child-rearing practices differ from social class to social class — for instance, middle class vs. working class. He indicates three factors as the major contributors to this difference: conditions of life, values, and behavior. This chapter clearly illustrates some of the ways in which social structure affects behavior.

The family is a basic social institution that fulfills human needs, preserves social norms, and maintains social stability. The adult is the cornerstone of the family social unit who radiates the warmth and emotional attachment for the members of the unit, provides material necessities and security, and transmits human experience to the younger generations. Many writers have argued that rapid social change has greatly modified the functions and qualities of the modern family. Have traditional family functions deteriorated or has family function advanced to a more differentiated form? These questions, raised by William Ogburn and Talcott Parsons, are considered by Clark Vincent in Chapter 36. Vincent sees evidence of viability for modern family functioning in the adaptive capacity of each family to meet the internal needs of its members and the external demands of the society.

References

1. Irving Sarnoff, *Personality Dynamics and Development.* New York: Wiley, 1962, pp. 402–03.

2. Erik H. Erikson, *Childhood and Society.* New York: Norton, 1950, p. 231.

3. Ibid.

4. Irving Sarnoff, *Personality Dynamics and Development.* New York: Wiley, 1962, p. 435.

5. William J. Lederer and Don D. Jackson, *The Mirages of Marriage.* New York: Norton, 1968.

6. Ibid., p. 42.

A. INTRAPERSONAL SYSTEM

CHAPTER 25

Adult Personality: Toward a Psychology of the Life Cycle

BERNICE L. NEUGARTEN

A psychology of the human life cycle has been slow in making its appearance. From one point of view, biological and sociological perspectives have not yet been integrated into an overarching theory of human behavior, nor have they been combined even in describing a meaningful context against which to view psychological change over the life cycle. From a different point of view, the primary problem is that we lack a developmental psychology of adulthood in the sense that we have a developmental psychology of childhood. Because the term "development" has been used with such a wide variety of philosophical as well as scientific meanings, it will be strategic for purposes of the present discussion to avoid the awkward juxtaposition of the terms "adult" and "development," and to speak of the need for a psychology of adulthood in which investigators are concerned with the orderly and sequential changes that occur with the passage of time as individuals move from adolescence through adulthood and old age, with issues of consistency and change in personality over relatively long intervals of time, and with issues of antecedent-consequent relationships.

Using this definition, the field of adult psychology and adult personality remains an underpopulated one among psychologists. The effect has been, to speak metaphorically, that as psychologists seated under the same circus tent, some of us who are child psychologists remain seated too close to the entrance and are missing much of the action that is going on in the main ring. Others of us who are gerontologists remain seated too close to the exit. Both groups are missing a view of the whole show.

One of our problems lies in the fact that we are as yet without sufficient systematic data on adults. A few sets of data have been reported in which individuals have been studied from childhood into adulthood (Havighurst et al., 1962; Hess, 1962; Honzik and McFarlane, 1966; Kagan and Moss, 1962); but these studies are few in number, and despite the growing recognition of the importance of longitudinal research, there have been as yet no major longitudinal studies of men and women as they move from youth to middle age, or from middle age to old age. There have been even few carefully designed and well-controlled cross-sectional studies of adult personality in which age differences, to say nothing of age *changes,* have constituted a central axis of investigation (Kuhlen, 1964).

Not only is there a paucity of data, but more important we are without a useful theory. Personality theorists have not for the most part faced the questions of stability and change over the entire life cycle. Attention has been focused primarily upon the first two but not on the last five sevenths of life. Although Erikson's formulation of the stages of ego development is a notable exception (Erikson, 1950); and although Jung (1933), Buhler (1933, 1935, 1962), Fromm (1941), Maslow (1954), Peck (1955), and White (1963) have made important contributions, there is no integrated body of theory that encompasses the total life span.

At the same time we are aware that changes in personality occur in adulthood and that the personality is by no means fixed once the organism becomes biologically mature (Worchel and Byrne, 1964). There is evidence on all sides: the changes that occur in adults undergoing psychotherapy or religious conversions or brainwashings, or who live in concentration camps or prisons or ghettos or institutions for the aged. Nor need we look to such extreme situations. We are impressed with — although so far as I know, no one has yet systematically studied — the changes in personality that accompany motherhood in young women, for example, or that accompany career success in middle-aged men.

To the psychologist, of course, as to any other scientist, much more is know than can be easily demonstrated. Confronted with the need to produce systematic evidence, the investigator who turns his attention to the study of adult personality is faced, first, with the problem of delineating those personality processes that are the most salient at successive periods in adulthood; then with the problem of describing those processes in terms that are appropriate; then to distinguish those changes that relate to increasing age from those that relate, say, to illness on the one hand or to social and cultural change on the other; then — and only then — to interpret his findings in light of the question: Which of these processes and changes are orderly and sequential, and which are not? At the same time, because theory and observation must proceed simultaneously, he must always be concerned with the construction of a body of theory that will help account for his findings.

The present paper is addressed primarily to conceptual problems in relating childhood to adulthood, and in relating biological to social pacemakers of change in personality. In this connection we shall turn attention, first, to the delineation of salient issues in adult personality and the types of personality change that are measurable in middle-aged and old adults, mainly to illustrate some of the conceptual problems involved in predicating continuity or discontinuity over the life cycle. Second, in describing an age-status system as one of the social contexts for viewing time-related changes, we shall comment upon the presence of social as well as biological time clocks.

The Salient Issues in Adulthood

First, with regard to the delineation of the salient issues, the criticism is sometimes made that psychologists focus on different phenomena and make use of different explanatory concepts for studying different age levels; and that they sometimes seem to regard children, adolescents, adults, and old people as members of different species. One implication is that if we were fortunate enough to have longitudinal data, the life span could be seen in more continuous and more meaningful terms and antecedent-consequent relations could more readily be investigated.

While this may be true, it is also true that longitudinal studies have thus far had a child-centered or what might be called a "childomorphic" orientation. The variables selected for study have been those particularly salient in childhood; or else those measured retrospectively from the data gathered when the subjects were children. In either instance, the investigator is confined to data and to concepts which may be only of secondary relevance when he attempts to explain the varieties or sequences in adult behavior. In this respect there are countless studies in the child-development literature that deal with dependence, aggression, cognition, the fate of the Oedipal personality issues which, when projected into adulthood, lose much of their compelling quality.[1]

[1] Perhaps this point can be made more clearly and with less controversy if we move outside the area usually delineated as personality. A recent review of research on the relationship between college grades and adult achievement is summarized by the author's statement that "present evidence strongly suggests that college grades bear little or no relationship to any measures of adult accomplishment" (Hoyt, 1965).

It is widely acknowledged that school grades are a salient issue to the child or adolescent himself; and at the same time, they are important to the psychologist as an index of the child's intellectual progress and of his success in relating to the adult world. It is known also that school grades are excellent predictors of later school and college grades; furthermore, that there is a close relationship between grades and measures of intelligence in children. Finally, knowledge of the distribution of intelligence test scores and of school grades in large populations of children has led to constant modifications in our theories of the nature and growth of intelligence.

Yet when we move to adulthood, these relationships and the interpretations to be drawn

What, then, are the salient issues of adulthood? At one level of generality, it might be said that they are issues which relate to the individual's use of experience; his structuring of the social world in which he lives; his perspectives of time; the ways in which he deals with the major life themes of work, love, time, and death; the changes in self-concept and changes in identity as he faces the successive contingencies of marriage, parenthood, career advancement and decline, retirement, widowhood, illness, and personal death.[2]

To be more specific: One of our studies is based on lengthy interviews with 100 men and women aged 45 to 55, selected because they have been visibly successful in career or in civic participation. In this study, we moved toward more and more naturalistic-phenomenological-type data as we listened to what our subjects were saying and as we encouraged their introspection. As a result, we have delineated certain issues that seem to be typical of middle adulthood and which, if they appear at all, take quite different form in younger or older persons.

There is, for instance, the middle-ager's sensitivity to the self as the instrument by which to reach his goals — his sense of "self-utilization" as contrasted to the "self-consciousness" of the adolescent.

There is the shift in body cathexis and its relations to the self-concept. "Body monitoring" is the term we use to encompass the large variety of protective strategies described by middle-aged people as techniques for

---

from them become relatively useless. We cannot predict from school to later achievement; indeed, we have no well-established criterion of adult achievement. Next, lacking a criterion by which to standardize a test, we have no test of intelligence that will help us — except in the grossest sense — to predict adult accomplishment. Finally, of course, we have no theory of adult intelligence that is useful, in this sense, beyond the very first steps in the work career, namely *selection* of occupation.

We are even worse off when we come back to the field of personality and when we attempt to apply what we think we know about children to what we would like to know about adults.

[2] In commenting upon some of these problems I am drawing upon a set of studies that I and various of my colleagues have been carrying out in the Committee on Human Development at the University of Chicago over the past decade — studies of personality, of adaptational patterns, of career lines, of age norms and age-appropriate behaviors in adults, and of attitudes and values across generational lines. Many of these studies have been reported earlier; some are just now being completed. While this line of inquiry does not involve longitudinal research on the same subjects (except for one group of 200 older persons who were followed over a seven-year period), it represents a related set of investigations in which the total number of adults who have participated now totals something over 2,000. Each study is based upon a relatively large sample of normal people, none of whom were volunteers. In some of the studies, samples have been drawn by probability techniques from the metropolitan community of Kansas City; in other instances, quota samples have been drawn from the metropolitan area of Chicago.

maintaining the body at given levels of performance and in combating the new sense of physical vulnerability.

There is the striking change in time perspective. Middle-aged adults restructure time in terms of time-left-to-live rather than time-since-birth. . . . They personalize death. There is also in middle age a "rehearsal for widowhood," which is more characteristic of women than of men, and an elaboration of the parenting-sponsoring theme, with regard to young associates as well as with regard to one's children. What we have called the "creation of social as well as biological heirs" appears to be the manifestation of what Erikson calls "generativity."

There is the heightened self-understanding that comes to the middle-aged person from observing the aging parent, on the one hand, and the young adult child, on the other. One perceptive woman described it in these terms: "It is as if I'm looking at a three-way mirror. In one mirror I see part of myself in my mother who is growing old, and part of her in me. In the other mirror, I see part of myself in my daughter. I have had some dramatic insights, just from looking in those mirrors. . . . It is a set of revelations that I suppose can only come when you are in the middle of three generations."

There is the sense of expertise that accompanies middle-adulthood. One man said:

> I believe in making decisions. . . . They may appear to others to be snap decisions, for I make them quickly and I don't look back and worry about what might have been if it had been done another way. I've had enough experience of my type. . . . I've been through it fifty times, so when I make decisions that seem to come out of the clear blue sky, they just represent a lot of experience. I've been over the ground before and I have the ready answer. . .

There is also the sense that accomplishment is not only appropriate but is to be expected. Those who fail to achieve recognition have failed not to achieve the *extra*ordinary, but merely the ordinary. As one forty-five-year-old put it, "Middle age is when you're not really considered young anymore, by anyone. It's considered special when recognition and reward are given to the young businessmen, those in their twenties or thirties. But anybody who makes it after forty — well, that's expected. He's no longer the young genius; he's just done what is par for the course."

In pondering the data on these highly articulate men and women, we have been impressed (as with the findings from some of the earlier studies in this series) with the central importance of what might be called the executive processes of personality: self-awareness, selectivity, manipulation and control of the environment, mastery, competence, the wide array of cognitive strategies.

We are impressed, too, with the heightened importance of introspection in the mental life of middle-aged persons: the stock-taking, the increased reflection, and above all, the structuring and restructuring of experience — that is, the processing of new information in the light of experience; the use of this knowledge and expertise for the achievement of desired ends; the handing over to others or guarding for oneself the fruits of one's experience.[3]

It is perhaps evident that these psychological issues are ones to which the investigator comes unprepared, as it were, from his studies of children and adolescents. It is perhaps evident also why it is that to psychologists of adulthood most of the existing personality theories seem inadequate. Neither psychoanalytic theory nor learning theory nor social-role theory are sufficiently embracing of the data. Where, except perhaps to certian ego psychologists who use terms such as "competence," "self," and "effectance," can we look for concepts to describe the incredible complexity shown in the behavior of a business executive, age fifty, who makes a thousand decisions in the course of a day? What terms shall we use to describe the strategies with which such a person manages his time, buffers himself from certain stimuli, makes elaborate plans and schedules, sheds some of his "load" by delegating some tasks to other people over whom he has certain forms of control, accepts other tasks as being singularly appropriate to his own competencies and responsibilities, and in the same twenty-four-hour period, succeeds in satisfying his emotional and sexual and aesthetic needs?

It is the incongruity between existing psychological concepts and theories, on the one hand, and the transactions that constitute everyday adult behavior, on the other, to which I am drawing attention.

Changes in Personality in the Second Half of Life

Although change and consistency in adult personality is a problem area which has thus far attracted relatively few psychologists, evidence is nevertheless beginning to accumulate that systematic and measurable changes occur in the second half of life. While the studies carried out at Chicago have not been longitudinal, we have begun to delineate processes of change that are characteristic of individuals as they move from middle to old age (Neugarten et al., 1964).

[3] If confronted with the question of whether or not there are any "inherent" or "inevitable" changes in personality that accompany adulthood, there is at least one that would come at once to mind: the conscious awareness of past experience in shaping one's behavior. Psychologists are accustomed to the idea that experience is registered in the living organism over time; and that behavior is affected accordingly. But in the case of the human organism, it is not merely that experience is recorded; it is that the *awareness* of that experience becomes increasingly dominant.

As already implied, the middle years of life — probably the decade of the fifties for most persons — represent an important turning point, with the restructuring of time and the formulation of new perceptions of self, time, and death. It is in this period of the life line that introspection seems to increase noticeably and contemplation and reflection and self-evaluation become characteristic forms of mental life. The reflection of middle age is not the same as the reminiscence of old age; but perhaps it is its forerunner.

Significant and consistent age differences are found in both working-class and middle-class people in the perceptions of the self vis-a-vis the external environment and in coping with impulse life. Forty-year-olds, for example, seem to see the environment as one that rewards boldness and risk-taking and to see themselves as possessing energy congruent with the opportunities perceived in the outer world. Sixty-year-olds, however, perceive the world as complex and dangerous, no longer to be reformed in line with one's wishes, and the individual as conforming and accommodating to outer-world demands.

Important differences exist between men and women as they age. Men seem to become more receptive to affiliative and nurturant promptings; women, more responsive toward and less guilty about aggressive and egocentric impulses. Men appear to cope with the environment in increasingly abstract and cognitive terms; women, in increasingly affective and expressive terms. In both sexes older people move toward more egocentric, self-preoccupied positions and to attend increasingly to the control and satisfaction of personal needs.

With increasing old age, ego functions are turned inward, as it were. With the change from active to passive modes of mastering the environment, there is also a movement of energy away from an outer-world to an inner-world orientation.[4]

Whether or not this increased "interiority" has inherent as well as reactive qualities cannot yet be established. It may be that in advanced old age, biologically based factors become the pacemakers of personality changes; but this is a question which awaits further disentangling of the effects of illness from the effects of aging, effects which are presently confounded in most older persons who are the subjects of psychological research.

[4] To this increased "interiority" of personality, we once gave the term psychological "disengagement," a term that accurately reflects the quality of some of these processes. The term has since become associated, in the field of gerontology, with issues of social role behavior, optimum patterns of aging, and even with the value systems of investigators as well as social policy makers who are concerned with the position of the aged in American society. Because the word has now taken on such a wide variety of meanings I prefer now to substitute the phrase "increased interiority of the personality."

Another important finding in this series of studies is that, in the age range fifty to eighty, and in relatively healthy individuals, age does not emerge as a major variable in the goal directed, purposive qualities of personality. In other words, while consistent age differences occur in covert processes (those not readily available to awareness or to conscious control and which have no direct expression in overt social behavior), they do *not* appear on those variables which reflect attempted control of the self and of the life situation.[5] Age-related changes appear earlier and more consistently, then, in the internal than in the external aspects of personality.

Nor is this all the evidence that exists of dynamic changes in personality in the second half of the life span. Lieberman, for instance, following a somewhat different line of inquiry, has found measureable changes in psychological functioning at the very end of life; changes which seem to be independent of illness and which seem to be timed, not by chronological age or distance from birth, but by distance from death (Lieberman, 1965). Similarly, Butler posits the universal occurrence in old people of an inner experience that he calls the life review; a process that perhaps accounts for the increased reminiscence of the aged and which often leads to dramatic changes in personality (Butler, 1963).

It might be pointed out, parenthetically, that awareness of approaching death should perhaps not be viewed as a signal for the dissolution of the personality structure, but instead as the impetus for a new and final restructuring; an event that calls for a major readaptation, and which leads, in some individuals, to constructive, and in others, to destructive reorientations.

There is at least some evidence, then, that personality change can occur all along the life span; and that any personality theory which is to be useful to us in comprehending the life cycle must take account of changes in advanced old age as well as in other periods of life.

Relations Between Biological and Psychological Change

This leads us back to questions of theory. Sometimes the very theories of developmental psychology hinder us in constructing a psychology of the life cycle. In this respect, the changes observable in behavior with the passage of short intervals of time — a month, six months, a year — are dramatic when one regards a young biological organism; and no less compelling are the overall regularities of biological change. It has been tempting for students of behavior to draw parallels between biological

[5] This differentiation reminds us of Brewster Smith's two personality subsystems — one which he has called adaptive or "external"; the other, the internal (Smith, 1959).

phenomena and psychological, and on this basis to establish a developmental psychology of childhood and adolescence. It has been relatively easy, if not always accurate, first to assume, then to look for ways of describing sequential and orderly progressions in psychological and social behavior. It has been easy to take growth as the model, to borrow from the biologist the concepts of increasing differentiation and integration, and the concept of an end point toward which change is necessarily directed. It has been understandable, therefore, that we have used the biological clock as a frame of reference, looking to biological changes as the pacemakers of psychological changes, and taking for granted the intimate relationship between these two classes of phenomena.

The difficulty with this approach, however, can be illustrated from one of our investigations in which we proposed to study the psychological correlates of the biologic climacterium in middle-aged women. Along with other psychologists, we have been impressed with the changes in behavior that accompany puberty; and with the interpretation that the personality differences that are measurable in adolescents relate to major *developmental* components of biologic changes. We reasoned by analogy that there should also be biologically based developmental components in the personality differences observable in middle-aged women — in short, that if puberty is an important developmental event in the psychology of females, so, probably, is the climacterium.

Accordingly we selected a sample of 100 normal women aged forty-three to fifty-three from working-class and middle-class backgrounds, and obtained data on a large number of psychological and social variables. Using menopausal status as the index of climacteric status (presence or absence of observed changes in menstrual rhythm, or cessation of menses), we found climacteric status to be unrelated to our wide array of personality measures. Furthermore, there were very few significant relationships between severity of somatic and psychosomatic symptoms attributed to climacteric changes and these variables.

Granted that the question would be better pursued by a longitudinal rather than a cross-sectional method, nevertheless the negative findings were more largely due, we believe, to the fact that we were pursuing a certain parallelism between childhood and adulthood, a parallelism that probably does not exist.

This is not to say that the menopause is a meaningless phenomenon in the lives of women; nor that biological factors are of no importance in adult personality. Instead these comments are intended to point out: (1) that the menopause is not necessarily the important event in understanding the psychology of middle-aged women that we might have assumed it to be, from a biological model or from psychoanalytic theory — not as important, seemingly, as illness; or worry over possible illness; or even as worry over possible illness that might occur in one's husband rather than in oneself; (2) that the timing of the biological event, the climacterium — at least to the extent that we could perceive it — did not produce order in our data; and (3) above all, that psychologists should proceed cautiously in assuming the same intimate relationships between biological and psychological phenomena in adulthood that hold true in childhood.

The Age-Status System

As already suggested, the biological components of human development take a certain precedence in viewing personality change in childhood and early adolescence, and perhaps also in the very last part of the life span when biological decrement may overwhelm other components in the personality. There remains, however, the span of the adult years — a period of now approximately fifty years, beginning when the organism reaches biological maturity around age twenty and extending to approximately seventy. In this long part of the life span the biological model is obviously insufficient, and we need a social framework for understanding the timing patterns that occur.

Psychologists have, of course, already looked at personality in one after another social context — the family, the school, the community, as well as in relation to social-structural variables in the society at large. Thus, social class groups, ethnic and religious groups, and now again racial groups are studied with regard to the ways in which these groups create different subcultures and, through processes of socialization, the ways they produce both similarities and differences in personalities. One major context has been neglected, however: the age structure of the society, the network of age norms and age expectations that govern behavior, and the ways in which different age groups relate to each other.[6]

[6] An analogy is appropriate: in sociology social stratification is a well-delineated field of study, and social class is recognized as one of the basic dimensions of social organization. Social classes, in turn, have been described as subcultures; and a tradition has developed of studying the behavior of both children and adults in relation to the social class to which they belong and in the context of the social-status structure of the community in which they live.

In similar fashion, some of us began several years ago to focus attention on the factor of age as another of the basic dimensions of social organization and the ways age groups relate to each other in the context of different social institutions. By a dimension of social organization, I refer to the fact that in all societies an age-status system exists, a system of implicit and explicit rules and expectations that govern the relations between persons of different ages. Certain behaviors come to be regarded as appropriate or inappropriate for each age group; and the relations between age groups are based upon dimensions of prestige, power, and deference. Older children have prestige in the eyes of younger; adolescents, in at least many ways, recognize the power of adults; and both children and adults, in at least some ways, show deference to the old. Each age group occupies a given status; and the system as a whole undergoes alterations in line with other social, economic, and political changes in the society.

Expectations regarding age-appropriate behavior form an elaborated and pervasive system of norms governing behavior and interaction, a network of expectations that is imbedded throughout the cultural fabric of life for the adult as much as for the child. There is a prescriptive timetable for the ordering of life events: a time when men and women are expected to marry, a time to raise children, a time to retire. This pattern is adhered to, more or less consistently, by most persons in the society, and even though the actual occurrences of these events are influenced by various life contingencies, it can easily be demonstrated that norms and actual occurrences are closely related.

Age norms and age expectations operate as prods and brakes upon behavior, in some instances hastening an event, in others delaying it. Men and women are not only aware of the social clocks that operate in various areas of their lives, but are also aware of their own timing and readily describe themselves as "early," "late," or "on time" with regard to family and occupational events.

Whether we identify ourselves as developmental psychologists or as social psychologists or as personality psychologists, we have perhaps taken the social dimension of age status so for granted that, like the air we breathe, we pay it no attention. As soon as it is pointed out we agree, of course, that the society is organized by age — rights, duties and obligations are differentially distributed according to age; and the relations between age groups change with historical time, as is so evident now, for instance, with the growth of political protest among college-age youth and with the changing age base for political and social responsibility in our society.

Every encounter between individuals is governed, at least to some extent, by their respective ages. When we meet a stranger — indeed, when we first glance at any person — we think first in categories based on age and sex. We notice, first, "this is a male child," or "a young female," or "an old man"; and we pattern our behavior accordingly — according, that is, to our own age and sex in relation to the other person's. So automatic and so immediate is this regulation of behavior that only when the cues are ambiguous do we give it full awareness — as, for instance, when we face an adult of indeterminate age and when we fumble about in discomfort lest we make a blunder. Age, then, provides one of the basic guidelines to social interaction.

At the same time that psychologists have neglected the age-status dimension of social behavior, we have also neglected its relevance for providing an understanding of the various phenomena of personality internal to the individual. For example, although we always report the ages of the children whom we are studying, we seldom investigate directly the way in which the child himself thinks of his age or the way his perception of age colors his relations to other people. We have, for instance, talked at length about sex-role identity, but not about changing age-role identity. We have given very little systematic study to the way in which children, let alone adults, internalize age norms and age expectations. Nor have we even attended to the ages of the adults who deal with the child. In most of the studies of child-rearing no mention at all is to be found of the age of the parent, to say nothing of the way parental age might regularly be built into research designs as an important variable.

The saliency of age and age norms in influencing the behavior of adults is no less than in influencing the behavior of children. The fact that the social sanctions related to age norms take on psychological reality can be readily granted. One has only to think of the young woman who, in 1940, was not yet married by age twenty-five; or who, in 1966, was not yet married by twenty-one; and who, if present trends continue over another few decades, will soon lie awake nights worrying over spinsterhood by the time she is eighteen. The timing of life events provides some of the mostpowerful cues to adult personality.

In recent investigations we have perceived of age norms and of the age-status system as forming a backdrop or cultural context against which the behavioral and personality differences of adults should be viewed. Early in this line of inquiry, we began to study the outlines of the age-status system and the extent to which there was consensus in the minds of adults. We began with middle-aged men and women in the belief that, by virtue of their age, they would have relatively accurate perceptions of an age-status system (if such perceptions could be elicited at all); and we asked them a series of questions, such as:

> "What would you call the periods of life that people go through after they are grown up?" For instance, we usually refer to people first as babies, then as children, then as teen-agers.
>
> "After a person is grown up, what periods does he go through?"
>
> "At what age does each period begin, for most people?"
>
> "What are the important changes from one period to the next?" [Neugarten and Peterson, 1957.]

We discovered a commonly held set of perceptions with regard to adult age periods, each with a distinguishing set of transition points and a distinguishing set of psychological and social themes. Thus, our respondents seemed to share a view that adulthood can be divided into periods of young adulthood, maturity, middle age, and old age, each with its distinctive characteristics and each with its own psychological flavor.

At the same time there were sex differences in these perceptions, and differences between members of different social classes. The timing of middle age and old age occurs earlier to working-class than to middle-class men and women.

For instance, the typical upper-middle-class man, a business executive or a professional, divides the lifeline at thirty, at forty, and at sixty-five. He considered a man "mature" at forty; at the "prime of life" and as "having the greatest confidence in himself" at age forty. A man in not "middle-aged" until almost fifty; nor is he "old" until seventy.

The unskilled worker, on the other hand, saw the life line as being paced more rapidly. For him, the major dividing points were placed at twenty-five, thirty-five, and fifty. In his view a man is "middle-aged" by forty; "old" by sixty.

In the eyes of upper-middle-class people, the period of young adulthood extending into the thirties is a time of exploration and groping; a period of "feeling one's way," of trying out and getting adjusted to jobs and careers, to marriage, to one's adult roles; a period of experimentation. By contrast, the working-class man regards young adulthood as the period, not when issues are explored, but when issues are settled. They may be settled by giving up a certain type of autonomy — "when responsibility is *hung* on you," with an undertone of regret; or the issues may be settled by establishing one's independence — the "now you are a man" refrain — but in either case, young adulthood to the working man is the time when one gives up youth with a note of finality and takes over the serious business of life — job, marriage, children, responsibilities.

In these first studies the gross outlines of an age-status structure that seems to crosscut various areas of adult life were delineated. While their view of the life span was more implicit than explicit in the minds of many respondents, it seemed nevertheless to provide a frame of reference by which the experiences of adult life were seen as orderly and rhythmical. More important, progression from one age-level to the next was conceived primarily in terms of psychological and social changes rather than in terms of biological changes.

We also asked questions regarding age-appropriate and age-linked behaviors:

"What do you think is the best age for a man to marry? To finish school? To hold his top job? To retire?"

"What age comes to your mind when you think of a young man? An old man?"

There was widespread consensus on such items. Among the middle-class middle-aged, for instance, 90 percent said the best age for a man to marry was from twenty to twenty-five. Nor was the consensus limited to middle-aged persons or to persons residing in a particular region of the U.S. When responses to the same set of questions were obtained from other middle-class groups, essentially the same patterns emerged in each set of data.

When we looked at life history data with regard to the ages at which major events had actually occurred, we found striking regularities within social-class groups, with the actual occurrences following the same patterns described above. The higher the social class, the later each of the following events was reported by our respondents: age at leaving school, age at first job, age at marriage, age at parenthood, age of top job, grandparenthood — even, in women, their reported age of menopause.

The perceptions, the expectations, and the actual occurrences of life events, then, are closely related; and the regularities within social class groups indicate that these are *socially* regulated.

Given this view that age norms and age expectations operate in the society as a system of social control, we undertook still another study of their psychological correlates, asking, How do members of the society vary in their perception of the strictures involved in age norms, or in the degree of *constraint* they perceive with regard to age-appropriate behaviors? We devised a questionnaire using items that relate to occupational career, some that relate to the family cycle, and some that refer to recreation, appearance, and consumption behaviors. For instance:

Would you approve a couple who likes to dance the "twist" when they are age twenty? Would you approve if they are age thirty? If they are age fifty-five?

What about a woman who decides to have another child when she is forty-five? When she is thirty-seven? When she is thirty?

A man who is willing to move his family from one town to another to get ahead in his company, when he's forty-five? When he's thirty-five? Twenty-five?

A couple who move across country so they can live near their married children — when they are forty? Fifty-five? Seventy?

We devised a score which reflected the degree of refinement with which the respondent makes age discriminations; then used this instrument with a group of 400 middle-class men and women divided into young (20-29), middle-aged (35-55), and old (65+) (Neugarten, Moore, Lowe, 1965). We found a significant increase in scores with age — that is, an increase in the extent to which respondents ascribe importance to age norms and place constraints upon adult behavior in terms of age appropriateness. The middle-aged and the old seem to have learned that age is a reasonable criterion by which to evaluate behavior; that to be "off time" with regard to life events or to show other age-deviant behavior brings with it social and psychological sequelae that cannot be disregarded. In the young, especially the young male, this view is only partially accepted; and there seems to be a certain denial of age as a valid dimension by which to judge behavior.

This age-related difference in point of view is reflected in the response of a twenty-year-old who, when asked what he though of marriage between seventeen-year-olds, said, "It would be OK if the boy got a job and if they loved each other. Why not?" While a forty-five-year-old said, "At that age, they'd be foolish. He couldn't support a wife and children. Kids who marry that young will suffer for it later."

We have begun also to study the correlates in personality and behavior of being "on time" or "off time." Persons who regard themselves as early or late with regard to a major life event describe ways in which the off-timeness has other psychological and social accompaniments. We are pursuing this line of inquiry now with regard to career timing. Thus, in a study of Army officers (the Army is one of the most clearly age-graded occupations available for study) the men who recognize themselves as being too long in grade — or late in career achievement — are also distinguishable on an array of social and psychological attitudes toward work, family, community participation, and personal adjustment.

These studies have relevance for a theory of the psychology of the life cycle, in two ways: First, in indicating that the age structure of a society, the internalization of age norms, and age-group identifications are important dimensions of the social and cultural *context* in which the course of the individual life line must be viewed.

Second, because these concepts point to at least one additional way of structuring the passage of time in the life span of the individual, and in *providing a time clock* that can be superimposed over the biological clock so that together they help us to comprehend the life cycle. The major punctuation marks in the adult life line (those, that is, which are orderly and sequential) tend to be more often social than biological — or, if biologically based, they are often biological events that occur to significant others rather than to oneself, like grandparenthood or widowhood.

If psychologists are to discover order in the events of adulthood, and if they are to discover order in the personality changes that occur in all individuals as they age, we should look to the social as well as to the biological clock, and certainly to social definitions of age and age-appropriate behavior.

Conclusion

In conclusion I would like to return once again to the problem of a personality theory that will encompass the life-cycle.

In commenting upon the salient issues of adulthood and in illustrating in particular from middle age, I have tried to illustrate the need for a theory which will emphasize the ego or executive functions of the personality; one which will help account for the growth and maintenance of cognitive competence and creativity, one that will help explain the conscious use of past experience.

In illustrating from our findings regarding differences between intrapsychic and adaptive aspects of personality as persons move from middle to old age, and from our findings with regard to biological climacterium in women, I have tried to illustrate some of the problems involved in any theory of personality that is based primarily upon a biological model of the life span.

Finally, in describing some of our studies regarding the age-status system and the pervasive quality of age norms in influencing the psychology of adulthood, I have suggested at least one of the sociological components that, in addition to the biological, provides a view of the orderliness of change that underlies the total life cycle.

References

Buhler, Charlotte. *Der Menschliche Lebenslauf als psychogiches Problem.* S. Hirzel, Leipzig, 1933.

Buhler, Charlotte. The curve of life as studied in biographies. *J. appl. Psychol.*, 1935, *19*, 405–9.

Buhler, Charlotte. Genetic aspects of the self. *Ann. N.Y. Acad. Sci.*, 1962, *96*, 730–64.

Butler, R. N. The life review: an interpretation of reminiscence in the aged. *Psychiat.*, 1963, *26*, 65–76.

Erikson, Erik H. *Childhood and Society.* New York, Norton, 1950.

Erikson, Erik H. Identity and the life cycle: Selected papers. *Psych. Issues,* 1959, I.

Fromm, E. *Escape from Freedom.* New York: Rinehart, 1941.

Havighurst, R. J., Bowman, P. H., Liddle, G. P., Matthews, C. V., & Pierce, J. V. *Growing Up in River City.* New York: John Wiley, 1962.

Hess, R. D. High school antecedents of young adults, performance (mimeo).

Hoyt, D. P. The relationship between college grades and adult achievement: a review of the literature. *ACT Research Reports,* September, 1965, No. 7. American College Testing Program: P.O. Box 168, Iowa City, Iowa.

Jung, C. G. *Modern Man in Search of a Soul.* New York: Harcourt, 1933.

Kagan, J., & Moss, H. A. *Birth to maturity.* New York: John Wiley, 1962.

Kuhlen, R. G. Personality change with age. In P. Worchel & D. Byrne (eds.), *Personality Change.* New York: John Wiley, 1964, pp. 524–55.

Lieberman, M. A. Psychological correlates of impending death: some preliminary observations. *J. of Geron.,* 1965, *20*, 181–90.

Maslow, A. H. *Motivation and Personality.* New York: Harper, 1954.

Neugarten, Bernice L., & Peterson, W. A. A study of the American age-grade system. *Proceedings of the Fourth Congress of the International Association of Gerontology,* Merano, Italy, 1957. Vol. 3, 497–502.

Neugarten, Bernice L., & Associates. *Personality in middle and late life.* New York: Atherton, 1964.

Neugarten, Bernice L., Moore, Joan W., & Lowe, J. C. Age-norms, age constraints, and adult socialization. *Amer. J. Soc.,* 1965, *70*, 710–17.

Peck, R. Psychological developments in the second half of life. In J. E. Anderson (ed.), *Psychological Aspects of Aging.* Washington, D.C.: A.P.A., 1956.

Smith, M. B. Research strategies toward a conception of positive mental health. *Amer. Psychol.,* 1959, *14*, 673–81.

White, R. W. Ego and reality in psychoanalytic theory. *Psych. Issues,* 1963, *3* (3).

Worchel, P., & Byrne, D. *Personality change.* New York: John Wiley, 1964.

CHAPTER 26

The Concept of Competence

ROBERT W. WHITE

When parallel trends can be observed in realms as far apart as animal behavior and psychoanalytic ego psychology, there is reason to suppose that we are witnessing a significant evolution of ideas. In these two realms, as in psychology as a whole, there is evidence of deepening discontent with theories of motivation based upon drives. Despite great differences in the language and concepts used to express this discontent, the theme is everywhere the same: Something important is left out when we make drives the operating forces in animal and human behavior.

The chief theories against which the discontent is directed are those of Hull and of Freud. In their respective realms, drive-reduction theory and psychoanalytic instinct theory, which are basically very much alike, have acquired a considerable air of orthodoxy. Both views have an appealing simplicity, and both have been argued long enough so that their main outlines are generally known. In decided contrast is the position of those who are not satisfied with drives and instincts. They are numerous, and they have developed many pointed criticisms, but what they have to say has not thus far lent itself to a clear and inclusive conceptualization. Apparently there is an enduring difficulty in making these contributions fall into shape.

In this paper I shall attempt a conceptualization which gathers up some of the important things left out by drive theory. To give the concept a name I have chosen the word *competence*, which is intended in a broad biological sense rather than in its narrow everyday meaning. As used here, competence will refer to an organism's capacity to interact effectively with its environment. In organisms capable of but little learning, this capacity might be considered an innate attribute, but in the mammals and especially man, with their highly plastic nervous systems, fitness to interact with the environment is slowly attained through prolonged feats of learning. In view of the directedness and persistence of the behavior that leads to these feats of learning, I consider it necessary to treat competence as having a motivational aspect, and my central argument will be that the motivation needed to attain competence cannot be wholly derived from sources of energy currently conceptualized as drives or instincts. We need a different kind of motivational idea to account fully for the fact that man and the higher mammals develop a competence in dealing with the environment which they certainly do not have at birth and certainly do not arrive at simply through maturation. Such an idea, I believe, is essential for any biologically sound view of human nature.

Competence and the Play of Contented Children

A backward glance at our survey shows considerable agreement about the kinds of behavior that are left out or handled poorly by theories of motivation based wholly on organic drives. Repeatedly we find reference to the familiar series of learned skills which starts with sucking, grasping, and visual exploration and continues with crawling and walking, acts of focal attention and perception, memory, language and thinking, anticipation, the exploring of novel places and objects, effecting stimulus changes in the environment, manipulating and exploiting the surroundings, and achieving higher levels of motor and mental coordination. These aspects of behavior have long been the province of child psychology, which has attempted to measure the slow course of their development and has shown how heavily their growth depends upon learning. Collectively they are sometimes referred to as adaptive mechanisms or as ego processes, but on the whole we are not accustomed to cast a single name over the diverse feats whereby we learn to deal with the environment.

I now propose that we gather the various kinds of behavior just mentioned, all of which have to do with effective interaction with the environment, under the general heading of competence. According to Webster, competence means fitness or ability, and the suggested synonyms include capability, capacity, efficiency, proficiency, and skill. It is therefore a suitable word to describe such things as grasping and exploring, crawling and walking, attention and perception, language and thinking, manipulating and changing the surroundings, all of which promote an effective — a competent — interaction with the environment. It is true, of course, that maturation plays a part in all these developments, but this part is heavily overshadowed by learning in all the more complex accomplishments like speech or skilled manipulation. I shall argue that it is necessary to make competence a motivational concept; there is a *competence motivation* as well as competence in its more familiar sense of achieved capacity. The behavior that leads to the building up of effective grasping, handling, and letting go of objects, to take one example, is not random behavior produced by a general overflow of energy. It is directed, selective, and persistent, and it is continued not because it serves primary drives, which indeed it cannot serve until it is almost perfected, but because it satisfies an intrinsic need to deal with the environment.

No doubt it will at first seem arbitrary to propose a single motivational conception in connection with so many and such diverse kinds of behavior. What do we gain by attributing motivational unity to such a large array of activities? We could, of course, say that each developmental sequence, such as learning to grasp or to walk, has its own built-in bit of motivation — its "aliment," as Piaget (1952) has expressed it. We could go further and say that each item of behavior has its intrinsic motive — but this makes the concept of motivation redundant. On the other hand, we might follow the lead of the animal psychologists and postulate a limited number of broader motives under such names as curiosity, manipulation, and mastery. I believe that the idea of a competence motivation is more adequate than any of these alternatives and that it points to very vital common properties which have been lost from view amidst the strongly analytical tendencies that go with detailed research.

In order to make this claim more plausible, I shall now introduce some specimens of playful exploration in early childhood. I hope that these images will serve to fix and dramatize the concept of competence in the same way that other images — the hungry animal solving problems, the child putting his finger in the candle flame, the infant at the breast, the child on the toilet, and the youthful Oedipus caught in a hopeless love triangle — have become memorable focal points for other concepts. For this purpose I turn to Piaget's (1952) studies of the growth of intelligence from its earliest manifestations in his own three children. The examples come from the first year of life, before language and verbal concepts begin to be important. They therefore represent a practical kind of intelligence which may be quite similar to what is developed by the higher animals.

As early as the fourth month, the play of the gifted Piaget children began to be "centered on a result produced in the external environment," and their behavior could be described as "rediscovering the movement which by chance exercised an advantageous action upon things" (1952, p. 151). Laurent, lying in his bassinet, learns to shake a suspended rattle by pulling a string that hangs from it. He discovers this result fortuitously before vision and prehension are fully coordinated. Let us now observe him a little later when he has reached the age of three months and ten days.

> I place the string, which is attached to the rattle, in his right hand, merely unrolling it a little so that he may grasp it better. For a moment nothing happens. But at the first shake due to chance movement of his hand, the reaction is immediate: Laurent starts when looking at the rattle and then

> violently strikes his right hand alone, as if he felt the resistance and the effect. The operation lasts fully a quarter of an hour, during which Laurent emits peals of laughter. [Piaget, 1952, p. 162.]

Three days later the following behavior is observed.

> Laurent, by chance, strikes the chain while sucking his fingers. He grasps it and slowly displaces it while looking at the rattles. He then begins to swing it very gently, which produces a slight movement of the hanging rattles and an as yet faint sound inside them. Laurent then definitely increases by degrees his own movements. He shakes the chain more and more vigorously and laughs uproariously at the result obtained. [Piaget, 1952, p. 185.]

Very soon it can be observed that procedures are used "to make interesting spectacles last." For instance, Laurent is shown a rubber monkey which he has not seen before. After a moment of surprise, and perhaps even fright, he calms down and makes movements of pulling the string, a procedure which has no effect in this case, but which previously has caused interesting things to happen. It is to be noticed that "interesting spectacles" consist of such things as new toys, a tin box upon which a drumming noise can be made, an unfolded newspaper, or sounds made by the observer such as snapping the fingers. Commonplace as they are to the adult mind, these spectacles enter the infant's experience as novel and apparently challenging events.

Moving ahead to the second half of the first year, we can observe behavior in which the child explores the properties of objects and tries out his repertory of actions upon them. This soon leads to active experimentation in which the child attempts to provoke new results. Again we look in upon Laurent, who has now reached the age of nine months. On different occasions he is shown a variety of new objects — for instance a notebook, a beaded purse, and a wooden parrot. His carefully observing father detects four stages of response: (*a*) visual exploration, passing the object from hand to hand, folding the purse, *etc.*; (*b*) tactile exploration, passing the hand all over the object, scratching, *etc.*; (*c*) slow moving of the object in space; (*d*) use of the repertory of action: shaking the object, striking it, swinging it, rubbing it against the side of the bassinet, sucking it, *etc.*, "each in turn with a sort of prudence as though studying the effect produced" (1952, p. 255).

Here the child can be described as applying familiar tactics to new situations, but in a short while he will advance to clear patterns of active experimentation. At 10 months and 10 days Laurent, who is unfamiliar with bread as a nutritive substance, is given a piece for examination. He manipulates it, drops it many times, breaks off fragments and lets them fall. He has often done this kind of thing before, but previously his attention has seemed to be centered on the act of letting go. Now "he watches with great interest the body in motion; in particular, he looks at it for a long time when it has fallen, and picks it up when he can." On the following day he resumes his research.

> He grasps in succession a celluloid swan, a box, and several other small objects, in each case stretching out his arm and letting them fall. Sometimes he stretches out his arm vertically, sometimes he holds it obliquely in front of or behind his eyes. When the object falls in a new position (for example on his pillow) he lets it fall two or three times more on the same place, as though to study the spatial relation; then he modifies the situation. At a certain moment the swan falls near his mouth; now he does not suck it (even though this object habitually serves this purpose), but drops it three times more while merely making the gesture of opening his mouth. [Piaget, 1952, p. 269.]

These specimens will furnish us with sufficient images of the infant's use of his spare time. Laurent, of course, was provided by his studious father with a decidedly enriched environment, but no observant parent will question the fact that babies often act this way during those periods of their waking life when hunger, erotic needs, distresses, and anxiety seem to be exerting no particular pressure. If we consider this behavior under the historic heading of psychology we shall see that few processes are missing. The child gives evidence of sensing, perceiving, attending, learning, recognizing, probably recalling, and perhaps thinking in a rudimentary way. Strong emotion is lacking, but the infant's smiles, gurgles, and occasional peals of laughter strongly suggest the presence of pleasant affect. Actions appear in an organized form, particularly in the specimens of active exploration and experimentation. Apparently the child is using with a certain coherence nearly the whole repertory of psychological processes except those that accompany stress. It would be arbitrary indeed to say that one was more important than another.

These specimens have a meaningful unity when seen as transactions between the child and his environment, the child having some influence upon the environment and the environment some influence upon the child. Laurent appears to be concerned about what he can do with the chain and rattles, what he can accomplish by his own effort to reproduce and to vary the entertaining sounds. If his father observed correctly, we must add that Laurent seems to have varied his actions systematically, as if testing the effect of different degrees of effort upon the bit of environment represented by the chain and rattles. Kittens make a similar study of parameters when delicately using their paws to push pencils and other objects ever nearer to the edge of one's desk. In all such examples it is clear that the child or animal is by no means at the mercy of transient stimulus fields. He selects for continuous treatment those aspects of his environment which he finds it possible to affect in some way. His behavior is selective, directed, persistent — in short, motivated.

Motivated toward what goal? In these terms, too, the behavior exhibits a little of everything. Laurent can be seen as appeasing a stimulus hunger, providing his sensorium with an agreeable level of stimulation by eliciting from the environment a series of interesting sounds, feelings, and sights. On the other hand we might emphasize a need for activity and see him as trying to reach a pleasurable level of neuromuscular exercise. We can also see another possible goal in the behavior: the child is achieving knowledge, attaining a more differentiated cognitive map of his environment and thus satisfying an exploratory tendency or motive of curiosity. But it is equally possible to discern a theme of mastery, power, or control, perhaps even a bit of primitive self-assertion, in the child's concentration upon those aspects of the environment which respond in some way to his own activity. It looks as if we have found too many goals, and perhaps our first impulse is to search for some key to tell us which one is really important. But this, I think, is a mistake that would be fatal to understanding.

We cannot assign priority to any of these goals without pausing arbitrarily in the cycle of transaction between child and environment and saying, "This is the real point." I propose instead that the real point is the transaction as a whole. If the behavior gives satisfaction, this satisfaction is not associated with a particular moment in the cycle. It does not lie solely in sensory stimulation, in a bettering of the cognitive map, in coordinated action, in motor exercise, in a feeling of effort and of effects produced, or in the appreciation of change brought about in the sensory field. These are all simply aspects of a process which at this stage has to be conceived as a whole. The child appears to be occupied with the agreeable task of developing an effective familiarity with his environment. This involves discovering the effects he can have on the environment and the effects the environment will have on him. To the extent that these results are preserved by learning, they build up an increased competence in dealing with the environment. The child's play can thus be viewed as serious business, though to him it is merely something that is interesting and fun to do.

Bearing in mind these examples, as well as the dealings with environment pointed out by other workers, we must now attempt to describe more fully the possible nature of the motivational aspect of competence. It needs its own name, and in view of the foregoing analysis I propose that this name be *effectance*.

Effectance

The new freedom produced by two decades of research on animal drives is of great help in this undertaking. We are no longer obliged to look for a source of energy external to the nervous system, for a consummatory climax, or for a fixed connection between reinforcement and tension-reduction. Effectance motivation cannot, of course, be conceived as having a source in tissues external to the nervous system. It is in no sense a deficit motive. We must assume it to be neurogenic, its "energies" being simply those of the living cells that make up the nervous system. External stimuli play an important part, but in terms of "energy" this part is secondary, as one can see most clearly when environmental stimulation is actively sought. Putting it picturesquely, we might say that the effectance urge represents what the neuromuscular system wants to do when it is otherwise unoccupied or is gently stimulated by the environment. Obviously there are no consummatory acts; satisfaction would appear to lie in the arousal and maintaining of activity rather than in its slow decline toward bored passivity. The motive need not be conceived as intense and powerful in the sense that hunger, pain, or fear can be powerful when aroused to high pitch. There are plenty of instances in which children refuse to leave their absorbed play in order to eat or to visit the toilet. Strongly aroused drives, pain, and anxiety, however, can be conceived as overriding the effectance urge and capturing the energies of the neuromuscular system. But effectance motivation is persistent in the sense that it regularly occupies the spare waking time between episodes of homeostatic crisis.

In speculating upon this subject we must bear in mind the continuous nature of behavior. This is easier said than done; habitually we break things down in order to understand them, and such units as the reflex arc, the stimulus-response sequence, and the single transaction with the environment seem like inevitable steps toward clarity. Yet when we apply such an analysis to playful exploration we lose the most essential aspect of the behavior. It is constantly circling from stimulus to perception to action to effect to stimulus to perception, and so on around; or, more properly, these processes are all in continuous action and continuous change. Dealing with the environment means carrying on a continuing transaction which gradually changes one's relation to the environment. Because there is no consummatory climax, satisfaction has to be seen as lying in a considerable series of transactions, in a trend of behavior rather than a goal that is achieved. It is difficult to make the word "satisfaction" have this connotation, and we shall do well to replace it by "feeling of efficacy" when attempting to indicate the subjective and affective side of effectance.

It is useful to recall the findings about novelty: the singular effectiveness of novelty in engaging interest and for a time supporting persistent behavior. We also need to consider the selective continuance of transactions in which the animal or child has a more or less pronounced effect upon the environment — in which something happens as a consequence of his activity. Interest is not aroused and sustained when the stimulus field is so familiar that it gives rise at most to reflex acts or automatized habits. It is not sustained when actions produce no effects or changes in the stimulus field. Our conception must therefore be that effectance motivation is aroused by stimulus conditions which offer, as Hebb (1949) puts it, difference-in-sameness. This leads to variability and novelty of response, and interest is best sustained when the resulting action affects the stimulus so as to produce further difference-in-sameness. Interest wanes when action begins to have less effect; effectance motivation subsides when a situation has been explored to the point that it no longer presents new possibilities.

We have to conceive further that the arousal of playful and exploratory interest means the appearance of organization involving both the cognitive and active aspects of behavior. Change in the stimulus field is not an end in itself, so to speak; it happens when one is passively moved about, and it may happen as a consequence of random movements without becoming focalized and instigating exploration. Similarly, action which has effects is not an end in itself, for if one unintentionally kicks away a branch while walking, or knocks something off a table, these effects by no means necessarily become involved in playful investigation. Schachtel's (1954) emphasis on focal attention becomes helpful at this point. The playful and exploratory behavior shown by Laurent is not random or casual. It involves focal *attention* to some object — the fixing of some aspect of the stimulus field so that it stays relatively constant — and it also involves the focalizing of *action* upon this object. As Diamond (1939) has expressed it, response under these conditions is "relevant to the stimulus," and it is change in the *focalized* stimulus that so strongly affects the level of interest. Dealing with the environment means directing focal attention to some part of it and organizing actions to have some effect on this part.

In our present state of relative ignorance about the workings of the nervous system it is impossible to form a satisfactory idea of the neural basis of effectance motivation, but it should at least be clear that the concept does not refer to any and every kind of neural action. It refers to a particular kind of activity, as inferred from particular kinds of behavior. We can say that it does not include reflexes and other kinds of automatic response. It does not include well-learned, automatized patterns, even those that are complex and highly organized. It does not include behavior in the service of effectively aroused drives. It does not even include activity that is highly random and discontinuous, though such behavior may be its most direct forerunner. The urge toward competence is inferred specifically from behavior that shows a lasting focalization and that has the characteristics of exploration and experimentation, a kind of variation within the focus. When this particular sort of activity is aroused in the nervous sytem, effectance motivation is being aroused, for it is characteristic of this particular sort of activity that it is selective, directed, and persistent, and that instrumental acts will be learned for the sole reward of engaging in it.

Some objection may be felt to my introducing the word *competence* in connection with behavior that is so often playful. Certainly the playing child is doing things for fun, not because of a desire to improve his competence in dealing with the stern hard world. In order to forestall misunderstanding, it should be pointed out that the usage here is parallel to what we do when we connect sex with its biological goal of reproduction. The sex drive aims for pleasure and gratification, and reproduction is a consequence that is presumably unforeseen by animals and by man at primitive levels of understanding. Effectance motivation similarly aims for the feeling of efficacy, not for the vitally important learnings that come as its consequence. If we consider the part played by competence motivation in adult human life we can observe the same parallel. Sex may now be completely and purposefully divorced from reproduction but nevertheless pursued for the pleasure it can yield. Similarly, effectance motivation may lead to continuing exploratory interests or active adventures when in fact there is no longer any gain in actual competence or any need for it in terms of survival. In both cases the motive is capable of yielding surplus satisfaction well beyond what is necesary to get the biological work done.

In infants and young children it seems to me sensible to conceive of effectance motivation as undifferentiated. Later in life it becomes profitable to distinguish various motives such as cognizance, construction, mastery, and achievement. It is my view that all such motives have a root in effectance motivation. They are differentiated from it through life experiences which emphasize one or another aspect of the cycle of transaction with environment. Of course, the motives of later childhood and of adult life are no longer simple and can almost never be referred to a single root. They can acquire loadings of anxiety, defense, and compensation, they can become fused with unconscious fantasies of a sexual, aggressive, or omnipotent character, and they can gain force because of their service in producing realistic results in the way of income and career. It is not my intention to cast effectance in the star part in adult motivation. The acquisition of motives is a complicated affair in which simple and sovereign theories grow daily more obsolete. Yet it may be that the satisfaction of effectance contributes significantly to those feelings of interest which often sustain us so well in day-to-day actions, particularly when the things we are doing have continuing elements of novelty.

The Biological Significance of Competence

The conviction was expressed at the beginning of this paper that some such concept as competence, interpreted motivationally, was essential for any biologically sound view of human nature. This necessity emerges when we consider the nature of living systems, particularly when we take a longitudinal view. What an organism does at a given moment does not always give the right clue as to what it does over a period of time. Discussing this problem, Angyal (1941) has proposed that we should look for the general pattern followed by the total organismic process over the course of time. Obviously this makes it necessary to take account of growth. Angyal defines life as "a process of self-expansion"; the living system "expands at the expense of its surroundings," assimilating parts of the environment and transforming them into functioning parts of itself. Organisms differ from other things in nature in that they are "self-governing entities" which are to some extent "autonomous." Internal processes govern them as well as external "heteronomous" forces. In the course of life there is a relative increase in the preponderance of internal over external forces. The living system expands, assimilates more of the environment, transforms its surroundings so as to bring them under greater control. "We may say," Angyal writes, "that the general dynamic trend of the organism is toward an increase of autonomy. . . . The human being has a characteristic tendency toward self-determination, that is, a tendency to resist external influences and to subordinate the heteronomous forces of the physical and social environment to its own sphere of influence." The trend toward increased autonomy is characteristic so long as growth of any kind is going on, though in the end the living system is bound to succumb to the pressure of heteronomous forces.

Of all living creatures, it is man who takes the longest strides toward autonomy. This is not because of any unusual tendency toward bodily expansion at the expense of the environment. It is rather that man, with his mobile hands and abundantly developed brain, attains an extremely high level of competence in his transactions with his surroundings. The building of houses, roads and bridges, the making of tools and instruments, the domestication of plants and animals, all qualify as planful changes made in the environment so that it comes more or less under control and serves our purposes rather than intruding upon them. We meet the fluctuations of outdoor temperature, for example, not only with our bodily homeostatic mechanisms, which alone would be painfully unequal to the task, but also with clothing, buildings, controlled fires, and such complicated devices as self-regulating central heating and air conditioning. Man as a species has developed a tremendous power of bringing the environment into his service, and each individual member of the species must attain what is really quite an impressive level of competence if he is to take part in the life around him.

We are so accustomed to these human accomplishments that it is hard to realize how long an apprenticeship they require. At the outset the human infant is a slow learner in comparison with other animal forms. Hebb (1949) speaks of "the astonishing inefficiency of man's first learning, as far as immediate results are concerned," an inefficiency which he attributes to the large size of the association areas in the brain and the long time needed to bring them under sensory control. The human lack of precocity in learning shows itself even in comparison with one of the next of kin: as Hebb points out, "the human baby takes six months, the chimpanzee four months, before making a clear distinction between friend and enemy." Later in life the slow start will pay dividends. Once the fundamental perceptual elements, simple associations, and conceptual sequences have been established, later learning can proceed with ever increasing swiftness and complexity. In Hebb's words, "learning at maturity concerns patterns and events whose parts at least are familiar and which already have a number of other associations."

This general principle of cumulative learning, starting from slowly acquired rudiments and proceeding thence with increasing efficiency, can be illustrated by such processes as manipulation and locomotion, which may culminate in the acrobat devising new stunts or the dancer working out a new ballet. It is especially vivid in the case of language, where the early mastery of words and pronunciation seems such a far cry from spontaneous adult speech. A strong argument has been made by Hebb (1949) that the learning of visual forms proceeds over a similar course from slowly learned elements to rapidly combined patterns. Circles and squares, for example, cannot be discriminated at a glace without a slow apprenticeship involving eye movements, successive fixations, and recognition of angles. Hebb proposes that the recognition of visual patterns without eye movement "is possible only as the result of an intensive and prolonged visual training that goes on from the moment of birth, during every moment that they eyes are open, with an increase in skill evident over a period of 12 to 16 years at least."

On the motor side there is likewise a lot to be cumulatively learned. The playing, investigating child slowly finds out the relationships between what he does and what he experiences. He finds out, for instance, how hard he must push what in order to produce what effect. Here the S-R formula is particularly misleading. It would come nearer the truth to say that the child is busy learning R-S connections — the effects that are likely to follow upon his own behavior. But even in this reversed form the notion of bonds or connections would still misrepresent the situation, for it is only a rare specimen of behavior that can properly be conceived as determined by fixed neural channels and a fixed motor response. As Hebb has pointed out, discussing the phenomenon of "motor equivalence" named by Lashley (1942), a rat which has been trained to press a lever will press it with the left forepaw, the right forepaw, by clinging upon it, or by biting it; a monkey will open the lid of a food box with either hand, with a foot, or even with a stick; and we might add that a good baseball player can catch a fly ball while running in almost any direction and while in almost any posture, including leaping in the air and plunging forward to the ground. All of these feats are possible because of a history of learnings in which the main lesson has been the effects of actions upon the stimulus fields that represent the environment. What has been learned is not a fixed connection but a flexible relationship between stimulus fields and the effects that can be produced in them by various kinds of action.

One additional example, drawn this time from Piaget (1952), is particularly worth mentioning because of its importance in theories of development. Piaget points out that a great deal of mental development depends upon the idea that the world is made up of objects having substance and permanence. Without such an "object concept" it would be impossible to build up the ideas of space and casuality and to arrive at the fundamental distinction between self and external world. Observation shows that the object concept, "far from being innate or ready made in experience, is constructed little by little." Up to 7 and 8 months the Piaget children searched for vanished objects only in the sense of trying to continue the actions, such as sucking or grasping, in which the objects had played a part. When an object was really out of sight or touch, even if only because it was covered by a cloth, the infants undertook no further exploration. Only gradually, after some study of the displacement of objects by moving, swinging, and dropping them, does the child begin to make an active search for a vanished object, and only still more gradually does he learn, at twelve months or more, to make allowance for the object's sequential displacements and thus to seek it where it has gone rather than where it was last in sight. Thus it is only through cumulative learning that the child arrives at the idea of permanent substantial objects.

The infant's play is indeed serious business. If he did not while away his time pulling strings, shaking rattles, examining wooden parrots, dropping pieces of bread and celluloid swans, when would he learn to discriminate visual patterns, to catch and throw, and to build up his concept of the object? When would he acquire the many other foundation stones necessary for cumulative learning? The more closely we analyze the behavior of the human infant, the more clearly do we realize that infancy is not simply a time when the nervous system matures and the muscles grow stronger. It is a time of active and continuous learning, during which the basis is laid for all those processes, cognitive and motor, whereby the child becomes able to establish effective transactions with his environment and move toward a greater degree of autonomy. Helpless as he may seem until he begins to toddle, he has by that time already made substantial gains in the achievement of competence.

Under primitive conditions survival must depend quite heavily upon achieved competence. We should expect to find things so arranged as to favor and maximize this achievement. Particularly in the case of man, where so little is provided innately and so much has to be learned through experience, we should expect find highly advantageous arrangements for securing a steady cumulative learning about the properties of the environment and the extent of possible transactions. Under these circumstances we might expect to find a very powerful drive operating to insure progress toward competence, just as the vital goals of nutrition and reproduction are secured by the powerful drives, and it might therefore seem paradoxical that the interests of competence should be so much entrusted to times of play and leisurely exploration. There is good reason to suppose, however, that a strong drive would be precisely the wrong arrangement to secure a flexible, knowledgeable power of transaction with the environment. Strong drives cause us to learn certain lessons well, but they do not create maximum familiarity with our surroundings.

This point was demonstrated half a century ago in some experiments by Yerkes and Dodson (1908). They showed that maximum motivation did not lead to the most rapid solving of problems, especially if the problems were complex. For each problem there was an optimum level of motivation, neither the highest nor the lowest, and the optimum was lower for more complex tasks. The same problem has been discussed more recently by Tolman (1948) in his paper on cognitive maps. A cognitive map can be narrow or broad, depending upon the range of cues picked up in the course of learning. Tolman suggests that one of the conditions which tend to narrow the range of cues is a high level of motivation. In everyday terms, a man hurrying to an important business conference is likely to perceive only the cues that help him to get there faster, whereas a man taking a stroll after lunch is likely to pick up a substantial amount of casual information about his environment. The latent learning experiments with animals, and experiments such as those of Johnson (1953) in which drive level has been systematically varied in a situation permitting incidental learning, give strong support to this general idea. In a recent contribution, Bruner, Matter, and Papanek (1955) make a strong case for the concept of breadth of learning and provide additional evidence that it is favored by moderate and hampered by strong motivation. The latter "has the effect of speeding up learning at the cost of narrowing it." Attention is concentrated upon the task at hand and little that is extraneous to this task is learned for future use.

These facts enable us to see the biological appropriateness of an arrangement which used periods of less intense motivation for the development of competence. This is not to say that the narrower but efficient learnings that go with the reduction of strong drives make no contribution to general effectiveness. They are certainly an important element in capacity to deal with the environment, but a much greater effectiveness results from having this capacity fed also from learnings that take place in quieter times. It is then that the infant can attend to matters of lesser urgency, exploring the properties of things he does not fear and does not need to eat, learning to gauge the force of his string-pulling when the only penalty for failure is silence on the part of the attached rattles, and generally accumulating for himself a broad knowledge and a broad skill in dealing with his surroundings.

The concept of competence can be most easily discussed by choosing, as we have done, examples of interaction with the inanimate environment. It applies equally well, however, to transactions with animals and with other human beings, where the child has the same problem of finding out what effects he can have upon the environment and what effects it can have upon him. The earliest interactions with members of the family may involve needs so strong that they obscure the part played by effectance motivation, but perhaps the example of the well fed baby diligently exploring the several features of his mother's face will serve as a reminder that here, too, there are less urgent moments when learning for its own sake can be given free rein.

In this closing section I have brought together several ideas which bear on the evolutionary significance of competence and of its motivation. I have sought in this way to deepen the biological roots of the concept and thus help it to attain the stature in the theory of behavior which has not been reached by similar concepts in the past. To me it seems that the most important proving ground for this concept is the effect it may have on our understanding of the development of personality. Does it assist our grasp of early object relations, the reality principle, and the first steps in the development of the ego? Can it be of service in distinguishing the kinds of defense available at different ages and in providing clues to the replacement of primitive defenses by successful adaptive maneuvers? Can it help fill the yawning gap known as the latency period, a time when the mastery of school subjects and other accomplishments claim so large a share of time and energy? Does it bear upon the self and the vicissitudes of self-esteem, and can it enlighten the origins of psychological disorder? Can it make adult motives and interests more intelligible and enable us to rescue the concept of sublimation from the difficulties which even its best friends have recognized? I believe it can be shown that existing explanations of development are not satisfactory and that the addition of the concept of competence cuts certain knots in personality theory. But this is not the subject of the present communication, where the concept is offered much more on the strength of its logical and biological probability.

Summary

The main theme of this paper is introduced by showing that there is widespread discontent with theories of motivation built upon primary drives. Signs of this discontent are found in realms as far apart as animal psychology and psychoanalytic ego psychology. In the former, the commonly recognized primary drives have proved to be inadequate in explaining exploratory behavior, manipulation, and general activity. In the latter, the theory of basic instincts has shown serious shortcomings when it is stretched to account for the development of the effective ego. Workers with animals have attempted to meet their problem by invoking secondary reinforcement and anxiety reduction, or by adding exploration and manipulation to the roster of primary drives. In parallel fashion, psychoanalytic workers have relied upon the concept of neutralization of instinctual energies, have seen anxiety reduction as the central motive in ego development, or have hypothesized new instincts such as mastery. It is argued here that these several explanations are not satisfactory and that a better conceptualization is possible, indeed that it has already been all but made.

In trying to form this conceptualization, it is first pointed out that many of the earlier tenets of primary drive theory have been discredited by recent experimental work. There is no longer any compelling reason to identify either pleasure or reinforcement with drive reduction, or to think of motivation as requiring a source of energy external to the nervous system. This opens the way for considering in their own right those aspects of animal and human behavior in which stimulation and contact with the environment seem to be sought and welcomed, in which raised tension and even mild excitement seem to be cherished, and in which novelty and variety seem to be enjoyed for their own sake. Several reports are cited which bear upon interest in the environment and the rewarding effects of environmental feedback. The latest contribution is that of Woodworth (1958), who makes dealing with the environment the most fundamental element in motivation.

The survey indicates a certain unanimity as to the kinds of behavior that cannot be successfully conceptualized in terms of primary drives. This behavior includes visual exploration, grasping, crawling and walking, attention and perception, language and thinking, exploring novel objects and places, manipulating the surroundings, and producing effective changes in the environment. The thesis is then proposed that all of these behaviors have a common biological significance: they all form part of the process whereby the animal or child learns to interact effectively with his environment. The word *competence* is chosen as suitable to indicate this common property. Further, it is maintained that competence cannot be fully acquired simply through behavior instigated by drives. It receives substantial contributions from activities which, though playful and exploratory in character, at the same time show direction, selectivity, and persistence in interacting with the environment. Such activities in the ultimate service of competence must therefore be conceived to be motivated in their own right. It is proposed to designate this motivation by the term effectance, and to characterize the experience produced as a *feeling of efficacy*.

In spite of its sober biological purpose, effectance motivation shows itself most unambiguously in the playful and investigatory behavior of young animals and children. Specimens of such behavior, drawn from Piaget (1952), are analyzed in order to demonstrate their constantly transactional nature. Typically they involve continuous chains of events which include stimulation, cognition, action, effect on the environment, new stimulation, *etc*. They are carried on with considerable persistence and with selective emphasis on parts of the environment which provide changing and interesting feedback in connection with effort expended. Their significance is destroyed if we try to break into the circle arbitrarily and declare that one part of it, such as cognition alone or active effort alone, is the real point, the goal or the special seat of satisfaction. Effectance motivation must be conceived to involve satisfaction — a feeling of efficacy — in transactions in which behavior has an exploratory, varying, experimental character and produces changes in the stimulus field. Having this character, the behavior leads the organism to find out how the environment can be changed and what consequences flow from these changes.

In higher animals and especially in man, where so little is innately provided and so much has to be learned about dealing with the environment, effectance motivation independent of primary drives can be seen as an arrangement having high adaptive value. Considering the slow rate of learning in infancy and the vast amount that has to be learned before there can be an effective level of interaction with surroundings, young animals and children would simply not learn enough unless they worked pretty steadily at the task between episodes of homeostatic crisis. The association of interest with this "work," making it play and fun, is thus somewhat comparable to the association of sexual pleasure with the biological goal of reproduction. Effectance motivation need not be conceived as strong in the sense that sex, hunger, and fear are strong when violently aroused. It is moderate but persistent, and in this, too, we can discern a feature that is favorable for adaptation. Strong motivation reinforces learning in a narrow sphere, whereas moderate motivation is more conducive to an exploratory and experimental attitude which leads to competent interactions in general, without reference to an immediate pressing need. Man's huge cortical association areas might have been a suicidal piece of specialization if they had come without a steady, persistent inclination toward interacting with the environment.

References

Angyal, A. *Foundations for a science of personality*. New York: Commonwealth Fund, 1941.

Bruner, J. S., Matter, J., & Papanek, M. L. Breadth of learning as a function of drive level and mechanization. *Psychol. Rev.*, 1955, 62, 1-10.

Diamond, S. A neglected aspect of motivation. *Sociometry*, 1939, 2, 77-85.

Hebb, D. O. *The organization of behavior*. New York: Wiley, 1949.

Johnson, E. E. The role of motivational strength in latent learning. *J. Comp. Physiol. Psychol.*, 1953, 45, 526-530.

Lashley, K. S. The problem of cerebral organization in vision. In H. Kluver, *Visual mechanisms*. Lancaster, Pa.: Jaques Cattell, 1942. Pp. 301-322.

Piaget, J. *The origins of intelligence in children*. (Trans. by M. Cook.) New York: International Univer. Press, 1952.

Schachtel, E. G. The development of focal attention and the emergence of reality. *Psychiatry*, 1954, 17, 309-324.

Tolman, E. C. Cognitive maps in rats and men. *Psychol. Rev.*, 1948, 55, 189-208.

Woodworth, R. S. *Dynamics of behavior*. New York: Holt, 1958.

Yerkes, R. M. & Dodson, J. D. The relation of strength of stimulus to rapidity of habit-formation. *J. Comp. Neurol. Psychol.*, 1908, 18, 459-482.

CHAPTER 27

Work as a Sublimation

KARL A. MENNINGER

Work and the Instinct

Sublimation, which has been described as "the keystone of our culture,"[1] is a psychological process for which Freud introduced this term in his celebrated *Three Contributions to the Theory of Sex*, published in 1905. At that time he believed that the chief function of the ego was the direction and control of the sexual impulses and their adaptation to reality and social law. His idea was that some sexual energy was "sublimated" or converted into desexualized activities which were socially acceptable. He was not entirely original in this. The same idea had been suggested as long ago as Ovid (who advised that "you who seek a termination of your passion, attend to your business; . . . soon will voluptuousness turn its back on you"),[2] and by many others since Ovid.

Dr. Harry B. Levey has made a very careful study of the changes in Freud's formulation subsequent to 1905 and the interpretations made of it by other analysts, and concludes his excellent survey with the comment that "the concept of sublimation is an improved recapitulation of empirically known facts [but] . . . we have neglected to revise it in accordance with metapsychological standards."[3] As his review shows, Freud never revised his theory of sublimation in accordance with his revision of the instinct theory.

[1] Levey, Harry B. A Critique of the Theory of Sublimation. *Psychiatry*, 2: 239-70, May 1939.

[2] Heroides, trans. Riley. London, Bohn's Libraries, 1912. Quoted by W. S. Taylor, *A Critique of Sublimation in Males.* Genetic Psychology Monographs, no. 13, p. 9, January 1933.

[3] Levey, *A Critique of the Theory of Sublimation.*

I think this revision in the concept of sublimation is now clearly indicated. It is generally accepted, even by those who do not subscribe to the instinct theory as a whole, that the ego has to manage not only the sexual impulses but also the aggressive tendencies, whatever their source. Those of us for whom the dual instinct theory of constructive and destructive impulses best explains the phenomena, conceive of the life instinct (the erotic impulses) as the chief influence whereby the destructive impulses are harnessed and redirected. In a primitive civilization, it is unnecessary to modify or to restrain either the aggressive or the erotic impulses, but in a civilized world it is necessary to restrain both, and this restraint is accomplished by fusion. If the erotic impulse sufficiently dominates, the result is constructive behavior; if the aggressive impulses dominate, the result is more or less destructive behavior. We could summarize the fate of the aggressive energy in the theoretically normal person thus: some of it is completely repressed, some is expressed directly in self-defense or in the protection of others; some is internalized as conscience, and some is sublimated, that is, converted into useful but essentially aggressive activity. In the less normal individual, we must allow for the portion which is directly expressed against others as cruelty, crime, provocativeness, etc., and that which is directed back upon the self as depression, inhibition, illness, neurosis, and suicide.

Of all the methods available for absorbing the aggressive energies of mankind in a useful direction, work takes first place. It may not be the oldest, it certainly is not the most pleasant. But it has a certain realistic quality which makes it seem the most practical and obvious of all sublimations. Almost everyone would accept it as self-evident that we must work in order to live; not everyone believes that we must play in order to live, or that we must have something to believe in or to love.

Furthermore, the connections of work with the destructive instinct are close and clear. It is easy to see that all work represents a fight against something, an attack upon the environment. The farmer plows the earth, he harrows it, tears it, pulverizes it; he pulls out weeds, or cuts them, or burns them; he poisons insects and fights against drought and floods. To be sure, all this is done in order to create something, for which reason we can call it work and not rage. The destructiveness is, so to speak, specialized or selectively directed, and a net "product" is obtained. But even this product must be torn or cut from its producing matrix by dint of more labor, transported by labor to other places for storage or consumption, prepared for use as food or clothing by still further destructive energy. It is cut, crushed, scorched, torn, burned, boiled, twisted, combed, exposed to the sun and air for desiccation. The construction of sheltering granaries requires the destruction of trees or the splitting and avulsion of rocks and the reassembly of the fragments into an artificial structure. The forging of tools or weapons, the making of clothes, even the fabrication and hoarding of money, the symbol of value, require the expenditure of aggressive energy.

One could carry the illustration further into the realms of mastering the sea and the air, or forcing earth, clay, or metal into molds, or creating energy by burning carbon, or slaughtering animals for food and clothing. In every instance it is the same: destructive energy is applied to a constructive goal through discriminating between the desirable and the undesirable, and encouraging the former by eliminating or suppressing or altering the latter.

The reader may feel a little taken in by the selection of these examples; he may think rather of such illustrations as the work of the housepainter, the seamstress, the lawyer, or the banker — which do not seem to be so obviously destructive. I admit that there is room for argument at this point. Some of my colleagues[4] feel that we must distinguish another element in work, best described as an impulse toward mastery — controlling, re-forming, organizing, directing, etc. But to me this urge to master something, whether it be a mechanical puzzle or a complicated accounting problem or a recalcitrant horse, seems to be indistinguishable in its *essence* from the aggressive, destructive impulse, purposively and expediently directed. To the extent that something is mastered, some kind of resistance is broken down or overcome. Of course the destructiveness of the whaler is different from that of the lumberman, and the destructiveness of the lumberman is different from that of the miner, and that of the miner is different from that of the surgeon; but all of them are working *against* something in an effort to master a situation or a material and to produce something in the end. It is the modification of this destructive energy in such a way as to achieve this creation of something that distinguishes work from wanton destructiveness.

It may also be objected by some that it is not always love that modifies the aggressive energies in such a way as to make them useful and fruitful. Hunger, the need for protection against the elements, the fear of approaching enemies — these would seem to be the more immediate determinants of the labors described. Man must eat to live and he must work to eat, we are reminded. But this objection loses sight of the broader concept of love that is implied in this discussion, just as the other objection takes too narrow a view of aggression. This is the old stumbling block in the psychonalytic theory of twenty years ago, a stumbling block that for many years confused even Freud's thinking. For one cannot fruitfully divide the

[4] Ives Hendrick, "Work and the Pleasure Principle," read before the 44th meeting of the American Psychoanalytic Association, May 18, 1942; and S. Bernfeld, *The Psychology of the Infant*, Brentano, 1929.

function of love as applied to the fostering of one's own life and love as applied to the fostering of another's. Love is the reflection of the instinct of life, and the love of ourselves is of the same texture as the love of others; it is the love of life itself. What man does to survive is dictated by the same love that dictates his compulsion to continue the race. The immediate emotional stimulus may be fear, or anger, or curiosity, or cupidity, or the desire for warmth and peace and caresses, but the instinct is the same.

So we shall not linger over these objections. The essential point is that in work, as contrasted with purposeless destruction, the aggressive impulses are molded and guided in a constructive direction by the influence of the creative (erotic) instinct.

Savage man's first work was killing. He hunted and killed men and beasts. Closely allied to this work was worship, which consisted of sacrifice and efforts to propitiate the gods and to forestall *their* aggressions. With the development of social organization, certain important changes took place. Mutual agreements had to be made between individuals for purely selfish reasons, to impose some restrictions upon indiscriminate destruction. According to Freud's concept, it was the father of the primitive horde who ended the rule of complete individualism and by means of his prohibitions directed the destructive energies toward constructive ends. These ends might be considered constructive primarily from *his*, the ruler's, standpoint; they were enforced through his power and for his purposes. But they also contributed to the personal advantage of the individual members of the group since they enabled them to concentrate destructive energies against foes. But dicipline, obedience, the postponement of personal gain for communal good, and the restriction of destruction to certain specified objects were probably initiated by force, and while they afforded personal advantages to the individual they were accepted only with resentment. This resentment against the power of authority is still to be seen in the way in which work tends to be regarded as a necessary evil. Work became dissociated from pleasure to the extent that it became dissociated from individual initiative. The conception of work as drudgery which everyone experiences to some extent, and which some persons experience in a very high degree, is bound up with this resistance to authority. In the course of time, with the development of society, this authority became introjected, and work — which had formerly been a free outpouring of aggressive energies — became surrounded with a certain sentimental, sacrificial halo. It was described as ennobling and considered an end in itself. But it was still *"work."*

Aversion to work does not spring solely from resentment against its necessity. If we ask ourselves just what it is in work that makes for drudgery, some of us might say it is monotony; others might say that it is the entailment of pain or muscular exertion — or the impossibility of seeing tangible results — or lack of connection with any creed or purpose. All these factors are probably present.[5] It would be interesting to examine how it came about that some people must do continuously what seems chiefly drudgery, while other people are able to do what seems to be pleasurable and even delightful work, if indeed it can be called work at all. Some anthropologists explain this by the theory that it was originally a question of physical power: men preferred the labor of hunting, and they were stronger than women, so they made the women (and the slaves) do the cooking, the firebuilding, and the other things that they (the men) didn't "like" to do. "Work only tires a woman, but it ruins a man," runs an old African proverb. Others explain the specialization as the consequence of some sort of instinctual predilection. Marxians, of course, explain it on the basis of economic determinism and the power of accumulated capital (or, prior to the capitalistic period, some other form of organized power). Psychologists in the past explained it on the basis of intellectual capacity. This theory was very flattering, and — while it remained current — certain stupid but financially successful persons made the most of it. Probably all of these factors have entered into the differentiations in quality of specialized labor. Thorstein Veblen and many others have analyzed this problem in recent years with varying conclusions.

The very multiplicity of theories to explain the painfulness of work suggests that none of them is adequate. We may venture, therefore, to add another interpretation in line with our psychological theories. Perhaps we could define drudgery as that form or aspect of work in which the satisfaction of the aggressive element is not combined with sufficient erotization to give some degree of conscious satisfaction in the work itself. The satisfaction in work may be related to the product, as for example the pleasure an artisan receives from making a beautiful vase or an author from writing a good book. Or it may be related to the approval received from a superior, or the feeling that the work has been done for his sake. Or the pleasure may be derived chiefly from a sense of cooperation, companionship, *esprit de corps*, brotherhood. Finally it may be derived from some erotization of the actual techniques involved in the performance of the labor itself.[6] I remember a charwoman who used to get a most evident pleasure out of her daily floor-scrubbing. It was not only that she took pride in the thoroughness of her cleansing, or in the compliments she received from people who praised her assiduousness: she actually *liked* to swash around in the suds and would attack the dirty floors with something

[5] For a discussion of factors improving and impairing productivity in work, see A.C. Ivy, "The Physiology of Work," *Journal of the A.M.A.*, 118: 569-73, February 21, 1942.

[6] Bernfeld *(Psychology of the Infant)* and Hendrick *(Work and the Pleasure Principle)*.

of the vehement pleasure portrayed in the advertisements for Old Dutch Cleanser. Yet what could be worse drudgery for the average person than floor-scrubbing? All of the above are ways in which the erotic instinct can actually neutralize the destructive elements in the work sublimation.

If work is done only by compulsion, external or internal, if it gives none of the pleasures just mentioned, it is felt to be drudgery, and it is not a complete sublimation. Take, for instance, the housewife's occupation: cleaning and so on certainly bear witness to her desire to make things pleasant for others and for herself, and as such are a manifestation of love for other people and for the things she cares for. But at the same time she also gives expression to her aggression in destroying the enemy, dirt, which in her unconscious mind has come to stand for 'bad' things. The original hatred and aggression derived from the earliest sources may break through in women who make life miserable for the family by continuously 'tidying up'; there the hatred is actually turned against the people she loves and cares for. In other words, the sublimation breaks down.

Although these principles might seem to be almost self-evident, it is astonishing to discover what widely divergent opinions prevail on the subject of work. In an age when billions of dollars are spent in promoting efficiency and increased production, an age when work has been exalted to being the justification of existence, it is almost unbelievable how consistently the basic requirements of work are overlooked. The psychological prerequisite of successful labor is dimly perceived and poorly manipulated even by those for whom it is a problem of paramount importance. Consider the stupid and costly blunders of industry in relation to manpower. The organization of labor against its employers furnished workers with a closely knit brotherhood and a formidable enemy upon whom they can project all their aggressions. The men in power see this as a menace and attempt to make their position more secure by fighting; but in doing so they are forced to yield ground because of the losses they incur in decreased production. The stiffer the opposition, the stronger waxes the battle, until much of the energy which would otherwise go into production of goods is deflected into personal warfare. That some of this waste is unavoidable is probably true, for the aggressive instinct seems never to be satisfied with entirely impersonal objectives. But that it may be redirected toward other ends is shown by the way in which labor difficulties become unpopular during war when employers and employees unite against a common foe. This is what has happened in England and in Russia; it is beginning to develop belatedly in our own country. The feeling of brotherhood among workers and employers is intensified and their hostilities are directed against national enemies.

Personal psychology has been so discredited in the entire setup of industry, and the profit motive has been assumed to be so powerful and all-controlling, that satisfaction in work has come to be assessed entirely in terms of wages and hours, not only by employers but by the employees themselves. Yet, from the standpoint of psychology, the recurring industrial and economic depressions might seem, in some degree, the result of profound dissatisfactions and disappointments in work, on the part of both employers and employees. When psychology has been applied to industry, it has too often been patronizingly used as an aggressive tool to perpetuate or feebly patch certain obvious evils. We pay only lip service to the idea of increasing production and profits by constructive reorganization of workers and redirection of aggressions, and both sides look hopefully to the government for redress and support.

An extraordinarily important study of a small group of workers in the Western Electric Company was made by some psychologists. For five years productivity was checked daily and compared with all sorts of conditions affecting the employees — their health, their working conditions, their home conditions, their attitudes, and their general morale. Of the many conclusions the most pertinent to the present theme is the fact that no matter what changes were made in working conditions, in management techniques, in hours of labor, etc., if the change was made with the ostensible purpose of benefiting the workers, efficiency and production were immediately and markedly stimulated. Reducing this to the simple terms of our thesis, we would say that, prosaic though it may sound, fostering the affection of workers for an employer is more important than any specific concession or regulation, because it permits more complete sublimation of the aggressive impulses which otherwise "leak out" in the form of resentment against the employer.

Economists do not deny that disturbances in industry are based on fundamental human conflicts. The dynamic nature of these conflicts is a legitimate study for psychiatrists, but these have confined themselves largely to the study of individuals and to neurotic attitudes toward work in the individual. The inability to play brings some persons to psychiatrists; the inability to work brings many more. It is from the study of these that the general principles enunciated above were derived.

The average person may suppose that when patients come to a sanitarium for the treatment of what appear to be relatively mild maladjustments, the great problem is how to keep them busy, how to fill their leisure time. It will probably astonish many, therefore, when I say that the great problem in every well-conducted sanitarium is how to get the patients to do *anything!* The place may offer ten times as much to do as the average patient can bring himself even to attempt; but, despite all the efforts of the medical director, the medical staff, the nursing staff, and the therapists — despite schedules and regulations and exhortations — the patients manage to evade with an uncanny skill every opportunity for amusement, recreation, exercise, constructive craftwork, and all the other devices so carefully planned for them. At the moment I am thinking of a rather characteristic example — the son of the president of a large manufacturing concern. Although the patient had an older brother, he expected ultimately to succeed his father as head of this business. However, he was thirty-five years old and he had never stuck consistently at any task. Several times he had worked briefly in the executive offices of his father's plant, but he soon became irritated at what he regarded as his father's bad management and either engaged in sharp arguments with his father or abandoned his duties for an alcoholic debauch or an extended loaf in California or Florida. He came to the sanitarium for treatment because of severe headaches which were recognized by his physicians to be emotional in origin and related to his terrific rage against his father — a rage which, while not entirely unjustified at times, was scarcely becoming in this indolent son.

It was easily recognizable that his indolence, like his headaches, was the result of an emotional reaction of fear and hate directed toward his father. But it was interesting to see how this reaction was carried out during his sanitarium residence. Although he was very anxious to get well and worked earnestly and cooperatively at any treatment that was immediately supervised by the physician — including, for example, his psychoanalytic interviews — everything else about the sanitarium he considered negligible. Time after time he reiterated his eagerness to get well and to go home. For many months during his treatment it was only rarely that he would turn his hand to doing anything constructive, even when urged by the physicians to do so for his own health. He would walk five or ten miles a day for exercise but would not lift a spadeful of dirt or make a single puppet; he would play tennis occasionally but he would never roll the tennis courts. He would play bridge until he had a blinding headache, but he would not set up a bridge table or manage a tournament for the pleasure of the other guests. Although his education needed supplementing in many particulars, he refused to attend any classes. During the first ten months of his stay he spent altogether only four hours in the craft and carpentry shops.

Toward the end of his treatment, when the energy locked up in his inhibitions began to be released, he began to take increasing interest in the very things he had so long avoided. It is a frequently observed paradox that when a patient really begins to take full advantage of the opportunities of psychiatric treatment he is approaching the time when he is able to give it up and return home.

Perhaps three-fourths of the patients who come to psychiatrists are suffering from an incapacitating impairment of the satisfaction in work or their ability to work. In many it is their chief complaint.

The harnessing of aggressive energy in work may break down so that the aggressive energies threaten to emerge directly and destructively; this then requires the erection of emergency defenses — costly, last-minute barriers against exploding. We recognize these defense measures as symptoms of illness; the more significant "illness" is the inability to work, to sublimate; and the genuine, deeper illness is the uncontrollable excess of hostility.

Vocational Choice and Psychology

Another index of our lack of scientific thinking in regard to the function of labor is our colossal ignorance and neglect of the problem of vocational choice. Here is one of the momentous decisions that cast the lives of human beings in fixed though diverse channels. Perhaps next to the choice of a marital partner, it is the most important and far-reaching decision made by the individual. (A place of residence is important, as are a standard of living and a selection of friends, but these things often depend immediately upon the mate and the vocation chosen.) Yet the subject of vocational choice has had all too little scientific investigation and is at best only the field of a few specialists. Both in literature and in psychology there appears to be a tacit assumption that one's vocation is something almost foreordained, something determined by chance and circumstances — or, to give them their more sophisticated titles, by economics and the social pattern. One inherits a family interest in a business or a profession, or acquires one by inspiration, or happens upon an opportunity which he exploits.

It is singular how little support psychiatry and psychoanalysis have given to those commendable efforts in the direction of vocational education which a few specialists have put forth.[7] If one has occasion to observe in an adolescent about to be graduated from high school his struggles over a choice of college and, particularly, over his course of study in that college, one cannot but be grateful to those who have made some effort to put at his disposal a survey of the complicated activities of life in which he will soon be forced to participate in some capacity or other. It would seem as if there were some taboo on the subject that makes us so loath to accept it as a necessary part of education. Not only we, but also the educators, are at fault in this respect. A survey of educational and vocational opportunities is seldom afforded a student until too late to be of much practical value. It was only a few years ago that colleges introduced orientation courses, and for the most part these served to orient the student with regard to knowledge only but not with regard to practical activities. Most of the high schools of the country have no courses in vocational guidance, and in none of them, so far as I know, is such a course compulsory. Even in those where it is offered, it seems to be presented in such a way as to be of value to only a

[7] See the following books: Arthur J. Jones, *Principles of Guidance,* McGraw-Hill, 1934; Donald G. Paterson, *Men, Women and Jobs: A Study in Human Engineering,* University of Minnesota, 1936; Donald G. Paterson, *Student Guidance Techniques,* McGraw-Hill, 1938; Edmund G. Williamson, *Student Personnel Work: An Outline of Clinical Procedure,* McGraw-Hill, 1937.

small percentage of the students. Of course, no vocational-guidance course in high school or even in college can hope to explore *unconscious* reasons for vocational choice, either in general or in particular. Furthermore, as one wise teacher put it to me, how can one expect the secluded and protected teachers in such institutions to present even the known realities about worldly professions and occupations?

Of course, there is some extenuation for this lack. One might assume that to high school students who are unable to afford further education the question of vocation is largely a question of opportunity and chance. On the other hand, for those who can afford to go to college, the entire college career might be looked upon as an orientation in vocational possibilities. Sometimes it works out this way, but all of us have seen cases in which a vocation must be decided upon before the student has gone far enough in college to know much about the world, past or present.

What is there about the practice of medicine, for example, that appeals to the young man as yet undecided about how he will exchange his efforts for other worldy values? The college or high school student has no very accurate idea about what the practice of medicine is, and he makes his decision largely on the basis of certain conceptions that are more or less illusory. Among the advantages that attract the young man or woman to medicine is probably the traditional dignity and social rank of the medical man. Some of this is perhaps unrealistic, but for the most part it is based upon a long tradition and a sound one. There are certain financial and social accruals that can reasonably be expected if one has obtained a legal prerogative for selling professional advice and service. Furthermore, this promised economic security is of an independent variety — "the doctor is his own boss," so they mistakenly think.

Such features of a profession probably have a wide though very superficial appeal. Banking has, or at least did have, more social prestige and better financial rewards than medicine and requires much less training and, theoretically, less intelligence. The practice of law permits an equal amount of independence and has more opportunities for personal exploitation.

Then, in certain instances, there are the purely fortuitous circumstances of family tradition or special opportunity. In many countries it is the custom for the son to follow the trade or profession of his father, and we know that even in America there is some tendency in this direction. How much this is to be ascribed to the special practical advantages which it makes possible, such as the inheritance of a practice, and how much to the psychological factors relating to the son's attitude toward the father must vary in different instances.

The wish to please one's parents, to live up to their ideal or ambition, is frequently a strong conscious determinant in the selection of life work. The mother who wants her son to be a minister, the father who wants his son to follow in his footsteps, are familiar contemporary figures. The son may comply because he desires to please his parents or because he fears to displease them. If his psychological maturity is somewhat greater, he will comply or not comply because of more external and objective reasons, not excluding the actual inspiration that he may have found in the profession of his father or that of some father substitute.

But we know that beneath the conscious and therefore more superficial determinants there are unconscious motives which strongly influence any decision. Among these, in the case of vocational choice, one must undoubtedly include the unconscious reaction of the son to his father's attitudes. Where the conscious identification with the father in the selection of the father's profession will appear to be positive, there will be negative valences in the unconscious and vice versa. In other words, a son may select his fathers's profession, or one that the father wishes him to follow, ostensibly because it flatters and pleases the father; but unconsciously such a son will often be motivated strongly by the repressed impulse to compete with, eclipse, or supersede his father. Similarly many a son who disappoints his father by what appears to be an aggressive rejection of the parental hopes is unconsciously deterred by love of the father, or by the fear of entering into competition with him. It reminds one of that parable of Jesus about the two sons, one of whom said quickly and politely, "I go, sir. I do your bidding," but went not; while the other said, "I will not, I refuse to obey," but did.

In a somewhat different way the unconscious attitude toward the mother, particularly the mother who has definite opinions as to the preferable vocation for her son, likewise influences his choice by very reason of her attitudes sometimes in one direction, sometimes in another.

All vocational choice ought therefore to be considered from three preliminary standpoints: (1) Have the parents indicated any preference? (2) Is the son inclined toward or away from the particular preference of the parents? (3) If the father's preference is something other than his own profession, is the son's inclination toward either the father's own profession or his preferred profession, or is it opposed to (away from) both? These questions should underlie the more usual investigation of personal qualities: preparation, prospective financial return, etc.

In this digression concerning the motives that determine the selection of various types of work, we may seem to have lost sight of our original thesis; namely, that one function of work is the sublimation of the aggressive impulses. The unconscious gratifications outlined as important for the selection of medicine or any other vocation represent not the energy, aggressive or otherwise, which is vocationally invested, but rather the internal criteria determining the direction in which the energy is turned. If a hostile force attacks a city, the citizens will spring to arms in defense of it. Even the criminals of the community might be willing to join in. One could say then that the aggressive impulses of the citizens of that community had been externally stimulated by the feeling of need for security, the defending of their city. However, a similar belligerent defense might be stimulated merely by more or less groundless fears of the possible approach of such an external menace. This would be an aggressive reaction in response to an internal stimulus. Similarly, when I say that the need for a feeling of greater security is one of the motives impelling the selection of a certain profession, I mean that this is one of the ways in which the aggressive energy is stimulated and indicates the objective toward which it is, in a refined and disguised way, directed.

It will be a long time before we arrive at a comprehensive analysis of the unconscious motives involved in all the various specialties of human labor, but what I have outlined here is sufficient to illustrate the possibilities, based on the original postulate; namely, that the destructive instinct may be modified by the sublimating effects of the erotic instinct into the constructive activity of work. When the doctor administers quinine for malaria or arsenic for syphilis, he is using a refined type of aggression, displaced from its original unconscious aim and directed toward an actual, dangerous foe. And just as the doctor combats disease and the agents of disease, so the teacher combats ignorance, the lawyer crime, the economist poverty, and the minister vice. Nor does this leave out of account the creative artist who combats ugliness, monotony, and boredom. These, too, are enemies of mankind. And in all of these activities the worker, by using his aggressive energies to save others, is saving himself.

In any circumstances, therefore, work is necessary and work does us good; but does work give us pleasure? Is it, as Marcus Manilius claimed, "a pleasure in itself"? Around this question revolve problems of world importance — problems of labor legislation, labor organization, vocational choice, public policy, personal adjustment. It is certainly no such open-and-shut question as many blandly assume. It is all very well for Tertullian to say, "Where our work is, there let our joy be" — for Carlyle to ask, "What is the use of health, or of life, if not to do some work therewith? . . . Blessed is he who has found his work; let him ask no other blessedness" — for Emerson to record that "When I go into my garden with a spade, and dig a bed, I feel such an exhilaration and health that I discover that I have been defrauding myself all this time in letting others do for me what I should have done with my own hands." The fact remains that Tertullian, Carlyle, and Emerson were not compelled to spade up gardens or dig ditches or lift stones or plow furrows; they did as much of these things as they enjoyed and under no external compulsion; they ceased when they became weary; they accounted to no one for the product and to no one for their time. It is all very easy to say in theory that work is one of our pleasures; human experience refutes this as often as it confirms it. Those who rhapsodize about the joy of labor are likely to be persons who are not obliged to do much of it.

There is no evidence that work is in itself pleasurable; the question is rather: In what circumstances is it pleasurable? These circumstances include certain external conditions and certain internal conditions. *Externally* there must be a minimum of compulsion, and opportunity for comfortable group feeling with fellow workers, absence of intense discomfort or fatigue in the performance of the work, proper provision for interspersed rest and recreation periods, a realization of pride in the product and a conviction that the work is useful and appreciated. *Internally*, there must be relative freedom from guilt feelings connected with pleasure and from neurotic compulsions either to work or not to work. The latter are carried over from the childhood era when work is a method of dealing with reality, not elected by the child but acquired by him from his parents. So long as his thinking is governed primarily by the pleasure principle, the child sees no necessity for work except the parental compulsion. If his introduction to reality has been accomplished with sufficient smoothness and grace, he will appreciate those tools which enable him to deal with it productively, but the technique of teaching a child to work is something we know very little about. We only know that most children are so clumsily taught that they seem to have learned more about how not to work than about how to work. Not methods so much as attitudes need to be taught.

B. INTERPERSONAL RELATIONS

CHAPTER 28

Family Systems

DAVID C. SPEER

In an article in which he discusses the relationships between man, society, and the environment from the social systems standpoint, Vickers (36) asked the question, "Is adaptability enough?" He answered his own question as follows:

> I have no doubt that adaptability, as commonly understood, is not enough as a goal or even as an explanation of our striving. We must assume also a "nature" which men and societies are striving to realize and which gives force and direction to their adaptations and sets limits to their adaptability. I believe that this nature is limited but not given by the genetic constitution of mankind: it is something which we may be said to both "make" and to "discover." At any given moment of history, it consists not so much in the mass of partly inconsistent relations which we are set to seek and shun as in the *idea of ourselves* which guides our valuations and hence our compromises. It is an active force in our evolution, none the less so for being its chief product. It will always be tentative, never final, never sacrosant; yet it will slowly come to hold more of what history has to tell us about what is worthwhile to try to be.

Although we have hopefully moved beyond the issue of the sufficiency of adaptability, those concerned with the family, particularly family systems, are at a point where an analogous question can be asked about the concept of homeostasis as it pertains to social groups in general and family groups in particular.

In light of the continuing interest in the conjoint family therapy approach and the impact it has had on clinical practice and research, we are approaching a point where it might be well to examine its conceptual underpinnings. Although it is premature to speak of a family systems *theory* at this time, it is appropriate to make explicit the conceptual framework of current family systems thinking and to consider at least one alternative that is compatible with the basic emphasis on communication.

Before going on, it might be well to point out that now is a particularly appropriate time to re-examine the axiomatic basis of the conjoint therapy approach. First, a simple fact of life is that the longer a conceptual system exists and the greater the social superstructure that is built upon it, the more crystallized and resistant to change it becomes. Secondly, because of its development within the area of social pathology and the resulting conceptual isolation from the fields of systems analysis and management in industry, clinical psychology, family sociology, and the social psychology of small groups, there has been a lack of the exchange and cross-fertilization which is necessary to the development of a viable and critical body of endeavor. Thirdly, at a time when there is an ever increasing interest and conceptual investment in interpersonal growth, changing basic social structures and institutions, social innovation, and creativity, there is something paradoxical and incongruent about a family systems approach based on change-resistant or change-minimizing concepts such as homeostasis, cybernetic machines, and negative feedback. Fourthly, as will become apparent, there is room for reasonable doubt about whether the homeostatic systems framework is sufficiently general to be able to incorporate thinking about the family processes of asymptomatic families, and unusually fully developed, growth-oriented, and creative families.

It should be noted here that although the focus of this paper is on the homeostatic assumption, the advocates of the conjoint therapy approach place considerable practical importance on the concepts of levels of communication, the definition and structure of relationships, self-concepts and self-esteem, *and on disequilibrating techniques*. It is when we look at families from a general theoretical and empirical point of view (and not at just the dysfunctional sector) or in terms of projecting new forms of family education and service programs that the implications of the homeostatic premise take on salience and importance.

The Homeostasis Concept in Conjoint Family Therapy

Nonetheless, it is not difficult to document the centrality of the organismic-homeostatic premise in the conceptual underpinnings of conjoint therapy. Jackson (17) and Jackson and Yalom (19) have used the phrase *family homeostasis* and speak of families as representing closed information systems. Haley (11) views family organization as "a special kind of system," and postulates that the schizophrenic symptoms of a family member serve a homeostatic function in the family. Specifically, he offers the family of the schizophrenic as a model in which the family is viewed as a cybernetic, error-activated, self-correcting system. Bateson (3), too, speaks of the family as a biosocial, feedback-governed, error-activated system. Jackson and Weakland (18) assert frankly that the basic function of the homeostatic, steady-state mechanisms of the family system is the restoration of the status quo.

Haley in 1962 reiterated his basic premise that "Any one family is a stable system." He goes on to state what he calls the First Law of Relationships: "When an organism indicates a change in relation to another, the other will act upon the first so as to diminish and modify that change." Here he emphasizes the cybernetic, deviation counteracting nature of family processes and that these processes are primarily concerned with maintaining the stability of the system. Family organization is viewed as maintaining a limited range of behavior in the service of stability and homeostasis. Similarly, Riskin (31) states his theoretical point of view as follows:

> The family is viewed as an ongoing system. It tends to maintain itself around some point of equilibrium which has been established as the family evolves. The system is a dynamic, not a static, one. There is a continuous process of input into the system, and thus a tendency for the system to be pushed away from the equilibrium point. . . . Over a period of time, the family develops certain repetitive, enduring techniques or patterns of interaction for maintaining its equilibrium when confronted by stress; this development tends to hold whether the stress is internal or external, acute or chronic, trivial or gross. These techniques, which are assumed to be characteristic for a given family, are regarded as homeostatic mechanisms.

Satir (32), too, has spoken of family homeostasis and the apparent need of families to maintain balance in relationships.

Even as recently as 1968, Lederer and Jackson espouse the family as a homeostatic, error-activated, negative feedback-governed system characterized by constant action and counteracting reaction oriented to regaining balance. "When one spouse becomes 'discordent' . . . there must be a compensatory mood or action (usually in the other spouse) or both the individuals and the system will get out of balance (p. 91)." Again, they state that once established, a system tends to remain in homeostasis.

The conjoint therapy writers who, to this writer's knowledge, come closest to departing from the narrower centralistic view of family homeostasis are Watzlawick, Beavin, and Jackson (38). Although they subscribe to Hall and Fagan's (16) "steady state" conceptualization of systems, state their preference for the concept of stability to the concept of negative feedback-mediated homeostasis, and discuss the role of positive (deviation amplifying) feedback in family growth and change, their exact attitude toward and degree of departure from the homeostasis concept is not clear. Watzlawick's illustrative reference to the furnace thermostat, the concepts of calibration and step functions, and the "implicit premise of some fundamental stability of variation," suggest a conceptual allegiance to organismic general system theory (37) that may be intrinsically untenable when applied to social, group, or interpersonal phenomena. The basic difficulty here may reduce simply to whether one chooses to emphasize homeostasis maintaining or nonhomeostatic process principles in one's conceptual approach to families in general.

The Sociocultural Systems Approach

The sociologist Buckley has recently (7, 8) reviewed the systems literature and has attempted to develop a general framework for approaching the investigation of sociocultural phenomena from a systems point of view. This conceptual approach is tantalizing both with respect to the concepts and premises which it has in common with conjoint family therapy thinking and with respect to those ideological areas, specifically the homeostasis premise, where the two frameworks part company. The reader is referred to Buckley (8) for his historical overview of systems concepts with special emphasis on social systems and to Buckley (7) for the detailed development of his "complex, adaptive, sociocultural systems" framework. Here we will limit the discussion to a rather gross comparison of similarities and differences of the two approaches, with particular emphasis on homeostasis and the possible implication of this concept for the family.

To begin with, Buckley develops his framework by trichotomizing the universe into 3 overlapping entities: the system, the system's components or constituent members, and the system's significant environment. Basically, he views a system as a complex of elements or components (for our purposes, individual family members, diads, coalitions, etc.) which are directly or indirectly related in a mutually causal network, "such that at least some of the components are related to some others in a more or less stable way *at any one time*" (Buckley, 7, p. 41). In Gestaltist fashion, Buckley subscribes to the premise of non-summativity of parts; that is, the effects of system membership on individual or system behavior are greater than a simple summation of the behavior tendencies or characteristics of the individuals comprising the system. The identifying characteristic of a system is the organizing network of these relatively causal and constraining interrelationships of the components. In the case of open social systems, these interrelationships are thought of as relatively fluid and flexible. From Buckley's point of view the overriding quality of the system's environment (physical and social) is intrinsic change, lack of constancy, and constant variation of input into the system.

Buckley uses the terms system *organization*, system *structure,* and *interrelations of components* interchangeably and synonymously. All refer to the system's primary function of mediating between both internal and external changes, or pressures to change, and the operations or behavior of the system and its members. To reiterate, the system *is* its organization or structure, and organization or structure *is* the network of relationships among the component members of the system. If a component does not causally influence and/or manifest some constraint on the range of possible alternative behaviors of another component either directly or indirectly, the two members are not related and display no organization, structure, or interaction.

It is when we turn to Buckley's conceptualization of the interrelations of components or members that we find the basic common ground between his views of social systems and the conceptual framework of conjoint family therapy. In a nutshell, he defines system structure or relationships between members in terms of interaction that is mediated by *communication*, exchange of information, and transmission of meaning. In addition to the premise of constant change and variation in the environment, Buckley assumes that one of the system's vital functions is to develop networks of operations for coping with and relating to the changeable and varied input ("variety") from the environment in order to remain viable and to survive. He calls this process a "mapping" of environmental variety. Successful mapping is a process of making available to the system and its components information about change and variation among the events, conditions, and component relationships of the environment. (For example, economic or racial changes in the neighborhood, changes in educational philosophy or teaching methods in the schools, etc.) Mapping, then, is a learning process whereby information about the meaning of changes in environmental events are internally represented, stored, subject to recall and synthesis with other information. Such information, then, becomes the basis for system and component operations both in relation to the environment and with respect to the relationships between members of the system.

Members and subsystems of social or family systems are also assumed to be characterized by : (1) diverse attributes and unique characteristics; (2) the direct receipt of environmental input; and (3) the potential for change and variation independent of impetus from the environment or other system components. Their intrasystemic relationships are assumed to be relatively flexible, unstable, and subject to change by small forces emanating from either the environment, the system, or the components themselves. Returning to intercomponent relationships (systems structure or organization), components interact, are related, or display structure only if there is communication or transfer of information or meaning between them.

Communication of information or meaning can be of several sorts. First, it can be of the type suggested by the terms signals, messages, or symbols, whereby certain acts refer to the same event, object, or condition for two or more people. From an individual's or system's standpoint this can be thought of as a match between some internal change or variation and some external change, and thereby conveys meaning. Secondly, communication can also be thought of as effecting changes in the probabilities or likelihood of certain present or future response tendencies or propensities for action. In either case, communication is assumed to have intrinsic causal *and* constraining effects on the communicating individuals and their relationship. That is, communication implicitly alters some aspect of at least one of the components and/or their relationships *and* implicitly limits the range of possible behavior or action within the relationship. To reiterate, the intercomponent relationship *is* the communicational process between the components and without some degree of this mutual causality and constraint the components do not manifest systemic structure or organization.

What we have been talking about here are essentially feedback processes, which Buckley believes are centrally characteristic of social systems. Parenthetically, because he views feedback processes as being integral social systems functions, Buckley also views social systems as purposeful, goal-oriented organizations. Unlike traditional cybernetic theorizing, and like Deutsch (10), he believes that social systems are "open" internally as well as externally. That is, social systems use as a partial basis for their operations, functioning and programming, mismatch information from both within and outside the system. Traditionally, cybernetic or negative feedback has been thought of as input information from outside the system indicating a discrepancy, incongruence, or divergence between the system's behavior and some preprogrammed environmental goal state. This information, or error data, about the effects of the system's behavior or actions acts as input data which the system uses as a basis for altering its operations and reducing its divergence from the internally represented environmental goal, or the system's value hierarchy. This type of information is Deutsch's *primary messages.* In addition, however, the social system or the family receives feedback information about the status or condition of its members, components, subsystems, and their relationships, which again signal incongruence or disparity between component states and the system's criteria values. This internal feedback, or Deutsch's *secondary messages* (a form of systemic "self-awareness"), is again used as input on the basis of which the system may alter its operations, its organization, or its criteria or goal values.

Buckley, following Maruyama (23), further categorizes feedback into *negative* (error-activated, deviations counteracting) and *positive* (also deviation-activated, but deviation *amplifying*) feedback processes. It is the latter which will lead us to the area of divergence between conjoint family thinking and the social systems framework. The primary difference between positive and negative feedback process is the nature of the systemic changes following the input of internal or external error or deviation information. The negative feedback process, or morphostatic function, is essentially that described above as the traditional cybernetic type of feedback. The main point is that, after receiving mismatch information, deviation counteracting operations are triggered so as to bring the individual's or system's behavior or status back into congruence or convergence with the extant internal standards or the system's governing criterion values. For example, parents of a conservative political persuasion might respond to their adolescent son's socialistic or militant views with prolonged and impassioned discourse on the benefits of the free enterprise system and the evils of anarchy. The interchange might contain sufficient subtle but intense personal acceptance messages to force the young person to accept his parents' view in order to insure their acceptance and support. The negative feedback process (consisting of receipt of information about system behavior, intrasystem transmission of this data, comparison with the internal standard or criterion, and error-correcting effector operations) then is basically a "sameness," status quo, homeostasis-maintaining, and a change-resistant set of operations.

The positive feedback or deviation amplifying process also begins with error or mismatch information resulting from a comparison of data about behavior with internal standards or criteria. The difference is that the subsequent effector operations do not act to reduce the discrepancy but rather act to increase the divergence between the system's or member's status and the original goal or standard values. The most frequently cited examples of positive feedback process are economic inflationary spirals, evolutionary genetic mutations, and the development of culturally labeled deviant subgroups such as gangs of delinquents (see Wilkins, 39, for a model of drug addiction based on the deviation amplifying concept). An example involving the family might be the reaction of parents with strong negative feelings about associating with minority group members to their grade school daughter's friendship with a minority child. Such parents saying, "We cannot feel comfortable with your friend, but we are very pleased that you are and hope that you do not develop the hang-up we have about minority people" would be an example of deviation amplification within the family system.

It should be pointed out that both Buckley and Maruyama, among others, view positive feedback process as being constructive, as system enhancing, and as centrally contributing to the maintenance of system viability, as well as having potentially destructive outcomes. Buckley believes that the positive feedback processes are *the* vehicles by which social systems "grow," create, and innovate, and consequently views them as *morphogenic* processes (literally, form- or structure-changing processes).

In the context of the assumption of intrinsic environmental and system component change and variation, Buckley flatly postulates that social systems *must* be capable of morphogenesis or of changing their basic structure, organization, and values in order to survive and remain viable. In contrast to the organismic-homeostasis criterion of system survival, Buckley believes that social systems must be "ultra-stable" (Caldwallader, 9); that is, must be capable of enduring basic changes of structures (the preschooler entering kindergarten, the adolescent getting married, changes in religious beliefs, etc.). Furthermore, he belives that the impetus for such change can come from within the system and its components (as, for example, might occur by spontaneous or intrinsic basic change in the state of a component), as well as from the environment (as in the sense of adaptation to environmental forces and pressures). Morphogenesis can take the form of change in inter-component relations (greater autonomy and decreased constraint); qualitative changes in the nature of relationships (such as between a sexually maturing son and his mother); change of governing values, purposes, and criterion standards (moral values, occupational goals); basic changes in internal and external effector operations (the increased earning power of the adolescent); and the ascendence of components or subsystems with different properties and attributes in the governing or management of the system (such as graduate student offspring still living at home).

A corollary of the morphogenic or positive feedback principle is that the organization of social systems tends to increase in *complexity* and flexibility with increased viability, variability or change within the system, and/or variability or change in environmental events or conditions. *Viability* is used here in the sense of the definition provided by Webster's Seventh New Collegiate Dictionary: capable of living, capable of growing or developing; workable. Two characteristics of complexity of system structure are: an increase in the autonomy of the components and an increase in the extent to which the system or its components determines system behavior, rather than being primarily reactive to input from the environment.

Thus, complexity of inter-component relationships does *not* necessarily imply structural rigidity and maximal constraint in relationships. As Pringle (30) has suggested, degree of constraint or order within the system and system conplexity are seen as essentially *independent* variables. It would seem tenable to postulate an inverse relationship between structural rigidity and complexity in social systems.

Buckley also subscribes to Deutsch's view that the feedback processes of social systems are governed to a large extent by "efficient causes"; that is, immediate, current, "here and now" inputs, conditions, and variations, rather than more abstract and/or temporally remote purposes, goal states, and objectives.

Another concept integrally related to the idea of positive feedback and morphogenesis, suggested by both Buckley (8) and Cadwallader (9), is the necessity of a constant flow and backlog of "variety"; a wide range and quantity of varied information, experience, and input into the system. Such a family would be willing to hear and at least consider a wide range of, for example, political or moral views even though such conflicted with their own values. Cadwallader postulated that the capacity for systemic innovation, and presumably morphogenesis, cannot exceed the quantity and variety of information available to the system. He also postulated that creative, viable systems must contain a mechanism for preventing the "freezing" or locking in of old or currently extant operations and programs.

It should be made explicit to the reader, in light of the above, that Buckley does *not* completely reject the concept and function of negative feedback and self-correcting processes in social systems. He views social systems as having both morphostatic, self-correcting processes *and* morphogenic, self-directing properties. He does reject homeostasis maintaining, negative feedback processes, however, as sufficient for the survival of social systems. Buckley prefers the more general term *viability* to homeostasis, as the governing criterion principle of social systems.

Equilibrial, Homeostatic, and Social Systems

Let us now briefly contrast the adaptive social systems model to the two major earlier systems models. *Equilibrial*, or mechanical systems, are exemplified by the inorganic chemical and mechanical systems which constitute the domain of investigation for physicists and engineers. Such systems, by definition, are relatively closed and entropic (tending toward a random or equal distribution of energy and an absence of structure or organization). These systems are affected only by external forces or influences and have no internal sources of change other than entropy. Their components tend to be simple in their own organization and are related to each other by transfer of energy (rather than information). Because they tend to be relatively closed systems, they have no feedback processes, or other self-regulating or adaptive mechanisms beyond the basic tendency toward a state of entropy.

Homeostatic or organismic systems on the other hand are exemplified by the living organisms constituting the subject matter of biology. These systems are open and negentropic (tending toward the generation and specialized structuring of energy). For our purposes, the principle characteristic of organic systems is maintaining the energy level within controlled limits and the structure or behavior of the system within pre-established limits. Although many such systems possess feedback loops with the environment, and possibly information as well as energy exchange with the environment, they are oriented primarily to self-regulation (morphostasis) rather than adaptation (morphogenesis).

Complex, adaptive systems (species, psychological, and sociocultural systems) are also open and negentropic, but they are open internally as well as externally. That is, the interaction of the components may result in significant changes in the nature of components themselves and as a consequence the system as a whole. Energy levels are subject to wide fluctuations and the limits on variation of central or vital system parameters are fluid, variable, and generally broad. Internal and external interaction are mediated by information transfer. Feedback control loops exist which make possible, not only self-regulation, but self-direction, internally instigated change and elaboration of basic systemic structure. Alteration of structure, or interrelations of components, is assumed to be a necessary condition of system survival or viability.

Thus, while equilibrium is the fundamental principle of inorganic, chemical, and mechanical systems, and homeostasis is the basic principle of lower and higher biological and organismic systems, viability with the implication of inherent capacities for growth and self-directed change is the criterion principle for social systems.

Discussion

To recapitulate, sociocultural systems are assumed to be organized, internally and externally "open," purposeful, self-regulating, *and* self-directing systems which are dependent on communication or information or information exchange as their basic process and which are governed by the principle of maximizing viability. A maximally viable social system is characterized by complex structural relationships (communication and interaction) between its components and subsystems, highly flexible organization, a minimum of rigid constraint in inter-component relationships, highly autonomous components, and considerable intra-system determinism and causality of system and component behavior. To become maximally viable a system must be ultra-stable, must possess effective morphogenic or positive feedback processes, and must have available to it a constant flow of varied and novel information input. Specifically, the social systems framework rejects the conjoint family therapy premise that homeostasis-maintaining morphostatic, negative feedback and efficient, self-regulating processes are sufficient explanatory principles for social phenomena.

Although not generated specifically by the sociocultural systems approach, the findings of several studies are pertinent to the postulated relationship between family system viability and flexibility, autonomy, and lack of rigid constraint in the interactional structure of the system. The pertinence of this evidence is dependent upon accepting the absence of significant psychiatric or behavioral symptoms as a gross, operational definition of viability.

In a study of the relationship between family disturbance and degree of repetitive interactional patterning, Haley (13) defined family organization in terms of limitations and constraints on flexible, spontaneous, minimally structured interaction among the members. He postulated that disturbed (symptomatic) families would display more rigid patterning and less flexible or quasi random interaction on the "who speaks after whom" variable than would normal (asymptomatic) family triads. This writer is aware of five studies in which this prediction was investigated (Haley, 13; Haley, 14; Winter and Ferreira, 40; Haley, 15; and Murrell and Stackowiak, 28). Although only two of the studies (Haley, 13, and Winter and Ferreira, 40) found reliable functional-dysfunctional family differences, four of the five studies (Murrell and Stackowiak, 28, being the exception) reported data indicating that normal and abnormal families nonetheless rank-ordered themselves in the predicted fashion with asymptomatic families displaying more equal participation in family discussions, less rigid patterning of interaction, and thus less order and limiting organization than did the symptomatic families.

Contrary to their expectations, Mishler and Waxler's (27) study of schizophrenic and normal family interaction revealed that asymptomatic families displayed markedly and reliably *more* direct and indirect disruptions of communication, and fragmentation and disturbances of speech patterns (intrusive laughter, sarcasm, joking, pausing, repetitions, and incomplete sentences) than did families with a schizophrenic male offspring. The investigators interpreted these findings as reflecting the general normal family flexibility and spontaneity which permits a greater freedom for members to insert and interject variation and potentially change relevant input into the family problem-solving process than is available to families with a schizophrenic offspring. Their findings that normal families also surpassed the symptomatic families on amount of expressiveness and "responsiveness" to others (acknowledgment that the previous speaker had been heard) indicates that considerable communication and transfer of information occurred in spite of these disruptions and intrusions, thus supporting the presumption that these asymptomatic families represented relatively viable systems.

Similarly, Minuchin et al. (26), in their study of multiple delinquency families, found that control mothers (from families without a delinquent member but matched on several SES variables) displayed considerably *more* disruptive verbal behavior in a family task situation than did mothers of delinquent children. This finding again fits with the greater flexibility and lesser rigid constraint and order postulated by the viability principle. Becker and Iwakami (4) have also recently reported finding reliably higher incidences of simultaneous speech, interruptions, disagreements, and aggressive verbalizations in the problem-solving interaction of *nonclinic* families than in families attending a child psychology clinic.

Although the sociocultural systems framework departs from the conjoint family therapy approach on the sufficiency and centrality of homeostasis as an organizing principle, there are those within the family therapy field who have also asserted a need for new concepts of family change and organization. Although still viewing the family as a self-regulating, homeostasis-maintaining "organism," Brody (5) has stated his belief that we need new concepts of family growth, self-evolution, and "restructuring." In his application of cybernetic principles to family systems, he emphasizes the necessity of families being able to be constructively responsive to change, able to change with change, and capable of learning to give up "obsolescent constraints." He believes that this is necessary for the maintenance of family stability in a society undergoing rapid social and technological change. One of the challenges confronting parents today, he believes, is teaching children to be able to cope with unthought-of, and as a result unprogrammable, situational and environmental changes. Brody views families which are unresponsive to change as closed, maximally constrained systems with limited evolutionary power. In such families, the members or components grow to fit predetermined or preprogrammed norms, behavior patterns, roles, and abstractions; change cannot occur except in predetermined ways which, by definition, are insensitive to new, novel, and unanticipated situational circumstances and environmental input.

Apropos of the hypothesized need for a wide range of varied and novel information and input, Brody (6, p. 101) says:

> We need to learn new ways of learning, of developing our senses to take in new information. The old information we can leave to the machines.
> The animal that perceives as much as possible of what exists around him, and can discard by choice what is of no interest to him, has in this capacity for choice an advantage over other creatures whose design for living remains more circumspect. This is the value system I propose.
> Then a family teaches contact with what exists within it and around it and outside it — and thus acknowledges the larger field of choice — broadens its fields of growth by teaching discovery.

Sonne (34) states his views about family homeostasis and growth rather succinctly. He believes that any psychic or social system "which is not growing is as if dead." This author believes that family systems at equilibrium or homeostasis are not growing, have no room for differentiation over time, and have no room for change. "Only sick or disturbed families are in equilibrium or homeostasis." This view is consonant with a conceptualization of functional and dysfunctional families offered by Miller (25). Miller suggests that the end points on the family functionality-dysfunctionality continuum be defined in terms of morphostatic and homeostatic processes at the dysfunctional end, and in terms of morphogenic or growth processes at the functional terminal.

Thus, what are some of the implications of the morphogenesis-viability concept for how we view and approach families? First and foremost has to do with the basic assumptions we make about what we would like to see families be like, and the related ideas we hold about what normal, effectively functioning, fully developed families are like. Theoretically, there would appear to be quite a difference between approaching troubled families with the assumption that their means of maintaining homeostasis and balance simply need improving in order to mitigate their difficulties, and the assumption that the fact that they are attempting to maintain homeostasis may be central to their difficulties. That is, family operations and processes oriented primarily to the goal state of maintaining homeostasis and minimizing changes in basic relationships and values may prove to be one form of system dysfunction and nonviability.

Related to this is the sad fact that we know almost nothing about the satisfaction, closeness, meaning-achieving, autonomy, problem-solving, communication, change, and basic relationship-organizing processes of exceptionally well-functionaing, broadly and deeply satisfied, fulfilled families. To this writer's knowledge there is no evidence available to indicate whether the fundamental system operations of such families are primarily morphostatic or morphogenic in nature. Following Miller, it may be that more knowledge about the relationship processes of these families will provide much needed evidence apropos of morphostasis vs. morphogenesis, or homeostasis vs. viability.

Another major implication is that of workers in the mental health field addressing themselves to the basic role and function of change in relation to the conditions for growth and development, and the general approach to life that we propose both to our clients and to society. We are sorely in need of new, daring, and creative clinical and educational approaches and techniques that will mitigate the resistance to and fear of change on the part of the public and social institutions. If we set higher goals for ourselves than just attempting to improve the self-regulating operations and communication patterns of troubled families and relationships, then it seems appropriate to address ourselves not only to our service-requesting clientele but to other social systems, such as the schools and the church, which influence people's assumptions and expectations about what family life and interpersonal relationships "should" or "ought" to be like.

If we can accept an isomorphism between the constant flow and backlog of "variety" or varied input information presumably necessary to maximal viability, and the cumulative and varying human *experiences* of families and their members suggested by Brody, another implication is suggested. That is, serious consideration might be given to greater preventive, pretreatment, and/or treatment use of experientially oriented primary and secondary feedback techniques such as those currently being used in encounter, sensitivity, and communication training groups and laboratories. To this writer's knowledge the potential for feedback "variety" of these awareness-of-self and other techniques has been little explored clinically. The usual basic assumption of such growth or "development of human potential" practitioners is that most people, asymptomatic as well as dysfunctional, have closed themselves off from large segments of basic human experiences, have attempted to minimize experiential "variety," and, as a result, have contributed substantially to their own human dilemmas and have diminished their own and their social and family systems viability (e.g., Bach, 1 and 2; Jourard, 20; Maslow, 22; Perls et al., 29; Schultz, 33; May, 24; and Stoller, 35). Although there has been little written about or done to ascertain the potentially harmful effects of such groups and techniques, the writer and his colleagues have been singularly impressed by the impact that such techniques have on rigidified and "locked in," obsolescent, self-defeating, and interpersonally alienating rules, operations, and communication patterns among married couples and parents. The least that can be said for encounter and sensitivity training groups is that they usually provided participants with a much broader range of "variety" and varied basic human experiences than most have allowed themselves previously.

While on this subject, the writer has been impressed with the sterility, superficiality, and unproductiveness of orthodox, didactic, and bibliographic approaches to family life education and premarital counseling programs. Although there is nothing new about a plea for more meaningful and practical approaches, the combination of the availability of new experiences and awareness-expanding techniques (sources of variety) and a somewhat different view of social and family systems offer the opportunity for innovative and creative preventive work in the family area. Although there is no plethora of research or documented experimentation in this area, Miller and Nunnally at the University of Minnesota have research underway in which they are investigating the effects of a family developmental communication training program on the communication and interpersonal behavior patterns of engaged couples. Obviously this is a much needed and very important area of endeavor.

One implication for clinical management and treatment planning will be a more judicious concern about subsequent "symptom substitution" generated by the homeostasis, equilibrium or balance-maintenance assumption, if the whole family system is not treated. Combining the three premises of (1) greater intrinsic family-system flexibility and autonomy in the relationships among family members than previously assumed, (2) positive family change and reorganization stimulated by influences from within the family, and (3) family members being subject to considerable influence by simultaneous membership in nonfamily systems, it is possible to view symptoms as not necessarily family-system generated, nor requiring total family or even symptom-bearer involvement in treatment. Examples might be: a grade school youngster who has to spend thirty hours a week with a hostile, dysfunctional teacher; a preschooler reacting to the power-struggle propensities or double-level messages of his nursery school teacher; a teenage girl struggling with rigid and punitive school policies and rules bearing on her unique wishes and characteristics; a teenage boy accepting the valid and true social grievances of his activist peers; and a teenager being subject to hostile shaming and depreciation by a juvenile court judge or probation officer. If we assume a perspective greater than the dimensions of our consulting rooms, we might consider investing our efforts in attempts to change the teachers, assistant principals, judges, etc., and the systems of which they are members.

Another possibliity suggested by the morphogenesis concept is that measures of past and present change of attitudes, behavior, family rules, and family operations, and/or measures of receptivity or reaction to discrepant ideas, facts, information, and input (along the lines of the Revealed Differences Technique, or Ferreira and Winter's family questionnaire) might be potentially useful prognostic, response to treatment, or research measures.

References

1. Bach, G. R. "The Marathon Group: Intensive Practice of Intimate Interaction." *Psychological Reports*, 18: 995-1002, 1966.

2. Bach, G. R. and Wyden, P. *The Intimate Enemy.* New York, Morrow, 1969.

3. Bateson, G. "The Biosocial Integration of Behavior in the Schizophrenic Family," in Ackerman, N. W.; Beatman, F. L.; and Sherman. S. N., eds., *Exploring the Base for Family Therapy.* New York, Family Service Association of America, 1961.

4. Becker, J., and Iwakami, E. "Conflict and Dominance Within Families of Disturbed Children." *J. Abnorm. Psychol.,* 74: 330-35, 1969.

5. Brody, W. M. "A Cybernetic Approach to Family Therapy," in Zuk, G. H., and Boszormenyi-Nagy, I., eds., *Family Therapy and Disturbed Families.* Palo Alto, Science and Behavior Books, 1967.

6. Brody, W. M. *Changing the Family.* New York, Clarkson M. Potter, 1968.

7. Buckley, W. *Sociology and Modern Systems Theory.* Englewood Cliffs, N.J., Prentice Hall, 1967.

8. Buckley, W., ed., *Modern Systems Research for the Behavioral Scientist.* Chicago, Aldine, 1968.

9. Cadwallader, M. "The Cybernetic Analysis of Change in Complex Social Organizations." *Am. J. Sociol.,* 65: 154-57, 1959.

10. Deutsch, K. W. "Some Notes on Research on the Role of Models in the Natural and Social Sciences." *Synthese,* 7: 506-533, 1948/49.

11. Haley, J. "The Family of the Schizophrenic: A Model System." *J. Nerv. Ment. Dis.,* 129: 357-74, 1959.

12. Haley, J. "Family Experiments: A New Type of Experimentation." *Fam. Proc.,* 1: 265-93, 1962.

13. Haley, J. "Research on Family Patterns: An Instrument Measurement." *Fam. Proc.,* 3: 41-65, 1964.

14. Haley, J. "Speech Sequences of Normal and Abnormal Families with Two Children Present." *Fam. Proc.,* 6: 81-97, 1967.

15. Haley, J. "Experiment with Abnormal Families: Testing Done in a Restricted Communication Setting." *Arch. Gen. Psychiat.,* 17: 53-63, 1967.

16. Hall, A. D., and Fagan, R. E. "Definition of System," in Hall, A. D., and Fagan, R. E., eds., *General Systems, I.* New York, Bell Telephone Laboratories, 1956.

17. Jackson, D. D. "The Question of Family Homeostasis." *Psychiat. Quart. (Suppl.),* 31: 79-90, 1957.

18. Jackson, D. D., and Weakland, J. H. "Conjoint Family Therapy: Some Considerations on Theory, Technique, and Results." *Psychiatry,* 24 (Suppl. to no. 2): 30-45, 1961.

19. Jackson, D. D., and Yalom, E. "Conjoint Family Therapy as an Aid to Intensive Psychotherapy," in Benton, A., ed., *Modern Psychotherapeutic Practice,* Palo Alto, Science and Behavior Books, 1965.

20. Jourard, S. M. *The Transparent Self.* Princeton, N.J., Van Nostrand, 1964.

21. Lederer, W. J., and Jackson, D. D. *The Mirages of Marriage.* New York, Norton, 1968.

22. Maslow, A. H. *Toward a Psychology of Being.* Princeton, N.J., Van Nostrand, 1968.

23. Maruyama, M. "The Second Cybernetics: Deviation-Amplifying Mutual Causal Processes." *American Scientist,* 51: 164, 1963.

24. May, R. *Psychology and the Human Dilemma,* Princeton, N.J., Van Nostrand, 1967.

25. Miller, S. L. "Family Crisis, Intervention, Adjustment, and Growth." Unpublished paper, Sociology Department, University of Minnesota, 1969.

26. Minuchin, S.; Montalvo, B.; Guerney, B. G., Jr.; Rosman, B. L.; and Shumer, F. *Families of the Slums.* New York: Basic Books, 1967.

27. Mishler, E. G., and Waxler, N. E. *Interaction in Families.* New York, Wiley, 1968.

28. Murrell, S. A., and Stachowiak, J. G. "Consistency, Rigidity, and Power in the Interaction Patterns of Clinic and Non-Clinic Families." *J. Abnorm. Psychol.,* 72: 265-72, 1967.

29. Perls, F.; Hefferline, R.; and Goodman, P. *Gestalt Therapy.* New York, Dell, 1964.

30. Pringle, J. W. S. "On the Parallel Between Learning and Evolution." *Behaviour,* 3: 174-215, 1951.

31. Riskin, J. "Methodology for Studying Family Interaction." *Arch. Gen. Psychiat.,* 8: 343-48, 1963.

32. Satir, V. *Conjoint Family Therapy.* Palo Alto, Science and Behavior Books, 1964.

33. Schutz, W. C. *Joy.* New York, Grove Press, 1967.

34. Sonne, J. C. "Entropy and Family Therapy: Speculations on Psychic Energy, Thermodynamics, and Family Interpsychic Communication," in Zuk, G. H., and Boszormenyi-Nagy, I., eds. *Family Therapy and Disturbed Families.* Palo Alto, Science and Behavior Books, 1967.

35. Stoller, F. H. "Accelerated Interaction: A Time-Limited Approach Based on the Brief, Intensive Group." *Int. J. Group Psychother.,* 18: 220-35, 1968.

36. Vickers, G. "Is Adaptability Enough?" *Behav. Sci.,* 4: 219-34, 1959.

37. Von Bertalanffy, L. *General Systems Theory.* New York, George Braziller, 1968.

38. Watzlawick, P.; Beavin, J. H.; and Jackson. D. D. *Pragmatics of Human Communication.* New York, Norton, 1967.

39. Wilkins, L. F. "A Behavioral Theory of Drug Taking." *Howard Journal,* 11: 6, 1965.

40. Winter, W. D., and Ferreira, A. J. "Interaction Process Analysis of Family Decision-Making." *Fam. Proc.,* 6: 155-72, 1967.

CHAPTER 29

The Resolution of Role Conflict Within the Family

JOHN P. SPIEGEL

In an investigation of the relations among cultural value conflict, family conflict, and the emotional adjustment of the individual, in which I am participating with Florence R. Kluckhohn and a number of co-workers, the concept of social role is being used to observe and analyze the details of behavior which is functional or dysfunctional for the family as a whole. The social role concept is useful for this purpose because it facilitates observation of the way the individual members of the family become involved in the family as a superordinate system of behavior. (See, for example, Ackerman, 1951; Ackerman and Sobel, 1950; Parsons and Bales, 1955; Pollak, 1952; Spiegel, 1954.) It helps to describe not only the interaction of two members as they adjust to each other, but also the transactions of a plurality of members as they interweave in the special type of compulsiveness or control which a going system always imposes on its members. (See Bentley, 1950; Dewey and Bentley, 1949; Kluckhohn and Spiegel, 1954; Spiegel, 1956.) Since the uniquely compulsive elements of the family system leave a characteristic stamp upon the personality development of the child, it is important to have a way of tearing apart the rather subtle elements of which it is composed.

In studying a group of families of emotionally disturbed children and comparing them with families in which the children are free of clinically manifest disturbance, we have found evidence of what promises to be a consistent difference between the two groups. In the first group, the children inevitably become involved in a conflict or disequilibrium situation which exists between the parents. Most frequently neither the child nor the parents are aware of this fact, nor are they aware of the ways in which it comes about. In the second group of families, although there may be sources of tension between the parents, the children are minimally involved in it. In order to avoid excessive variability in our two sets of families, we have kept them similar with respect to size, ethnic, regional, and class variables. Nevertheless, the sources of tension can be related in every case to differences and incompatibilities in cultural value orientations and, as a corollary, in definitions of social role expectations. These incompatibilities have a pronounced bearing upon the object relations and unconscious psychodynamics of the transacting members of the family. However, this is not the place to deal with the origin of the cultural value conflict or its direct relation to the intrapsychic process. These connections will be reported in subsequent communications. In this paper they will be assumed to underlie the role conflict in the family, and our attention will be centered rather on the ways in which the role conflict[1] is handled.

While we were studying the ways in which parents unwittingly involve one or more children in their own conflicts, it became clear that this process, so ably reported by Adelaide Johnson and her co-workers (1952, 1953), could be described in the usual psychodynamic terms. Through identification with the unconscious wishes of the parent the child acts out the parent's unconscious emotional conflict. The acting-out serves as a "defense" for the parent, making it unnecessary for him to face his own conflicts. This vocabulary is adequate for most purposes, and besides, confers a kind of credibility upon the description because of long usage and ready acceptance in the mind of the user. Nevertheless, it left us unsatisfied. Even with the qualifications of the term "unconscious," the description sounds too planned, too much under the control of one or more individuals. A constant observer of the family — or of any other persistent group process — has a somewhat contrary impression that much of what occurs in the way of behavior is not under the control of any individual or even set of individuals, but is rather the upshot of complicated processes beyond the ken of anyone involved. Something in the group process itself takes over as a steering mechanism, and brings about results which no one anticipates or wants, consciously or unconsciously. Or the steering mechanism may bring about a completely unexpected pleasant effect. On the basis of numerous observations, we were struck with the fact that so

[1]The expression "role conflict" has been used in two different ways. In the first, and perhaps more common, usage it refers to a situation in which ego is involved in a difficult or impossible choice between two different roles toward two different alters. No matter what decision he makes, he is in trouble with one or the other of his role partners in the situation. In the second usage, ego and alter have conflicting or incompatible notions of how to play their reciprocal role. The conflict is not over which of several possible roles to take, but rather how to enact the role they have both decided to take. It is the second definition which is used in this paper. Settlement of the terminology problem should not prove too difficult, but will have to be postponed for the present.

often what is functional for one member of the family group may be dysfunctional for the family as a whole. The opposite also holds: What is functional for the family as a whole may have very harmful effects on an individual. Then phenomena take place unwittingly not only because of the unconscious dynamics within the individual but also because of the operations of the system of relations in which the members of the family are involved.

To describe the characteristics of a system of relations within a group accurately over a considerable span of time is no small task. The most successful attempt to do this known to us is the method of interaction process analysis devised by Parsons and Bales (Chap. V, 1953). However, the categories of interaction used by these workers are at too high a level of abstraction for our purposes. We decided, therefore, to use their basic concepts of behavior occurring within role systems of ego and alter, or any number of alters, but to devise our own set of categories for observing the roles involved. Thus the basic concept used in analyzing the family as a system consists of describing the behavior of an individual in terms of his role in transaction with a role partner or partners. A role is defined as a goal-directed pattern or sequence of acts tailored by the cultural process for the transactions a person may carry out in a social group or situation. It is conceived that no role exists in isolation but is always patterned to gear in with the complementary or reciprocal role of a role partner (alter). Thus all roles have to be learned by the individuals who wish to occupy them in accordance with the cultural (or subcultural) values of the society in which they exist. If that society is fairly homogeneous and well integrated, then the roles will be patterned in such a way that their complementary structure is obvious and stable.

The roles pertinent to the family as a system consist of husband and wife, mother and father, son and daughter, brother and sister. This is not an exhaustive list, and refers to the nuclear rather than the extended family. But if one compares these roles on any axis of variation, such as ethnic or class affiliation, it is apparent that they are defined differently, and their complementary structure varies according to the particular mode of family organization characteristic of that class or ethnic group. It is true that even within a class or ethnic group there is considerable variation of pattern. Nevertheless, one mode of organization tends to be typical or dominant compared to the others. For example, an American middle-class wife tends to expect her husband to treat her as an equal. She expects of her husband a good deal of independence, initiative, and planning for future success in his occupation, but in his relations with her and with the children, she expects cooperation, sharing of responsibility, and individual consideration. Reciprocally, the husband expects his wife to help in his plans for future economic and social success, notably, by putting his success goals above any personal career or occupational goals of her own, and by developing the social and domestic skills suitable to his particular occupational status. There is evidence that these complementary role expectations may not be precisely reciprocal — that is, there may be some built-in strain — but on the whole they fit with each other fairly well.

By way of comparison, it is illuminating to select some of the complementary role patterns in the lower-class Italian family. Here the wife has no wish that her husband spend a great deal of time thinking or planning about occupational success. She expects her husband to work steadily and do his best to bring in enough money to satisfy the needs of the family, but she doesn't expect his economic or social status to change. She expects rather to have a large number of children who will soon join the husband in trying to increase the economic intake of the family, but in the meantime there is always help to be expected from relatives and friends if there is real need. On the other hand, she does expect him to spend a lot of time keeping up contacts with the extended and complicated networks of relatives and friends in order to keep their own position secure. At home she doesn't expect to be treated as an equal. Rather she expects to be relieved of responsibility through his making the chief decisions so that she can tend to the needs of the large brood of children. For his part, the husband expects submission but also a good deal of nurturant care from his wife. He wants her to be chiefly concerned with his children. Everything else is secondary. For both of them there is an accent on enjoyment and a sense of festivity in family life which is of greater importance than hard work and planning for the sake of social ambition. Again, although definite strains can be noted here and there, the roles of the family members vis-a-vis each other are characterized by a complementarity of expectations which fit each other in fairly smooth and systematic ways.

The Equilibrium-Disequilibrium Balance

I hope these all too brief examples of contrasting husband-wife role patterns illustrate how complementarity can be maintained in spite of variation in goals, values, and concrete sequences of acts within the role systems. The principle of complementarity is of the greatest significance because it is chiefly responsible for that degree of harmony and stability which occurs in interpersonal relations. Because so many of the roles in which any of us are involved are triggered off by cultural cues in a completely complementary fashion, we tend not to be aware of them. We enact them automatically, and all goes well. This automatic function of role systems has significance for psychological economy of effort. We are spared the necessity of coming to decisions about most of the acts we perform because we know our parts so well. This saves our efforts for those acts which occur in less stabilized role systems. In this way role reciprocity confers spontaneity upon human behavior. Self-consciousness and self-guarding enter the scene along with role conflict which sharply raises the number of decisions which have to be made with respect to any sequence of acts. As long as complementarity is maintained at high levels of equilibrium,[2] decisions are decentralized, so to speak. They are taken care of by the system of role relations rather than by the individual acting in a self-conscious manner.

However, it is a part of the human condition that high levels of equilibrium figured by precise complementarity of roles are seldom maintained for long. Sooner or later disharmony enters the picture. Complementarity fails; the role systems characterizing the interpersonal relations move toward disequilibrium. The role partners disappoint each other's expectations. The failure of complementarity feeds back into the awareness of the participants in the form of tension, anxiety or hostility, and self-consciousness. If the process continues without change, it will end in the disruption of the system. This process is so familiar and inevitable that it seems to merit no further comment. Yet, it has appeared to us that it may contain some general elements which can throw light on family behavior, if it were to be subjected to critical scrutiny. The key to its analysis would consist of a study of the conditions leading to the breakdown of complementarity and to its subsequent restoration. Although this study has not been carried as far as I would like, our current experience indicates that there are at least five causes for failure of complementarity in role systems within the family. Because of limitations of time and space, I will review them here very briefly, without the extended discussion and illustration which they deserve.

Cognitive Discrepancy

One or both individuals involved in the role system may not know or have sufficient familiarity with the required roles. This is especially likely to occur with respect to age roles, and therefore frequently characterizes sources of disequilibrium between parents and children. When the pattern of acts constituting the role is not clearly mastered or not cognitively mapped or internalized, complementarity can be maintained only with difficulty. Cues are misinterpreted, and misunderstanding reduces complementarity of expectations. Both participants must have a relatively high tolerance of frustration and failure, and both must alternatively

[2]In this context "equilibrium" does not denote a rigid, static state, but rather a balancing of process in a moving or changing state. The phrase "moving equilibrium" might, perhaps, be a better name.

assume informally the roles of teacher and learner alternately. This alternation and reversal of roles will be discussed later in connection with the mechanisms of restoration of complementarity. In our culture cognitive discrepancy is a characteristic problem between adolescents and the adult world. It also occurs between husband and wife at various developmental crises, or with respect to any sudden, new situation. For example, the wedding and immediate postnuptial situation requires much new learning of roles. So does the birth of the first child, the first severe illness, and so forth.

Discrepancy of Goals

Roles are patterns of acts directed toward immediate or ultimate goals. The goal of ego, interlocking with the goal of alter, determines the motivational principle behind the individual's taking of the role. Some goals serve the purpose of gratification, while others are chosen for the sake of defense. The same goal may serve either purpose, but if there is a shift in motivation, there is usually a shift in the definition of the role. For example, in one of our "sick" Italian families an eleven-year-old daughter, the middle one of three girls, repeatedly made demands upon her father for gifts of all sorts. Her motive was originally desire for gratification, but it was mixed with a defensive need to test whether she were being rejected or not. At first the father gratified her demands intermittently and inconsistently. He gave when he felt like it and at other times refused. Both giving and refusing represented satisfactions for him, and he included rewarding and withholding as legitimate goals in his conception of the father's role. However, the daughter gradually defined his withholding as confirmation of her fear of rejection and tested more intensively by increasing her demands. The father defined this as "pestering" and responded with increased withholding and disapproval while claiming that he was trying his best to satisfy her. This claim was not true since he consistently rewarded the older sister more than this middle girl. But now the goal of withholding had become defensive against the implied meaning of her demands — that he actually preferred the older sister. In this complicated transaction, the defense was accomplished on the father's side through defining the daughter's motivation as coercive and pinning this down in the informal role "pest," while giving himself the informal role of "victim." Although a tenuous complementarity was maintained by the defensive establishment of the informal "pest-victim" relation, actually their goals became more and more discrepant. This discrepancy of goals was one of the chief reasons why the family brought the girl into the psychiatric clinic for treatment. The parents verbalized the failure of complementarity by characterizing the girl to our interviewers as a bad and disobedient daughter. They had tried their best to teach her "right" from "wrong" but she was unable to "learn." It is significant of the defensive problem in this family that her behavior was ascribed to a cognitive and value discrepancy — that she couldn't "learn" the correct behavior — when actually it was due to a motivational problem concerning unavowed goals.

Another source of discrepancy in goals is biologically determined, rather than of motivational origin. Fatigue, illness, and lack of maturation are accompanied by a *restricted capacity for goal attainment.* Other biological limitations such as deficiency of intelligence have the same effect. Such limitations produce disequilibrium when one of the role partners is unable to accommodate through a change in level of expectancy of goals as rewards, for example, the parent who can't accept the limited intelligence of his child.

Allocative Discrepancy

In any particular social situation there is a question of the individual's right to the role he wishes to occupy. There are four principal ways in which roles are sorted out among those who contend for them.

(1) Some roles, such as age and sex roles, are *ascribed* (Linton, 1936). This means that they are universally expected and the individual has practically no leeway: he is not free to decide to change his sex or age role. If a man tries to change his sex role, as in transvestism, he is likely to invoke intense criticism. The same is true, through to a lesser extent, of age roles. The child who tries to act like an adult usually produces a critical response, and the same thing holds for the reverse situation.

(2) Some roles, such as occupational and some domestic roles, have to be *achieved* (Linton, 1936). As an allocative principle, achievement involves effort, the satisfaction of prerequisites, and some form of ceremonial recognition such as licensure, contract, conferring of a diploma, appointment, and so forth. There is more leeway than in the case of ascribed roles, but strong sanctions will be invoked if an achieved role is simply taken without observing the required formalities.

(3) Some roles, in the main of an informal character, can be taken simply through *adoption.* No one has to ask permission to take an adopted role, although there may not always be approval for it. For example, the father in the Italian family just discussed adopted the role of "victim." He could have responded to his daughter's demands with some other role activity. He could have treated them as childish antics and laughed them off in the role of amused "spectator." This was actually a tack he frequently took when his feelings were not so intensely involved. By adopting the role of "victim," however, he *assigned* her the complementary role of "pest." The assignment was implicit rather than explicit. This is to say that it was

concealed or masked, and that on the whole he treated her as if she had spontaneously adopted the role of pest toward him. Thus adoption-assignment describes for role transactions what is denoted for the individual by the concepts of introjection-projection. If he had been able to laugh off her demands, he would have treated her behavior as essentially playful.

(4) Playfulness is the sign of the last allocative principle, which is based on *assumption.* Assumed roles are not serious. They are taken in games or play, and are held to be at some distance from "reality." The child who plays "mother" is not really confusing herself with her mother. Thus there are no sanctions invoked for assumed roles, provided the individual has emitted the culturally appropriate cue indicating the assumption of a role. The facial configuration referred to in the expression "smile when you say that" is such a cue. It is obvious that assumed roles are of the greatest importance to the development and socialization of the child. But they are of equal importance to adults, not only for the sake of recreation and informality, but also to escape from a disequilibrium situation. The formula "I was only kidding" changes an adopted or achieved role into an assumed one, and thus establishes a new type of complementarity when the old one was threatened with failure. In this connection, withholding a cue indicating whether a role is adopted or assumed is frequently used to conceal or mask motivation. Alter is left in the dark or misinterprets whether ego was serious or not.

The most common sources of allocative discrepancy leading to a failure of complementarity are (*a*) use of a culturally invalid or inappropriate allocative principle; (*b*) withholding of a cue indicating the allocative principle being used; and (*c*) emission of a misleading cue which gives alter the impression that one allocative principle is in use when in fact another one is actually present. For example, in the Italian family that I have been discussing, the mother was angry about the favoritism and excessive attention which the father showed toward their oldest daughter. In her eyes his behavior was largely seductive. At the same time she was ambivalent about his behavior, and unable to express the full range of her feelings. She preferred to attack him on the grounds that he was not a typical American daddy. She reproached him for showing favoritism, for being unfair to the other children, saying nothing about the competitive feelings toward her daughter which his behavior stimulated in her. His response was to deny anything inappropriate in his behavior toward his daughter, and to accuse his wife of being irritable and unduly apprehensive in this situation. Actually neither of them wanted to push the situation to the full extent of their feelings. There was an implicit agreement to avoid it and to substitute in its place their cooperative concern with the excessive demands and "disobedience" of the middle daughter.

An analysis of the allocative principles involved in this source of disequilibrium between the parents reveals that (*a*) the mother defines the father's role as invalid. In her eyes he acts like a lover to his daughter and this is doubly inappropriate. It is not a part of his ascribed role as a father, nor of his achieved role as a husband. He has no right to this role. (*b*) The father agrees with the mother's view of the allocative principles but denies that he has taken a lover's role. But since both the accusation and the denial are implicit — that is, they are only hinted at, not directly verbalized — we have to look for the operations through which the potentially explosive aspects of this situation are avoided. This occurs by a mutually unconscious shift of the dispute to the ground of a cultural value discrepancy — the father's failure to be a typical American rather than a misguided Italian daddy. At the same time, according to the observations of the interviewers who are studying the family, there is an ill-defined but quite intense intimacy between the father and daughter. It is hard to decide whether it is merely a playful aspect of filial attention and devotion, or whether it is something more than this. At times the daughter seems actually to take the mother's role toward the father. The cue distinguishing this as an assumed, adopted, achieved, or ascribed role is missing. But the father's direct description of his activity (how he perceives his behavior) on being questioned is that it is merely a part of his generally ascribed role as father. He even goes so far as to deny to the interviewer that he shows any favoritism, claiming that he treats all his children alike.

Withholding allocative cues or emitting misleading cues are in part attempts to avert the full denouement of failure of complementarity with its accompanying intense disequilibrium. Insofar as they have this function they will be discussed below in connection with that step in the restoration of equilibrium for which I will propose the term *masking*. It is probably obvious that these are general processes occurring in transactions at all levels of the social system. Withholding allocative cues universally produces a masked or ambiguous situation favorable for the "reading in" or projection of intentions. Emitting misleading cues is also a familiar device, whether in the hands of spies, at the international level, or confidence men on home territory. Be that as it may, their connection with failure of complementarity is this: that at the point at which the situation becomes unmasked, the allocative discrepancy is revealed in all its starkness. The disequilibrium is characterized by disillusionment ("You deceived me!"), protest ("You have no right to do what you did!"), alarm ("I've been robbed!"), and various similar phrasings in the vocabulary of victimization.

Instrumental Discrepancy

A review of the origins of failure to maintain complementarity in role relations cannot neglect the fact that nonhuman events and objects form part of the context of all behavior. Insofar as role activities require technical instruments, equipment, furniture, props, costumes, climate and other appropriate physical facilities (including money!), a deprivation or insufficiency of these instrumental prerequisites interferes with role transactions. The point is so obvious that it is represented in various traditional and contemporary maxims, of somewhat dubious accuracy. When equestrian skills were at a premium, instrumental discrepancy was pictured as "For want of a nail, the shoe was lost. For want of a shoe, the horse was lost. For want of a horse, the battle was lost. . . ." Today, in a less heroic cultural climate, one frequently hears, "There's nothing wrong with him that money won't cure!"

Despite the therapeutic oversimplification, such sentiments underscore the potential for severe frustration inherent in instrumental discrepancy. In addition to legitimate and actual deprivation, instrumental descrepancy easily assumes displaced or symbolic functions. For example, in our Italian family, the father complained that he did not have the money to buy the things that his family demanded. Actually he tried desperately to earn more money by taking extra jobs in addition to his main employment. These frenzied efforts defined him as a failure in the dominant American cultural pattern of occupational and economic success because he was unable to plan, budget, or save any money. On the other hand, this strenuous activity relieved him of the potential accusation of neglect — of not caring for his family's welfare. Yet the need to neglect underlay much of his overcompensatory striving. Unconsciously he resented having to take the role of the father, the provider, and would have preferred to compete with his children as the recipient of parental care and concern. This source of role discrepancy, however, had to be hidden from his conscious awareness and its energy had to be partly displaced into other types of activity or passive avoidance of activity.

Unconsciously contrived instrumental deficiency admirably served this purpose. The family suffered from protean forms of equipment failure. The screens had holes, the cellar frequently flooded, the car broke down, the icebox was constantly in need of repair, fuses blew, pipes broke, paint peeled. In the midst of this chaos, the father gave the impression of much activity, rushing about to attend to the latest crisis, accompanied by strident advice from his wife. Actually, he neglected repairing obvious defects until it was too late. The result of the neglect was painful to the wife, who had high standards of housekeeping. He met all criticism from her with the attitude "What can I do? I'm doing my best!"

From this description, it is apparent that instrumental discrepancy can be consciously or unconsciously motivated. To the extent that this is true, it is closely related to goal discrepancy. It must be kept in mind, however, that it can occur quite fortuitously, as in the case of accidental loss or deprivation by fire, robbery, or some other external agent.

Discrepancy in Cultural Value Orientations

As was said before, roles are patterned in accordance with the value orientations of a culture or subculture. In mixed marriages, in families that have moved suddenly from one culture to another as in emigration, and in families that are moving up or down the social class ladder, the possibilities of confusion or outright conflict in cultural values are very great. However, even in families not involved in such dramatic transitions, there is a possibility of discrepancy of cultural value orientations. This is especially true in the United States, because of the extreme mixture of values beneath the surface layers of apparent uniformity of the social system. In this country, cultural traditions are so various and so frequently at odds with each other that almost any individual will have internalized some degree of cultural conflict.

In our project we are using the scheme of variation in cultural value orientations proposed by Florence Kluckhohn (1953, 1957) to keep track of the cultural attitudes which can give rise to conflict. This has proved very useful, but it is too detailed and involved to set forth here. However, the way in which cultural value discrepancies can give rise to disequilibrium can be illustrated again in the case of the Italian family discussed above. The mother was born in this country of native Italian parents. The father was born in Italy and did not come to the United States until he was eight years old. Consequently, the mother considers herself, correctly, to be more Americanized than the father. In both of them there is a great deal of conflict and confusion over the transition to the American patterns, but on any specific issue between them, she is always closer to the American middle-class cultural orientation. She would like to cook only American food, but he insists on Italian dishes. She would like to get away from the home, visit with friends and ultimately obtain a job, but he insists that she stay home and care for the children constantly. She would like her husband to show more initiative and independence though she has the capacity for making decisions and solving problems. He backs away from responsibility and is unable to discipline the children. She would like to plan for their future and the future of the children, but he is occupied with present concerns and he can't get his eyes on the future as a good American would.

These discrepancies in cultural values are associated with incompatible definitions of their roles as husband and wife, mother and father. Thus the complementarity of their role relations is always somewhat strained. The strain would be reduced if the father were moving, culturally, in the direction the mother wants to go. But her activity toward him makes it impossible for him to utilize what potentials for movement he possesses, since he is continuously defined as a failure in terms of the American patterns. He defends himself by pleading incapacity, by claiming that he is "trying" as hard as he can, and by asking that she accept as culturally adequate substitutes other informal roles. One of these is the role of comedian, which he plays with great skill, offering entertainment in the place of successful performance. However, his position vis-a-vis the value of discrepancy is essentially destructive to his self-esteem. He takes his revenge on his wife through his seductive relations with his oldest daughter. In this way a value discrepancy, in which he is the loser, is compensated by an allocative and goal discrepancy in which he is the victor. Since these complicated transactions represent attempts to stabilize or restore equilibrium through *masking* and *compromise,* their further discussion will be postponed until we take up the discussion of these processes.

It is apparent that in discussing the varieties of failure of complementarity in any concrete empirical focus it is virtually impossible to avoid discussing simultaneously the efforts occurring in the system of transactions to compensate or re-establish equilibrium. Failure of complementarity is so disruptive that it is almost always accompanied by processes of restoration for which I would like to use the term *re-equilibration.* In any ongoing system of relations such as a family, then, one can observe re-equilibration occurring whenever the balance of equilibrium to disequilibrium in the state of the system moves too close to the disequilibrium pole. It seems to me that it is the empirical admixture of these three processes — that is, of equilibrium (high complementarity), disequilibrium (low complementarity), and re-equilibration — that has made the processes involved in the stabilization or healthy internal adjustment of the system so difficult to recognize.

Re-equilibrium

The restoration of equilibrium, once complementarity is threatened with failure, is itself an extremely complicated process. I have distinguished eleven steps in the process which I will here describe briefly. I believe these steps have a temporal order and that this order has a kind of internal logic. Unfortunately, I am unable to discern the basis of the order and must therfore leave the presentation in an excessively descriptive and *ad hoc* condition. The description has heuristic value, though it will not leave the reader free of the suspicion that it is arbitrary and incomplete. I am myself dissatisfied with it, but at least it is a method of systematically noting processes in the family which are subtle and difficult to observe.

With respect to the problem of internal logic or the underlying process connecting the various categories, one thing can be said. The eleven categories fall into two groups which are basically different. The first five categories belong together, as do the last five. The sixth forms a connecting link between the two groups. The difference between the two groups is concerned with the method by means of which the role conflict is handled and the equilibrium restored. In the first group the resolution is affected by means of a unilateral decision. Ego resolves the discrepancy by giving in to alter, or vice versa. One or the other parties to the conflict agrees, submits, goes along with, becomes convinced, or is persuaded in some way. For this group, therefore, I would propose the term *role induction*.[3] The net effect, whatever the particular step may be, is that alter is induced to take the complementary role which will restore the equilibrium with ego. Ego's role, on the other hand, does not essentially change. The techniques of induction are dealt with in the classical tradition of rhetoric and have been given a contemporary analysis by Kenneth Burke (1950). They have also been considered in contemporary studies of propaganda devices. I am very much indebted to Burke for his detailed and illuminating studies of the relation between persuasion and discrepancy.

In the second group of categories, re-equilibration is accomplished through a change in roles of both ego and alter. Complementarity is re-established on a mutually new basis. Because of the novel solution of the conflict, I suggest for this group the term *role modification*. The change in role expectations is bilateral and the modification techniques are based on interchanges and mutual identifications of ego with alter. Although the distinction may be somewhat vague, induction techniques are founded on manipulative and instrumental procedures, while modification techniques are based on insight and communicative procedures.

Role Induction

(1) *Coercing* holds first place as the most universally available induction technique. It may hold its primacy either on biological or cultural grounds or both. It can be defined as the manipulation of present and future punishments. Thus it ranges from overt attack to threats of attack in the future, and from verbal commands to physical force. It varies in intensity from mildly aversive manipulations to cruel and unusual torture. It owes its universality to its connection with the hostile-aggressive patterns of behavior in the individual. The reverse is, of course, also true. This is to say that the hostile-aggressive behavior would have no biologically useful

[3]H. S. Sullivan used the term "induction" for the process through which anxiety in the parent elicits anxiety in the child. However, Sullivan applied the word only to the transmission of anxiety; in this paper it refers to a variety of interpersonal influences. A further distinction is that Sullivan regarded the process as somewhat mysterious — a unitary phenomenon, incapable of analytic penetration. For further details, see Sullivan (1953).

function if coercion did not exist as a culturally patterned mode of settling role conflicts. It exists in every family we have studied, and it is probably safe to say that it is present in every enduring social system, no matter how much it may be veiled. If it is successful, the role conflict is settled through submission in which ego accepts the complementary role enforced by alter. However, none of the induction techniques can guarantee success. They may all be met in one of two ways: either by a specific neutralizing technique or by a counterinduction. The specific neutralizing technique for coercing is *defying*. The counterinductions may vary from retaliatory coercion to any of the other re-equilibration categories.

(2) *Coaxing* is in second place not because it is less universal than coercion but because it seems somewhat less readily available as an induction technique. It is probably not the best term for this category, though it specifies the basic principle involved in it. Coaxing can be defined as the manipulation of present and future rewards. Thus it includes asking, promising, pleading, begging, and tempting. Ego accedes to alter's request in order to gain alter's reward, just as in coercing ego submits in order to escape alter's punishment. The child who says, "Please!" by word or gesture rewards his mother with love or compliance when she gratifies his request. In tempting, bribing, or seducing, alter's rewards are likely to be more concrete! However, in seduction, the behavior is invaded by masking — insofar as the seducer conceals his actual motives — and consequently this is probably not a pure case.

Coaxing owes it universality (and its irresistibility) to the fact that it expresses ego's wish for gratification and stimulates a wish to gratify in alter. It epitomizes desire. In spite of its power, it is no guarantee of success in resolving role conflict. As with the other induction techniques, it can misfire if ego responds with a specific neutralizing technique or a counterinduction. The specific neutralization for coaxing is *refusing* or *withholding*. All specific neutralizing techniques are essentially without affect. The affective neutrality occurs because the response is simply a technical way of meeting a persuasion. However, the neutrality may be hard to maintain, and some degree of affective response may creep into it. To the extent that this happens, the response becomes transformed into a counterinduction. For example, defying is simply a holding out against threat and is not in itself affectively toned. But, if ego feels anxiety over the success of defying as a way of warding off threat, then he is likely either to submit, or to become hostile and respond with countercoercion. Similarly, refusing is merely a way of warding off the pressure of coaxing, but if ego is anxious about its effect — for example, if he feels guilty — he may respond by coercing or postponing, or some other induction.

(3) *Evaluating* operates upon the role conflict in a somewhat more derived way than coercing and coaxing. In the usual case it follows upon

them, and therefore is likely to be a counterinduction, though this is not necessary or inevitable. In evaluating, alter responds to ego's behavior by identifying or categorizing it in a value context. Thus it includes such activities as praising, blaming, shaming, approving, and disapproving. For example, if alter tries to resolve the role conflict through coercion, ego may evaluate his behavior by saying, "Stop behaving like a fool!" or "Quit trying to act like a little Hitler!" The "stop" and "quit" signal defiance, but ego clearly is responding as if defiance were not enough either to express the degree of affect mobilized in him or to neutralize the degree of coercion emitted by alter.

The effect of this kind of induction is based upon the manipulation of reward and punishment. It differs from coercing and coaxing in that the reward or punishment is generalized, categorized, and thus placed in a class of value judgments — either positive or negative — linked by verbal and visual imagery to the category. When ego says "acting like a fool," he is linking alter's behavior to a class of punished or devalued activities symbolized by the figure of the fool. He establishes an identity between alter and all other fools. If alter accepts the identity, then he will define his having coerced as punishable or noneffective and will terminate or extinguish his coercive activity. He may then substitute some other induction, such as coaxing, to resolve the role conflict. However, he may not accept the identity employed in the evaluation, and if so, he may use the specific neutralizing technique to be employed against evaluation. This is *denying.* For example, alter may respond to ego's evaluating by saying, "I am *not* behaving like a fool, and if you don't do what I've asked you to do, you'll have to suffer the consequences!" After denying, ego returns to coercing, showing the circular pattern characteristic of any protracted quarrel.

The same mechanisms hold true for positive evaluating such as praising. Of course, positive evaluating is more likely to be accepted, since it is a reward, though it may not be so interpreted as in the case of what is held to be unwarranted flattery. The case of flattery, however, is another example of a compound induction because it is likely to be mixed with various degrees of masking. Alter is apt to perceive ego's flattering as concealing a hidden motive. Apart from masking, there are still good reasons for denying positive evaluating. Since the motive behind ego's positive evaluation is to induce alter to take the complementary role which will restore equilibrium, alter may deny in order to ward off this outcome. This is certainly what happens in the case of praise, encouragement or support, if alter is resisting the induction process. A mother, attempting to encourage her reluctant son to go to school for the first time, may say, "Johnny, I'm sure you'll enjoy school. You'll have a good time, and Mommy will be proud of you, just like she is of Freddy [older brother]."

First the mother coaxes, by holding out the promise of future reward (enjoyment) and then she reinforces with a positive evaluation, putting Johnny with Freddy in a class of rewarded objects (pride). Such an inducement can easily backfire. Johnny bursts into tears and says, "No! I don't wanna go. I won't have a good time." (Refusal of coaxing.) "And I don't care about Freddy. I'm not *like* him!" (Denial of identity and of evaluation.) This leaves the discrepancy of goals about where it started, at high disequilibrium, and the mother may now try coercing, or she may postpone the settlement of the conflict until Daddy comes home, or until tomorrow when Johnny's resistance may be lowered.

(4) *Masking* is another universal induction technique, more indirect than the three discussed so far. It may be defined as the withholding of correct information or the substitution of incorrect information pertinent to the settlement of the conflict. It includes such behavior as pretending, evading, censoring, distorting, lying, hoaxing, deceiving, etc. These words are taken from ordinary usage and are apt to have a negative connotation. However, it is not my intention to give masking (nor any other induction technique) either a positive or negative value. It occurs universally in the course of organism-environment transactions, and has its biological and cultural aspects. The tiger stalking its prey is masking, as is the camouflaged bird sitting on its nest. Every culture has its patterned ways of concealing information and its criteria for determining what information may or may not be revealed, with or without distortion. In studying masking my intention is merely to determine its *function* for the way the system is working. I believe it is as significant to the function of the social system, large or small, as is *repression* to the function of the personality as a system. Repression is universal as an intrapsychic process, and it means that information available to certain components of the personality is either completely unavailable to another component or reaches it only in disguised form. Repression has a biological basis in the function of the organism, but the content of what is repressed is related to the content of what is masked in the social system. This is a point which Sullivan (1953) repeatedly stressed in calling attention to the significance of interpersonal relations to the function of the personality. However, Sullivan tended to see only the negative side of masking. He noted how it produces obstacles to successful communication which the individual internalized, but he was not interested in its function for the social system itself.

Masking is so complex and so intrinsic to re-equilibrating processes in the family that it is impossible to discuss it adequately in this small compass. "Little white lies" and minor disguises of motives take place so automatically that they are scarcely noticeable. For example, displacement and substitution of roles between parent and child are ubiquitous. A child bumps itself on a chair, and the mother says, "Naughty chair!" assigning the chair a human activity and then evaluating that activity as if it were part of a coercive induction. Why does she do this? Pain produces anger and in order to avoid the potential role conflict which may be precipitated between herself and her child, she involves the child in a make-believe conflict with the chair, with herself in the role of referee. Furthermore, she denies thereby the potential negative evaluation of herself as insufficiently protective of the child, by displacing the carelessness to the chair. This preserves equilibrium between herself and the child and thus is functional for that role system. But one can ask whether what is functional for their role system may not be dysfunctional for the child's ability to test reality. She conceals the important information that pain and accidents can occur without motive and need to be endured in the inevitable process of maturation and acquistion of autonomy by the child. Thus her masking ties the child to her in a dependent relation in which she plays the role of protector. She conceals both from herself and her child information about her resentment at the growing independence of the child, which, if it were available as a message, would read, "If you're going to act so independent, you ought to be punished. But I don't want you to know that I think this, so I'll pretend that it's not your behavior I resent but the chair's. You will understand that the world is full of hostile chairs, and you need me to protect you from them." If the child does not see through this masking, he will take the complementary dependent role which his mother desires for him.

In studying the family it is often difficult to disentangle the significance of minor masking, such as the example just discussed, from major transactions in which the masking is very dramatic. For example, in the Italian family discussed above both the cultural value discrepancy and the sexual goal discrepancy between the parents were masked and the role conflict displaced to the middle daughter, who was explicitly defined by both parents as the major source of all their difficulties. The test of significance is to discover what happens when the induction technique is unmasked. *Unmasking* is the specific neutralizing technique for masking. The role partners confront each other with what has been concealed or disguised. Where the masking has averted a major disequilibrium, unmasking can be extremely explosive. As a result of therapy with the mother, father, and middle child in this Italian family, the mother began to displace less of her role conflict to the middle child and to pay more attention to the father's relation to the oldest daughter. The change was registered in a violent scene in which the mother openly voiced her resentment to the father, who then lost his temper and threw a lighted cigarette at his wife, denying all the while the truth of her accusation. This unmasked the sexual situation, but left the cultural discrepancy still concealed, that is, not directly stated as a source of role conflict between the parents. It is our hunch that when this conflict opens up, the violence in their feelings will be even greater.

(5) *Postponing* may seem to fit uneasily as an induction technique since it appears to be merely a negative or passive way of dealing with role conflict. Nevertheless, it is undertaken with the expectation in both ego and alter that "in the interval he will change his mind." The process by which the conflict is to be settled is deferred in the hope of change of attitude. Indeed, this is very likely to be successful since the intrapsychic process always tends to work toward a resolution of conflict. The implied instruction "think it over," or "I'll sleep on it," often achieves the desired effect. Most role conflicts in the family are not settled at the moment, but are deferred and taken up afresh, time and time again. From the point of view of persuasion, the question between ego and alter is: Who has the most to gain from postponing? If ego considers that he has very little to gain, he may attempt the specific neutralizing technique when alter attempts to postpone. This is *provoking*. If ego is afraid of postponement, he may provoke or incite the conflict to appear in full force.

(6) *Role reversal*[4] is a transitional re-equilibration midway between role induction and role modification. It can be defined in G. H. Mead's (1934) sense as the process of taking the role of alter. Ego proposes that alter put himself in ego's shoes, try to see it through his eyes. Or ego initiates the reversal in the hope that alter will do the same. Ego may say, "Well, I think I'm beginning to see your point, but . . ." or "It doesn't make too much sense to me, but I think I see what you mean." Insofar as this is a nonmanipulative approach, it can't be classified as an induction, and it therefore requires no specific neutralizing technique. On the one hand, if alter responds to role reversal with an induction, then ego may give up the attempt to reverse roles, and the whole process will revert to inductive and counterinductive maneuvers. On the other hand, the role reversal may well kick the process of re-equilibration toward role modification and a novel resolution. It is this ambivalent position that makes it impossible to classify role reversal as belonging to either group; it is really transitional between both of them.

Whether role reversal is effective or not depends in large part on the intensity of masking procedures in the family relations. The more energy in disequilibrium being defended by masking, the less likely is role reversal to take effect. In our Italian family, the interviewers, seeing the parents, tried repeatedly to test their ability to reverse roles with Joanne, the middle daughter. For example, the mother's interviewer would say, "Do you think Joanne is sort of feeling left out in the family? Maybe she feels she isn't

[4] Like "role conflict," the expression "role reversal" has also been used in two different ways. In one usage it refers to a situation in which ego and alter permanently exchange roles.

getting enough attention." To this sort of approach, the mother, for a long time, would respond with the statement "But how could she? We try so hard to treat them all the same!" The same sort of thing tended to happen with respect to Joanne's stealing. Joanne was not given an allowance or permitted to baby-sit in order to earn some money. This was always defended on the basis of the evaluative induction: Joanne steals. She's not reliable. We can't trust her, etc. The interviewer asked how Joanne could ever learn to take responsibility if she were not given some. After a while this role reversal "took" with the mother, who started to treat Joanne as if she were not an irretrievably deviant daughter This coincided with an intensive role reversal program between the interviewer and the mother in which the interviewer tried continuously to understand how the mother was feeling. The double-barreled procedure moved Joanne out of the masking process in which she had been held as if in a vise. In turn this led to the unmasking of the sexual conflict between the mother and father with respect to Rosemarie, the oldest daughter.

For example, a husband and wife settle on an arrangement in which the husband stays home and looks after the house and children while the wife takes a job and earns the income for the family. In the second usage, the phrase refers to a process in which ego and alter temporarily exchange roles, in action or in imagination, for the sake of gaining insight into each other's feelings and behavior. This definition has been extensively used by J. L. Moreno and his associates, to whom I am much indebted. For examples, see Moreno (1955).

Role Modification

(7) *Joking* is an outgrowth of role reversal. It is the first sign that role modification is in progress. The role partners, having successfully exchanged places with each other and thus having obtained some insight into each other's feelings and perceptions, are now able to achieve some distance from their previous intense involvement in the conflict. They are able to laugh at themselves and each other. The laughing proceeds in part, as Freud pointed out, from the saving of psychic energy coincident with the partial solution of the conflict. The jokes also permit the expression in sublimated form of some of the induction techniques which are about to be relinquished — such as coercing and evaluating. The joking process moves the allocative base of the transaction to a whole set of assumed roles, and thus introduces playfulness into what was previously a tense set of achieved or adopted roles. In play the role partners try on for size a series of weird or impossible solutions, out of which is gradually fabricated the substance of the possible solution.

(8) *Referral to a third party.* Role reversal and joking may not of themselves create a role modification. They are helpful but not necessarily

sufficient for this type of re-equilibration. Therefore, ego or alter may suggest that the conflict be referred to a third person (or organization) for help in its solution. The assumption is that the third party is less intensely involved in the conflict and has information or skills not available to ego or alter. Thus he can visualize a solution with greater ease. There are two difficulties which may arise from this re-equilibrating procedure. First, the third party chosen may steer the process back to a manipulative procedure and thereby restimulate the induction process. Secondly, and coincidental with this, the third person may form a coalition with ego against alter, or vice versa. In families, the attempted solution through referral frequently gets grounded on the rocks of a coalition. The third person, who was to have taken the role of impartial judge or referee, actually teams up with either ego or alter. This triadic situation has been studied by Simmel (1950) and more recently by Mills (1953, 1954) in artificially composed groups. However, the process involved in it needs much more extensive investigation. In our Italian family, third party referral always seemed to end in a coalition. At the outset, the parents were allied against Joanne. As unmasking proceeded, the father and Rosemarie were revealed in a coalition against the mother. There was evidence that the youngest child and the mother were in alliance against the father. These shifting triadic relations are among the most difficult transactions to unearth and keep track of in the family. Yet they are of the greatest importance to the dynamics of role conflict and thus to the way in which the family system is organized and functions.

Referral is invoked whenever a family comes to a community agency for help and is inevitably associated with the role of the psychiatrist or other mental health worker. Implicitly or explicitly, the helper is asked to judge, referee, or take sides. The interviewers seeing our families are inevitably pitted against each other in a semi-coalition with the particular member of the family they are seeing. This process is neutralized by our team approach in which the interviewers exchange information continuously with each other. If there is delay in the collaborative interchange between interviewers, then the coalitions are apt to get out of hand. The high level of communication between the interviewers permits all of them to obtain a balanced view of the overall family process. In this way excessive identification with the member they are seeing is avoided. It seems a good working rule that the more information available to the person taking the role of the third party, the easier it is to avoid getting entangled in a coalition. As a corollary to this proposition, the more information available to the third person, the easier it is for him to help the role partners to a novel solution and to avoid a manipulated solution of the conflict.

(9) *Exploring* is the next step in role modification. Ego and alter probe and test each other's capacity to establish a novel solution. This process was

already initiated in the joking phase but now it is undertaken more seriously. If a third party has been able to avoid becoming entangled in a coalition, he can be of great help in promoting exploration. To a considerable extent this describes the activity of the psychiatrist, case worker, nurse, or whoever is involved in the solution of a family problem. It is almost always accompanied by temporary relapses to an induction procedure, but once initiated, it tends to be self-steering. Ego and alter propose and reject possible solutions. This is accomplished not so much through verbal formulations as through actual behavior, though both paths toward the solution are probably necessary.

(10) *Compromising.* After a sufficient amount of exploration, ego and alter come to see that restoration of equilibrium involves some change in the goals each desired or in the values by which they were guided. Thus they must settle for somewhat different complementary roles than those with which they started. If the process of re-equilibration has involved a successful referral, the third person takes very little part in the actual compromise solution. His role has accomplished its function when exploration moves re-equilibration to the threshold of compromise.

(11) *Consolidating* is the last step and it is required because the compromise solution, which is characterized by novelty and cognitive strain, is still present. Even though ego and alter establish a compromise, they must still learn how to make it work. To put the matter somewhat differently, compromise can be defined as the adjustment and redistribution of goals. Then consolidating is associated with the adjustment and redistribution of rewards. The roles are modified through the redistribution of goals. The new roles still have to be worked through and internalized by ego and alter as they discover how to reward each other in playing the new roles.

Conclusion

The study of how the family functions and maintains itself as a going system is greatly facilitated by the observation of role transactions concerned with equilibrium (high complementarity of roles), disequilibrium (low complementarity of roles), and re-equilibration (restoration of complementarity). I suspect that these same processes occur in other small-scale social systems such as a factory or a mental hospital. To what extent they can be detected in large-scale social systems, such as a total society, I do not know. In a small-scale system like the family, most of the process which can be seen by the observer is concerned with equilibrium. Complementarity of roles is high, decision-making is low, and most events take place automatically, leaving a considerable degree of spontaneity to the individuals in transaction with each other. This is the "routine," the way the system usually works. However, there are inevitable strains in any such system, and these give rise to disequilibrium. The strains can be analyzed in terms of the cognitive, goal, allocative, instrumental, and value structures of the roles. A strain represents a discrepancy in the expectations of any ego and alter with respect to these role structures. Thus it can be described in terms of role conflict. Strain gives rise to anxiety because, if left unchecked, it will lead to a rupture of the role relations, and thus to a disruption of the system. Without discussing the origin of this anxiety in the basic structure and function of the intrapsychic process in the individual, it can be said that the role conflict gives rise to defensive processes both in the individual and in the family system. For the family system this reactive process can be described as an attempt to restore the threatened complementarity of roles. The process itself can be called re-equilibration since its effect is to restore the equilibrium which has been shattered.

Re-equilibration can be analyzed as an eleven-step process. The first five of these steps are manipulative. Ego attempts to persuade or get alter to comply with his expectations. If compliance is achieved and alter takes the necessary complementary role, then equilibrium is restored. For this reason, these steps are grouped together as a process called role induction. The last five steps are based on mutual insight rather than manipulation. They lead to a novel solution of the role conflict underlying the disequilibrium. These steps are grouped together in a process called role modification. The sixth step is intermediate between the two groups since it can lead to either induction or modification.

If modification is successful, then the new solution of the role conflict sinks into the normal "routine" of the family. The "problem" has disappeared. In this way modification differs from induction. Induction is primarily defensive. It wards off the disequilibrium but it is always likely to crop up again. It is an unsettled problem to the system and the resolution of the strain is more apparent than real. In this way it becomes internalized by the members of the family where it is likely to be productive either of a neurotic symptom or of difficulties in interpersonal relations. In dealing with emotionally disturbed individuals, whether in office practice or in a mental institution, one observes new versions of the old, unsettled family role conflict appear. Therefore, it is fruitful to examine the role systems which the patient re-creates in these settings to see in what way they reproduce the defensive, inductive procedures which were experienced in the family. Also, it is necessary to discover in what way the new institutional settings may have elaborated role conflicts and inductive re-equilibrations — because of their own internal organization — which resemble the original strain in the family.

References

Ackerman, N. W., and Sobel, R. Family diagnosis: an approach to the pre-school child. *American Journal of Orthopsychiatry,* 1950, 20, 744–53.

Ackerman, N. W. "Social role" and total personality. *American Journal of Orthopsychiatry,* 1951, 21, 1-17.

Bentley, A. F. Kennetic inquiry. *Science,* 1950, 112, 775-83.

Burke, K. *A rhetoric of motives.* New York: Prentice-Hall, 1950.

Dewey, J., and Bentley, A. F. *Knowing and the known.* Boston: Beacon Press, 1949.

Johnson, Adelaide M., and Szurek, S. A. The genesis of antisocial acting out in children and adults. *Psychoanalytic Quarterly,* 1952, 22, 323-43.

Johnson, Adelaide M. Factors in the etiology of fixations and symptom choice. *Psychoanalytic Quarterly,* 1953, 22, 475-96.

Kluckhohn, Florence R. Dominant and variant value orientations. In Kluckhohn, C.; Murray, H. A.; and Schneider, D. M., eds., *Personality in nature, society, and culture.* New York: Knopf, 1953.

Kluckhohn, Florence R. *Variants in value orientations.* Evanston, Ill.: Row-Peterson, 1957.

Kluckhohn, Florence R., and Spiegel, J. P. *Integration and conflict in family behavior* (Report N. 27). Topeka, Kansas: Group for the Advancement of Psychiatry, 1954.

Linton, R. *The study of man.* New York: Appleton-Century, 1936.

Mead, G. H. *Mind, self, and society.* Chicago: University of Chicago Press, 1934.

Mills, T. M. Power relations in three-person groups. In Cartwright, D., and Zander, A., eds., *Group Dynamics.* Evanston, Ill.: Row-Peterson, 1953.

Mills, T. M. The coalition pattern in three-person groups. *American Sociological Review,* 1954, 19, 657-67.

Moreno, J. L. The discovery of the spontaneous man — with special emphasis on the technique of role reversal. *Group Psychotherapy,* 1955, 8, 103-29.

Parsons, T.; Bales, R. F.; and Shils, E. A. *Working papers in the theory of action.* Glencoe, Ill.: Free Press, 1953.

Parsons, T.; Bales, R. F. *Family, socialization, and interaction process.* Glencoe, Ill.: Free Press, 1955.

Pollak, O. *Social science and psychotherapy for children.* New York: Russell Sage Foundation, 1952.

Simmel, G. Quantitative aspects of groups. In Wolff, K. J., trans. and ed., *The sociology of Georg Simmel.* Glencoe, Ill.: Free Press, 1950.

Spiegel, J. P. The social roles of doctor and patient in psychoanalysis and psychotherapy. *Psychiatry,* 1954, 17, 369-76.

Spiegel, J. P. A model for relationships among systems. In Grinker, R. R., ed., *Toward a unified theory of human behavior.* New York: Basic Books, 1956.

Sullivan, H. S. *The interpersonal theory of psychiatry.* New York: Norton, 1953.

CHAPTER 30

Counseling with Sexually Incompatible Marriage Partners

WILLIAM H. MASTERS and VIRGINIA E. JOHNSON

At least one result of the cultural relaxation of sexual taboos has been of major consequence. Today, more — many more — marital partners are seeking professional assistance when sexual incompatibility threatens their marriage. Anyone exposed professionally to the emotional anguish and disrupted marriages caused by such clinical problems as impotence and frigidity will look upon this help-seeking trend with considerable satisfaction.

Most of the sexually distressed people are bringing their problems to their family physicians. Although the individual or combined efforts of psychiatrists, psychologists, marriage counselors, social workers and/or clergymen may be needed in addition to those of the chosen physician to solve some problems of sexual inadequacy, it is the family physician, taking advantage of initial rapport and established confidence, who ordinarily overcomes any patient reluctance or embarrassment and builds motivation for further treatment.

Unfortunately, until recently the physician has been hampered in treatment by three major stumbling blocks:

First, there has been a long-standing and widespread medical misconception that a patient will not reveal sex history background with sufficient accuracy and in adequate detail for effective therapy.

Second, in the past the physician has been provided with very little basic information in sexual physiology upon which to develop any effective treatment of sexual inadequacy.

Third, many physicians have been convinced that since most sexual problems are psychogenic in origin, only a specialized psychopathologist can treat them effectively.

Increasingly large numbers of physicians are demonstrating clinically that none of the obstacles now have much substance in fact.

Almost ten years of investigation in the broad areas of human sexual response has brought conviction to the writers that if the interviewing physician can project sincere interest in the patient's problem and, even more important, exhibit no personal embarrassment in an open sexual discussion, almost any individual's sexual history will be reported with sufficient accuracy and in adequate detail for treatment purposes. Others, such as Eisenbud,[1] who have worked with human sexual problems, also believe that patients are usually very ready to talk freely about their disturbed sexual behavior patterns once they have gathered their courage to a degree sufficient to seek professional guidance.

While it is true that the amount of research in sexual physiology has in the past been meager indeed, this situation is rapidly being corrected.[2, 6] Some of this recent material is synthesized in the latter part of this chapter and quite possibly may provide a minimal baseline for the more adequate clinical treatment of frigidity or impotence.

With regard to the third stumbling block — that of requisite referral to the psychopathologist of problems of sexual incompatibility — two things should be noted. First, there is ample clinical evidence for the observation that sexual imbalance or inadequacy is not confined to individuals who have been identified with major psychoses or even severe neuroses. Secondly, long-maintained individually oriented psychotherapy for sexual inadequacy frequently places irreversible strains on the marital state. While the psychopatholgist is working with one marriage partner or the other toward the resolution of his or her individual sexual inadequacy, the marriage itself may be deteriorating. One or two years of therapy directed specifically toward the impotent male or frigid female frequently leaves the unsupported marital partner in a state of severe frustration. Not only are unresolved sexual tensions of the nontreated spouse of major moment, but frequently no significant attempt is made by the therapist to keep the supposedly adequate partner apprised of his or her mate's fundamental problems and/or the specifics of therapeutic progress. Such situations of spouse neglect not only are sure to increase the performance pressures on the sexually inadequate partner, but obviously may lead to many other areas of marital strife and, for that matter, stimulate extramarital interests.

As the result of these observations, the conviction has grown that the most effective treatment of sexual incompatibility involves the technique of working with both members of the family unit. The major factor in effective diagnosis and subsequent productive counseling in sexual problems lies in gaining access to and rapport with both members of the family unit. This community approach not only provides direct therapy for the sexually inadequate partner, but provides something more. An indirect therapeutic gain results from enlisting the complete cooperation and active participation of the adequate spouse (the husband of the frigid woman or the wife of the impotent male). It is virtually impossible for the mate of the sexually distressed partner to remain isolated from or uninvolved in his or her partner's concern for adequate sexual performance. Therefore, most of these individuals can and will be most cooperative in absorbing the necessary material of both physiologic and psychologic background necessary to convert them into active members of the therapy team.

As in so many other areas of medical practice, treating sexual incompatibility involves, first, recognizing the nature of the patient's problem; second, determining the type and degree of the incompatibility; and third, developing and activating the therapeutic approaches applicable to the particular clinical involvement.

Recognizing the Sexual Problem

The patient with sexual distress defines the problem directly with increasing frequency during this era of marked change in our cultural attitudes toward sexual material. However, many women initially may discuss such symptoms as fatigue, "nerves," pelvic pain, headaches or any other complaint for which specific pathology cannot be established. The physician-interviewer must anticipate conscious vocal misdirection when Victorian concepts of sexual taboos still exist, or where there is a personal demand to fix blame on the marital partner.

If, for example, the female is the partner experiencing major dissatisfaction with her marriage, for any one of a number of reasons, she purposely may obscure her basic personal antipathies by describing gross sexual irregularities on the part of her marital partner. Sometimes when it is the husband who wishes to end the marriage, he often employs the pressure of partial sexual withdrawal, or even complete sexual refusal. At this point, medical consultation is sought solely to justify condemnation of what is termed the mate's unfair, inadequate or perverted sexual behavior.

Actually, the marital incompatibility which brings the couple to the physician usually is not primarily of sexual origin. Sexual incompatibility may well be the secondary result of marital disagreement over such problems as money, relatives, or child care. Such areas of dispute easily may undermine any poorly established pattern of sexual adjustment. Frequently, witholding of sexual privileges is used as punishment in retaliation for true or fancied misdeeds in other areas. If the preliminary history reveals such a situation of secondary sexual incompatibility, the physician must decide whether he wishes to carry the full, time-consuming burden of total marriage counseling or if referral is in order. In the latter case, he still may wish to retain an active clinical role in the psychosexual aspects of the problems involved.

However, once the problem is established as primarily sexual in nature and as the cause and not the effect of the marital incompatibility, the complaint should be attacked directly and with the same sense of medical urgency with which clinical complaints of either a medical or a surgical background are investigated. Otherwise, permanent impairment of the marital relationship may be inevitable.

The Sexual History

The need to acquire accurate and detailed sexual histories is basic to determining the type and the degree of the incompatibility of the members of the distressed family unit.

Sex histories must reflect accurately details of early sexual training and experience, family attitudes toward sex, the degree of the family's demonstrated affection, personal attitude toward sex and its significance within the marriage, and the degree of personal regard for the marital partner. While the actual nature of the existing sex difficulty may be revealed during an early stage of history taking, the total history, as it discloses causation and subsequent effect, provides the basis for the most effective means of therapy.

The first step in the team approach to diagnosis and treatment has been to see the husband and wife together as a complaining unit during the initial interview.[7, 9] Procedures and philosophies are explained to them. If the family unit desires to continue after the investigative concepts have been outlined, the couple is separated for individual interrogation after both marital partners are assured that similar background material will be covered simultaneously by the two interviewers.

The knowledge that both unit members are undergoing similar interrogative procedures, that essentially the same background material will be investigated, and that all areas of professed concern will be probed in depth, produces an atmosphere that encourages honest reporting and an unusual amount of patient attention to detail.

Finite details of past and present sexual behavior may be obtained during the initial interview with the facility and integrity anticipated for the recording of a detailed medical history. Encouraged by a receptive climate, controlled, brief questioning and a nonjudgmental attitude, the patient is just as free to discuss the multiple facets of, for example, a homosexual background, as he might be to present the specific details of an attack of chronic chololithiasis in a medical history.

It should be noted particularly that in the process of acquiring a detailed sexual history, the usual basic physical and social histories of medical and behavioral significance also are recorded.

For the rapid diagnosis and treatment of sexual incompatability, a male-female therapy team approach has been developed as reported elsewhere.[9]

This approach involves the male marriage partner being interviewed first by the male member of the therapy team. Simultaneously, material from the female partner of the involved marital unit is acquired by the female member of the therapy team. Prior to the second investigative session, members of the therapy team exchange pertinent details of the marital unit's reported sexual distress. During the second session the female partner of the complaining couple is reinterviewed by the male member of the therapy team. Meanwhile, the husband of the distressed unit is evaluated by the female therapist. At the third interview, the therapy team and the distressed family unit meet as a committee of the whole to review the positive features of the prior interrogative sessions and to discuss in detail the active degree of the sexual incompatibility.

While the male-female therapy approach has been found to be eminently satisfactory, obviously this techniques usually is not possible in the typical physician's practice. However, the broad general steps toward diagnosis and evaluation which are outlined here can be adapted by the individual physician. For instance, the advantage of honest reporting obtained by simultaneous interviews of members of the marital unit can be retained by interrogating family unit members consecutively.

The Therapeutic Process

Once the background of the individual couple's sexual imbalance has been defined, and the clinical picture delineated and presented to their satisfaction and understanding, a discussion of therapeutic procedure is developed for the family unit.

In general terms, the psychotherapeutic concepts and physiologic techniques employed to attack the problems of frigidity and impotence are explained without reservation. Specific plans are outlined for the therapeutic immediacies and a pattern for long-range support is described. With this specific information available, a decision must be reached as to whether there is sufficient patient need or interest for active participation in the therapeutic program. The decision obviously is based not only on a joint evaluation of the quality of the marriage and the severity of the sexual distress, but also on a review of the individual abilities to cooperate fully with the program. If doubt exists, on the part of either member of the investigative team or either partner of the sexually incompatible family unit, as to real interest in marital unity exposure to remedial techniques or ability to cooperate fully as a unit, the couple is directed toward other sources of clinical support.

Since the two major sexual incompatibilities are frigidity and impotence, treatment for these problems will be discussed in detail.

Impotence

Three major types of impotence ordinarily are encountered in the human male. They are:

1. Failed erection. Penile erection cannot be achieved.

2. Inadequate erection. Full penile erection either cannot be achieved or, if accomplished, is maintained fleetingly and lost, usually without ejaculation.

3. Non-emissive erection. Full penile erection is achieved, but ejaculation cannot be accomplished with the penis contained within the vagina.

Note. Premature ejeculation, ejaculation before, during or immediately after mounting is accomplished, while not considered a form of impotence, is discussed in this chapter due to the similarity of therapeutic approach.

Impotence is rarely, if ever, the result of lesions of the posterior urethra. Eliminating the possibility of spinal cord disease or certain endocrinopathies, such as hypogonadism or diabetes insipidus, the total history should be scrutinized for the omnipresent signs of psychogenic origin for the specific type of male impotence reported.

In the case of the male with failed or inadequate erection, history-taking should stress the timetable of symptom onset. Has there always been difficulty, or is loss of erective power of recent origin? If recent in origin, what specific events inside or outside the marriage have been associated with onset of symptoms? Are there any masturbatory difficulties? Is there a homosexual background of significance?

Further questioning should define the male's attitude toward his sexual partner. Is there rejection not only of the marriage partner, but also of other women as well? Are the female partner's sexual demands in excess of his levels of sexual interest or ability to comply? Is there a sexual disinterest that may have resulted from the partner's physical or personal traits, such as excessive body odor or chronic alcoholism?

In the case of a patient with premature ejeculation, questions should be concentrated in a different area. Does this rapid ejaculatory pattern date from the beginning of his sexual activity? Has he been exposed to prostitute demand for rapid performance during his teen-age years? Does he come from a level of society where the female sexual role is considered to be purely one of service to male demand?

When working with the male with a non-emissive erection still other questions are more appropriate. Has the male always been unable to ejaculate during intercourse or has this difficulty been confined to exposure to his marital partner? Are nocturnal emissions frequent, especially after heterosexual encounters? Is there an active homosexual history?

Actually, the fundamental therapeutic approach to all problems of impotence is one of creating and sustaining self-confidence in the patient. This factor emphasizes the great advantage in training the wife to be an active member of the therapeutic team. All pertinent details of the anatomy, physiology and psychology of male impotence should be explained to her satisfaction. The rationale of treatment, together with an explanation of the specific stimulative techniques most effective in dealing with the specific type of impotence distressing her husband, must be made clear to her.

In the early stages of treating failed or non-erective impotence, it is wise to avoid emphasizing the demand that intercourse be the end of all sexual play. Frequently, the male's inability to meet just such a repetitive female demand is already one of the primary factors in his impotence. Some males find release from fear of performance when they are given to understand that sexual play need not necessarily terminate in intercourse. They are then able to relax, enjoy and participate freely in the sexually stimulative situations created by their clinically oriented wives to a point where erection does occur. After several such occasions of demand-free spontaneous erections, the males may even initiate the mounting procedure and complete the sexual act. This casual mating may well be the beginning of release from their chronic or acute failed or non-erective impotence.

In most cases, manual penile manipulation varying in degree of intensity and duration probably will be necessary. This controlled penile stimulation must be provided by his previously trained female partner. The male with inadequate erection syndrome should be exposed to long and regularly recurrent periods of manual stimulation in a sensitive, sexually restrained, but firmly demanding fashion.

In the opposite vein, the male with the difficult problem of premature ejaculation should be manually stimulated for short, controlled periods with stimulation withheld at his own direction as he feels ejaculation is imminent. The shaft of the penis should be well lubricated to reduce cutaneous sensation. This technique will fail frequently and ejaculation will occur. However, the family unit should be encouraged to return to the technique repetitively until the male's obviously improved control leads to the next therapeutic step. This will be a female superior mounting which can later be converted to a nondemanding lateral resting position. The progressive control techniques emphasize the family unit approach to the problem of sexual inadequacy and from here on psychogenic support and family cooperation certainly will reclaim many of those males who were formerly sexually quite inadequate.

The problem of the male with non-emissive erection is somewhat different. His is largely an infertility problem rather than one of sexual incompatibility. In these cases, reassuring both husband and wife that the problem is of little clinical consequence provides the basic therapy.

Sometimes the infertility concern connected with this variant can be overcome by artificially inseminating the wife with her husband's seminal fluid obtained by manipulation. Since psychotherapy has produced so few positive results with this type of impotence, providing clinical reassurance and conceptive information may have to suffice in these cases.

Frigidity

There is a great deal of misunderstanding over the connotation of the word "frigidity." It is often used in a context which presumes an irrevocable lack of sexuality on the part of a female sexual partner. Misconceptions occur too frequently when overdependency is placed on this word as a diagnostic term.

From a therapeutic point of view, the maximal meaning of the word should indicate no more than a prevailing inability or subconscious refusal to respond sexually to effective stimulation. A woman is not necessarily lacking in sexual responsiveness when she does not experience an orgasm. Therefore, the achievement of orgasmic response should not be considered the end-all of sexual gratification for the responding female. Unhappily, many women, unable to achieve an orgasmic level of sexual response in the past, have been labeled frigid not only by their marital partner but also by the physician they may have consulted.

The free use of this term frequently does great psychologic damage. Frigidity is a term that should rarely by employed in the presence of the sexually inadequate female for it may well add shame, and/or fear of inadequate performance to whatever other psychologic problems she may have.

It is true that there are a number of women who experience a persistently high degree of sexual tension, but, for unidentified reasons, are not able to achieve a satisfactory means of tension release. In evaluating this problem, initial exploration should be concentrated in two areas of psychosexual withdrawal. The first is to determine the presence or absence of psychologic inability to respond to effective sexual stimulation. The second is to define the possible existence of sexual incompatibility caused by misunderstandings resulting from a difference in the sex tension demands of the marital partners.

As describe elsewhere,[7] three positive indications of female psychosexual inadequacy can be developed by careful history taking:

1. Attitude toward sex and its significance within the marriage.
2. Degree of personal regard for the marital partner.
3. Fear of pregnancy.

In investigating the attitude toward sex, existing negative concepts should be pursued by careful interrogation. Questioning should explore early sexual training and experience, exposure to lack of demonstrated parental affection, history of homosexual experience, if any, and/or any traumatic sex-oriented incidents which might have affected natural sexual responsiveness.

When exploring the area of personal regard for the marital partner, the female partner's disinterest or lack of cooperation with the consulting physician may be an interesting clinical symptom of itself. When essential indifference toward a marital partner has been exposed, the existence of a basically unwanted marriage or marriage undertaken without intelligent preparation or emotional maturity is a real possibility. Perhaps, in these cases, referral to a marriage counselor or undertaking marriage counseling in the more general frame of reference, is in order, rather than concentrating on the sexual aspects of the problem.

When there is any indication of fear of pregnancy the therapeutic approach is obvious to the counseling physician. Actually, satisfactory results are ordinarily more easily achieved in pregnancy phobia situations than in either of the other two areas of psychosexual withdrawal.

After the background of the female's sexual unresponsiveness has been established, and the marital unit has accepted the conclusions presented during the diagnostic sessions, therapy may begin. Female sexual responsiveness may well depend upon the successful orientation to the following framework of therapeutic approach:

1. The possibility of anatomic or physiologic abnormalities that can contribute to varying shades of dyspareunia should be eliminated. Orientation to male and female sexual anatomy, directly if necessary, should be accomplished.
2. Affirmation that sexual expression represents an integral basis for sharing within the marriage should be emphasized.
3. A mutually stimulative sexual pattern should be developed and adapted to the individual psychosocial backgrounds of the marriage partners.
4. Gentleness, sensitivity and technical effectiveness in the male partner's approach to sexual encounter should be encouraged.
5. Emphasis should be placed on the fact that female orgasm is not necessarily the end-all of every sexual encounter.

With regard to pelvic abnormalities, it might be noted that a history indicating actual pain or any other physical displeasure during sex play or coition certainly suggests the need for an adequate physical examination. If physiologic variants, such as pelvic endometriosis, causing severe, recurrent dyspareunia with deep penile penetration, are revealed, subsequent medical and/or surgical adjustments may be indicated.

However, it should be noted that sometimes the simple clinical expedient of teaching the family unit proper positioning for coital activity may remove the female partner's distress despite existent pelvic pathology.

A high percentage of psychologically based problems of inadequate female response begin as the result of rejection of, or ignorance of, effective sex techniques by either or both marital partners. The physician may also be called upon to provide reassurance as to the propriety of variants of stimulative sexual behavior. Although the number of patients who are sexually incompatible as the result of the wife's or husband's total lack of sexual experience before marriage may well be declining, patients with this type of problem are seen occasionally. Moreover, many women have been taught that only certain specifics of sexual stimulation or certain coital positions are acceptable. These women do not readily accept any deviation from what they consider "right and proper" regardless of the interests of their marital partners. Victorianism, although vanishing from the American social scene, leaves a residual influence that may well require attention for at least another fifty years.

Teaching the sexually inadequate woman and her partner the basic rudiments of sexual anatomy may be extremely important. Many males, however experienced in coition, are unaware of the importance of adequate techniques for clitoral area stimulation. Few are aware that it is the gentle friction of the mons area or of the clitoral shaft rather than the clitoral glans that provides the most effective stimulation for the female partner. Moreover, many females as well as males are not aware of the basic physiology of sexual response and of the fact that physiologic orgasm takes place within the vagina and in the clitoris, regardless of where sensation is perceived by the female or initiated by the male.

In the development of a mutually stimulative sexual pattern it is important that the marital unit's move toward maximal female sexual responsiveness should be accompanied by the female's vocalizing such things as: specific sexual preferences, desired zones of erogenous stimulation, choice of coital positioning and, particularly, the fact of her approaching orgasm. The family unit must be taught to consider moments of individual preference for sexual encounter. Experimentation with varieties of time, place and sexual techniques should be made in order to achieve the necessary mood conducive to the female's successful sexual response. It is well to bear in mind that the two basic deterrents to female sexual responsiveness are *fatigue* and *preoccupation*.

The item in the therapeutic framework emphasizing gentleness and sensitivity needs little elaboration. But it should be noted that the male's approach — his ability to project both security and affection to the female — may be an absolute essential to any improvement in the female's sexual responsiveness. A reevaluation of the male's attitudes toward sex and toward women may be as important to the progress of therapy as the attention paid to his education in specific sexual techniques.

The second major interrogative direction (area two) in the treatment of frigidity is concerned with the possible difference in the degree of basic sexual tension demonstrated by the wife of the pair as opposed to that indicated by the husband. In analyzing this area of the husband-wife relationship, it should be emphasized that an impression of low level female sexual demand should only be established in relative comparison to a higher tension partner. A lower level of demand does not necessarily connote either inability to respond adequately to effective heterosexual stimulation or homosexual tendency. Yet, when such a divergence in sexual interest is encountered, there are inevitable misunderstandings between the marital partners. In some cases there may be a conscious sexual withdrawing by the lower response partner, developing from a sense of personal inadequacy or from a wish to punish what is considered as excessive demand. Conscious sexual withdrawal also may develop from a deep resentment or a sense of rejection felt by the partner wishing a higher degree of sexual participation.

The marital unit's understanding and acceptance of a difference in sexual tension demand is far more important than its causation and the determination of a specific spouse role-playing. A higher level of demand may well belong to either partner. This is evident in marriages between younger partners as well as in many marriages between older individuals. Feelings of sexual inadequacy, distrust, or withdrawal may be corrected by education of each mate to the other partner's individual, highly personal, sexual requirements. Thereafter, the problem becomes one of adjusting acknowledged differences in sexual tension to a mutally accepted plan for effective release of the higher level of demand. It has been noted frequently that the relief of inhibitions of the lower tension partner (once the family unit problem is understood) may be marked by a more receptive, or even anticipatory participation in family unit sexual activity, even though there is no permanent elevation of the lower level partner's own sexual tensions.[9]

As emphasized many times previously, the individual or combined interests of psychiatrists, psychologists, medical specialists, marriage counselors, social workers and clergymen may be needed to solve severe problems of sexual incompatibility. However, the advice of the initially consulted family physician frequently will be the most important step in relief of marital sexual maladjustments. The physician's forthright guidance and intial reassurance, whether he refers to other professionals or treats the patients himself, provide the best foundation for the solution of problems of sexually incompatible marriages.

References

1. Eisenbud, J. A psychiatrist looks at the report. In *Problems of Social Behavior*, pp. 20-27. Social Hygiene Association, New York, 1948.

2. Masters, W. H. The sexual response cycle of the human female: I. Gross anatomic considerations. *West. J. Surg.*, 68: 57-72, 1960.

3. Masters, W. H. The sexual response cycle of the human female: II. Vaginal lubrication. *Ann. New York Acad. Sc.*, 83: 301-317, 1959.

4. Masters, W. H., and Johnson, V. E. The physiology of vaginal reproductive function. *West J. Surg.*, 69: 105-120, 1961.

5. Masters, W. H., and Johnson, V. E. The sexual response cycle of the human female: III. The clitoris: anatomic and clinical considerations. *West. J. Surg.*, 70: 248-57, 1962.

6. Masters, W. H., and Johnson, V. E. The sexual response cycle of the human male: I. Gross anatomic considerations. *West. J. Surg.*, 71: 85-95, 1963.

7. Johnson, V. E. and Masters, W. H. Treatment of the sexually incompatible family unit. *Minnesota Med.*, 44: 466-71, 1961.

8. Johnson, V. E., and Masters, W. H. Sexual incompatibility: diagnosis and treatment. In *Human Reproduction and Sexual Behavior*, ed. Charles W. Lloyd, pp. 474-89. Lea & Febiger, Philadelphia, 1964.

9. Johnson, V. E., and Masters, W. H. A team approach to the rapid diagnosis and treatment of sexual incompatibility. *Pac. Med. & Surg.* (formerly *West. J. Surg.)*, 72: 371-75, 1964.

C. CULTURAL VARIATIONS

CHAPTER 31

Marital Sexuality in Four Cultures of Poverty

LEE RAINWATER

Oscar Lewis has asserted that "poverty becomes a dynamic factor which affects participation in the larger national culture and creates a subculture of its own. One can speak of a culture of the poor, for it has its own modalities and distinctive social and psychological consequences for its members . . . (it) cuts across regional, rural-urban and even national boundaries." He sees similarities between his own findings about the Mexican poor and findings of others in Puerto Rico, England, and the United States. This paper deals with such similarities in one area of interpersonal relationships, namely, in the attitudes and role behaviors which characterize the sexual relationships of lower-class husbands and wives in these four countries. This paper seeks to demonstrate that, in spite of important differences in the cultural forms of these four areas, there are a number of striking similarities in the ways husbands and wives act sexually and in the ways they regard their actions. The concern is thus not with the simplest level of sexual description — "Who has intercourse with whom, how and how often?" — but with the meanings these experiences have for their participants and for the assimilation of heterosexual with other family roles.

The materials from which this comparison is drawn are rather varied. For Mexico, there is Lewis's work on Tepoztlan (1); for Puerto Rico, J. M. Stycos' investigation of family and fertility in the northeastern area of the island (2) and David Landy's study of child-rearing in an eastern cane-raising area (3). For England, there are several studies of English working-class life, among which those of Spinley (4), Madeline Kerr (5), and Slater and Woodside (6) deal specifically with sexual relations. For the United States, there are the Kinsey studies (7) and this author's exploratory work on lower-class marital sexuality (8).

Methodologically, these studies represent two kinds of approaches: Lewis, Landy, Kerr, and Spinley depended on both anthropological field techniques and some systematic interviewing of respondents; this author's work and that of Stycos and of Slater and Woodside are based only on interviews. In none of these studies were sexual relations a primary focus of the study. Given these differences in method and focus, it is perhaps surprising that there should be as much compatibility in findings as in fact exists.

The Central Sexual Norm: "Sex Is a Man's Pleasure and a Woman's Duty"

That sexual relations exist for the pleasure of the man and that enjoyment for the woman is either optional or disapproved is specifically noted in each of these four cultures. Women are believed either not to have sexual desires at all or to have much weaker sexual needs, needs which do not readily come to the fore without stimulation from a man. In Tepoztlan, ". . . women who are passionate and 'need' men are referred to as *loca* (crazy) . . ." (9); in the other three areas, women are likely to be regarded as immoral if they show too much interest in sexual relations with their husbands. In Gorer's study of English character (10), one set of questionnaire items deals with the nature of women's sexuality. Gorer reports that the poor were more likely than the more affluent to agree that women "don't care much about the physical side of sex" and "don't have such an animal nature as men" and to disagree that women enjoy sex as much or more than men.

Man's nature demands sexual experience; he cannot be happy without it; and if he is not satisfied at home, it is understandable that he looks elsewhere. That the wife might look elsewhere is a common fantasy of men in these areas, but neither men nor women so often say that dissatisfaction with sexual relations is likely to lead the wife to stray. The husband's anxiety that his wife's "unnatural" impulses could lead her to look for a lover, however, is often given as a reason for not stimulating her too much or developing her sensual capacities through long or elaborated lovemaking (11). Stycos notes that Puerto Rican men expect such more elaborated sexual experiences in their relations with prostitutes or other "bad" women (12).

In all four areas, it is not considered appropriate for parents to devote attention to the sexual education of their children. Boys may be encouraged, either overtly or covertly, to acquire sexual experience. This seems most fully institutionalized in Puerto Rico (13). Elsewhere, the boy seems to be left more to his own devices. In any case, there is recognition that boys will acquire a fair amount of knowledge about sexual relations and that they probably will have intercourse with available women. In Puerto Rico and Tepoztlan, these women are seen as very much in the status of prostitutes or "loose women," and the boy feels he must be careful about approaching a more respectable girl. These lines seem more blurred in England and the United States — in these countries, in the context of group or individual dating, boys seem to feel freer about forcing their attentions and less vulnerable to repercussions from the girl's family (14). The Latin pattern of sharp separation of "loose women" and the virginal fiancee seems a highly vulnerable one in any case and quickly breaks down under the pressures of urbanization in a lower-class environment (15).

Girls, on the other hand, are supposed not to learn of sexual relations either by conversation or experience. Mothers in all four cultures do not discuss sex with their daughters and usually do not even discuss menstruation with them. The daughter is left very much on her own in this area, with only emergency attention from the mother — e.g., when the girl proves unable to cope with the trauma of onset of menses or begins to seem too involved with boys. Later, women will say that they were completely unprepared for sexual relations in marriage, that no one had ever told them about this, and that they had only the vaguest idea of what this shrouded part of their marital responsibility involved. Girls tend to be trained to a prudish modesty in relation to their bodies (even though they may also elaborate their dress or state of undress to attract boys) — in England, for example, Spinley notes that girls will not undress in front of each other or their mothers (16). Modesty in the two Latin cultures is, of course, highly elaborated.

The sexual stimulation that comes in all of these cultures from the close living together of children and adults is apparently systematically repressed as the child grows older. The sexual interests stimulated by these and other experiences are deflected for the boys onto objects defined as legitimate marks (loose women, careless girls, prostitutes, etc.) and for the girls are simply pushed out of awareness with a kind of hysterical defense (hysterical because of the fact that later women seem to protest their ignorance too much).

In these cultures, therefore, marriage is hardly made attractive from the point of view of providing sexual gratification. The girls are taught to fear sex and most often seem to learn to regard it in terms of the nonerotic gratifications it may offer. The boys learn that they may expect fuller sexual experiences from other, less respectable objects, and in some groups (Puerto Rico most overtly, 17), because of their identification of the wife as a "second mother," men have very potent reasons for not regarding the wife hopefully as a sexual object. Yet both boys and girls know that they will marry; the girls are anxious to do so, and the boys feign resistance more than maintain it.

For the girls, the transition to the married state often takes place via a period of high susceptibility to romantic love. The girl becomes involved with notions of falling in love with a man — this being but vaguely defined in her mind and oriented to an idealized conception of love and marriage. Stycos sees this as a "psychological mechanism intervening between . . . rebellion (against the cloistered life imposed by parents) and elopement, providing the dynamism by which this radical move can be made" (18). Spinley notes that girls fall in love more with love than with their particular boy friends (19). Arnold Green has described a similar pattern for a lower-class Polish group in the United States (20). Lewis notes that participation in courtship is one of the main gratifications of the adolescent period, albeit one flavored with many risks (21). The girl is both pushed toward marriage by her desire to get away from home (where demands for work and/or support tend to be made increasingly as she gets older) and *pulled* in that direction by her knowledge that the only appropriate role for a woman in her culture is that of wife and mother, that if she does not marry soon she runs the risk of being regarded as an immoral, "loose" woman or a ridiculous old maid.

For the young man, too, marriage looms large as representing the final transition to adult status. In Tepoztlan, only married men may hold responsible positions; without marriage, one is still tied to one's own father. In Puerto Rico, the situation is similar, and in addition, one cannot be considered truly masculine until one has fathered children, preferably sons. If one is to establish himself as an independent adult, he needs a woman to wash and cook and take care of him as his mother would. Sex, too, plays a part in the man's desire for marriage. Because in reality women of easy virtue are not as available as the norm has it, the man may want to be married to have sexual relations whenever he wants — as one American said, "It's nice to have it ready for me when I get home." Also, some American men express the desire to have a woman whom they know is "safe" and "clean," i.e., does not have a venereal disease.

But the assertion of independent adult status which marriage represents in these groups meets with considerable opposition from parents, more so for the girl than for the young man. Lewis notes that in the mid-forties as many as 50 percent of unions took place by elopement (22), and Stycos indicates a similarly high percentage for Puerto Rico (23). Because girls are not cloistered in the United States and England as they are in these Latin cultures, clear-cut elopement is not so much the pattern, but feigned surprise and anger are common parental responses, and it is not unusual for a premarital pregnancy to be used as a final argument to the parents to accept the marriage of their daughter to a man whom they like to feel is "not good enough for her." The wife in particular, then, is launched into marriage during a period of overt strain in her relations with her parents. Although later relationships with relatives may come to be central to her integration socially, during this period she is often more "alone" than at any other time in her life.

The "Honeymoon Trauma"

The adjustment to sexual relations is observed to be difficult for many women in all four groups. Stycos notes that in a majority of cases, the wedding night was traumatic for the woman, with trembling, weeping, and speechlessness being frequently reported (24). Several women were so frightened that they managed to delay the first intercourse for several days. Lewis reports a similar pattern for Tepoztlan (25), although he feels that the less sheltered girls of today are less likely to be so resistant. Lewis also notes that the fact that couples often start their marriage sleeping in the same room with the husband's family imposes additional constraints. Slater and Woodside indicate that many of their English lowerclass women reported unpleasant wedding nights but claimed to have overcome their initial fear and repugnance (26), a pattern also noted in the United States (27). It should be noted that the wife's modesty and reticence are not necessarily disapproved by her new husband; he may value them as an indication that she is still a virgin and that he is not "oversexed." Even so, he is confronted with a problem: while he does not wish his wife to be desirous independent of his initiation, he does need her cooperation, and he does not like to be made to feel guilty by her protests and fright.

Individual Behavior and the Norm

The effect of early socialization processes and later experiences, then, seems to be to establish the husband as the one to whom sexual relations are really important and the wife as the unwilling vehicle for his gratification. Given this cultural statement of the nature of men and women, what are the actual patterns of sexual gratification in these four cultures? For Tepoztlan, Lewis reports only what is presumably the majority pattern: ". . . much of the women's expressed attitudes toward sexual relations with their husbands dwell upon its negative aspects and reveal feelings of self-righteousness which border on martyrdom. Women speak of submitting to their husbands' 'abuse' because it is their obligation to do so" (28). For the husband's part, he reports, "Husbands do not expect their wives to be sexually demanding or passionate, nor are these viewed as desirable traits in a wife. Husbands do not complain if their wives are not eager for or do not enjoy sexual intercourse. . . . Some husbands deliberately refrain from arousing their wives sexually, because they do not want them to 'get to like it too much.' . . . Few husbands give attention to the question of their wives' sexual satisfaction. In general, sexual play is a technique men reserve for the seduction of other women." (Furthermore, a passionate wife may be considered a victim of black magic.) Perhaps some wives of Tepoztlan do enjoy sexual relations with their husbands, and the husbands do not object, but Lewis apparently did not find this pattern frequent enough to warrant comment (29).

Stycos reports a similar pattern for Puerto Rico: most women say they do not enjoy sexual relations; for them, sex is a duty and their emotional stance a continuation of the premarital rejection of sex as an appropriate interest for a woman (30). Women report a sense of disgust and revulsion about this necessary role, or they communicate a sense of detachment and minor irritation. Some women say they deceive the husband into believing that they enjoy sexual relations somewhat — perhaps to keep them from feeling too guilty, perhaps to allay any suspicion that they have a lover. The woman in this case seeks a balance of apparent enjoyment and reticence in which she communicates to her husband that her interest is solely due to her love for him, that her enjoyment is secondary to his right. Women use various excuses to cut down on the frequency of intercourse —they feign sleep, or illness, argue about the danger of becoming pregnant, welcome menstruation, seek to prolong postpartum abstinence — but they feel that such a course is risky because it may make the husband violent or suspect infidelity. Stycos also notes that over one-third of the women in his sample indicated real enjoyment of sexual relations, but he does not discuss how these women differ from the majority who to some extent reject sexual relations.

The patterns described by Stycos for Puerto Rico are also apparent in data from lower-class American families currently being studied by the author (31). Table 1 presents a tabulation of wives' attitudes toward sexual relations with their husbands for 195 middle- and lower-class women. The women's responses were categorized into three gross patterns: *highly accepting of sexuality* (referring to positive statements of interest in, desire for, and enjoyment of sexual relations with the husband and explicit or implicit indications that sexual relations were highly significant in the marital relationship), *moderately accepting of sexuality* (referring to positive statements about sexual relations with the husband, but without glowing testimony to the importance of, or gratification in, sexual relations, and often with an effort to place sexuality in proper perspective in relation to other activities and gratifications in marriage), and *lack of acceptance of sexuality* (in which the wife indicates that sexual relations are for the husband's gratification, not hers). In the middle class (no significant differences between upper-middles and lower-middles), only 14 percent of the women indicate lack of acceptance of sexuality; in the upper-lower class, this proportion rises to 31 percent; and in the lower-lower class, 54 percent of the women do not show acceptance of themselves as sexually interested and do not indicate enjoyment of sexual relations (32). The women who do not find sexual relations enjoyable range in their attitude toward the necessity to have intercourse: a good many try to neutralize the unpleasantness they feel, others are overtly hostile to the husband about his demands, but the latter pattern seems a difficult one to maintain since it has ready repercussions on the marital relationship, and generally the women fear their husbands will stray or desert them. Some of these women report with pride that they never directly refuse their husbands, although they use the same devices reported by Stycos to reduce the frequency of their husbands' demands. (One device reported by these women but not mentioned by Stycos is precipitating an argument with the husband so that he will go out to a tavern.)

TABLE 1. SOCIAL CLASS AND WIFE'S ENJOYMENT OF SEXUAL RELATIONS

	Middle Class (58)	Upper-Lower Class (68)	Lower-Lower Class (69)
Highly accepting: very positive statements about enjoyment	50%	53%	20%
Moderately accepting: enjoyment not emphasized	36	16	26
Lack of acceptance: avoidant or rejecting attitudes expressed	14	31	54
	100%	100%	100%

$X^2=26.48$, df=4, $P < .005$.

For England, Spinley reports only that the most common pattern is for sex to be only the man's pleasure (33), but Slater and Woodside supply some idea of the frequency of the wife's enjoyment of sexual relations. They report that only a minority of women find real gratification in sexual relations and that about half indicate that they do not participate of their own wish (34).

Women in all of these areas sometimes justify holding back from emotional participation in sexual relations by saying that they are less likely to become pregnant if they do not have orgasm. This is perhaps related to the general tendency, observed in the English and American reports at least, for the sexual relationship to become less and less involved, more automatic, after the first few years of marriage. There is not only a decrease in frequency, but also a tendency to relegate intercourse more and more to the category of satisfying the husband's biological need, and for whatever sense of mutuality has existed to wither. Several American women who reject sex comment that earlier in marriage they had sometimes enjoyed intercourse but that now, with many children and other preoccupations, would just as soon do without it. Slater and Woodside also note tht the longer-married women tend to be the more dissatisfied (35).

Kinsey's data on educational level and sexual behavior are by and large congruent with the patterns outlined here (36). He finds that although men of lower educational status are much more likely than men with more education to have premarital relations, women of this status are less likely to do so. He finds that for women, erotic arousal from any source is less common at the lower educational levels, that fewer of these women have ever reached orgasm, and that the frequency for those who do is lower. For men, he reports that foreplay techniques are less elaborated at the lower educational levels, most strikingly so with respect to oral techniques. In positional variations in intercourse, the lower educational levels show somewhat less versatility, but more interesting is the fact that the difference between lower and higher educational levels increases with age because variations among lower status men drop away rapidly with age, while the drop among more educated men is much less. The same pattern characterizes nudity in marital coitus.

One final aspect of marital sexuality can be considered: the prevalence of extramarital relations. As noted, the sexual norms of all of these cultures or subcultures treat extramarital relations on the part of men as understandable, sometimes as to be expected, while such relations by wives are strongly disapproved. Kinsey finds that men of below college level are more likely than others to have extramarital relations, but the lowest educational level is not the most frequent participant in such relationships. Also, the differential between college and grammar-school-only men disappears with age (by the 36-40 age period, college-level men show a slightly higher incidence than those of grammar school levels). Extramarital intercourse, except in the early married years, does not seem as highly class-bound as does premarital intercourse. Since no comparable data are available for the other three societies, it can only be said that extramarital relations by the husband — so long as there is no marked interference with the life of the couple — are not heavily condemned. Indeed, in Puerto Rico and Tepoztlan, at least, they are to be expected.

For women, considerable unclarity exists in the reports for all four areas. It is known that the partners for erring husbands are usually prostitutes or single, separated, or widowed women, but this is not always the case. It is not clear who the married women who participate in these affairs are, or whether they do so out of sexual desire or from other motives — to get even with the husband, to receive attention and presents from another man because the husband ignores them, or what. Kinsey's data indicate clearly that few lower-status women have extramarital relations and that the proportion is lower than among more educated women, especially for marriages of longer duration. It seems likely that in these lower-class groups, the concern about the wife's extramarital relations is more a manifestation of the husband's concern over her taking revenge for his domination than a prevalent pattern of deviance which he must realistically guard against.

Cultures mold the expression of sexual drives, the manifestations of male potency and female receptivity, in varied ways to conform to the requirements of particular social and cultural systems. Each individual in the system responds sexually not only, or even primarily, in terms of sexual drive, but in terms of the interpersonal implications which such action has (37). What, then, are the characteristics of the social systems and processes of socialization to which the patterns of sexual behavior and attitudes outlined above represent accommodations?

In all of these lower-class subcultures, there is a pattern of highly segregated conjugal role relationships. Men and women do not have many joint relationships; the separation of man's work and woman's work is sharp, as is the separation of man's and woman's play. Stycos indicates that half of the women in his sample do not report common activities with their husbands outside the home, and many of the remaining cases report only infrequent or limited outside activities (38); a similar pattern seems characteristic of Tepoztlan (39); and in both groups, the necessity for respect toward the husband reduces joint activities. This has been the traditional pattern in England and the United States, although trends toward less segregation are observable (40).

The low value placed on mutuality in sexual relations can be seen, then, as in part an extension of a more generalized pattern of separateness in the marital relationship. It is not difficult to understand that husbands and wives do not think of sexual relations as a way of relating intimately when they have so few other reasons for doing so. The role segregation of which the pattern of sexual relationships seems a part has as one consequence a considerable difficulty in communication between husbands and wives on matters not clearly defined in terms of traditional expectations. It is difficult for such couples to cope with problems which require mutual accommodation and empathy. This has been noted as one reason couples in these groups are not able to practice birth control effectively (41) and seems also to be involved in the distance husbands and wives feel with respect to sexual relations.

In these groups, then, husbands and wives tend to be fairly isolated from each other. They do not seem to be dependent on each other emotionally, though each performs important services for the other. In the traditional social systems of these groups, social integration is somewhat separate for the husband and wife. Each participates in relatively closed social networks to which he can look for a sense of stability and continuity in his life (42). For women, social relations often are organized about relations with kin; the woman regards herself as most importantly a person embedded in a network of kin extending upward to maternal figures and downward to her children. The importance of "Mum" for the English lower class has been noted by many observers, and Mum in turn is the center of a network of kin and neighbors which absorbs many of the emotional demands of the wife (43). In Tepoztlan, the young wife traditionally orients herself to her husband's mother or to her own at a later date. The tie with the grandmother also seems important emotionally. In the United States, this pattern of kin relating is not so sharply defined as in other areas, but some evidence exists that lower-class people maintain kin relationships more fully than do middle-class people (44). It is not clear from the Puerto Rican data how much wives orient themselves to kin networks, but it is clear that whatever their social relationships, these do not depend on joint relating by the couple to others (45).

The husband's social network is not as dependent on kindred as that of the wife, although his too tends to be closed in the sense that the men he relates to tend also to relate to each other. His status in the home, and among the wife's kin network, tends to be tangential. Though he may be defined as the final authority, by default he usually has less influence on what goes on from day to day than his wife or the maternal figure to whom she looks for guidance. His important social relationships are outside the home, with other men — some of whom may be relatives, others not. His performance as a husband and father is more influenced by the standards they set than by his wife's desires, just as her behavior is more influenced by the standards and expectations of her kin-based network (46).

What all of this suggests, in short, is that in a system characterized by closed social networks, the impact in the direction of highly segregated conjugal roles makes close and mutually gratifying sexual relations difficult because neither party is accustomed to relating intimately to the other. Further, a close sexual relationship has no particular social function in such a system since the role performances of husband and wife are organized on a separate basis, and no great contribution is made by a relationship in which they might sharpen their ability for cooperation and mutual regulation (47). It is possible that in such a system, a high degree of intimacy in the marital relationship would be antagonistic to the system since it might conflict with the demands of others in ego's social network (48).

However, not all lower-class couples can be said to be caught up in closed social networks of the kind just discussed. Where residence is neolocal and geographical movement breaks up both kin and lifelong community relationships, the lower-class couple finds itself either isolated or participating in loose social networks in which there is not so great an opportunity for relationships with others to take up the emotional slack of a segregated relationship between husband and wife. In these situations, there is a tendency for husband and wife to be thrown more on each other for meaningful standards and emotional support, a push in the direction of joint organization and joint role relationships (49). The impact of such a disruption of previous social networks is probably greatest on the wife. Since she does not have the work situation as a ready base for forming a new network of relationships, her network probably remains denuded for a longer period of time. This is apparent in some of the United States data (50) and can be inferred for Puerto Rico. In this situation, the stage is set for sexual relations to assume a more important role in the couple's relationship. The wife wishes her husband to be more affectionate and to spend more time with her. She may try to overcome her resistance to sex. The outcome will depend on the husband's ability to adapt to the new situation, in which his sexual need is not the sole factor, and in which he is expected to moderate his demandingness in the service of a mutually gratifying relationship. The relationship becomes one in which, as one American woman said, "He takes and I give, but I take, too." Given the cultural norms to which such couples have been socialized, however, such a delicate accommodation is not easily achieved (51).

The socialization experiences of husbands and wives, of course, provide the motivational basis for the role behaviors discussed above and account for some of the resistance to change which individuals show when the network of relationships changes (52). As has been noted, girls growing up are not encouraged to internalize a role as interested sexual partner, but are taught instead a complex of modesty, reticence, and rejection of sexual interests which continues into marriage. Girls are not rewarded, either prospectively or after marriage, by the significant others in their social networks for tendencies in the direction of passionate wife. Instead, the direction of proffered goals is toward functioning as a mother to children and husband, and perhaps continuing as daughter to a maternal figure. Although during late adolescence the girl may break from this pattern via romance and elopement, she finds that her greatest security comes from a return to the fold (if it is available). Finally, since her father has not been an integrated part of the household during her childhood, she has never developed the early psychic basis for a close relationship with a man which could be transferred to her husband. (Or the relationship with the father has come to have incestuous overtones which make a transfer difficult.)

The socialization experiences of the boy make greater allowance for a sexual role, but primarily in a narcissistic way. The boy learns to regard sex as a kind of eliminative pleasure for himself, and there is little emphasis on the sexual relationship as a social relationship. The aggressive component in sex is strongly emphasized, in the form of seduction and fantasies of raping, fantasies which in England (53) and the United States (54) are sometimes acted out in sham fashion as part of individual or group "dating" behavior. The masculinity of the *macho* pattern in the Latin cultures incorporates an aggressive pride in relations with men and with women; with respect to women, it stresses both having many partners and being insistent on taking one's own pleasure in each relationship. In the English and American lower class, there is no comparable name for this pattern, but the behavior encouraged is very similar. This exaggerated masculinity may be viewed as an overcompensation for the difficulty the boy has in developing a masculine identity. His early life is spent very much in the company of women; he tends to identify more with his mother than with his father since the latter is not much in the home and tends not to interact closely with his children (55). The heavy emphasis on man's pleasure and indifference to the woman serves to ward off feelings of inadequacy stemming from both past difficulties in identification and present marginal status in the family (56). Given this vulnerability in his sense of masculine competence, it is not difficult to understand that the husband in these groups does not want to complicate his functioning as a sexual partner by having to take into account, or by stimulating, his wife's needs and demands. Nor does he want to feel that he has to compete with other men to keep her affections.

Thus, even though the current social situation may encourage joint role organization and greater dependence of husband and wife on each other, the legacy of socialization directed toward a system in which men and women orient themselves to different social networks and sharply segregated conjugal roles makes change difficult, and reduces the frequency with which couples develop sexual relations involving mutual gratification.

Summary

This paper began with an examination of similarities in the patterns of marital sexuality in four "cultures of the poor," all part of, or strongly dominated by, the overall Western European culture (57). It was shown that among the lower classes of certain communities in Mexico, Puerto Rico, England, and the United States, there are significant similarities in the sentiments expressed by husbands and wives concerning sexual experiences and in their expectations about sexual role performances by the two marital partners. The paper has concluded with an explanatory hypothesis which seems adequate to account for the data from these four cultures, but which perhaps has wider applicability (58). It will occur to the reader, for example, that middle-class marriage according to the "Victorian" model was and is similarly marked by a lack of mutuality in sexual relations; in place of mutuality, a variety of repressive and mistress-lover patterns have developed (59). A more general hypothesis can be advanced which very likely is relevant to these situations also. This is that *in societies where there is a high degree of segregation in the role relationships of husbands and wives, the couple will tend not to develop a close sexual relationship, and the wife will not look upon sexual relations with her husband as sexually gratifying* (although she may desire such relations as signifying the continuing stability of the relationship). This leaves open the question of whether in other cultures having such a role segregation pattern, the wife may commonly seek other relationships for sexual gratification, as on Truk (60), nor does it take into account the complexities introduced in societies in which polygamy is common. Should such a hypothesis have wider validity, it would represent an additional step toward understanding patterns of sexual relations in different societies as more than anthropological *curiosa.*

References

1. Oscar Lewis, *Life in a Mexican Village: Tepoztlán Restudied.1 Urbana: University of Illinois Press, 1951.*

2. *J. Mayone Stycos, Family and Fertility in Puerto Rico: A Study of the Lower Income Group.* New York: Columbia University Press, 1955.

3. David Landy, *Tropical Childhood: Cultural Transmission and Learning in a Rural Puerto Rican Village.* Chapel Hill: University of North Carolina Press, 1959.

4. B. M. Spinley, *The Deprived and the Privileged: Personality Development in English Society.* London: Routledge & Kegan Paul, 1953.

5. Madeline Kerr, *The People of Ship Street.* London: Routledge & Kegan Paul, 1958.

6. Eliot Slater and Moya Woodside, *Patterns of Marriage: A Study of Marriage Relationships in the Urban Working Classes.* London: Cassell, 1951.

7. Alfred C. Kinsey, et al. *Sexual Behavior in the Human Male.* Philadelphia: W. B. Saunders, 1948; and *Sexual Behavior in the Human Female.* Philadelphia: W. B. Saunders, 1953.

8. Lee Rainwater, *And the Poor Get Children: Sex, Contraception and Family Planning in the Working Class.* Chicago: Quadrangle Books, 1960.

9. Lewis, *Life in a Mexican Village,* p. 326.

10. Geoffrey Gorer, *Exploring English Character.* New York: Criteron Books, 1955, pp. 115–16.

11. Lewis, *Life in a Mexican Village,* p. 326; Slater and Woodside, *Patterns of Marriage,* p. 172.

12. Stycos, *Family and Fertility in Puerto Rico,* p. 143.

13. Ibid., pp. 42-43; Landy, *Tropical Childhood,* pp. 108, 159-60.

14. Spinley, *The Deprived and the Privileged,* p. 87.

15. Cf. Oscar Lewis, *Five Families: Mexican Case Studies in the Culture of Poverty.* New York: Basic Books, 1959; and *The Children of Sánchez: Autobiography of a Mexican Family.* New York: Random House, 1961.

16. Spinley, *The Deprived and the Privileged,* pp. 62-63.

17. Stycos, *Family and Fertility in Puerto Rico,* p. 142.

18. Ibid., p. 99.

19. Spinley, *The Deprived and the Privileged,* p. 87.

20. Arnold W. Green, "The 'Cult of Personality' and Sexual Relations." *Psychiatry,* vol. 4, 1941, pp. 343-44.

21. Lewis, *Life in a Mexican Village,* pp. 399-405.
22. Ibid., p. 407.
23. Stycos, *Family and Fertility in Puerto Rico,* pp. 91-97.
24. Ibid., pp. 134-35.
25. Lewis, *Life in a Mexican Village,* pp. 326-27.
26. Slater and Woodside, *Patterns of Marriage,* p. 173.
27. Rainwater, *And the Poor Get Children,* pp. 60-64.
28. Lewis, *Life in a Mexican Village,* p. 326.
29. Lewis' studies of urban Mexican couples suggest that a more complex sexual relationship is common (*Five Families, The Children of Sánchez*).
30. Stycos, *Family and Fertility in Puerto Rico,* pp. 134-42.
31. Rainwater, *Family Design: Marital Sexuality, Family Size and Contraception.* Chicago: Aldine, 1964.
32. These data are taken from a study currently in progress concerning family size desires and family planning success based on interviews with 150 couples (husband and wife interviewed separately) and 100 individual husbands and wives. An analysis of the latter 100 interviews has been presented in Rainwater, *And The Poor Get Children;* the discussion of sexual relations is given on pp. 92-121. A fuller analysis, based on the total sample, of variations in sexual relations by social class status is presented in Rainwater, *Family Design.* The lower class sample tabulated in Table 1 includes both Negroes and whites. However, at each class level, the differences in acceptance of sexual relations between Negro and white wives are very small.
33. Spinley, *The Deprived and the Privileged,* p. 61.
34. Slater and Woodside, *Patterns of Marriage,* pp. 168-69.
35. Slater and Woodside note that women frequently shift from "he" to "they" when discussing sexual relationships with the husband, a usage apparent also in the American interviews. Apparently, a good many women find it difficult in connection with sex to think of the husband as other than a representative of demanding men as a type.
36. The Kinsey data suggest that late in marriage (36-40 age period) at the lowest educational level, men have extramarital relations four times as often as women; at the intermediate level slightly more than twice as often; and at the highest educational level, only 50 percent more often.
37. Margaret Mead, *Male and Female.* New York: Morrow, 1949, pp. 201-222.
38. Stycos, *Family and Fertility in Puerto Rico,* p. 149.
39. Lewis, *Life in a Mexican Village,* pp. 319-25.
40. The literature on the English and American working class is quite extensive; in addition to the works cited in the discussion of sexual relations, there are several English studies (Raymond Firth, *Two studies of Kinship in London.* London: Athlone Press, 1956; J. M. Mogey, *Family and Neighborhood.* London: Oxford University Press, 1956; Richard Hoggart, *The Uses of Literacy.* London: Chatto & Windus, 1957; Michael Young and Peter Willmott, *Family and Kinship in East London.* London: Routledge & Kegan Paul, 1957; Peter Willmott and Michael Young, *Family and Class in a London Suburb.* New York: Humanities Press, 1961), and several American studies (Allison Davis, Burleigh B. Gardner, and Mary R. Gardner, *Deep South: A Social and Anthropological Study of Caste and Class.* Chicago: University of Chicago Press, 1941; Allison Davis and Robert J. Havighurst, *Father of the Man.* New York: Houghton Mifflin, 1947; Allison Davis, *Social Class Influences upon Learning.* Cambridge: Harvard University Press, 1952; Martin B. Loeb, *Social Class as Evaluated Behavior,* unpublished Ph.D. dissertation, University of Chicago, 1957; Lee Rainwater, Gerald Handel, Richard P. Coleman, *Workingman's Wife: Her Personality, World and Life Style.* New York: Oceana Publications, 1959; Jerome K. Myers and Bertram H. Roberts, *Family and Class Dynamics in Mental Illness.* New York: John Wiley, 1959;

Bennett M. Berger, *Working Class Suburb: A Study of Auto Workers in Suburbia.* Berkeley: University of California Press, 1960; Herbert Gans, *Urban Villagers,* Glencoe, Ill.: Free Press, 1962), which discuss family relationships in a way pertinent to this paper.

41. Reuben Hill, J. M. Stycos, and Kurt W. Back, *The Family and Population Control.* Chapel Hill: University of North Carolina Press, 1959.

42. Elizabeth Bott, *Family and Social Network.* London: Tavistock, 1957.

43. Kerr, *The People of Ship Street;* Firth, *Two Studies of Kinship;* Young and Willmott, *Family and Kinship.*

44. Davis, Gardner, and Gardner, *Deep South;* Rainwater, Handel, and Coleman, *Workingman's Wife.*

45. Various Puerto Rican sources (Stycos, *Family and Fertility in Puerto Rico;* Julian Steward, ed., *The People of Puerto Rico.* Urbana: University of Illinois Press, 1956; Landy, *Tropical Childhood*) suggest that there is some tendency for the wife to retain a tie with a maternal figure after marriage, but not in the same organized way that is traditional in Tepoztlan or common with the "Mum" system in England.

46. Elizabeth Bott *(Family and Social Network),* from a sample not confined to the lower class, finds a correlation between the value placed by a couple on sexual relations and the degree of joint organization in their role relationships and "looseness" of their social network. She notes, in the one case of a highly segregated conjugal role relationship associated with a closed network, that the wife "felt physical sexuality was an intrusion on a peaceful domestic relationship . . . as if sexuality was felt to be basically violent and disruptive" (p. 73). Among families having a joint conjugal role relationship associated with a loose-knit network, ". . . a successful sexual relationship was felt . . . to be very important for a happy marriage . . . to prove that all was well with the joint relationship, whereas unsatisfactory relations were indicative of a failure in the total relationship" (p. 83). And among couples with intermediate segregation of role relationships and medium-knit networks, "In general, the greater the importance attached to joint organization and shared interests, the greater the importance attached to sexual relations" (p. 88). In the present author's current research on American lower-class couples referred to above, a similar pattern emerges. Among lower-class wives with highly segregated role relationships, 60 percent indicated a lack of acceptance of sexual relations; among those with less segregated role relationships, only 28 percent indicated such a lack of acceptance. Phrased another way, 77 percent of the wives in highly segregated role relationships indicated less enjoyment of sexual relationships than did their husbands; among wives in less segregated relationships, only 36 percent were less interested in, or enjoyed sex less than, their husbands.

47. Erik H. Erikson, *Childhood and Society.* New York: Norton, 1950.

48. See the closely related discussions by William J. Goode ("The Theoretical Importance of Love," *American Sociological Review,* vol. 24, 1959, pp. 38-47) and Max Gluckman (*Custom and Conflict in Africa.* Oxford: Blackwell, 1955), in which these authors argue that romantic love between a couple tends to interfere with other important solidarities in a society and that therefore societies tend to operate in ways that keep love from disrupting existing social arrangements. While the efforts of society to control love are clearest in connection with mate selection, these efforts continue after marriage also and have some bearing on the kind of sexual relationship that commonly exists among couples in a society.

49. Bott, *Family and Social Network.*

50. Rainwater et al., *Workingman's Wife.*

51. In a study of Detroit wives' feeling about their marriages, Robert O. Blood and Donald M. Wolfe (*Husbands and Wives: The Dynamics of Married Living.* Chicago: Free Press, 1960, pp. 221-35), it was found that the wife's satisfaction with the "love and affection" she receives from her husband increases steadily with social status. This seems related to the fact that there is a greater degree of sharing, communication, and joint participation in social relations at the higher- than on the lower-status levels.

52. Melford Spiro, "Social Systems, Personality and Functional Analysis," in *Studying Personality Cross-Culturally,* ed. Bert Kaplan. Evanston, Ill.: Row, Peterson, 1961.

53. Spinley, *The Deprived and the Privileged,* p. 87.

54. Green, "The 'Cult of Personality' and Sexual Relations," pp. 343-44.

55. Spinley, *The Deprived and the Privileged,* pp. 81-82.

56. Kerr, *The People of Ship Street,* p. 88.

57. Jamaica represents a fifth "culture of poverty" that has been the subject of a number of studies which touch on sexual life (Judith Black, *Family Structure in Jamaica: The Social Context of Reproduction.* Glencoe, Ill.: Free Press, 1961; Fernando Henriques, *Family and Colour in Jamaica.* London: Eyre & Spottiswoode, 1953; Yehudi A. Cohen, *Social Structure and Personality: A Casebook.* New York: Holt, Rinehart & Winston, 1961, pp. 71-81, 167-81). However, the specific discussion of attitudes and feelings toward sexuality is both conflicting and somewhat offhand. None of the studies contains the detail of the references for the four subcultures discussed in this paper. On balance, the Jamaican studies suggest that marital sexuality there has much in common with the pattern outlined here, but this can be asserted only very tentatively.

58. Cf. Margaret Mead's *Growing Up in New Guinea* (New York: Morrow, 1930, pp. 101-102) for a description of marital sexual relations among the Manus: "Unrelieved by romantic fictions or conventions of wooing, untouched by tenderness, unbulwarked by cooperativeness and good feeling as between partners, unhelped by playfulness, preliminary play or intimacy, sex is conceived as something bad, inherently shameful, something to be relegated to the darkness of the night. . . . Most women welcome children because it gives their husbands a new interest and diverts their unwelcome attentions from themselves." The pattern of rejection of her sexual role by a wife can be viewed as a rather extreme case of the lack of attachment to a social role in spite of commitment to it, a commitment which, in Erving Goffman's sense, has fateful consequences for the performer. As Goffman notes (*Encounters: Two Studies in the Sociology of Interaction.* Indianapolis: Bobbs-Merrill, 1961, pp. 88-89), social scientists have tended to "neglect the many roles that persons play with detachment, shame or resentment."

59. Richard Lewinsohn, *A History of Sexual Customs.* London: Longmans, Green, 1958. Also see Joseph A. Banks's *Prosperity and Parenthood: A Study of Family Planning Among the Victorian Middle Classes.* (London: Routledge & Kegan Paul, 1954), in which the discussion is congruent with the view presented here, although without the desirable empirical detail. Clellan S. Ford and Frank A. Beach's *Patterns of Sexual Behavior* (New York: Harper, 1959) is a cross-cultural study of sexual behavior which unfortunately contains almost nothing about the particular aspect of sexual behavior under discussion here, being instead concerned mainly with sexual techniques, rules for sexual mateships and liaisons, and prevalence of particular kinds of sexual behavior.

60. Marc J. Swartz, "Sexuality and Aggression on Romonum, Truk." *American Anthropologist,* vol. 60, no. 3, June 1958, pp. 467–86.

CHAPTER 32

The Inexpressive Male: A Tragedy of American Society

JACK O. BALSWICK and CHARLES W. PEEK

The problem of what it means to be "male " and "female" is a problem which is faced and dealt with in its own way in every society. Through cross-cultural research one now surmises that culture rather than "nature" is the major influence in determining the temperamental differences between the sexes. It may be no accident that a woman, Margaret Mead, did the classic study demonstrating that temperamental differences between the sexes are explained very little in terms of innateness, but rather in terms of culture. In her book *Sex and Temperament*, Mead reported on the differences in sex roles for three New Guinea societies. Using ethnocentric Western standards in defining sex roles, she found that the ideal sex role for both the male and female was essentially "feminine" among the Arapesh, "masculine" among the Mundugumor, and "feminine" for the male and "masculine" for the female among the Tchambuli. The Tchambuil represents a society that defines sex roles in a complete reversal of the traditional distinctions made between masculine and feminine roles in the United States.

It is the purpose of this paper to consider a particular temperament trait that often characterizes the male in American society. As sex role distinctions have developed in America, the male sex role, as compared to the female sex role, carries with it prescriptions which encourage inexpressiveness. In some of its extreme contemporary forms, the inexpressive male has even come to be glorified as the epitome of a real man. This will be discussed later in the paper when two types of inexpressive male are examined.

The Creation of the Inexpressive Male

Children, from the time they are born, both explicitly and implicitly are taught how to be a man or how to be a woman. While the girl is taught to act "feminine" and to desire "feminine" objects, the boy is taught how to be a man. In learning to be a man, the boy in American society comes to value expressions of masculinity and devalue expressions of femininity. Masculinity is expressed largely through physical courage, toughness, competitiveness, and aggressiveness, whereas femininity is, in contrast, expressed largely through gentleness, expressiveness, and responsiveness. When a young boy beings to express his emotions through crying, his parents are quick to assert, "You're a big boy and big boys don't cry." Parents often use the term, "He's all boy," in reference to their son, and by this term usually refer to behavior which is an expression of aggressiveness, getting into mischief, getting dirty, etc., but never use the term to denote behavior which is an expression of affection, tenderness, or emotion. What parents are really telling their son is that a real man does not show his emotions and if he is a real man he will not allow his emotions to be expressed. These outward expressions of emotion are viewed as a sign of femininity, and undesirable for a male.

Is it any wonder, then, that during the most emotional peak of a play or movie, when many in the audience have lumps in their throats and tears in their eyes, that the adolescent boy guffaws loudly or quickly suppresses any tears which may be threatening to emerge, thus demonstrating to the world that he is above such emotional feeling?

The Inexpressive Male as a Single Man

At least two basic types of inexpressive male seem to result from this socialization process: the cowboy and the playboy. Manville (1969) has referred to the *cowboy type* in terms of a "John Wayne neurosis," which stresses the strong, silent, and two-fisted male as the 100 percent American he-man. For present purposes, it is especially in his relationship with women that the John Wayne neurosis is particularly significant in representing many American males. As portrayed by Wayne in any one of his many type-cast roles, the mark of a real man is that he does not show any tenderness or affection toward girls because his culturally acquired male image dictates that such a show of emotions would be distinctly unmanly. If he does have anything to do with girls, it is on a "man to man" basis: the girl is treated roughly (but not sadistically), with little hint of gentleness or affection. As Manville puts it:

> The on-screen John Wayne doesn't feel comfortable around women. He does like them sometimes — God knows he's not *queer*. But at the right time, and in the right place — which he chooses. And always with his car/horse parked directly outside, in/on which he will ride away to his more important business back in Marlboro country [1969, 111].

Alfred Auerback, a psychiatrist, has commented more directly (1970) on the cowboy type. He describes the American male's inexpressiveness with women as part of the "cowboy syndrome." He quite rightly states that "the cowboy in moving pictures has conveyed the image of the rugged 'he-man,' strong, resilient, resourceful, capable of coping with overwhelming odds. His attitude toward women is courteous but reserved." As the cowboy equally loved his girl friend and his horse, so the present day American male loves his car or motorcycle and his girl friend. Basic to both these descriptions is the notion that the cowboy does have feelings toward women but does not express them, since ironically such expression would conflict with his image of what a male is.

The *playboy type* has recently been epitomized in *Playboy* magazine and by James Bond. As with the cowboy type, he is resourceful and shrewd, and interacts with his girl friend with a certain detachment which is expressed as "playing it cool." While Bond's relationship with women is more in terms of a Don Juan, he still treats women with an air of emotional detachment and independence similar to that of the cowboy. The playboy departs from the cowboy, however, in that he is also "non-feeling." Bond and the playboy he caricatures are in a sense "dead" inside. They have no emotional feelings toward women, while Wayne, although unwilling and perhaps unable to express them, does have such feelings. Bond rejects women as women, treating them as consumer commodities; Wayne puts women on a pedestal. The playboy's relationship with women represents the culmination of Fromm's description of a marketing-oriented personality in which a person comes to see both himself and others as persons to be manipulated and exploited. Sexuality is reduced to a packageable consumption item which the playboy can handle because it demands no responsibility. The woman in the process becomes reduced to a playboy accessory. A successful "love affair" is one in which the bed was shared, but the playboy emerges having avoided personal involvement or a shared relationship with the woman.

The playboy, then, in part is the old cowboy in modern dress. Instead of the crude mannerisms of John Wayne, the playboy is a skilled manipulator of women, knowing when to turn the lights down, what music to play on the stereo, which drinks to serve, and what topics of conversation to pursue. The playboy, however, is not a perfect likeness; for unlike the cowboy, he does not seem to care for the women from whom he withholds his emotions. Thus, the inexpressive male as a single man comes in two types: the inexpressive feeling man (the cowboy) and the inexpressive non-feeling man (the playboy).

The Inexpressive Male as a Married Man

When the inexpressive male marries, his inexpressiveness can become highly dysfunctional to his marital relationship *if* he continues to apply it across the board to all women, his wife included. The modern American family places a greater demand upon the marriage relationship than did the family of the past. In the typical marriage of one hundren or even fifty years ago, the roles of both the husband and the wife were clearly defined as demanding, task-oriented functions. If the husband successfully performed the role of provider and protector of his wife and family and if the wife performed the role of homemaker and mother to her children, chances were the marriage was defined as successful, from both a personal and a societal point of view. The traditional task functions which in the past were performed by the husband and wife are today often taken care of by indviduals and organizations outside the home. Concomitant with the decline of the task functions in marriage has been the increase in the importance of the companionship and affectionate function in marriage. As Blood and Wolfe (1960, 172) concluded in their study of the modern American marriage, "companionship has emerged as the most valued aspect of marriage today."

As American society has become increasingly mechanized and depersonalized, the family remains as one of the few social groups where what sociologists call the primary relationship has still managed to survive. As such, a greater and greater demand has been placed upon the modern family and especially the modern marriage to provide for affection and companionship. Indeed, it is highly plausible to explain the increased rate of divorce during the last seventy years, not in terms of a breakdown in marriage relationships, but instead as resulting from the increased load which marriage has been asked to carry. When the husband and wife no longer find affection and companionship from their marriage relationship they most likely question the wisdom of attempting to continue in their conjugal relationship. When affection is gone, the main reason for the marriage relationship disappears.

Thus, within the newly defined affectively oriented marriage relationship male inexpressiveness toward *all* women, wife included, would be dysfunctional. But what may happen for many males is that through progressively more serious involvements with women (such as going steady, being pinned, engagement, and the honeymoon period of marriage), they begin to make some exceptions. That is, they may learn to be *situationally rather than totally inexpressive*, inexpressive toward women in most situations but not in all. Like the child who learns a rule and then, through further experience, begins to understand the exceptions to it, many American males may pick up the principle of inexpressiveness toward women, discovering its exceptions as they become more and more experienced in the full range of man-woman relationships. Consequently, they may become more expressive toward their wives while remaining essentially inexpressive toward other women; they learn that the conjugal relationship is one situation that is an exception to the cultural requirement of male inexpressiveness. Thus, what was once a double *sexual* standard, where men had one standard of sexual conduct toward their fiancée or wife and another toward other women, may now be primarily a double *emotional* standard, where men learn to be expressive toward their fiancée or wife but remain inexpressive toward women in general.

To the extent that such situational inexpressiveness exists among males, it should be functional to the maintenance of the marriage relationship. Continued inexpressiveness by married males toward women other than their wives would seem to prohibit their forming meaningful relationships with these women. Such a situation would seem to be advantageous to preserving their marital relationships, since "promiscuous" expressiveness toward other women could easily threaten the stability of these companionship-oriented marital relationships.

In short, the authors' suggestion is that situational inexpressiveness, in which male expressiveness is essentially limited to the marital relationship, may be one of the basic timbers shoring up many American marriages, especially if indications of increasing extramarital sexual relations are correct. In a sense, then, the consequences of situational inexpressiveness for marital relationships do not seem very different from those of prostitution down through the centuries, where prostitution provided for extramarital sex under circumstances which discouraged personal affection toward the female partner strong enough to undermine the marital relationship. In the case of the situationally inexpressive husband, his inexpressiveness in relations with women other than his wife may serve as a line of defense aginst the possible negative consequences of such involvement toward marital stability. By acting as the cowboy or playboy, therefore, the married male may effectively rob extramarital relationships of their expressiveness and thus preserve his marital relationship.

The inexpressiveness which the American male early acquires may be bothersome in that he has to partially unlearn it in order to effectively relate to his wife. However, if he is successful in partially unlearning it (or learning a few exceptions to it), then it can be highly functional to maintaining the conjugal relationship.

But what if the husband does not partially unlearn his inexpressiveness? Within the newly defined expressive function of the marriage relationship, he is likely to be found inadequate. The possibility of an affectionate and companionship conjugal relationship carries with it the assumption that both the husband and wife are bringing into marriage the expressive capabilities to make such a relationship work. This being the case, American society is ironically short-changing males in terms of their ability to fulfill this role expectation. Thus, society inconsistently teaches the male that to be masculine is to be inexpressive, while at the same time, expectations in the marital role are defined in terms of sharing affection and companionship which involves the ability to communicate and express feelings. What exists, apparently, is another example of a discontinuity in cultural conditioning of which Benedict (1938) spoke more than thirty years ago.

Conclusion and Summary

It has been suggested that many American males are incapable of expressing themselves emotionally to a woman, and that this inexpressiveness is a result of the way society socialized males into their sex role. However, there is an alternative explanation which should be explored, namely, that the learning by the male of his sex role may not actually result in his inability to be expressive, but rather only in his thinking that he is not supposed to be expressive. Granted, according to the first explanation, the male cannot express himself precisely because he was taught that he was not supposed to be expressive, but in this second explanation inexpressiveness is a result of present perceived expectations and not a psychological condition which resulted from past socialization. The male perceives cultural expectations as saying, "Don't express yourself to women," and although the male may be capable of such expressiveness, he "fits" into cultural expectations. In the case of the married male, where familial norms do call for expressiveness to one's wife, it may be that the expectations for the expression of emotions to his wife are not communicated to him.

There has been a trickle of evidence which would lend support to the first explanation, which stresses the male's incapacity to be expressive. Several studies (Balswick, 1970; Hurvitz, 1964; Komarovsky, 1962; Rainwater, 1965) have suggested that especially among the lowly educated, it is the wife playing the feminine role who is often disappointed in the lack of emotional concern shown by her husband. The husband, on the other hand, cannot understand the relatively greater concern and emotional expressiveness which his wife desires, since he does not usually feel this need himself. As a result of her research, Komarovsky (1962, 156) has suggested that "the ideal of masculinity into which . . . [men are] socialized inhibits expressiveness both directly, with its emphasis on reserve, and indirectly, by identifying personal interchange with the feminine role." Balswick (1970) found that males are less capable than females of expressing or receiving companionship support from their spouses. His research also supports the view than inadequacy of expressiveness is greatest for the less educated males. Although inexpressiveness may be found among males at all socioeconomic levels, it is especially among the lower-class males that expressiveness is seen as being inconsistent with their defined masculine role.

There may be some signs that conditions which have contributed toward the creation of the inexpressive male are in the process of decline. The deemphasis in distinctiveness in dress and fashions between the sexes, as exemplified in the "hippy" movement, can be seen as a reaction against the rigidly defined distinctions between the sexes which have characterized American society. The sexless look, as presently being advanced in high fashion, is the logical end reaction to a society which has superficially created strong distinctions between the sexes. Along with the blurring of sexual distinctions in fashion may very well be the shattering of the strong, silent male as a glorified type. There is already evidence of sharp criticisms of the inexpressive male and exposure of him as a constituting a "hangup." Marriage counselors, sensitivity group leaders, "hippies," and certainly youth in general are critical of inexpressiveness, and want candid honesty in interpersonal relations. Should these views permeate American society, the inexpressive male may well come to be regarded as a pathetic tragedy instead of the epitome of masculinity and fade from the American scene. Not all may applaud his departure, however. While those interested in more satisfactory male-female relationships, marital and otherwise, will probably gladly see him off, those concerned with more stable marital relationships may greet his departure less enthusiastically. Although it should remove an important barrier to satisfaction in all male-female relationships via an increase in the male's capacity for emotional response toward females, by the same token it also may remove a barrier against emotional entanglement in relations with females outside marital relationships and thus threaten the stability of marriages. If one finds the inexpressive male no longer present one of these days, then, it will be interesting to observe whether any gains in the stability of marriage due to increased male expressiveness *within* this relationship will be enough to offset losses in stability emanating from increasing displays of male expressiveness *outside* it.

References

Auerback, Alfred. The Cowboy Syndrome. Summary of research contained in a personal letter from the author, 1970.

Balswick, Jack O. The Effect of Spouse Companionship Support on Employment Success. *Journal of Marriage and the Family*, 1970, 32, 212-15.

Benedict, Ruth. Continuities and Discontinuities in Cultural Conditioning. *Psychiatry*, 1938, 1, 161-67.

Blood, Robert, and Donald Wolfe. *Husbands and Wives: The Dynamic of Married Living*. Glencoe, Ill.: Free Press, 1960.

Cox, Harvey. Playboy's Doctrine of Male. In Wayne H. Cowan, ed., *Witness to a*

Generation: Significant Writings from Christianity and Crisis (1941-1966) Indianapolis: Bobbs-Merrill, 1966.

Hurvitz, Nathan. Marital Strain in the Blue-Collar Family. In Arthur Shostak and William Gomberg, eds., *Blue-Collar World*. Englewood Cliffs, N.J.: Prentice-Hall, 1964.

Komarovsky, M. *Blue-Collar Marriage*. New York: Random House, 1962.

Mead, Margaret. *Sex and Temperament in Three Primitive Societies*. New York: Morrow, 1935.

Manville, W. H. The Locker Room Boys. *Cosmopolitan*, 1969, 166 (11), 110-15.

Popplestone, John. The Horseless Cowboys. *Trans-action*, 1966, 3, 25-27.

Rainwater, Lee. *Family Design: Marital Sexuality, Family Size, and Contraception.* Chicago: Aldine, 1965.

CHAPTER 33

Transition to Parenthood

ALICE S. ROSSI

The Problem

The central concern in this sociological analysis of parenthood will be with two closely related questions. **1.** What is involved in the transition to parenthood: what must be learned and what readjustments of other role commitments must take place in order to move smoothly through the transition from a childless married state to parenthood? **2.** What is the effect of parenthood on the adult: in what ways do parents, and in particular mothers, change as a result of their parental experiences?

To get a firmer conceptual handle on the problem, I shall first specify the stages in the development of the parental role and then explore several of the most salient features of the parental role by comparing it with the two other major adult social roles — the marital and work role. Throughout the discussion, special attention will be given to the social changes that have taken place during the past few decades which facilitate or complicate the transition to and the experience of parenthood among young American adults.

From Child to Parent: An Example

What is unique about this perspective on parenthood is the focus on the adult parent rather than the child. Until quite recent years, concern in the behavioral sciences with the parent-child relationship has been confined almost exclusively to the child. Whether a psychological study such as Ferreira's on the influence of the pregnant woman's attitude to maternity upon postnatal behavior of the neonate (1), Sears and Maccoby's survey of child-rearing practices (2), or Brody's detailed observations of mothering (3), the long tradition of studies of maternal deprivation (4) and more recently of maternal employment (5), the child has been the center of attention. The design of such research has assumed that, if enough were known about what parents were like and what they in fact did in rearing their children, much of the variation among children could be accounted for (6).

The very different order of questions which emerge when the parent replaces the child as the primary focus of analytic attention can best be shown with an illustration. Let us take, as our example, the point Benedek makes that the child's need for mothering is *absolute* while the need of an adult woman to mother is *relative* (7). From a concern for the child, this discrepancy in need leads to an analysis of the impact on the child of separation from the mother or inadequacy of mothering. Family systems that provide numerous adults to care for the young child can make up for this discrepancy in need between mother and child, which may be why ethnographic accounts give little evidence of postpartum depression following childbirth in simpler societies. Yet our family system of isolated households, increasingly distant from kinswomen to assist in mothering, requires that new mothers shoulder total responsibility for the infant precisely for that stage of the child's life when his need for mothering is far in excess of the mother's need for the child.

From the perspective of the mother, the question has therefore become: what does maternity deprive her of? Are the intrinsic gratifications of maternity sufficient to compensate for shelving or reducing a woman's involvement in non-family interests and social roles? The literature on maternal deprivation cannot answer such questions, because the concept, even in the careful specification Yarrow has given it (8), has never meant anything but the effect on the child of various kinds of insufficient mothering. Yet what has been seen as a failure or inadequacy of individual women may in fact be a failure of the society to provide institutionalized substitutes for the extended kin to assist in the care of infants and young children. It may be that the role requirements of maternity in the American family system extract too high a price of deprivation for young adult women reared with highly diversified interests and social expectations concerning adult life. Here, as at several points in the course of this paper, familiar problems take on a new and suggestive research dimension when the focus is on the parent rather than the child.

Background

Since it is a relatively recent development to focus on the parent side of the parent-child relationship, some preliminary attention to the emergence of this focus on parenthood is in order. Several developments in the behavioral sciences paved the way to this perspective. Of perhaps most importance have been the development of ego psychology and the problem of adaptation of Murray (9) and Hartmann (10), the interpersonal focus of Sullivan's psychoanalytic theories (11), and the life cycle approach to identity of Erikson (12). These have been fundamental to the growth of the human development perspective: that personality is not a stable given but a constantly changing phenomenon, that the individual changes along the life line as he lives through critical life experiences. The transition to parenthood, or the impact of parenthood upon the adult, is part of the heightened contemporary interest in adult socialization.

A second and related development has been the growing concern of behavioral scientists with crossing levels of analysis to adequately comprehend social and individual phenomena and to build theories appropriate to a complex social system. In the past, social anthropologists focused as purely on the level of prescriptive normative variables as psychologists had concentrated on intrapsychic processes at the individual level or sociologists on social-structural and institutional variables. These are adequate, perhaps, when societies are in a stable state of equilibrium and the social sciences were at early stages of conceptual development, but they become inadequate when the societies we study are undergoing rapid social change and we have an increasing amount of individual and subgroup variance to account for.

Psychology and anthropology were the first to join theoretical forces in their concern for the connections between culture and personality. The question of how culture is transmitted across the generations and finds its manifestations in the personality structure and social roles of the individual has brought renewed research attention to the primary institutions of the family and the schools, which provide the intermediary contexts through which culture is transmitted and built into personality structure.

It is no longer possible for a psychologist or a therapist to neglect the social environment of the individual subject or patient, nor is the "family" they are concerned with any longer confined to the family of origin, for current theory and therapy view the adult individual in the context of his current family of procreation. So too it is no longer possible for the sociologist to focus exclusively on the current family relationships of the individual. The incorporation of psychoanalytic theory into the informal, if not the formal, training of the sociologist has led to an increasing concern for the quality of relationships in the family of origin as determinants of the adult attitudes, values, and behavior which the sociologist studies.

Quite another tradition of research has led to the formulation of "normal crises of parenthood." "Crisis" research began with the studies of individuals undergoing traumatic experiences, such as that by Tyhurst on natural catastrophes (13), Caplan on parental responses to premature births (14), Lindemann on grief and bereavement (15), and Janis on surgery (16). In these studies attention was on differential response to stress — how and why individuals vary in the ease with which they coped with the stressful experience and achieved some reintegration. Sociological interest has been piqued as these studies were built upon by Rhona and Robert Rapoport's research on the honeymoon and the engagement as normal crises in the role transitions to marriage and their theoretical attempt to build a conceptual bridge between family and occupational research from a "transition task" perspective (17). LeMasters, Dyer, and Hobbs have each conducted studies of parenthood precisely as a crisis or disruptive event in family life (18).

I think, however, that the time is now ripe to drop the concept of "normal crises" and to speak directly, instead, of the transition to and impact of parenthood. There is an unconfortable incongruity in speaking of any crisis as normal. If the transition is achieved and if a successful reintergration of personality or social roles occurs, then crisis is a misnomer. To confine attention to "normal crises" suggests, even if it is not logically implied, successful outcome, thus excluding from our analysis the deviant instances in which failure occurs.

Sociologists have been just as prone as psychologists to dichotomize normality and pathology. We have had one set of theories to deal with deviance, social problems, and conflict and quite another set in theoretical analyses of a normal system — whether a family or a society. In the latter case our theories seldom include categories to cover deviance, strain, dysfunction, or failure. Thus, Parsons and Bales' systems find "task-leaders" oriented to problem solution, but not instrumental leaders attempting to undercut or destroy the goal of the group, and "sociometric stars" who play a positive integrative function in cementing ties among group members, but not negatively expressive persons with hostile aims of reducing or destroying such intragroup ties (19).

Parsons' analysis of the experience of parenthood as a step in maturation and personality growth does not allow for negative outcome. In this view either parents show little or no positive impact upon themselves of their parental role experience, or they show a new level of maturity. Yet many women, whose interests and values made a congenial combination of wifehood and work role, may find that the addition of maternal responsibilities has the consequence of a fundamental and undesired change in both their relationships to their husbands and their involvements outside the family. Still other women, who might have kept a precarious hold on adequate functioning as adults had they *not* become parents, suffer severe retrogression with pregnancy and childbearing, because the reactivation of older unresolved conflicts with their own mothers is not favorably resolved but in fact leads to personality deterioration (20) and the transmission of pathology to their children (21).

Where cultural pressure is very great to assume a particular adult role, as it is for American women to bear and rear children, latent desire and psychological readiness for parenthood may often be at odds with manifest desire and actual ability to perform adequately as parents. Clinicians and therapists are aware, as perhaps many sociologists are not, that failure, hostility, and destructiveness are as much a part of the family system and the relationships among family members as success, love, and solidarity are (22).

A conceptual system which can deal with both successful and unsuccessful role transitions, or positive and negative impact of parenthood upon adult men and women, is thus more powerful than one built to handle success but not failure or vice versa. For these reasons I have concluded that it is misleading and restrictive to perpetuate the use of the concept of "normal crises." A more fruitful point of departure is to build upon the stage-task concepts of Erikson, viewing parenthood as a developmental stage, as Benedek (23) and Hill (24) have done, a perspective carried into the research of Raush, Goodrich, and Campbell (25) and of Rhona and Robert Rapoport (26) on adaptation to the early years of marriage and that of Cohen, Fearing et al. (27) on the adjustments involved in pregnancy.

Role Cycle Stages

A discussion of the impact of parenthood upon the parent will be assisted by two analytic devices. One is to follow a comparative approach, by asking in what basic structural ways the parental role differs from other primary adult roles. The marital and occupational roles will be used for this comparison. A second device is to specify the phases in the development of a social role. If the total life span may be said to have a cycle, each stage with its unique tasks, then by analogy a role may be said to have a cycle and each stage in that role cycle, to have its unique tasks and problems of adjustment. Four broad stages of a role cycle may be specified:

1. Anticipatory Stage

All major adult roles have a long history of anticipatory training for them, since parental and school socialization of children is dedicated precisely to this task of producing the kind of competent adult valued by the culture. For our present purposes, however, a narrower conception of the anticipatory stage is preferable: the engagement period in the case of the marital role, pregnancy in the case of the parental role, and the last stages of highly vocationally oriented schooling or on-the-job apprenticeship in the case of an occupational role.

2. Honeymoon Stage

This is the time period immediately following the full assumption of the adult role. The inception of this stage is more easily defined than its termination. In the case of the marital role, the honeymoon stage extends from the marriage ceremony itself through the literal honeymoon and on through an unspecified and individually varying period of time. Raush (28) has caught this stage of the marital role in his description of the "psychic honeymoon": that extended postmarital period when, through close intimacy and joint activity, the couple can explore each other's capacities and limitations. I shall arbitrarily consider the onset of pregnancy as marking the end of the honeymoon stage of the marital role. This stage of the parental role may involve an equivalent psychic honeymoon, that post-childbirth period during which, through intimacy and prolonged contact, an attachment between parent and child is laid down. There is a crucial difference, however, from the marital role in this stage. A woman knows her husband as a unique real person when she enters the honeymoon stage of marriage. A good deal of preparatory adjustment on a firm reality-base is possible during the engagement period which is not possible in the equivalent pregnancy period. Fantasy is not corrected by the reality of a specific individual child until the birth of the child. The "quickening" is psychologically of special significance to women precisely because it marks the first evidence of a real baby rather than a purely fantasized one. On this basis alone there is greater interpersonal adjustment and learning during the honeymoon stage of the parental role than of the marital role.

3. Plateau Stage

This is the protracted middle period of a role cycle during which the role is fully exercised. Depending on the specific problem under analysis, one would obviously subdivide this large plateau stage further. For my present purposes it is not necessary to do so, since my focus is on the earlier anticipatory and honeymoon stages of the parental role and the overall impact of parenthood on adults.

4. Disengagement-Termination Stage

This period immediately precedes and includes the actual termination of the role. Marriage ends with the death of the spouse or, just as definitively, with separation and divorce. A unique characteristic of parental role termination is the fact that it is not clearly marked by any specific act but is an attenuated process of termination with little cultural prescription about when the authority and obligations of a parent end. Many parents, however, experience the marriage of the child as a psychological termination of the active parental role.

Unique Features of the Parental Role

With this role cycle suggestion as a broader framework, we can narrow our focus to what are the unique and most salient features of the parental role. In doing so, special attention will be given to two further questions: **1.** the impact of social changes over the past few decades in facilitating or complicating the transition to and experience of parenthood; **2.** the new interpretations or new research suggested by the focus on the parent rather than the child.

1. Cultural Pressure to Assume the Role

On the level of cultural values, men have no freedom of choice where work is concerned: They must work to secure their status as adult men. The equivalent for women has been maternity. There is considerable pressure upon the growing girl and young woman to consider maternity necessary for a woman's fulfillment as an individual and to secure her status as an adult (29).

This is not to say there are no fluctuations over time in the intensity of the cultural pressure to parenthood. During the depression years of the 1930's, there was more widespread awareness of the economic hardships parenthood can entail, and many demographic experts belive there was a great increase in illegal abortions during those years. Bird has discussed the dread with which a suspected pregnancy was viewed by many American women in the 1930's (30). Quite a different set of pressures were at work during the 1950's, when the general societal tendency was toward withdrawal from active engagement with the issues of the larger society and a turning in to the gratifications of the private sphere of home and family life. Important in the background were the general affluence of the period and the expanded room and ease of child rearing that go with suburban living. For the past five years, there has been a drop in the birth rate in general, fourth and higher-order births in particular. During this same period there have been increased concern and debate about women's participation in politics and work, with more women now returning to work rather than conceiving the third or fourth child (31).

2. Inception of the Parental Role

The decision to marry and the choice of a mate are voluntary acts of individuals in our family system. Engagements are therefore consciously considered, freely entered, and freely terminated if increased familiarity decreases, rather than increases, intimacy and commitment to the choice. The inception of a pregnancy, unlike the engagement, is not always a voluntary decision, for it may be the unintended consequence of a sexual act that was recreative in intent rather than procreative. Secondly, and again unlike the engagement, the termination of a pregnancy is not socially sanctioned, as shown by current resistance to abortion-law reform.

The implication of this difference is a much higher probability of unwanted pregnancies than of unwanted marriages in our family system. Coupled with the ample clinical evidence of parental rejection and sometimes cruelty to children, it is all the more surprising that there has not been more consistent research attention to the problem of *parental satisfaction*, as there has for long been on *marital satisfaction* or *work satisfaction*. Only the extreme iceberg tip of the parental satisfaction continuum is clearly demarcated and researched, as in the growing concern with "battered babies." Cultural and psychological resistance to the image of a non-nurturant woman may afflict social scientists as well as the American public.

The timing of a first pregnancy is critical to the manner in which parental responsibilities are joined to the marital relationship. The single most important change over the past few decades is extensive and efficient contraceptive usage, since this has meant for a growing proportion of new marriages, the possiblity of and increasing preference for some postponement of childbearing after marriage. When pregnancy was likely to follow shortly after marriage, the major transition point in a woman's life was marriage itself. *This transition point is increasingly the first pregnancy rather than marriage.* It is accepted and increasingly expected that women will work after marriage, while household furnishings are acquired and spouses complete their advanced training or gain a foothold in their work (32). This provides an early marriage period in which the fact of a wife's employment presses for a greater egalitarian relationship between husband and wife in decision-making, commonality of experience, and sharing of household responsibilities.

The balance between individual autonomy and couple mutuality that develops during the honeymoon stage of such a marriage may be important in establishing a pattern that will later affect the quality of the parent-child relationship and the extent of sex-role segregationof duties between the parents. It is only in the context of a growing egalitarian base to the marital relationship that one could find, as Gavron has (33), a tendency for parents to establish some barriers between themselves and their children, a marital defense against the institution of parenthood as she describes it. This may eventually replace the typical coalition in more traditional families of mother and children against husband-father. Parenthood will continue for some time to impose a degree of temporary segregation of primary responsibilities between husband and wife, but, when this takes place in the context of a previously established egalitarian relationship between the husband and wife, such role segregation may become blurred, with greater recognition of the wife's need for autonomy and the husband's role in the routines of home and child rearing (34).

There is one further significant social change that has important implications for the changed relationship between husband and wife: the increasing departure from an old pattern of role-inception phasing in which the young person first completed his schooling, then established himself in the world of work, then married and began his family. Marriage and parenthood are increasingly taking place *before* the schooling of the husband, and often of the wife, has been completed (35). An important reason for this trend lies in the fact that, during the same decades in which the average age of physical-sexual maturation has dropped, the average amount of education which young people obtain has been on the increase. Particularly for the college and graduate or professional school population, family roles are often assumed before the degrees needed to enter careers have been obtained.

Just how long it now takes young people to complete their higher education has been investigated only recently in several longitudinal studies of college-graduate cohorts (36). College is far less uniformly a four-year period than high school is. A full third of the college freshmen in one study had been out of high school a year or more before entering college (37). In a large sample of college graduates in 1961, one in five were over 25 years of age at graduation (38). Thus, financial difficulties, military service, change of career plans, and marriage itself all tend to create interruptions in the college attendance of a significant proportion of college graduates. At the graduate and professional school level, this is even more marked: the mean age of men receiving the doctorate, for example, is 32, and of women, 36 (39). It is the exception rather than the rule for men and women who seek graduate degrees to go directly from college to graduate school and remain there until they secure their degrees (40).

The major implication of this change is that more men and women are achieving full adult status in family roles while they are still less than fully adult in status terms in the occupational system. Graduate students are, increasingly, men and women with full family responsibilities. Within the family many more husbands and fathers are still students, often quite dependent on the earnings of their wives to see them through their advanced training (41). No matter what the couple's desires and preferences are, this fact alone presses for more egalitarian relations between husband and wife, just as the adult family status of graduate students presses for more egalitarian relations between students and faculty.

3. Irrevocability

If marriages do not work out, there is now widespread acceptance of divorce and remarriage as a solution. The same point applies to the work world: we are free to leave an unsatisfactory job and seek another. But once a pregnancy occurs, there is little possibility of undoing the commitment to parenthood implicit in conception except in the rare instance of placing children for adoption. We can have ex-spouses and ex-jobs but not ex-children. This being so, it is scarcely surprising to find marked differences between the relationship of a parent and one child and the relationship of the same parent with another child. If the culture does not permit pregnancy termination, the equivalent to giving up a child is psychological withdrawal on the part of the parent.

This taps an important area in which a focus on the parent rather than the child may contribute a new interpretive dimension to an old problem: the long history of interest, in the social sciences, in differences among children associated with their sex-birth-order position in their sibling set. Research has largely been based on data gathered about and/or from the children, and interpretations make inferences back to the "probable" quality of the child's relation to a parent and how a parent might differ in relating to a first-born compared to a last-born child. The relevant research, directed at the parents (mothers in particular), remains to be done, but at least a few examples can be suggested of the different order of interpretation that flows from a focus on the parent.

Some birth-order research stresses the influence of sibs upon other sibs, as in Koch's finding that second-born boys with an older sister are more feminine than second-born boys with an older brother (42). A similar sib-influence interpretation is offered in the major common finding of birth-order correlates, that sociability is greater among last-borns (43) and achievement among first-borns (44). It has been suggested that last-borns use social skills to increase acceptance by their older sibs or are more peer-oriented because they receive less adult stimulation from parents. The tendency of first-borns to greater achievement has been interpreted in a corollary way, as a reflection of early assumption of responsibility for younger sibs, greater adult stimulation during the time the oldest was the only child in the family (45), and the greater significance of the first-born for the larger kinship network of the family (46).

Sociologists have shown increasing interest in structural family variables in recent years, a primary variable being family size. From Bossard's descriptive work on the large family (47) to more methodologically sophisticated work such as that by Rosen (48), Elder and Bowerman (49), Boocock (50), and Nisbet (51), the question posed is: what is the effect of growing up in a small family, compared with a large family, that is attributable to this group-size variable? Unfortunately, the theoretical point of departure for sociologists' expectations of the effect of the family-size variables is the Durkheim-Simmel tradition of the differential effect of group size or population density upon members or inhabitants (52). In the case of the family, however, this overlooks the very important fact that family size is determined by the key figures *within* the group, i.e, the parents. To find that children in small families differ from children in large families is not simply due to the impact of group size upon individual members but to the very different involvement of the parent with the children and to relations between the parents themselves in small versus large families.

An important clue to a new interpretation can be gained by examining family size from the perspective of parental motivation toward having children. A small family is small for one of two primary reasons: either the parents wanted a small family and achieved their desired size, or they wanted a large family but were not able to attain it. In either case, there is a low probability of unwanted children. Indeed, in the latter eventuality they may take particularly great interest in the children they do have. Small families are therefore most likely to contain parents with a strong and positive orientation to each of the children they have. A large family, by contrast, is large either because the parents achieved the size they desired or because they have more children than they in fact wanted. Large families therefore have a higher probability than small families of including unwanted and unloved children. Consistent with this are Nye's finding that adolescents in small families have better relations with their parents than those in large families (53) and Sears and Maccoby's finding that mothers of large families are more restrictive toward their children than mothers of small families (54).

This also means that last-born children are more likely to be unwanted than first- or middle-born children, particularly in large families. This is consistent with what is known of abortion patterns among married women, who typically resort to abortion only when they have achieved the number of children they want or feel they can afford to have. Only a small proportion of women faced with such unwanted pregnancies actually resort to abortion. *This suggests the possibility that the last-born child's reliance on social skills may be his device for securing the attention and loving involvement of a parent less positively predisposed to him than to his older siblings.*

In developing this interpretation, rather extreme cases have been stressed. Closer to the normal range, of families in which even the last-born child was desired and planned for, there is still another element which may contribute to the greater sociability of the last-born child. Most parents are themselves aware of the greater ease with which they face the care of a third fragile newborn than the first; clearly, parental skills and confidence are greater with last-born children than with first-born children. But this does not mean that the attitude of the parent is more positive toward the care of the third child than the first. There is no necessary correlation between skills in an area and enjoyment of that area. Searls (55) found that older homemakers are *more* skillful in domestic tasks but experience *less* enjoyment of them than younger homemakers, pointing to a declining euphoria for a particular role with the passage of time. In the same way, older people rate their marriages as "very happy" less often than younger people do (56). It is perhaps culturally and psychologically more difficult to face the possibility that women may find less enjoyment of the maternal role with the passage of time, though women themselves know the difference between the romantic expectation concerning child care and the incorporation of the first baby into the household and the more realistic expectation and sharper assessment of their own abilities to do an adequate job of mothering as they face a third confinement. Last-born children may experience not only less verbal stimulation from their parents than first-born children but also less prompt and enthusiastic response to their demands — from feeding and diaper-change as infants to requests for stories read at three or a college education at eighteen — simply because the parents experience less intense gratification from the parent role with the third child than they did with the first. The child's response to this might well be to cultivate winning, pleasing manners in early childhood that blossom as charm and sociability in later life, showing both a greater need to be loved and greater pressure to seek approval.

One last point may be appropriately developed at this juncture. Mention was made earlier that for many women the personal outcome of experience in the parent role is not a higher level of maturation but the negative outcome of a depressed sense of self-worth, if not actual personality deterioration. There is considerable evidence that this is more prevalent than we recognize. On a qualitative level, a close reading of the portrait of the working-class wife in Rainwater (57), Newsom (58), Komarovsky (59) Gavron (60), or Zweig (61), gives little suggestion that maternity has provided these women with opportunities for personal growth and development. So too, Cohen (62) notes with some surprise that in her sample of middle-class educated couples, as in Pavenstadt's study of lower-income women in Boston, there were more emotional difficulty and lower levels of maturation among multiparous women than primiparous women. On a more extensive sample basis, in Gurin's survey of Americans viewing their mental health (63) as in Bradburn's reports on happiness (64), single men are less happy and less active than single women, but among the married respondents the women are unhappier, have more problems, feel inadequate as parents, have a more negative and passive outlook on life, and show a more negative self-image. All of these characteristics increase with age among married women but show no relationship to age among men. While it may be true, as Gurin argues, that women are more introspective and hence more attuned to the psychological facets of experience than men are, this point does not account for the fact that the things which the women report are all on the negative side; few are on the positive side, indicative of euphoric sensitivity and pleasure. The possibility must be faced, and at some point researched, that women lose ground in personal development and self-esteem during the early and middle years of adulthood, whereas men gain ground in these respects during the same years. The retention of a high level of self-esteem may depend upon the adequacy of earlier preparation for major adult roles: men's training adequately prepares them for the primary adult roles in the occupational system, as it does for those women who opt to participate significantly in the work world. Training in the qualities and skills needed for family roles in contemporary society may be inadequate for both sexes, but the lowering of self-esteem occurs only among women because their primary adult roles are within the family system.

4. Preparation for Parenthood

Four factors may be given special attention on the question of what preparation American couples bring to parenthood.

a) *Paucity of preparation*. Our educational system is dedicated to the cognitive development of the young, and our primary teaching approach is the pragmatic one of learning by doing. How much one knows and how well he can apply what he knows are the standards by which the child is judged in school, as the employee is judged at work. The child can learn by doing in such subjects as science, mathematics, art work, or shop, but not in the subjects most relevant to successful family life: sex, home maintenance, child care, interpersonal competence, and empathy. If the home is deficient in training in these areas, the child is left with no preparation for a major segment of his adult life. A doctor facing his first patient in private practice has treated numerous patients under close supervision during his interneship, but probably a majority of American mothers approach maternity with no previous child-care experience beyond sporadic baby-sitting, perhaps a course in child psychology, or occasional care of younger siblings.

b) *Limited learning during pregnancy*. A second important point makes adjustment to parenthood potentially more stressful than marital adjustment. This is the lack of any realistic training for parenthood during the anticipatory stage of pregnancy. By contrast, during the engagement period preceding marriage, an individual has opportunities to develop the skills and make the adjustments which ease the transition to marriage. Through discussions of values and life goals, through sexual experimentation, shared social experiences as an engaged couple with friends and relatives, and planning and furnishing an apartment, the engage couple can make considerable progress in developing mutuality in advance of the marriage itself (65). No such headstart is possible in the case of pregnancy. What preparation exists is confined to reading, consultation with friends and parents, discussions between husband and wife, and a minor nesting phase in which a place and the equipment for a baby are prepared in the household (66).

c) *Abruptness of transition*. Thirdly, the birth of a child is not followed by any gradual taking on of responsibility, as in the case of a professional work role. It is as if the woman shifted from a graduate student to a full professor with little intervening apprenticeship experience of slowly increasing responsibility. The new mother starts out immediately on 24-hour duty, with responsibility for a fragile and mysterious infant totally dependent on her care.

If marital adjustment is more difficult for very young brides than more mature ones (67), adjustment to motherhood may be even more difficult. A woman can adapt a passive dependence on a husband and still have a successful marriage, but a young mother with strong dependency needs is in for difficulty in maternal adjustment, because the role precludes such dependency. This situation was well described in Cohen's study (68) in a case of a young wife with a background of co-ed popularity and a passive dependent relationship to her admired and admiring husband, who collapsed into restricted incapacity when faced with the responsibilities of maintaining a home and caring for a child.

d) *Lack of guidelines to successful parenthood*. If the central task of parenthood is the rearing of children to become the kind of competent adults valued by the society, then an important question facing any parent is what he or she specifically can do to create such a competent adult. This is where the parent is left with few or no guidelines from the expert. Parents can readily inform themselves concerning the young infant's nutritional, clothing, and medical needs and follow the general prescription that a child needs loving physical contact and emotional support. Such advice may be sufficient to produce a healthy, happy, and well-adjusted preschooler, but adult competency is quite another matter.

In fact, the adults who do "succeed" in American society show a complex of characteristics as children that current experts in child-care would evaluate as "poor" to "bad." Biographies of leading authors and artists, as well as the more rigorous research inquiries of creativity among architects (69) or scientists (70), do not portray childhoods with characteristics currently endorsed by mental health and child-care authorities. Indeed, there is often a predominance of tension in childhood family relations and traumatic loss rather than loving parental support, intense channeling of energy in one area of interest rather than all-round profile of diverse interests, and social withdrawal and preference for loner activities rather than gregarious sociability. Thus, the stress in current child-rearing advice on a high level of loving support but a low level of discipline or restriction on the behavior of the child — the "developmental" family type as Duvall calls it (71) — is a profile consistent with the focus on mental health, sociability, and adjustment. Yet the combination of both high support and high authority on the part of parents is most strongly related to the child's sense of responsibility, leadership quality, and achievement level, as found in Bronfenbrenner's studies (72) and that of Mussen and Distler (73).

Brim points out (74) that we are a long way from being able to say just what parent role prescriptions have what effect on the adult characteristics of the child. We know even less about how such parental prescriptions should be changed to adapt to changed conceptions of competency in adulthood. In such an ambiguous context, the great interest parents take in school reports on their children or the pediatrician's assessment of the child's developmental progress should be seen as among the few indices parents have of how well *they* are doing as parents.

System and Role Requirements: Instrumentality and Integration

Typological dichotomies and unidimensional scales have loomed large in the search by social scientists for the most economical and general principles to account for some significant portion of the complex human behavior or social organization they study. Thus, for example, the European dichotomy of *Gemeinschaft* and *Gesellschaft* became the American sociological distinction between rural and urban sociology, subfields that have outlasted their conceptual utility now that the rural environment has become urbanized and the interstices between country and city are swelling with suburban developments.

In recent years a new dichotomy has gained more acceptance in sociological circles — the Parsonian distinction between *instrumental* and *expressive*, an interesting dichotomy that is unfortunately applied in an indiscriminate way to all manner of social phenomena including the analysis of teacher role conflict, occupational choice, the contrast between the family system and the occupational system, and the primary roles or personality tendencies of men compared to women.

On a system level, for example, the "instrumental" occupational system is characterized by rationality, efficiency, rejection of tradition, and depression of interpersonal loyalty, while the "expressive" family system is characterized by nurturance, integration, tension-management, ritual, and interpersonal solidarity. Applied to sex roles within the family, the husband-father emerges as the instrumental rational leader, a symbolic representative of the outside world, and the wife-mother emerges as the expressive, nurturant, affective center of the family. Such distinctions may be useful in the attempt to capture some general tendency of a system or a role, but they lead to more distortion than illumination when applied to the actual functioning of a specific system or social role or to the actual behavior of a given individual in a particular role.

Take, for example, the husband-father as the instrumental role within the family on the assumption that men are the major breadwinners and therefore carry the instrumentality associated with work into their roles within the family. To begin with, the family is not an experimental one-task small group but a complex, ongoing 24-hour entity with many tasks that must be performed. Secondly, we really know very little about how occupational roles affect the performance of family roles (75). An aggressive courtroom lawyer or a shrewd business executive are not lawyers and businessmen at home but husbands and fathers. Unless shown to be in error, we should proceed on the assumption that behavior is role-specific. (Indeed, Brim [76] argues that even personality is role-specific.) A strict teacher may be an indulgent mother at home; a submissive wife may be a dominant mother; a dictatorial father may be an exploited and passive worker on the assembly line; or, as in some of Lidz's schizophrenic patients' families (77), a passive dependent husband at home may be a successful dominant lawyer away from home.

There is, however, a more fundamental level to the criticism that the dichotomous usage of instrumentality and expressiveness, linked to sex and applied to intrafamily roles, leads to more distortion than illumination. The logic of my argument starts with the premise that every social system, group, or role has two primary, independent, structural axes. Whether these axes are called "authority and support," as in Straus's circumplex model (78), or "instrumental and expressive" as by Parsons (79), there are tasks to be performed and affective support to be given in all the cases cited. There must be discipline, rules, and division of labor in the nation-state as in the family or a business enterprise *and* there must be solidarity among the units comprising these same systems in order for the system fo function adequately. *This means that the role of father, husband, wife, or mother each has these two independent dimensions of authority and support, instrumentality and expressiveness, work and love.* Little is gained by trying to stretch empirical results to fit the father role to the instrumental category, as Brim (80) has done, or the mother role to the expressive category, as Zelditch has done (81).

In taking a next logical step from this premise, the critical issue, both theoretically and empirically, becomes gauging the *balance* between these two dimensions of the system or of the role. Roles or systems could be compared in terms of the average difference among them in the direction and extent of the discrepancy between authority and support; or individuals could be compared in terms of the variation among them in the discrepancy between the two dimensions in a given role.

An example may clarify these points. A teacher who is all loving, warm support to her students and plans many occasions to evoke integrative ties among them but who is imcompetent in the exercise of authority or knowledge of the subjects she teaches would be judged by any school principal as an inadequate teacher. The same judgment of inadequacy would apply to a strict disciplinarian teacher, competent and informed about her subjects but totally lacking in any personal quality of warmth or ability to encourage integrative and cooperative responses among her students. Maximum adequacy of teacher performance requires a relatively high positive level on both of these two dimensions of the teacher role.

To claim that teachers have a basic conflict in approaching their role because they are required to be a "bisexual parent, permissive giver of love and harsh disciplinarian with a masculine intellectual grasp of the world," as Jackson and Moscovici (82) have argued, at least recognizes the two dimensions of the teacher role, though it shares the view of many sociologists that role *conflict* is inherent wherever these seeming polarities are required. Why conflict is predicted hinges on the assumed invariance of the linkage of the male to authority and the female to the expressive-integrative roles.

It is this latter assumed difference between the sexes that restricts theory-building in family sociology and produces so much puzzlement on the part of researchers into marriage and parenthood, sex-role socialization, or personality tendencies toward masculinity or femininity. Let me give one example of recent findings on this latter topic and then move on to apply the two-dimension concept to the parental role. Vincent (83) administered the Gough Femininity Scale along with several other scale batteries from the California Personality Inventory to several hundred college men and women. He found that women *low* on femininity were higher in the Class I scale which measures poise, ascendancy, and self-assurance, and men *high* in femininity were higher in dominance, capacity for status, and responsibility. Successful adult men in a technological society are rarely interested in racing cars, soldiering, or hunting; they are cautious, subtle, and psychologically attuned to others. So too, contemporary adult women who fear windstorms, the dark strange places, automobile accidents, excitement, crowded parties, or practical jokes (and are therefore high on femininity in the Gough scale) will be inadequate for the task of managing an isolated household with neither men nor kinswomen close by to help them through daily crises, for the assumption of leadership roles in community organizations, or for holding down supplementary breadwinning or cakewinning jobs.

When Deutsch (84) and Escalona (85) point out that today's "neurotic" woman is not an assertive dominant person but a passive dependent one, the reason may be found in the social change in role expectations concerning competence among adult women, not that there has been a social change in the characteristics of neurotic women. In the past an assertive, dominant woman might have defined herself and been defined by her analyst as "neurotic" because she could not fill the expectations then held for adequacy among adult women. Today, it is the passive dependent woman who will be judged "neurotic" because she cannot fill adequately the expectations now set for and by her. What is really meant when we say that sex role definitions have become increasingly blurred is that men are now required to show more integrative skills than in the past, and women more instrumental skills. This incurs potential sex-role "confusion" only by the standards of the past, not by the standards of what is required for contemporary adult competence in family and work roles.

Once freed from the assumption of a single bipolar continuum of masculinity-femininity (86), authority-integration, or even independence-dependence (87), one can observe increased instrumentality in a role with no implication of necessarily decreased integration, and vice versa. Thus, an increasing rationality in the care of children, the maintenance of a household, or meal planning for a family does not imply a decreasing level of integrative support associated with the wife-mother role. So, too, the increased involvement of a young father in playful encounters with his toddler carries no necessary implication of a change in the instrumental dimension of his role.

The two-dimensional approach also frees our analysis of parenthood on two other important questions. Brim has reviewed much of the research on the parent-child relationship (88) and noted the necessity of specifying not only the sex of the parent but the sex of the child and whether a given parent-child dyad is a cross-sex or same-sex pair. It is clear from his review that fathers and mothers relate differently to their sons and daughters: fathers have been found to be stricter with their sons than with their daughters, and mothers stricter with their daughters than with their sons. Thus, a two-dimensional approach to the parent role is more appropriate to what is already empirically known about the parent-child relationship.

Secondly, only on a very general overview level does a parent maintain a particular level of support and of discipline toward a given child: situational variation is an important determinant of parental response to a child. A father with a general tendency toward relatively little emotional support of his son may offer a good deal of comfort if the child is hurt. An indulgent and loving mother may show an extreme degree of discipline when the same child misbehaves. Landreth found that her four-year-olds gave more mother responses on a care item concerning food than on bath-time or bedtime care and suggests, as Brim has (89), that "any generalizations on parent roles should be made in terms of the role activities studies" (90).

Let me illustrate the utility of the two-dimensional concept by applying it to the parental role. Clearly there are a number of expressive requirements for adequate performance in this role: spontaneity and flexibility, the ability to be tender and loving and to respond to tenderness and love from a child, to take pleasure in tactile contact and in play, and to forget one's adultness and unself-consciously respond to the sensitivities and fantasies of a child. Equally important are the instrumental requirements for adequate performance in the parental role: firmness and consistency; the ability to manage time and energy; to plan and organize activities involving the child; to teach and to train the child in body controls, motor and language skills, and knowledge of the natural and social world; and interpersonal and value discriminations.

Assuming we had empirical measures of these two dimensions of the parental role, one could then compare individual women both by their levels on each of these dimensions and by the extent to which the discrepancy in level on the two dimensions was tipped toward a high expressive or instrumental dimension. This makes no assumptions about what the balance "should" be; that remains an empirical question awaiting a test in the form of output variables — the characteristics of children we deem to be critical for their competence as adults. Indeed, I would predict that an exhaustive count of the actual components of both the marital and parental roles would show a very high proportion of instrumental components in the parental role and a low proportion in the marital role and that this is an underlying reason why maternal role adjustment is more difficult for women than marital role adjustment. It also leaves as an open, empirical question what the variance is, among fathers, in the level of expressiveness and instrumentality in their paternal role performance and how the profile of fathers compares with that of mothers.

It would not surprise many of us, of course, if women scored higher than men on the expressive dimension and men scored higher on the instrumental dimension of the parental role. Yet quite the opposite might actually result. Men spend relatively little time with their children, and it is time of a particular kind: evenings, weekends, and vacations, when the activities and mood of the family are heavily on the expressive side. Women carry the major burden of the instrumental dimension of parenting. If, as Mabel Cohen (91) suggests, the rearing of American boys is inadequate on the social and sexual dimension of development and the rearing of American girls is inadequate on the personal dimension of development, then from the perspective of adequate parenthood performance, we have indeed cause to reexamine the socialization of boys and girls in families and schools. Our current practices appear adequate as preparation for occupational life for men but not women, and inadequate as preparation for family life for both sexes.

However, this is to look too far ahead. At the present, this analysis of parenthood suggests we have much to rethink and much to research before we develop policy recommendations in this area.

References

1. Antonio J. Ferreira, "The Pregnant Woman's Emotional Attitude and Its Reflection on the Newborn," *American Journal of Orthopsychiatry*, vol. 30, 1960, pp. 553–61.

2. Robert Sears, E. Maccoby, and H. Levin, *Patterns of Child-Rearing*. Evanston, Ill.: Row, Peterson, 1957.

3. Sylvia Brody, *Patterns of Mothering: Maternal Influences During Infancy*. New York: International Universities Press, 1956.

4. Leon J. Yarrow, "Maternal Deprivation: Toward an Empirical and Conceptual Re-evaluation." *Psychological Bulletin*. vol. 58, no. 6, 1961, pp. 459–90.

5. F. Ivan Nye and L. W. Hoffman, *The Employed Mother in America*. Chicago: Rand McNally, 1963; Alice S. Rossi, "Equality Between the Sexes: An Immodest Proposal," *Daedalus*, vol. 93, no. 2, 1964, pp. 607–52.

6. The younger the child, the more was this the accepted view. It is only in recent years that research has paid any attention to the initiating role of the infant in the development of his attachment to maternal and other adult figures, as in Ainsworth's research which showed that infants become attached to the mother, not solely because she is instrumental in satisfying their primary visceral drives, but through a chain of behavioral interchange between the infant and the mother, thus supporting Bowlby's rejection of the secondary drive theory of the infant's ties to his mother. Mary D. Ainsworth, "Patterns of Attachment Behavior Shown by the Infant in Interaction with His Mother," *Merrill-Palmer Quarterly*, vol. 10, no. 1, 1964, pp. 51–58; John Bowlby, "The Nature of the Child's Tie to His Mother," *International Journal of Psychoanalysis*, vol, 39, 1958, pp. 1–34.

7. Therese Benedek, "Parenthood as a Developmental Phase," *Journal of American Psychoanalytic Association,* vol. 7, no. 8, 1959, pp. 389–417.

8. Yarrow, "Maternal Deprivation."

9. Henry A. Murray, *Explorations in Personality*. New York: Oxford University Press, 1938.

10. Heinz Hartmann, *Ego Psychology and the Problem of Adaptation*. New York: International Universities Press, 1958.

11. Patrick Mullahy, ed., *The Contributions of Harry Stack Sullivan*. New York: Hermitage House, 1952.

12. E. Erikson, "Identity and the Life Cycle; Selected Papers," *Psychological Issues*, vol. 1, 1959, pp. 1–171.

13. J. Tyhurst, "Individual Reactions to Community Disaster," *American Journal of Psychiatry*, vol. 107, 1951, pp. 764–69.

14. G. Caplan, "Patterns of Parental Response to the Crisis of Premature Birth: A Preliminary Approach to Modifying the Mental Health Outcome," *Psychiatry*, vol. 23, 1960, pp. 365–74.

15. E. Lindemann, "Symptomatology and Management of Acute Grief," *American Journal of Psychiatry*, vol. 101, 1944, pp. 141–48.

16. Irving Janis, *Psychological Stress*. New York: Wiley, 1958.

17. Rhona Rapoport, "Normal Crises, Family Structure and Mental Health," *Family Process*, vol. 2, no. 1, 1963, pp. 68–80; Rhona Rapoport and Robert Rapoport, "New Light on the Honeymoon," *Human Relations*, vol. 17, no. 1, 1964, pp. 33–56; Rhona Rapoport, "The Transition from Engagement to Marriage," *Acta Scoiologica*, 8, fasc, 1–2, 1964, pp. 36–55; and Robert Rapoport and Rhona Rapoport, "Work and Family in Contemporary Society," *American Sociological Review*, vol, 30, no. 3, 1965, pp. 381–94.

18. E. E. Le Masters, "Parenthood as Crisis," *Marriage and Family Living*, vol. 19, 1957, pp. 352–55; Everett D. Dyer, "Parenthood as Crisis: A Re-Study," *Marriage and Family Living*, vol. 25, 1963, pp. 196–201; and Daniel F. Hobbs, Jr., "Parenthood as Crisis: A third Study," *Journal of Marriage and the Family*, vol. 27, no. 3, 1963, pp. 367–72. Le Masters and Dyer both report the first experience of parenthood involves extensive to severe crises in the lives of their young parent respondents. Hobbs's study does not show first parenthood to be a crisis experience, but this may be due to the fact that his couples have very young (seven-week-old) first babies and are therefore still experiencing the euphoric honeymoon stage of parenthood.

19. Parsons' theoretical analysis of the family system builds directly on Bales's research on small groups. The latter are typically comprised of volunteers willing to attempt the single task put to the group. This positive orientation is most apt to yield the empirical discovery of "sociometric stars" and "task leaders," least apt to sensitize the researcher or theorist to the effect of hostile nonacceptance of the group task. Talcott Parsons and R. F. Bales, *Family Socialization and Interaction Process*, New York: Free Press, Macmillan, 1955.

Yet the same limited definition of the key variables is found in the important attempts by Straus to develop the theory that every social system, as every personality, requires a circumplex model with two independent axes of authority and support. His discussion and examples indicate a variable definition with limited range: support is defined as High (+) or Low (-), but "low" covers both the absence of high support and the presence of negative support; there is love or neutrality in this system, but not hate. Applied to the actual families, this groups destructive mothers with low-supportive mothers, much as the non-authoritarian pole on the Authoritarian Personality Scale includes both mere non-authoritarians and vigorously anti-authoritarian personalities. Murray A. Straus, "Power and Support Structure of the Family in Relation to Socialization," *Journal of Marriage and the Family*, vol. 26, no. 3, 1964, pp. 318–26.

20. Mabel Blake Cohen, "Personal Identity and Sexual Identity," *Psychiatry*, vol. 29, no. 1, 1966, pp. 1–14; Joseph C. Rheingold, *The Fear of Being a Woman: A Theory of Maternal Destructiveness*. New York: Grune & Stratton, 1964.

21. Theodore Lidz, S. Fleck, and A. Cornelison, *Schizophrenia and the Family*, New York: International Universities Press, 1965; Rheingold, *Fear of Being a Woman.*

22. Cf. the long review of studies Rheingold covers in his book on maternal destructiveness, *Fear of Being a Woman.*

23. Benedek, "Parenthood as a Developmental Phase."

24. Reuben Hill and D. A. Hansen, "The Identification of a Conceptual Framework Utilized in Family Study," *Marriage and Family Living*, vol, 22, 1960, pp. 299–311.

25. Harold L. Raush, W. Goodrich, and J. D. Campbell, "Adaptation fo the First Years of Marriage," *Psychiatry*, vol. 26, no. 4, 1963, pp. 368–80.

26. Rapoport, "Normal Crises."

27. Cohen, "Personal Identity."

28. Raush, *Adaptation to the First Years of Marriage.*

29. The greater the cultural pressure to assume a given adult social role, the greater will be

the tendency for individual negative feelings toward that role to be expressed covertly. Men may complain about a given job but not about working per se, and hence their work dissatisfactions are often displaced to the non-work sphere, as psychosomatic complaints or irritation and dominance at home. An equivalent displacement for women of the ambivalence many may feel toward maternity is to dissatisfactions with the homemaker role.

30. Caroline Bird, *The Invisible Scar*. New York: McKay, 1966.

31. When it is realized that a mean family size of 3.5 would double the population in forty years, while a mean of 2.5 would yield a stable population in the same period, the social importance of witholding praise for procreative prowess is clear. At the same time, a drop in the birth rate may reduce the number of unwanted babies born, for such a drop would mean more efficient contraceptive usage and a closer correspondence between desired and attained family size.

32. James A. Davis, *Stipends and Spouses: The Finances of American Arts and Sciences Graduate Students*. Chicago: University of Chicago Press, 1962.

33. Hannah Gavron, *The Captive Wife*. London: Routledge & Kegan Paul, 1966.

34. The recent increase in natural childbirth, prenatal courses for expectant fathers, and greater participation of men during childbirth and postnatal care of the infant may therefore be a consequence of greater sharing between husband and wife when both work and jointly maintain their new households during the early months of marriage. Indeed, natural childbirth builds directly on this shifted base to the marital relationship. Goshen-Gottstein has found in an Israeli sample that women with a "traditional" orientation to marriage far exceed women with a "modern" orientation to marriage in menstrual difficulty, dislike of sexual intercourse, and pregnancy disorders and complaints such as vomiting. She argues that traditional women demand and expect little from their husbands and become demanding and narcissistic by means of their children, as shown in pregnancy by an over-exaggeration of symptoms and attention-seeking. Esther R. Goshen-Gottstein, *Marriage and First Pregnancy: Cultural Influences on Attitudes of Israeli Women*, London: Tavistock, 1966. A prolonged psychic honeymoon uncomplicated by an early pregnancy, and with the new acceptance of married women's employment, may help to cement the egalitarian relationship in the marriage and reduce both the tendency to pregnancy difficulties and the need for a narcissistic focus on the children. Such a background is fruitful ground for sympathy toward and acceptance of the natural childbirth ideology.

35. James A. Davis, *Stipends and Spouses;* James A. Davis, *Great Aspirations,* Chicago: Aldine, 1964; Eli Ginsberg, *Life Styles of Educated Women*, New York: Columbia University Press, 1966; Ginsberg, *Educated American Women: Self Portraits*, New York: Columbia University Press, 1967; National Science Foundation, *Two Years After the College Degree—Work and Further Study Patterns,* Washington, D.C.: U.S. Government Printing Office, NSF vol. 63, no. 26, 1963.

36. Davis, *Great Aspirations; Laure Sharp, "Graduate Study and Its Relation to Careers: The Experience of a Recent Cohort of College Graduates," Journal of Human Resources,* vol. 1, no. 2, 1966, pp. 41–58.

37. James D. Cowhig and C. Nam, "Educational Status, College Plans and Occupational Status of Farm and Nonfarm Youths," *U.S. Bureau of the Census Series ERS (P-27)*, no. 30, 1961.

38. Davis, *Great Aspirations.*

39. Lindsey R. Harmon, *Profiles of Ph.D.'s in the Sciences: Summary Report on Follow-up of Doctorate Cohorts, 1935–1960.* Washington, D.C.: National Research Council, Publication 1293, 1965.

40. Sharp, "Graduate Study and Its Relation to Careers."

41. Davis, *Stipends and Spouses.*

42. Orville G. Brim, "Family Structure and Sex-Role Learning by Children," *Sociometry*,

vol. 21, 1958, pp. 1–16; H. L. Koch, "Sissiness and Tomboyishness in Relation to Sibling Characteristics," *Journal of Genetic Psychology*, vol. 88, 1956, pp. 231–44.

43. Charles MacArthur, "Personalities of First and Second Children," *Psychiatry*. vol. 19, 1956, pp. 47–54; Stanley Schacter, "Birth Order and Sociometric Choice," *Journal of Abnormal and Social Psychology*, vol. 68, 1964, pp. 453–56.

44. Irving Harris, *The Promised Seed*, New York: Free Press, 1964; Bernard Rosen, "Family Structure and Achievement Motivation," *American Sociological Review*. vol, 26, 1961, pp. 574–85; Alice S. Rossi, "Naming Children in Middle-Class Families," *American Sociological Review*, vol. 30, no. 4, 1965, pp. 499–513; Stanley Schachter, "Birth Order, Eminence and Higher Education," *American Sociological Review*, vol. 28, 1963, pp. 757–68.

45. Harris, *Promised Seed*.

46. Rossi, "Naming Children in Middle-Class Families."

47. James H. Bossard, *Parent and Child*, Philadelphia: University of Pennsylvania Press, 1953; James H. Bossard and E. Boll, *The Large Family System*, Philadelphia: University of Pennsylvania, 1956.

48. Rosen, "Family Structure and Achievement Motivation."

49. Glen H. J. Elder and C. Bowerman, "Family Structure and Child Rearing Patterns: The Effect of Family Size and Sex Composition on Child-Rearing Practices," *American Sociological Review*, vol. 28, 1963, pp. 891–905.

50. Sarane S. Boocock, "Toward a Sociology of Learning: A Selective Review of Existing Research," *Sociology of Education*, vol. 39, no. 1, 1966, pp. 1–45.

51. John Nisbet, "Family Environment and Intelligence," in *Education, Economy and Society*, ed. Halsey et al. New York: Free Press, 1961.

52. Thus Rosen writes: "Considering the sociologist's traditional and continuing concern with group size as an independent variable (from Simmel and Durkheim to the recent experimental studies of small groups), there have been surprisingly few studies of the influence of group size upon the nature of interaction in the family" ("Family Structure and Achievement Motivation," p. 576).

53. Ivan Nye, "Adolescent-Parent Adjustment: Age, Sex, Sibling, Number, Broken Homes, and Employed Mothers as Variables," *Marriage and Family Living*, vol. 14, 1952, pp. 327–32.

54. Sears et al., *Patterns of Child-Rearing*.

55. Laura G. Searls, "Leisure Role Emphasis of College Graduate Homemakers," *Journal of Marriage and the Family*, vol, 28, no. 1, 1966, pp. 77–82.

56. Norman Bradburn and D. Caplovitz, *Reports on Happiness*. Chicago: Aldine, 1965.

57. Lee Rainwater, R. Coleman, and G. Handel, *Workingman's Wife*. New York: Oceana Publications, 1959.

58. John Newsom and E. Newsom, *Infant Care in an Urban Community*. New York: International Universities Press, 1963.

59. Mirra Komarovsky, *Blue Collar Marriage*. New York: Random House, 1962.

60. Gavron, *The Captive Wife*.

61. Ferdinand Zweig, *Women's Life and Labor*. London: Camelot Press, 1952.

62. Cohen, "Personal Identity and Sexual Identity."

63. Gerald Gurin, J. Veroff, and S. Feld, *Americans View Their Mental Health*. New York: Basic Books, Monograph Series no. 4, Joint Commission on Mental Illness and Health, 1960.

64. Bradburn and Caplovitz, *Reports on Happiness*.

65. Rapoport, "The Transition from Engagement to Marriage"; Raush et al., "Adaptation to the First Years of Marriage."

66. During the period when marriage was the critical transition in the adult woman's life rather than pregnancy, a good deal of anticipatory "nesting" behavior took place from the

time of conception. Now more women work through a considerable portion of the first pregnancy, and such nesting behavior as exists may be confined to a few shopping expeditions or baby showers, thus adding to the abruptness of the transition and the difficulty of adjustment following the birth of a first child.

67. Lee G. Burchinal, "Adolescent Role Deprivation and High School Marriage," *Marriage and Family Living*, vol. 21, 1959, pp. 378–84; Floyd M. Martinson, "Ego Deficiency as a Factor in Mariage," *American Sociological Review*, vol. 22, 1955, pp. 161–64; J. Joel Moss and Ruby Gingles, "The Relationship of Personality to the Incidence of Early Marriage," *Marriage and Family Living*, vol. 21, 1959, pp. 373–77.

68. Cohen, "Personal Identity and Sexual Identity."

69. Donald W. MacKinnon, "Creativity and Images of the Self," in *The Study of Lives,* ed. Robert W. White. New York: Atherton Press, 1963.

70. Anne Roe, *A Psychological Study of Eminent Biologists,* Psychological Monographs, vol. 65, no. 14, 1951; Anne Roe, "A Psychological Study of Physical Scientists," *Genetic Psychology Monographs*, vol. 43, 1951, pp. 121–239; Anne Roe, "Crucial Life Experiences in the Development of Scientists," in *Talent and Education*, ed. E. P. Torrance, Minneapolis: University of Minnesota Press, 1960.

71. Evelyn M. Duvall, "Conceptions of Parenthood," *American Journal of Sociology*, vol. 52, 1946, pp. 193–203.

72. Urie Bronfenbrenner, "Some Familial Antecedents of Responsibiliy and Leadership in Adolescents," in *Studies in Leadership,* ed. L. Petrullo and B. Bass, New York: Holt, Rinehart, & Winston, 1960.

73. Paul Mussen and L. Distler, "Masculinity, Identification and Father-Son Relationships," *Journal of Abnormal Psychology,* vol. 51, 1959, pp. 350–56.

74. Orville G. Brim, "The Parent-Child Relation as a Social System: I. Parent and Child Roles," *Child Development*, vol. 28, no. 3, 1957, pp. 343–64.

75. Miller and Swanson have suggested a connection between the trend toward bureaucratic structure in the oppupational world and the shift in child-rearing practices toward permissiveness and a greater stress on personal adjustment of children. Their findings are suggestive rather than definitive, however, and no hard research has subjected this question to empirical inquiry. Daniel R. Miller and G. Swanson, *The Changing American Parent*, New York: Wiley 1958.

The same suggestive but nondefinitive clues are to be found in von Mering's study of the contrast between professional and nonprofessional women as mothers. She shows that the professionally active woman in her mother role tends toward a greater stress on discipline rather than indulgence and has a larger number of rules with fewer choices or suggestions to the child: the emphasis is in equipping the child to cope effectively with rules and techniques of his culture. The nonprofessional mother, by contrast, has a great value stress on insuring the child's emotional security, tending to take the role of the clinician in an attempt to diagnose the child's problems and behavior. Faye H. von Mering, "Professional and Non-Professional Women as Mothers," *Journal of Social Psychology*, vol. 42, 1955, pp. 21–34.

76. Orville G. Brim, "Personality Development as Role Learning, in *Personality Development in Children*, ed. Ira Iscoe and Harold Stevenson. University of Texas Press, 1960.

77. Lidz et al., *Schizophrenia and the Family.*

78. Straus, "Power and Support Structure."

79. Parsons and Bales, *Family Socialization.*

80. Brim, "The Parent-Child Relation"

81. Parsons and Bales, *Family Socialization.*

82. Philip Jackson and F. Moscovici, "The Teacher-to-Be: A Study of Embryonic Identification with a Professional Role," *School Review*, vol. 71, no. 1, 1963, pp. 41–65.

83. Clark E. Vincent, "Implications of Changes in Male-Female Role Expectations for Interpreting M-F Scores," *Journal of Marriage and the Family*, vol. 28, no. 2, 1966, pp. 196–99.

84. Helene Deutsch, *The Psychology of Women: A Psychoanalytic Interpretation*, vol. 1. New York: Grune & Stratton, 1944.

85. Sibylle Escalona, "The Psychological Situation of Mother and Child upon Return from the Hospital," in *Problems of Infancy and Childhood: Transactions of the Third Conference*, ed. Milton Senn, 1949.

86. Several authors have recenty pointed out the inadequacy of social science usage of the masculinity-femininity concept. Landreth, in a study of parent-role appropriateness in giving physical care and companionship to the child, found her four-year-old subjects, particularly in New Zealand, made no simple linkage of activity to mother as opposed to father. Catherine Landreth, "Four-Year-Olds' Notions About Sex Appropriateness of Parental Care and Companionship Activities," *Merrill-Palmer Quarterly*, vol. 9, no. 3, 1963, pp. 175–82. She comments that in New Zealand "masculinity and femininity appear to be comfortably relegated to chromosome rather than to contrived activity" (p. 176). Lansky, in a study of the effect of the sex of the children upon the parents' own sexidentification, calls for devising tests which look at masculinity and femininity as two dimensions rather than a single continuum. Leonard M. Lansky, "The Family Structure Also Affects the Model: Sex-Role Identification in Parents of Preschool Children," *Merrill-Palmer Quarterly*, vol. 10, no. 1, 1964, pp. 39–50.

87. Beller has already shown the value of such an approach, in a study that defined independence and dependence as two separate dimensions rather than the extremes of a bipolar continuum. He found, as hypothesized, a very *low* negative correlation between the two measures. E. K. Beller, "Exploratory Studies of Dependency, *New York Academy of Science*, vol. 21, 1959, pp. 414–26.

88. Brim, "The Parent-Child Relation as a Social System."

89. *Ibid.*

90. Landreth, "Four-Year Olds' Notions About Sex Appropriateness," p.8.

91. Cohen, "*Personal Identity and Sexual Identity.*"

D. SOCIETAL STRUCTURE

CHAPTER 34

Women's Liberation and the Family

SYLVIA CLAVAN

The current efforts to win more equitable status on the part of many women in American society have emerged as a significant movement. Its potential for changing major social institutions, particularly that of the family, should not be underestimated. This paper is a statement on the present nature of the Women's Liberation Movement, the issues involved, and the implications it has for the future.

Transformation of the structure of the family as an institution in American society is under way. Changes in the role of the female, the role of the male, and the relationship between them are both cause and effect of the transformation. The potential for greater change in these areas is contained in the ideology and goals of the Women's Liberation Movement. An editorial from *Women: A Journal of Liberation* (vol. 1, no. 3) states, "Traditionally, women have been most oppressed by the institution of the family. . . . To be free, women must understand the source of their oppression and how to control it." Another editorial (vol. 1, no. 2) discussing the limited goals of earlier feminists states, "It is significant that the common phrase which describes the present women's movement is the word 'liberation.' This word implies a deep consciousness of the significance of our struggle: Women are asking for nothing less than the total transformation of the world."

Often it is the atypical or the deviant social phenomenon that points to future change. Jessie Bernard (1968, 6) suggests this when she writes:

> In discussing changes over time, it is important to remind ourselves of the enormous stability of social forms. The modal or typical segments of population show great inertia: they change slowly. . . . What does change, and rapidly, is the form the nontypical takes. It is the nontypical which characterizes a given time: that is, the typical, which tends to be stable, has to

> be distinguished from the characteristic or characterizing, which tends to be fluctuating. When we speak of the "silent generation" or the "beat generation" or the "anti-establishment generation," we are not referring to the typical member of any generation but to those who are not typical.

It is suggested that some of the various actions and ideas of the Women's Liberation Movement might be considered as possible "nontypical behavior anticipating the norm."

The available material on the Women's Liberation Movement mainly consists of articles and journals prepared by women in the movement, a few studies underway, and recent coverage by the mass media including national magazines and local newspapers. The earlier feminist movements are part of history, and source materials pertaining to them are more readily available.[1]

The literature dealing with the family does contain some reference to the relationship between a changing female role and possible changes in the present family structure. In general, these references have to do with the effects of role conflict on modern women in American society. Most often the role conflict is depicted as the outcome of antagonism between a female's needs and desires and the behavior expected of her. At other times the problem is presented as conflict inherent in the way modern women are socialized. That is, they are formally educated about the same as men, but are expected to assume the more traditional female roles of wife-mother-homemaker upon marrying. These types of references almost always apply to educated, middle-class women. They are mostly found in descriptive studies using general knowledge of role theory rather than in studies testing hypotheses. Dager's discussion (1964, 757–59) of sex-role identification point to factors frequently alluded to in these studies. Other studies of the female role treat the effects of the working wife or mother on the family.

[1] The primary sources of material on the current movement are the many articles written by members of the various organizations and distributed through them. *Women: A Journal of Liberation* is sporadically available in private book stores, particularly those serving university students. The author examined many bibliographies suggested by the movement literature as pertinent to understanding the modern feminist. Simone de Beauvior, *The Second Sex*, New York: Knopf, 1953 (originally 1949), and Betty Friedan, *The Feminine Mystique*, New York: W. W. Norton, 1963, are referred to consistently. Friedrich Engels, *The Origins of Family, Private Property, and the State*, New York: New World, originally 1884, is also mentioned often. For a historical background, Mary Beard, *Woman As Force in History*, New York: Macmillan, 1946; Eleanor Flexnor, *A Century of Struggle: The Woman's Rights Movement in the U.S.A.*, Harvard, 1959; and Aileen S. Kraditor, *Ideas of the Woman Suffrage Movement*, 1890–1920, New York: Columbia University, 1965, are helpful. For examples of journalistic coverage, see *The Atlantic*, March, 1970, 18–126; *The Saturday Review*, February 21, 1970, 27–30 and 55; "Sisterhood is Powerful," *New York Times Magazine*, March 15, 1970; and the special report "Women's Liberation: The War on 'Sexism,'" *Newsweek*, March 23, 1970.

One example would be Nye's (1967) discussion of possible changing trends in the family occurring because of women's increasing participation in the occupational world. Nye and Hoffman's (1963) *The Employed Mother in America* also looks at these effects. Goode's (1963, 373) *World Revolution and Family Patterns* touches on the possibility of the revolutionary idea of full equality for women, and he states: "We believe that it is possible to develop a society in which this would happen, but not without a radical reorganization of the social structure."

In general, the sociological literature does not seem to recognize any incipient and rapid change in the roles of women in American society as either an impetus for change or a predictor of change in other social sectors.

The Issues Involved

At the present point of its history, Women's Liberation Movement is an "umbrella name" covering a proliferation of women's groups, some more highly organized than others, but all dedicated to some aspect of improving women's status in this society. The current resurgence of active interest in the status of modern American women is often traced to the publication in 1963 of Betty Friedan's *The Feminine Mystique*. The book examined the post–World War II "back to the home" movement of American women and attacked as a myth the picture of the American woman as a fulfilled and happy housewife. Friedan established the National Organization for Women (NOW) in 1966 as a parliamentary style organization emphasizing improvement for women through legislative change.

Following the appearance of NOW, other feminist groups began to emerge. Among those frequently alluded to are the New York Radical Women, Women's International Terrorist Conspiracy from Hell (WITCH), Redstockings, the Feminists, and Female Liberation. Many of the young women comprising these groups came out of the civil rights and/or peace movements where, ironically, they found that they were treated as inferiors because of their sex.

For purpose of analysis, the Women's Liberation Movement may be conceptualized in two ways. First, it may be viewed as part of an ongoing process that results from industrialization or modernization of a society, of movement toward more equal status for men and women. Thus, in the United States, it may be seen as reactivation of the earlier feminist movement.

Second, the movement may be conceptualized in a more narrowly political sense as emerging out of the revolutionary spirit that characterized the 1960's. This view would hold that oppression of women is but one manifestation of a society that needs complete restructuring. Proponents of this view see female liberation as secondary to the primary focus of social revolution.[2] Although the ideology of the political approach is still nontypical in the American social world, many of its ideas, such as experiments with leaderless societies, rejection of traditional roles and institutions, self-determination, communal living, and shared responsibility for child-rearing, are pertinent to American family structure.

The aims of the different feminist groups are varied and often contradictory. The lack of an organized program can be attributed to newness of the movement, to organizational difficulties encountered at the outset of any new program, or probably most important, basic divisiveness as to an agreed-upon goal or goals. It is possible, however, to pick out several ideas common to most groups. All of the groups see the present conjugal family structure with its traditional division of labor as destructive to full female identity. Much of the focus has been on trying to alleviate the burdens of housework and to get help through free collective child care. The work world it attacked because of its sexual discriminatory practices. The traditional male-female relationship is viewed on a continuum ranging from a point that advocates some changes in sex roles to one that demands complete new definitions. For purposes of discussion, the areas generally considered by women liberationists to be in need of change fall roughly into economic and familial categories. Crossing both of these categories are questions of legal rights. Rights of women under law, however, have been and continue to be gained both worldwide and nationwide. Attempts to implement these rights are part of the feminist struggle.

The Economic Dimension of the Common Goal

Economic discrimination against women takes many forms, some overt and some covert. The argument for broader rights under the law overlaps the economic area, particularly for the single woman. The impetus for redress and amelioration is coming primarily from the highly educated, professional, semiprofessional, or upper-ranked occupational segment of women in the labor force. Their particular situation can best be described as that of second-class citizens in a world of men, if they are permitted to enter that world at all. Although their grievances might appear strange to a female factory worker, they are visible and real and, to some extent, have been documented (Rossi, 1970; Berman and Stocker, 1970). With an increasing number of women earning college degrees and seeking further professional training, action toward attaining a balance between men and women in the higher reaches of occupations will probably become intensified.

[2] See Roxanne Dunbar, "Female Liberation as the Basis for Social Revolution," article published by the New England Free Press, Boston, Mass., 1969.

Two structural factors have been important in the increase of interaction of women and the economy, industrialization and the conjugal family unit. Goode (1963, 369; 1964, 110) has noted that with industrialization the family structure of a society tends toward a conjugal system. He suggests that both modern industrialism and the conjugal structure offer women more economic freedom. Some substantiation for this can be based on the figures given in the President's Commission on the Status of Women (1965, 45) to the effect that in 1962 there were 23 million women in the labor force. In 1967, the Federal Bureau of Labor Statistics (Report No. 94) reported that the number of women working was 27 million, and in the June, 1970 issue of the BLS *Employment and Earnings,* that the figure had grown to 30,974,000. Approximately three out of five women workers are married and among married women, one in three is working. While it is possible to infer from this that discrimination and prejudice against women in the world of work have lessened, the reality is that traditional women's jobs are accorded lower status, earnings are lower than those for men in equivalent jobs, many industries use women as an expendable work force, and men are given preference in hiring where qualified women are available, to name but a few common discriminatory practices. The complete disregard of the female as a member of the work force in her own right is underlined when one notes that they are rarely mentioned, if at all, in the literature dealing with work and occupations. When they are considered, it is almost always within the context of the family structure, i.e., characterized as *still* single, the secondary jobholder in an *organized* family, or the major jobholder in a *disorganized* family. Caplow and McGee (1958, 95 and 194) succinctly speak to the point: ". . . Women tend to be discriminated against in the academic profession, not because they have low prestige but because they are outside the prestige system entirely. . . ." And: "Women scholars are not taken seriously and cannot look forward to a normal professional career. This bias is part of the much larger pattern which determines the utilization of women in our economy."

Johnstone (1968, 103), in considering what economic rights are for women, summarizes them as follows: ". . . The right of access to vocational, technical, and professional training at all levels; the right of economic life without discrimination and to advancement in work life on the basis of qualifications and merit; the right to equal treatment in employment, including equal pay; and the right to maternity protection."

And Rossi (1970, 99–102), exploring the problems of job discrimination, suggests that women ". . . have numerical strength, and a growing number of women's rights organizations to assist them in tackling all levels of discrimination in employment." She predicts that unless protections are forthcoming, there will be an increased militancy by American women. She suggests further that ". . . it must be recognized that such militant women will win legal, economic, and political rights for the daughters of today's traditionalist Aunt Bettys, just as our grandmothers won the vote that women can exercise today."

The Familial Dimension of the Common Goal

Although strengthening women's economic and legal positions would affect other changes in American society, social acceptance of Women's Liberation goals regarding the family has greater far-reaching implications. The assumptions underlying the President's Commission's Report on the Status of Women and its recommendations for improving women's economic, political, and legal positions presented only five years ago are summarized by Margaret Mead (1965, 183–84) in an epilogue to that report. In part, they assume "that both males and females attain full biological humanity only through marriage and the presence of children in the home. . . ." Americans feel that "a life that includes a legal and continuous sex relationship is the only good life." The typical woman is depicted as marrying early, having several children, and living many years after her children are grown. Mead goes on to say:

> Here it [the Report] makes the following assumptions: all women want to marry; marriage involves having (or at least, rearing) children; children are born (or adopted) early in marriage; the home consists of the nuclear family only; and special attention must be given to women not in the state assumed to be normal — the single, the divorced, and the widowed.

Women's Liberation questions and challenges these traditionally held ideas about the ideal American woman. Not all liberationists favor destruction of the conjugal family system, but most view the expected role structure of the husband as provider and the wife as homemaker and child's nurse as the basis of their oppression. It is possible that the movement heralds a revolutionary change in the American family. The movement's proposals generate many questions and reconceptualizations of existing family organization. Some of these are suggested below:

1. The conjugal family system has often been presented as congenial to the mode of life in modern industrialized societies. However, some have suggested that American society has entered a new socioeconomic era. For example, a decade ago Galbraith (1958) spoke of the need for less emphasis on production, and the society is frequently referred to as a post-modern society. Within this context, it is possible to speculate that the conjugal unit may be outmoded. If upper strata women can be considered to have reached a post-modern economic level, then the traditional family arrangements may no longer suit their needs. This may partially explain the phonemenon of the protest coming from this socioeconomic level.
2. Change has occurred in the American family since the early colonial period. There has been no serious suggestion, however, that child-rearing

be made a public rather than a private responsibility so that women could pursue their own goals. In the past, instances of interest in public child-rearing have been closely allied to political and/or economic goals. The most frequently cited examples are the Hebrew kibbutz, the Chinese commune, and the Russian system of state nurseries for children.[3] It may be argued that the change to public responsibility has been going on with the transfer of the educational, religious, and recreational functions from the family to other social institutions. However, the functions of the family are most often stated to be the socialization of the child and the psychological function of providing emotional support for its members. Seen in this way, extended public child care would seriously weaken the basis for maintenance of a conjugal system.

Seeking an alternative mode of child care as a means of freeing women bears an inherent conflict. The basic premise is that the job is oppressive and demeaning. It is accorded low status and attracting competent men and women to the job might prove difficult. It is possible, however, that what is considered oppressive in private family settings may not be so considered should the task be accorded some professional ranking. Women in the movement are aware that a satisfactory solution to the problem will not be easy.

3. The institutionalized normative expectations of the female in American society today require that she regard her wife-mother-homemaker role as primary. Any employment that she may participate in is considered to be secondary to her basic role. A consequence of these prescribed behaviors is that an increasing number of women, particularly those who have enjoyed a higher education, find marriage and the home a source of discontent and unhappiness. They conform to expectations of them as women, but find that there is no social acceptance of their desire to participate fully in the world of work. The result is guilt on the part of those who pursue careers or unhappiness on the part of those who continue to conform. If, as Women's Liberation proposes, the career role moved to primary position and were socially sanctioned, would those women who find their present traditional roles acceptable become the new disaffected group? Put in another way, would exchange of one set of prescribed and proscribed behaviors liberate one segment of the female population but inhibit another segment?

4. If nothing else, the aims of Women's Liberation necessitate some degree of restructuring of the traditional roles of the male and female and the relationship between them. Jessie Bernard (1968, 14) described this problem area as somewhat like a "zero-sum game." For instance, on a

[3] For a discussion of the relationship between economic and political factors and public child-rearing, see Jesse R. Pitts, The Structural-Functional Approach, in H. T. Christensen (Ed.), *Handbook of Marriage and the Family,* Chicago: Rand McNally, 1967, 110–11

material level when women are given rights such as the right to vote, men lose nothing. But when women are given property or employment rights, men are deprived of what was theirs. The same theory could be applied on a socio-psychological level. The emphasis on changes in the female role tends to hide the necessary corresponding changes in the male role. Bernard states, "For women, the relevant problems have to do with the implications of sexuality for equality; for men, with the implications of equality for sexuality." What is often referred to as the "emasculation of the male" has been given attention from time to time in both the professional literature and in fiction. This, of course, refers to emasculation within the context of the socio-cultural definition of masculinity. Margaret Mead (1935) suggested that different definitions exist when she detailed different cultural manifestations of sexuality such as the unaggressive males and females of the Arapesh tribe, the "male-like" males and females of the Mundugumor, and the reversal of sex attitudes that were found in the Tchambuli tribe, as compared to American sexual definitions.

The Women's Liberation Movement views men in varying ways. All of the views emphasize the present differences and the drive toward equality. This emphasis tends to obliterate the many attributes that men and women have in common, common needs, feelings, desires, emotions, etc. If society's prescriptions for approved female behavior have inhibited full realization of her potential, then those same prescriptions for the male have affected him similarly. The suggestion is that in a network of interdependent role relationships, it is unrealistic to emphasize one role over the other.

Where Will It Lead?

As women begin to attain some of the goals toward which their efforts are directed, changes in family structure can be anticipated. The nature of the changes cannot be predicted with any degree of certainty. In general, the movement has emphasized the necessity for change in social expectations of the female without giving much attention to the attendant effects on the family that such change would bring. The exception to this is the stand taken by the most extreme liberationists who hold that the present family structure is not acceptable. If, however, the nuclear family unit is still viewed as functional and desirable, then efforts can be made toward accommodating future changes. While attention is focused on the demands for change in sexual roles, an opportunity exists also for bringing into close correspondence the needs of both men and women for self-realization and the broader societal need for a healthy viable family form.

The kinds of action needed as steps toward this end open a new area of inquiry for those in the applied family services. It would seem that whatever the course the implementation takes, the underlying directional philosophy should be to permit choice. Oppression appears to exist where there is no choice of acceptable alternatives for the individual. It has already been suggested that freedom to choose a life style within marriage is an important indicator of happiness in that relationship (Orden and Bradburn, 1969; and Janeway, 1970). The freedom to choose alternative patterns of behavior in the other role relationships within the nuclear family unit may well prove to be a source of strength for that unit.

Industrialization, advanced technology, and higher levels of education may spell the end of traditional division of labor by sex. Societal recognition and acceptance of variation in sex role patterns would be living proof of a social revolution fashioned by the Women's Liberation Movement of the 70's.

References:

Beard, Mary. *Woman as Force in History*. New York: Macmillan, 1946.

Berman, J., and E. Stocker. Women's Lib in the American Psychological Association. *Women: A Journal of Liberation,* 1970, 1, 52–53.

Bernard, Jessie. The Status of Women in Modern Patterns of Culture. *The Annals,* 1968, 375, 3–14.

Caplow, Theodore, and R. J. McGee. *The Academic Marketplace*. Garden City, N.Y.: Doubleday, 1958.

Dager, Edward Z. Socialization and Personality Development in the Child. In H. T. Christensen (Ed.), *Handbook of Marriage and the Family*. Chicago: Rand McNally, 1967.

De Beauvior, Simone. *The Second Sex*. New York: Knopf, 1953 (originally 1949).

Dunbar, Roxanne. Female Liberation as the Basis for Social Revolution. Boston: New England Free Press, 1969.

Engels, Friedrich. *The Origins of Family, Private Property, and the State*. New York: International Publishers, original date 1884.

Flexnor, Eleanor. *A Century of Struggle: The Women's Rights Movement in the U.S.A.* Cambridge: Harvard, 1959.

Friedan, Betty. *The Feminine Mystique*. New York: Norton, 1963.

Galbraith, John K. *The Affluent Society*. Boston: Houghton Mifflin, 1958.

Goode, W. J. *The Family*. Englewood Cliffs, N.J.: Prentice-Hall, 1964.

Goode, W. J. *World Revolution and Family Patterns*. New York: Free Press, 1963.

Janeway, Elizabeth. Happiness and the Right to Choose. *The Atlantic,* March, 1970, 118–26.

Johnstone, Elizabeth. Women in Economic Life: Rights and Opportunities. *The Annals,* 1968, 375, 102–114.

Komisar, Lucy. The New Feminism. *Saturday Review,* February 21, 1970, 27–30 and 55.

Kraditor, Aileen S. *Ideas of the Woman Suffrage Movement, 1890–1920*. New York: Columbia University, 1965.

Mead, Margaret. *Sex and Temperament in Three Primitive Societies*. New York: Morrow, 1935.

Mead, Margaret, and F. Kaplan (Eds.). The Report of the President's Commission on the Status of Women and Other Publications of the Commission. *American Women*. New York: Scribner, 1965.

Nye, F. Ivan. Values, Family, and a Changing Society. *Journal of Marriage and the Family,* 1967, 29, 241–48.

Nye, F. Ivan, and Lois W. Hoffman. *The Employed Mother in America.* Chicago: Rand McNally, 1963.

Orden, S. R., and N. M. Bradburn. Working Wives and Marriage Happiness. *American Journal of Sociology,* 1969, 74, 392–407.

Pitts, Jesse R. The Structural-Functional Approach. In H. T. Christensen (Ed.), *Handbook of Marriage and the Family.* Chicago: Rand McNally, 1967.

Rossi, Alice S. Status of Women in Graduate Departments of Sociology, 1968–69. *American Sociologist,* 1970, 5, 1–12.

Rossi, Alice S. Job Discrimination and What Women Can Do About It. *Atlantic,* March, 1970, 225, 99–102.

U.S. Bureau of Labor Statistics. Employment and Earnings, June, 1970.

U.S. Bureau of Labor Statistics. Report no. 94, 1967.

Anon. Sisterhood Is Powerful. *New York Times Magazine,* March 15, 1970.

Anon. The War on "Sexism." *Newsweek,* March 23, 1970.

Anon. *Women: A Journal of Liberation.* Winter 1970, 1; Spring 1970, 2.

CHAPTER 35

Social Class and Parent-Child Relationships

MELVIN L. KOHN

This essay is an attempt to interpret, from a sociological perspective, the effects of social class upon parent-child relationships. Many past discussions of the problem seem somehow to lack this perspective, even though the problem is one of profound importance for sociology. Because most investigators have approached the problem from an interest in psychodynamics, rather than social structure, they have largely limited their attention to a few specific techniques used by mothers in the rearing of infants and very young children. They have discovered, *inter alia,* that social class has a decided bearing on which techniques parents use. But, since they have come at the problem from this perspective, their interest in social class has not gone beyond its effects for this very limited aspect of parent-child relationships.

The present analysis conceives the problem of social class and parent-child relationships as an instance of the more general problem of the effects of social structure upon behavior. It starts with the assumption that social class has proved to be so useful a concept because it refers to more than simply educational level, or occupation, or any of the large number of correlated variables. It is so useful because it captures the reality that the intricate interplay of all these variables creates different basic conditions of life at different levels of the social order. Members of different social classes, by virtue of enjoying (or suffering) different conditions of life, come to see the world differently — to develop different conceptions of social reality, different aspirations and hopes and fears, different conceptions of the desirable.

The last is particularly important for present purposes, for from people's conceptions of the desirable — and particularly from their conceptions of what characteristics are desirable in children — one can discern their objectives in child-rearing. Thus, conceptions of the desirable — that is, values (1) — become the key concept for this anaysis, the bridge between position in the larger social structure and the behavior of the individual. The intent of the analysis is to trace the effects of social class position on parental values and the effects of values on behavior.

Since this approach differs from analyses focused on social class differences in the use of particular child-rearing techniques, it will be necessary to re-examine earlier formulations from the present perspective. Then three questions will be discussed, bringing into consideration the limited available data that are relevant: What differences are there in the values held by parents of different social classes? What is there about the conditions of life distinctive of these classes that might explain the differences in their values? What consequences do these differences in values have for parents' relationships with their children?

Social Class

Social classes will be defined as aggregates of individuals who occupy broadly similar positions in the scale of prestige (2). In dealing with the research literature, we shall treat occupational position (or occupational position as weighted somewhat by education) as a serviceable index of social class for urban American society. And we shall adopt the model of social stratification implicit in most research, that of four relatively discrete classes: a "lower class" of unskilled manual workers, a "working class" of manual workers in semiskilled and skilled occupations, a "middle class" of white-collar workers and professionals, and an "elite," differentiated from the middle class not so much in terms of occupation as of wealth and lineage.

Almost all the empirical evidence, including that from our own research, stems from broad comparisons of the middle and working class. Thus we shall have little to say about the extremes of the class distribution. Furthermore, we shall have to act as if the middle and working classes were each homogeneous. They are not, even in terms of status considerations alone. There is evidence, for example, that within each broad social class, variations in parents' values quite regularly parallel gradations of social status. Moreover, the classes are heterogeneous with respect to other factors that affect parents' values, such as religion and ethnicity. But even when all such considerations are taken into account, the empirical evidence clearly shows that being on one side or the other of the line that divides manual from non-manual workers has profound consequences for how one rears one's children (3).

Stability and Change

Any analysis of the effects of social class upon parent-child relationships should start with Urie Bronfenbrenner's analytic review of the studies that had been conducted in this country during the twenty-five years up to 1958 (4). From the seemingly contradictory findings of a number of studies, Bronfenbrenner discerned not chaos but orderly change: there have been changes in the child-training techniques employed by middle-class parents in the past quarter-century; similar changes have been taking place in the working class, but working-class parents have consistently lagged behind by a few years; thus, while middle-class parents of twenty-five years ago were more "restrictive" than were working-class parents, today the middle-class parents are more "permissive"; and the gap between the classes seems to be narrowing.

It must be noted that these conclusions are limited by the questions Bronfenbrenner's predecessors asked in their research. The studies deal largely with a few particular techniques of child-rearing, especially those involved in caring for infants and very young children, and say very little about parents' over-all relationships with their children, particularly as the children grow older. There is clear evidence that the past quarter-century has seen change, even faddism, with respect to the use of breast-feeding or bottle-feeding, scheduling or not scheduling, spanking or isolating. But when we generalize from these specifics to talk of a change from "restrictive" to "permissive" practices — or, worse yet, of a change from "restricive" to "permissive" parent-child relationships — we impute to them a far greater importance than they probably have, either to parents or to children (5).

There is no evidence that recent faddism in child-training techniques is symptomatic of profound changes in the relations of parents to children in either social class. In fact, as Bronfenbrenner notes, what little evidence we do have points in the opposite direction: the over-all quality of parent-child relationships does not seem to have changed substantially in either class (6). In all probability, parents have changed techniques in service of much the same values, and the changes have been quite specific. These changes must be explained, but the enduring characteristics are probably even more important.

Why the changes? Bronfenbrenner's interpretation is ingenuously simple. He notes that the changes in techniques employed by middle-class parents have closely paralleled those advocated by presumed experts, and he concludes that middle-class parents have changed their practices *because* they are responsive to changes in what the experts tell them is right and proper. Working-class parents, being less educated and thus less directly responsive to the media of communication, followed behind only later (7).

Bronfenbrenner is almost undoubtedly right in asserting that middle-class parents have followed the drift of presumably expert opinion. But why have they done so? It is not sufficient to assume that the explanation lies in their greater degree of education. This might explain why middle-class parents are substantially more likely than are working-class parents to *read* books and articles on child-rearing, as we know they do (8). But they need not *follow* the experts' advice. We know from various studies of the mass media that people generally search for confirmation of their existing beliefs and practices and tend to ignore what contradicts them.

From all the evidence at our disposal, it looks as if middle-class parents not only read what the experts have to say but also search out a wide variety of other sources of information and advice: they are far more likely than are working-class parents to discuss child-rearing with friends and neighbors, to consult physicians on these matters, to attend Parent-Teacher Association meetings, to discuss the child's behavior with his teacher. Middle-class parents seem to regard child-rearing as more problematic than do working-class parents. This can hardly be a matter of education alone. It must be rooted more deeply in the conditions of life of the two social classes.

Everything about working-class parents' lives — their comparative lack of education, the nature of their jobs, their greater attachment to the extended family — conduces to their retaining familiar methods (9). Furthermore, even should they be receptive to change, they are less likely than are middle-class parents to find the experts' writings appropriate to their wants, for the experts predicate their advice on middle-class values. Everything about middle-class parents' lives, on the other hand, conduces to their looking for new methods to achieve their goals. They look to the experts, to other sources of relevant information, and to each other not for new values but for more serviceable techniques (10). And within the limits of our present scanty knowledge about means-ends relationships in child-rearing, the experts have provided practical and useful advice. It is not that educated parents slavishly follow the experts but that the experts have provided what the parents have sought.

To look at the question this way is to put it in a quite different perspective: the focus becomes not specific techniques nor changes in the use of specific techniques but parental values.

Values of Middle- and Working-Class Parents

Of the entire range of values one might examine, it seems particularly strategic to focus on parents' conceptions of what characteristics would be most desirable for boys or girls the age of their own children. From this one can hope to discern the parents' goals in rearing their children. It must be assumed, however, that a parent will choose one characteristic as more desirable than another only if he considers it to be both important, in the sense that failure to develop this characteristic would affect the child adversely, and problematic, in the sense that it is neither to be taken for granted that the child will develop that characteristic nor impossible for him to do so. In interpreting parents' value choices, we must keep in mind that their choices reflect not simply their goals but the goals whose achievement they regard as problematic.

Few studies, even in recent years, have directly investigated the relationship of social class to parental values. Fortunately, however, the results of these few are in essential agreement. The earliest study was Evelyn Millis Duvall's pioneering inquiry of 1946 (11). Duvall characterized working-class (and lower middle-class) parental values as "traditional" — they want their children to be neat and clean, to obey and respect adults, to please adults. In contrast to this emphasis on how the child comports himself, middle-class parental values are more "developmental" — they want their children to be eager to learn, to love and confide in the parents, to be happy, to share and co-operate, to be healthy and well.

Duvall's traditional-developmental dichotomy does not describe the difference between middle- and working-class parental values quite exactly, but it does point to the essence of the difference: working-class parents want the child to conform to externally imposed standards, while middle-class parents are far more attentive to his internal dynamics.

The few relevant findings of subsequent studies are entirely consistent with this basic point, especially in the repeated indications that working-class parents put far greater stress on obedience to parental commands than do middle-class parents (12). Our own research, conducted in 1956–57, provides the evidence most directly comparable to Duvall's (13). We, too, found that working-class parents value obedience, neatness, and cleanliness more highly than do middle-class parents, and that middle-class parents in turn value curiosity, happiness, consideration, and — most importantly — self-control more highly than do working-class parents. We further found that there are characteristic clusters of value choice in the two social classes: working-class parental values center on conformity to external proscriptions, middle-class parental values on *self*-direction. To working-class parents, it is the overt act that matters: the child should not transgress externally imposed rules; to middle-class parents, it is the child's motives and feelings that matter: the child should govern himself.

In fairness, it should be noted that middle- and working-class parents share many core values. Both, for example, value honesty very highly — although, characteristically, "honesty" has rather different connotations in the two social classes, implying "trustworthiness" for the working-class and "truthfulness" for the middle-class. The common theme, of course, is that parents of both social classes value a decent respect for the rights of others; middle- and working-class values are but variations on this common theme. The reason for emphasizing the variations rather than the common theme is that they seem to have far-ranging consequences for parents' relationships with their children and thus ought to be taken seriously.

It would be good if there were more evidence about parental values — data from other studies, in other locales, and especially, data derived from more than one mode of inquiry. But, what evidence we do have is consistent, so that there is at least some basis for believing it is reliable. Furthermore, there is evidence that the value choices made by parents in these inquiries are not simply a reflection of their assessments of their own children's deficiencies or excellences. Thus, we may take the findings of these studies as providing a limited, but probably valid, picture of the parents' generalized conceptions of what behavior would be desirable in their preadolescent children.

Explaining Class Differences in Parental Values

That middle-class parents are more likely to espouse some values, and working-class parents other values, must be a function of differences in their conditions of life. In the present state of our knowledge, it is difficult to disentangle the interacting variables with a sufficient degree of exactness to ascertain which conditions of life are crucial to the differences in values. Nevertheless, it is necessary to examine the principal components of class differences in life conditions to see what each may contribute.

The logical place to begin is with occupational differences, for these are certainly pre-eminently important, not only in defining social classes in urban, industrialized society, but also in determining much else about people's life conditions (14). There are at least three respects in which middle-class occupations typically differ from working-class occupations, above and beyond their obvious status-linked differences in security, stability of income, and general social prestige. One is that middle-class occupations deal more with the manipulation of interpersonal relations, ideas, and symbols, while working-class occupations deal more with the manipulation of things. The second is that middle-class occupations are more subject to self-direction, while working-class occupations are more subject to standardization and direct supervision. The third is that getting ahead in middle-class occupations is more dependent upon one's own actions, while in working-class occupations it is more dependent upon collective action, particularly in unionized industries. From these differences, one can sketch differences in the characteristics that make for getting along, and getting ahead, in middle- and working-class occupations. Middle-class occupations require a greater degree of self-direction; working-class occupations, in larger measure, require that one follow explicit rules set down by someone in authority.

Obviously, these differences parallel the differences we have found between the two social classes in the characteristics valued by parents for children. At minimum, one can conclude that there is a congruence between occupational requirements and parental values. It is, moreover, a reasonable supposition, although not a necessary conclusion, that middle- and working-class parents value different characteristics in children *because* of these differences in their occupational circumstances. This supposition does not necessarily assume that parents consciously train their children to meet future occupational requirements; it may simply be that their own occupational experiences have significantly affected parents' conceptions of what is desirable behavior, on or off the job, for adults or for children (15).

These differences in occupational circumstances are probably basic to the differences we have found between middle- and working-class parental values, but taken alone they do not sufficiently explain them. Parents need not accord pre-eminent importance to occupational requirements in their judgments of what is most desirable. For a sufficient explanation of class differences in values, it is necessary to recognize that other differences in middle- and working-class conditions of life reinforce the differences in occupational circumstances at every turn.

Educational differences, for example, above and beyond their importance as determinants of occupation, probably contribute independently to the differences in middle- and working-class parental values. At minimum, middle-class parents' greater attention to the child's internal dynamics is facilitated by their learned ability to deal with the subjective and the ideational. Furthermore, differences in levels and stability of income undoubtedly contribute to class differences in parental values. That middle-class parents still have somewhat higher levels of income, and much greater stability of income, makes them able to take for granted the respectability that is still problematic for working-class parents. They can afford to concentrate, instead, on motives and feelings — which, in the circumstances of their lives, are more important.

These considerations suggest that the differences between middle- and working-class parental values are probably a function of the entire complex of differences in life conditions characteristic of the two social classes. Consider, for example, the working-class situation. With the end of mass immigration, there has emerged a stable working class, largely derived from the manpower of rural areas, uninterested in mobility into the middle class, but very much interested in security, respectability, and the enjoyment of a decent standard of living (16). This working class has come to enjoy a standard of living formerly reserved for the middle class, but has not chosen a middle-class style of life. In effect, the working class has striven for, and partially achieved, an American dream distinctly different from the dream of success and achievement. In an affluent society, it is possible for the worker to be the traditionalist — politically, economically, and, most relevant here, in his values for his children (17). Working-class parents want their children to conform to external authority because the parents themselves are willing to accord respect to authority, in return for security and respectability. Their conservatism in child-rearing is part of a more general conservatism and traditionalism.

Middle-class parental values are a product of a quite different set of conditions. Much of what the working class values, they can take for granted. Instead, they can — and must — instill in their children a degree of self-direction that would be less appropriate to the conditions of life of the working class (18). Certainly, there is substantial truth in the characterization of the middle-class way of life as one of great conformity. What must be noted here, however, is that *relative to* the working class, middle-class conditions of life require a more substantial degree of independence of action. Furthermore, the higher levels of education enjoyed by the middle class make possible a degree of internal scrutiny difficult to achieve without the skills in dealing with the abstract that college training sometimes provides. Finally, the economic security of most middle-class occupations, the level of income they provide, the status they confer, allow one to focus his attention on the subjective and the ideational. Middle-class conditions of life both allow and demand a greater degree of self-direction than do those of the working class.

Consequences of Class Differences in Parents' Values

What consequences do the differences between middle- and working-class parents'values have for the ways they raise their children?

Much of the research on techniques of infant- and child-training is of little relevance here. For example, with regard to parents' preferred techniques for disciplining children, a question of major interest to many investigators, Bronfenbrenner summarizes past studies as follows: "In matters of discipline, working-class parents are consistently more likely to employ physical punishment, while middle-class families rely more on reasoning, isolation, appeals to guilt, and other methods involving the threat of loss of love" (19). This, if still true (20), is consistent with middle-class parents' greater attentiveness to the child's internal dynamics, working-class parents' greater concern about the overt act. For present purposes, however, the crucial question is not *which* disciplinary method parents prefer, but when and why they use one or another method of discipline.

The most directly relevant available data are on the conditions under which middle- and working-class parents use physical punishment. Working-class parents are apt to resort to physical punishment when the direct and immediate consequences of their children's disobedient acts are most extreme, and to refrain from punishing when this might provoke an even greater disturbance (21). Thus, they will punish a child for wild play when the furniture is damaged or the noise level becomes intolerable, but ignore the same actions when the direct and immediate consequences are not so extreme. Middle-class parents, on the other hand, seem to punish or refrain from punishing on the basis of their interpretation of the child's intent in acting as he does. Thus, they will punish a furious outburst when the context is such that they interpret it to be a loss of self-control, but will ignore an equally extreme outburst when the context is such that they interpret it to be merely an emotional release.

It is understandable that working-class parents react to the consequences rather than to the intent of their children's actions: the important thing is that the child not transgress externally imposed rules. Correspondingly, if middle-class parents are instead concerned about the child's motives and feelings, they can and must look beyond the overt act to why the child acts as he does. It would seem that middle- and working-class values direct parents to see their children's misbehavior in quite different ways, so that misbehavior which prompts middle-class parents to action does not seem as important to working-class parents, and vice versa (22). Obviously, parents' values are not the only things that enter into their use of physical punishment. But unless one assumes a complete lack of goal-directedness in parental behavior, he would have to grant that parents' values direct their attention to some facets of their own and their children's behavior, and divert it from other facets.

The consequences of class differences in parental values extend far beyond differences in disciplinary practices. From a knowledge of their values for their children, one would expect middle-class parents to feel a greater obligation to be *supportive* of the children, if only because of their sensitivity to the children's internal dynamics. Working-class values, with their emphasis upon conformity to external rules, should lead to greater emphasis upon the parents' obligation to impose constraints (23). And this, according to Bronfenbrenner, is precisely what has been shown in those few studies that have concerned themselves with the over-all relationship of parents to child: "Over the entire twenty-five-year period studied, parent-child relationships in the middle-class are consistently reported as more acceptant and equalitarian, while those in the working-class are oriented toward maintaining order and obedience" (24).

This conclusion is based primarily on studies of *mother*-child relationships in middle- and working-class families. Class differences in parental values have further ramifications for the father's role (25). Mothers in each class would have their husbands play a role facilitative of the child's development of the characteristics valued in that class: Middle-class mothers want their husbands to be supportive of the children (especially of sons), with their responsibility for imposing constraints being of decidedly secondary importance; working-class mothers look to their husbands to be considerably more directive — support is accorded far less importance and constraint far more. Most middle-class fathers agree with their wives and play a role close to what their wives would have them play. Many working-class fathers, on the other hand, do not. It is not that they see the constraining role as less important than do their wives, but that many of them see no reason why they should have to shoulder the responsibility. From their point of view, the important thing is that the child be taught what limits he must not transgress. It does not much matter who does the teaching, and since mother has primary responsibility for child care, the job should be hers.

The net consequence is a quite different division of parental responsibilities in the two social classes. In middle-class families, mother's and father's roles usually are not sharply differentiated. What differentiation exists is largely a matter of each parent taking special responsibility for being supportive of children of the parent's own sex. In working-class families, mother's and father's roles are more sharply differentiated, with mother almost always being the more supportive parent. In some working-class families, mother specializes in support, father in constraint; in others, perhaps in most, mother raises the children, father provides the wherewithal (26).

Thus, the differences in middle- and working-class parents' values have wide ramifications for their relationships with their children and with each other. Of course, many class differences in parent-child relationships are not directly attributable to differences in values; undoubtedly the very differences in their conditions of life that make for differences in parental values reinforce, at every juncture, parents' characteristic ways of relating to their children. But one could not account for these consistent differences in parent-child relationships in the two social classes without reference to the differences in parents' avowed values.

Conclusion

This paper serves to show how complex and demanding are the problems of interpreting the effects of social structure on behavior. Our inquiries habitually stop at the point of demonstrating that social position correlates with something, when we should want to pursue the question, "Why?" What are the processes by which position in social structure molds behavior? The present analysis has dealt with this question in one specific form: Why does social class matter for parents' relationships with their children? There is every reason to believe that the problems encountered in trying to deal with that question would recur in any analysis of the effects of social structure on behavior.

In this analysis, the concept of "values" has been used as the principal bridge from social position to behavior. The analysis has endeavored to show that middle-class parental values differ from those of working-class parents; that these differences are rooted in basic differences between middle- and working-class conditions of life; and that the differences between middle- and working-class parental values have important consequences for their relationships with their children. The interpretive model, in essence, is: social class — conditions of life — values — behavior.

The specifics of the present characterization of parental values may prove to be inexact; the discussion of the ways in which social class position affects values is undoubtedly partial; and the tracing of the consequences of differences in values for differences in parent-child relationships is certainly tentative and incomplete. I trust, however, that the perspective will prove to be valid and that this formulation will stimulate other investigators to deal more directly with the processes whereby social structure affects behavior.

References

1. "A value is a conception, explicit or implicit, distinctive of an individual or characteristic of a group, of the desirable which influences the selection from available modes, means, and ends of action" (Clyde Kluckhohn, "Values and Value Orientations," in Talcott Parsons and Edward A. Shils, eds., *Toward a General Theory of Action*. Cambridge: Harvard University Press, 1951, p. 395). See also the discussion of values in Robin M. Williams, Jr., *American Society: A Sociological Interpretation*. New York: Knopf, 1951, chap. 11, and his discussion of social class and culture on p. 101.

2. Williams, *American Society*, p. 89.

3. These, and other assertions of fact not referred to published sources, are based on research my colleagues and I have conducted. For the design of this research and the principal substantive findings see my "Social Class and Parental Values," *American Journal of Sociology*, vol. 44, January 1959, pp. 337-51; my "Social Class and the Exercise of Parental Authority," *American Sociological Review*, vol. 24, June 1959, pp. 352-66; and with Eleanor E. Carroll, "Social Class and the Allocation of Parental Responsibilities," *Sociometry*, vol. 23, December 1960, pp. 372-92. I should like to express my appreciation to my principal collaborators in this research, John A. Clausen and Eleanor E. Carroll.

4. Urie Bronfenbrenner, "Socialization and Social Class through Time and Space," in Eleanor E. Maccoby, Theodore M. Newcomb, and Eugene L. Hartley, eds., *Readings in Social Psychology*. New York: Holt, 1958.

5. Furthermore, these concepts employ *a priori* judgments about which the various investigators have disagreed radically. See, e.g., Robert R. Sears, Eleanor E. Maccoby, and Harry Levin, *Patterns of Child Rearing*, Evanston, Ill.: Row, Peterson, 1957, pp. 444-47; and Richard A. Littman, Robert C. A. Moore, and John Pierce-Jones, "Social Class Differences

in Child Rearing: A Third Community for Comparison with Chicago and Newton," *American Sociological Review*, vol. 22, December 1957, pp. 694-704, esp. p. 703.

6. Bronfenbrenner, "Socialization and Social Class," pp. 420-22 and 425.

7. Bronfenbrenner gives clearest expression to this interpretation, but it has been adopted by others, too. See, e.g., Martha Sturm White, "Social Class, Child-Rearing Practices, and Child Behavior," *American Sociological Review*, vol. 22, December 1957, pp. 704-712.

8. This was noted by John E. Anderson in the first major study of social class and family relationships ever conducted, and has repeatedly been confirmed. (*The Young Child in the Home: A Survey of Three Thousand American Families*. New York: Appleton-Century, 1936.)

9. The differences between middle- and working-class conditions of life will be discussed more fully later in this paper.

10. Certainly middle-class parents do not get their values from the experts. In our research, we compared the values of parents who say they read Spock, Gesell, or other books on child-rearing to those who read only magazine and newspaper articles, and those who say they read nothing at all on the subject. In the middle class, these three groups have substantially the same values. In the working class, the story is different. Few working-class parents claim to read books or even articles on child-reading. Those few who do have values much more akin to those of the middle class. But these are atypical working-class parents who are very anxious to attain middle-class status. One suspects that for them the experts provide a sort of handbook to the middle class; even for them, it is unlikely that the values come out of Spock and Gesell.

11. "Conceptions of Parenthood," *American Journal of Sociology*, vol. 52, November 1946, pp. 193-203.

12. Alex Inkeles has shown that this is true not only for the United States, but for a number of other industrialized societies as well ("Industrial Man: The Relation of Status to Experience, Perception, and Value," *American Journal of Sociology*, vol. 66, July 1960, pp. 20-21 and Table 9).

13. Kohn, "Social Class and Parental Values."

14. For a thoughtful discussion of the influence of occupational role on parental values, see David F. Aberle and Kaspar D. Naegele, "Middle-Class Fathers' Occupational Role and Attitudes Toward Children," *American Journal of Orthopsychiatry*, vol. 22, April 1952, pp. 366-78.

15. Two objections might be raised here. (1) Occupational experiences may not be important for a mother's values, however crucial they are for her husband's, if she has had little or no work experience. But even those mothers who have had little or no occupational experience know something of occupational life from their husbands and others, and live in a culture in which occupation and career permeate all of life. (2) Parental values may be built not so much out of their own experiences as out of their expectations of the child's future experiences. This might seem particularly plausible in explaining working-class values, for their high valuation of such stereotypically *middle-class* characteristics as obedience, neatness, and cleanliness might imply that they are training their children for a middle-class life they expect the children to achieve. Few working-class parents, however, do expect (or even want) their children to go on to college and the middle-class jobs for which a college education is required. This is shown in Herbert H. Hyman, "The Value Systems of Different Classes: A Social Psychological Contribution to the Analysis of Stratification," in Reinhard Bendix and Seymour Martin Lipset, eds., *Class, Status and Power: A Reader in Social Stratification*, Glencoe, Ill.: Free Press, 1953, and confirmed in unpublished data from our own research.

16. See, e.g., S. M. Miller and Frank Riessman, "The Working Class Subculture: A New View," *Social Problems*, vol. 9, Summer 1961, pp. 86-97.

17. Relevant here is Seymour Martin Lipset's somewhat disillusioned "Democracy and Working-Class Authoritarianism," *American Sociological Review*, vol. 24, August 1959, pp. 482-501.

18. It has been argued that as larger and larger proportions of the middle class have become imbedded in a bureaucratic way of life — in distinction to the entrepreneurial way of life of a bygone day — it has become more appropriate to raise children to be accommodative than to be self-reliant. But this point of view is a misreading of the conditions of life faced by the middle-class inhabitants of the bureaucratic world. Their jobs require at least as great a degree of self-reliance as do entrepreneurial enterprises. We tend to forget, nowadays, just how little the small- or medium-sized entrepreneur controlled the conditions of his own existence and just how much he was subjected to the petty authority of those on whose pleasure depended the survival of his enterprise. And we fail to recognize the degree to which monolithic-seeming bureaucracies allow free play for — in fact, require — individual enterprise of new sorts: in the creation of ideas, the building of empires, the competition for advancement.

At any rate, our data show no substantial differences between the values of parents from bureaucratic and entrepreneurial occupational worlds in either social class. But see Daniel R. Miller and Guy E. Swanson, *The Changing American Parent: A Study in the Detroit Area.* New York: Wiley, 1958.

19. Bronfenbrenner, "Socialization and Social Class," p. 424.

20. Later studies, including our own, do not show this difference.

21. Kohn, "Social Class and the Exercise of Parental Authority."

22. This is not to say that the methods used by parents of either social class are necessarily the most efficacious for achievement of their goals.

23. The justification for treating support and constraint as the two major dimensions of parent-child relationships lies in the theoretical argument of Talcott Parsons and Robert F. Bales, *Family, Socialization and Interaction Process*, Glencoe, Ill.: Free Press, 1955, esp. p. 45, and the empirical argument of Earl S. Schaefer, "A Circumplex Model for Maternal Behavior," *Journal of Abnormal and Social Psychology*, vol. 59, September 1959, pp. 226-34.

24. Bronfenbrenner, "Socialization and Social Class," p. 425.

25. From the very limited evidence available at the time of his review, Bronfenbrenner tentatively concluded: "Though the middle-class father typically has a warmer relationship with the child, he is also likely to have more authority and status in family affairs" (ibid., p. 422). The discussion here is based largely on subsequent research, esp. "Social Class and the Allocation of Parental Responsibilities."

26. Fragmentary data suggest sharp class differences in the husband-wife relationship that complement the differences in the division of parental responsibilities discussed above. For example, virtually no working-class wife reports that she and her husband ever go out on an evening or weekend without the children. And few working-class fathers do much to relieve their wives of the burden of caring for the children all the time. By and large, working-class fathers seem to lead a largely separate social life from that of their wives; the wife has full-time responsibility for the children, while the husband is free to go his own way.

CHAPTER 36

Familia Spongia: The Adaptive Function

CLARK E. VINCENT

The adaptive function is a vital but overlooked function of the family in all societies that are either highly industrialized or undergoing industrialization. This thesis, which the author is deliberately and provocatively writing to rather than attempting to test, could also be stated: The rapid and pervasive social changes associated with industrialization necessitate a family system that both structurally and functionally is highly adaptive externally to the demands of other social institutions and internally to the needs of its own members.

This thesis does not imply that the family is the cause or prime mover in social change. Nor does it imply that the adaptive function is performed exclusively by the family, or that the family is essentially passive in relation to other institutions. Other social institutions are deeply involved in social change, do respond to changing needs and demands of the family system; and the family system is selective in its adaptations. But the family, to a greater degree and more frequently than is true of the other major social institutions, facilitates social change by adapting its structure and activities to fit the changing needs of the society and other social institutions. A major reason for this is that the strategic socialization function of the family in preparing the individual for adult roles in the larger society is inseparable from the family's *mediation* function (1) whereby the changing requirements (demands, goals) of the society and its other social institutions are translated and incorporated into the ongoing socialization of both child and adult members of the family. A second reason is that the family as a social institution lacks an institutional spokesman or representative voice through which it might resist change.

In addressing this thesis the present paper is organized into four parts.

First is an abbreviated and highly selective review of some of the background issues and historical junctures in the literature on functions of the family. Second is a consideration of the adaptive function of the family in relation to the society and other social institutions. The third part considers the adaptive activities of the family in relation to its individual members and the fourth part raises the question of when adaptation becomes dysfunctional.

Background Issues and Historical Junctures

William F. Ogburn comes readily to mind when considering the functions of the family. His major interest in the processes of social change and his earlier writing on the impact of technology, inventions, and ideologies on the family provided the context for the massive empirical data he compiled in the late 1920's to emphasize the increasing transfer of economic, protective, recreational, educational, and religious activities from the family to outside agencies (2). His initial and more cautious interpretation that these increases in outside-the-home activities were indices of decreases in the *traditional* functions of the family was replaced in his later writings by assertions about the family's loss of functions (3).

Ogburn's initial interpretation was rarely given critical examination or tested in the textbooks and writings on the family in the 1930's and 1940's. Consequently, his observations and impressive statistical data concerning the decreases in the *traditional* (forms of) functions of the family became the basis or reference point for two widely held beliefs: (a) The family has lost many of its functions. (b) This loss of functions represents a decline (decay, disorganization) of the family.

Textbooks and journal articles published since the early 1930's have included a variety of data and illustrative materials interpreted to demonstrate that the family has lost many of its functions. Descriptions of an "ideal typical" pioneer or rural family needing only a few dollars a year for supplies it could not produce were contrasted with census data on, for example, the number of women in the labor force and the increasing number of restaurants, laundries, stores, etc., to show that the family was no longer a self-sustaining production unit economically. Loss of the educational function was illustrated with observations that sons were no longer apprenticed to their fathers, daughters learned cooking in home economics courses rather than at home, and the teaching hours and authority of the schools had increased constantly since the turn of the century. The loss of the protective function was illustrated with references to the duties of the policeman, truant officer, nurse, fireman, and the use of nursing homes and mental institutions. Support for the notion that the religious function was being transferred from the family was found, for example, in statistics reporting a decreasing proportion of families having daily devotions, reading the Bible, and saying grace before meals. In regard to recreational activities, it was noted that the family no longer produced its own recreation in the form of quilting parties, corn husking bees, and parlor games, and figures were given to show the marked increase in attendance at movies and spectator sports.

The Loss of Functions — A Myth?

It is interesting to speculate about what might have happened if students of the family: (a) had kept im mind Ogburn's central interest in social change, and (b) had emphasized that it was the traditional content and form of given functions, rather than functions *qua* functions, that were being performed decreasingly by the family.

Taking the latter possibility first, one can argue that in each case of a traditional function supposedly lost to the family as a social institution, the loss has in reality been but a *change in content and form*. For example, although the U.S. family is no longer an economic producing unit to the degree it was in the pioneer and rural America of several generations ago, it is an economic consuming unit. Is consumption by the family unit any less important an economic function in today's society than production by a family unit was in yesterday's society? To what degree does our current economy depend on the family *qua* family to "consume" houses, cars, boats, cereals, furniture, vacations, sterling silver, china, and pet food?

Similarly, one might argue that society is currently quite dependent on the family function of consuming recreation. It is quite possible (but almost impossible to measure) that today's family spends far more time not only in consuming but also in producing its own recreation than did the family of fifty or one hundred years ago. Here we think not only of the multimillion dollar sales annually of croquet, ping-pong, and badminton sets, cameras and home movies, family card games and barbecue equipment, but also of the family's expenditures for and use of swimming pools, rumpus rooms, camping equipment, summer homes, boats, television and hi-fi sets, etc.

Similar arguments can be made in relation to the purported loss of the educational, religious, and protective functions. That fewer families, for example, say grace and have daily Bible reading today than one hundred years ago (assuming this to be the case) does not demonstrate the loss of the religious function. For to omit grace before meals, nightly prayers, and daily Bible reading is one kind of religious instruction, albeit not the traditional kind. If the family has lost its educational and religious functions, why do the majority of children hold religious, political, and social class beliefs similar to those of their parents? Why are the asocial attitudes and immoral practices of the delinquent and the criminal traced to the family and not to the church? Why is it that the family in general and the parents in particular are considered to be key variables in determining how well and how far the child progresses in school? Why is the family, more so than the school system, blamed for dropouts? Did parents of one hundred, fifty, or even twenty years ago spend as much time as today's parents do in helping and prodding their children with homework? Did the pioneer parents who withdrew their children from school to work on the farm perform more of an educational function than today's parents who save, borrow, and mortgage to provide sixteen-plus years of schooling for their children?

The foregoing questions and examples grossly oversimplify the arguments and beg the question on many issues involved. In fact, many of these questions and examples would be irrelevant if Ogburn and the earlier family textbook writers had emphasized that the family had lost, for example, an *economic production* function but gained an *economic consumption* function. Instead, however, they emphasized that the family has lost its *economic* function. (Similarly, they emphasized the loss of the religious, educational, protective, and recreational functions.) They thereby precluded analysis of changes in the family's economic function and set the stage for the subsequent equating of the loss of functions with the decline of the family. Thus, the foregoing questions remain relevant and are intended to provoke some students of the family to critically examine the myth of the family's loss of functions. Hopefully, such students will focus their attention more on the structural changes, the sharing of functions among social institutions, and the changing content and form of the functions of the family.

What might have happened if Ogburn's observations and data had not been taken out of the context of his general interest in social change and his specific interest in tracing the causes of changes in the family? Would the family textbooks and literature of the past three decades still have had a predominant emphasis on the declining importance of the family? Probably! Because, since the earliest writings available, changes occurring in the institution of the family have been used and interpreted to support either an optimistic or a pessimistic premise concerning social change, and the pessimists have consistently outnumbered the optimists. As Goode has noted in his sound critism of the descriptions of the United States family of the past as a misleading stereotype of "the classical family of Western nostalgia," the same stereotype has been accepted as the baseline by those who view subsequent changes as progress as well as by those who interpret the subsequent changes as retrogression of the family (4).

Ogburn's stature as a sociologist, his considerable ability in mining and compiling impressive empirical data, and his delineation and naming of broad categories of functions purportedly lost by the family all combined to make his writings an important juncture for the family literature of the past three decades. The issues involved are of long standing, however, and space permits only passing reference to the existence of a much earlier and voluminous literature in which the differences and/or changes in the structure and functions of the family were interpreted to support quite different "theories" of social change. Some earlier writers with an optimistic premise of unilinear progress made considerable use of the "voyage literature" and "social Darwinism" to try to demonstrate a progressive evolution of the family from "primitive" to "modern" forms and from promiscuity to monogamy (5). Other earlier writers reflected a more pessimistic premise in attempting to show that changes from a previous form of social order represented decay, instability, or disorganization of the family (6).

To move quickly and briefly from Ogburn to the present, it was noted earlier that the purported loss of functions by the family was interpreted in the majority of family textbooks written during the 1930's and early 1940's to be evidence, if not the cause, of the decay, disorganization, and deterioration of the family as a social institution. Sorokin wrote: "The family as a sacred union of husband and wife, of parents and children will continue to disintegrate — the main socio-cultural functions of the family will further decrease until the family becomes a mere overnight parking place mainly for sex relationship" (7). But Sorokin was not alone. John B. Watson in psychology and Carle Zimmerman and Ruth Anshen in sociology were only a few among the many writers in the 1930's and 1940's who not only assumed a decay or decline of the family, but (a) attempted to explain how that decline had come about and (b) posited that the family was the prime mover or first cause of social change (8).

A More Optimistic Premise About Social Change

The assumption that the loss of functions was synonymous with a decline of the family began to share the limelight in the late 1940's with more optimistic interpretations. One interpretation, largely attributable to Burgess and Locke, emphasized that the changes in the family really represented progress in the form of a change from an institutional to a companionship orientation (9). In the late 1950's and early 1960's, a generally more optimistic view of changes in the institution of the family gained support from a number of writers who still accepted the premise of a loss of functions but who argued that the remaining functions had become more important. Straus (10), Kirkpatrick (11), and Rodman (12) have noted the increasing variety of conceptual labels used to convey this more optimistic view of the changing nature of the American family. As an alternative to Burgess and Locke's "companionship family," Miller and

Swanson (13) have proposed the "colleague family," and Farber has suggested the "permanent-availability model" (14).

Notable among the writers emphasizing that the remaining functions are more important is Parsons. His current interpretations, as Rodman has noted (15), represent a much more optimistic position than he had taken earlier. In his more recent writings, Parsons has emphasized that changes occurring in the family involve gains as well as losses and that when functions are lost by a particular unit in society, that unit is freer to concentrate upon other functions. "When two functions, previously imbedded in the same structure, are subsequently performed by two newly differentiated structures, they can *both* be fulfilled more intensively and with a greater degree of freedom" (16). Parsons has also emphasized increasingly that the contemporary American family is differentiated and not disorganized. "The family is more specialized than before, but not in any general sense less important, because the society is dependent *more* exclusively on it for the performance of *certain* of its various functions" (17).

Goode has also supported a more optimistic interpretation of changes in the family by emphasizing its *mediating* function. The idea that the family is a mediator (buffer, strainer, funnel) between the individual and the larger society has been both implicit and explicit in the family textbooks for several decades, but Goode is the first (to this writer's knowledge) to base the strategic significance of the family specifically on its mediating function (18).

The Adaptive Function and Society

The following discussion of the adaptive function of the family in relation to society and/or other social systems in that society is within the framework of what Mannheim called "relationism" (19) and what Goode and others have referred to as the "fit" between a given family system and the larger society (20). Thus, the present discussion is not dependent on an "organic analogy," or on the idea that there is some inherent or ideal function that the family "ought to perform" (21).

Superficially, the adaptive function of the family has some sponge-like characteristics that are evidenced by the family's absorption of blame for most social problems (mental illness, delinquency, drop-outs, alcoholism, suicide, crime, illegitimacy, etc.). And future studies of the scapegoat function within and among groups may have some applicability to the scapegoat function among social systems or institutions. Evidences of the family's adaptive function relevant to the present discussion may be illustrated with reference to the economic system.

Adaptation to the Economic System

The economic system of a highly industrialized society demands a mobile labor force as well as some professional, skilled, and semiskilled personnel who will work on holidays, Sundays, and at night. When the company employing father decrees that father shall move to another city, furtherance of the company's objectives is made possible by the adaptiveness (willingly or grudgingly) of the entire family; collectively and individually, the family members uproot themselves, adapt to a new city and neighborhood, enter different schools, and make new friends.

The varieties of family adaptation required by particular occupations have been illustrated in a number of studies, such as the early one by W. F. Cottrell, "Of Time and the Railroader" (22). The family of the railroad engineer, fireman, conductor, or porter might celebrate Christmas on the 23rd or the 27th of December, as dictated by the railroad schedule. The reader can supply many examples of jobs in transportation, communication, entertainment, and various professional services which require considerable adaptiveness in the schedules and patterns of the families involved. William Foote Whyte (23), among others, has described in some detail the degree to which the family and particularly the wives of management are required to adapt to the large corporation. Somewhat conversely, studies such as the one by Alvin W. Gouldner have shown how becoming a husband and father can influence the union leader's role performance on the job (24).

The adaptation of the family to occupational demands and economic pressures also includes the pattern observed in the Appalachian area where employment is more readily obtained by wives and where thousands of husbands have adapted to the role of homemaker.

It is true that the family breadwinner has a choice, and that the family can be selective in its adaptation. The breadwinner can change jobs, but this becomes only a choice of the manner in which one adapts. Rarely is the worker able to refuse to adapt to the demands of a job or position and still retain that position without a future adaptation of his family to less security and income than might have been forthcoming.

The adaptations required of the family by the educational systems are both minor and major. The minor adaptations may be found in such areas as P.T.A. pressures for parental attendance at monthly meetings, increased homework and the expected supervision of such homework by parents, funds for daily lunches, lack of control over the use of personality and achievement test results, categorizing of rapid and slow learners, and split or double shifts. Major adaptations are related to the increases in the number of years, the costs, and the specialization of formal education that frequently necessitate heavy family indebtedness.

The Reproductive Function and Reciprocal Adaptation

That the educational, religious, and economic systems in society adapt to the family is most evident in regard to the reproductive function of the family. Educational and religious institutions have had to expand their facilities considerably as a result of the rise in the birth rate in the middle and late 1940's. The business world has adapted its advertising and merchandising to the crest of the population wave — the initial boom in infant foods, children's toys, clothes catering to teen-agers' tastes and influence on family buying habits, the increasing market in automobiles for the 16-21-year-olds, the recent and current increase in sales of diamond rings and sterling silver, and the anticipated uptrend in housing for newlywed couples in the late 1960's.

At least three crucial points may be hypothesized concerning the reciprocal adaptation among various social systems: (1) Social institutions or systems other than the family adapt to the degree that such adaptation is in the interest of their respective goals. (2) If there is a conflict of interests or goals, it is the family which "gives in" and adapts. (3) The family adapts for lack of an alternative and in so doing serves the goals of other social systems and facilitates the survival of a society based on social change.

The plausibility of the first hypothesis is suggested by the fact that although the reproductive function would appear to be the one major function whereby the family "forces" adaptation from other social institutions, this is tolerated only to the degree that such adaptation furthers the ends or goals of the other institutions. The upswing in births in the 1940's was initially interpreted to represent more profits for business. In fact, the baby boom was equated with prosperity. The birth rate rise was also favorably viewed as meaning more potential converts for the churches, and higher wages and better job security for teachers, school administrators, and professors. However, in the late 1950's and early 1960's, the baby boom acquired another, almost opposite interpretation as the increasing number of teen-agers about to enter the labor market added to fears about the unemployment rate, and as high schools and colleges faced enrollments and building programs that necessitated sharp increases in tax monies. Equally, if not more important, has been the world-wide concern about depletion of natural resources and living space.

The subsequent and current concerted attack upon the problem of "conceptionitis," "birthquake," or "population explosion" provides a fascinating illustration of how even in regard to its traditional function of reproduction, the family adapts (gradually, and rarely through force) to the goals and interests of the society and of its other major social systems.

That it is the family system that gives in or adapts in cases of conflict of interests or goals was noted earlier in the discussion of the demands of the labor market, job position, and business corporation. The school system, the business world, and even church services are geared to time schedules that serve first the needs, interests, and efficiency of the school, the business, and the church. The family adapts its schedule accordingly. Even in times of war and armed conflict, the adaptations required of economic, educational, and religious institutions usually have some side-effects beneficial to those social institutions, whereas the family sacrifices most in the interests of winning the war.

That the family lacks an alternative to adaptation (although it may select among several patterns of adaptation) may be illustrated with reference to what Ogburn called the protective function. Within the past half decade, there has developed very rapidly a nationwide program to return mental patients to their families. Backed by multi-million dollars in federal funds, this program is intended to greatly reduce the number of patients in mental hospitals and institutions. Comprehensive community mental health centers will be built and staffed to provide outpatient, night-care, day-care and "half-way cottage" services. The family is expected to adapt to the return of its mentally ill or emotionally disturbed members, just as it was expected to adapt to the return of the parolee member of the family several decades ago. The family will also be expected to adapt to the intrusion of the mental health personnel concerned with the rehabilitation of the patient, just as it has adapted to the intrusion of the parole and probation officers, the judge of the juvenile court, and the social worker.

Why? Because the family has no realistic alternative. Given the mores of our society, how could the family maintain its ideological image if it refused to accept one of its members convalescing from mental illness or rehabilitating from crime or delinquency?

More importantly, who would be the spokesman for the family's refusal? The National Association for Mental Health has a powerful and effective lobby. The family has none. Almost every segment of the religious, educational, professional, recreational, political, and occupational worlds has strong and powerful spokesmen at local, state, and national levels. Each group of 20 physicians, 30 ministers, 40 school teachers, five manufacturers, or three union men in a given city can exert more influence and pressure directly or indirectly than can 5,000 families living in that same city.

Thus, no one asks: How will the family be affected by the return of a mentally ill member? What will double shifts at school do to the family? Will the regulations of ADC encourage husbands to desert the family? Will urban renewal disrupt the family and the network of extended family relationships? Would it be easier on the family to draft 45-year-old fathers for many service tasks prior to drafting 25-year-old fathers for those same tasks?

And even if such questions were asked, who would answer? The family system has no collective representative, no lobbyist, no official spokesman. Therefore, to observe that the family is the most adaptive of the several social systems in a rapidly changing society is perhaps only to recognize that it is the least organized.

Adaptation Within the Family

Adaptiveness would appear to be inversely related to the degree of organization and to the size or number of the group. The rigidity of the army, for example, is apparently positively related to its chain of command and its size. The size or number involved in what we refer to as the family system, however, tends to be the number in each individual family; and because the family system is *un*organized as a family system beyond each individual family, it is easily divided and its resistance conquered. Thus, in a given community, the organizational spokesmen for the teachers, the union, the clergy, or business can be and are heard and heeded much more clearly than are 50,000 *individual* families.

Its small numerical size and its lack of an organizational tie-in with all other families not only predisposes the individual family to adapt to the needs and demands of the other social systems, but facilitates its adaptation to the needs of its individual members. The highly individualized needs of each of 40 persons cannot possibly be heard or met to the same degree in the classroom, factory, office, or church as within the respective families of those 40 persons.

Much of the lament about the impersonalization, alienation, or dehumanization of human beings in the multiversity, factory, corporation, hospital, or large urban church obscures the lack of an alternative. The same individuals who may privately bemoan the *a*personal cashier in the supermarket, the tight-lipped teller in the bank, the hurried physician, the unavailable professor, or the uncommunicative dispenser of other professional services would strongly object to waiting in line for an extra hour while other customers and clients were being responded to warmly and personally on an individual basis. The Lilliputians, who thought that Gulliver's timepiece must be a God to require such frequent consultations, would justifiably infer that the citizens of highly industrialized societies not only worship but are governed and ruled by time. In such societies, the family becomes even more important as a flexible social unit wherein there is time and tolerance for expressing and acting out individual needs, and wherein being a few minutes late does not disrupt the production lines, board meetings, transportation schedules, and classroom lectures.

The time-scheduling demands which a technological society makes of the individual are perhaps minor in comparison with its demands for productive output, self-discipline, and emotional control. In combination, these demands increase the importance of what Goode has phrased as the family "task of restoring the input-output emotional balance of individualism . . ." (25).

Can a society undergo industrialization and/or remain highly industrialized without a family system that is highly adaptive to change, to the demands of other social systems or institutions, and to the needs of its individual members? Although a negative answer to this question was stated in positive form as a thesis at the beginning of this paper, the final answers will depend on considerable historical and cross-cultural research *in context*. Goode's recent work has provided some invaluable bench marks for such research. His comparative analysis of the family systems in quite disparate cultures provides sound support for the idea that ". . . we are witnessing a remarkable phenomenon: The development of similar family behavior and values among much of the world's population" (26).

Eufunctional or Dysfunctional?

The thesis of the present paper represents only one selected facet of Goode's much broader inquiry concerning whether and why there is an increasing world-wide similarity in family behavior and values. Thus far, our attempts to illustrate the merits of this thesis have interpreted the family's adaptiveness to other social systems and to its own members as predominately eufunctional. Is the adaptive function of the family at times dysfunctional? And if so, dysfunctional at what point? for whom? and in relation to what purposes?

One example of dysfunctional adaptation is provided by the Aid to Dependent Children (ADC) program. In adapting to the early regulations of ADC, an unknown proportion of fathers deserted their families, or perhaps in collusion with their wives simply disappeared from public view, to enable their wives and children to qualify for ADC funds. Current awareness that the early regulations may have encouraged such desertions, and the belief that such families need a father present (or the premise that an unemployed father in the home is better than no father or a series of adult males) have resulted in much discussion and some revisions of the regulations. Similarly, the family's adaptiveness (for lack of an alternative) to urban renewal may prove in some instances to have ill-served the interests of either the families forced to move or the city planners and taxpayers.

The fact that these two examples pertain to lower-income families illustrates the variability in the degree and form of adaptive activities among the family systems of various socio-economic and ethnic groups. In the United States, for example, it has been postulated that the nuclear family system of the middle class is more likely to manipulate the extended family, whereas the lower- and the upper-class family systems are more likely to be manipulated by or to adapt to the extended family (27).

An example or illustration which cuts across class lines is to be found in the internal adaptiveness of the family to its teen-age members. When familial adaptation to the needs and wants of its teen-age members reaches the point or degree where parental control is lost, such degree of adaptation becomes dysfunctional within the context of the socialization function of the family. That parental control is frequently lost or tenuously held at best is not surprising when we consider that: (a) a sizable proportion of the current generation of teen-agers was reared via a permissive philosophy that equated wants with needs; (b) teen-agers are highly organized in their selective translations to parents about what the teen-age peer group is allowed to do by other parents; and (c) parents are remarkably *un*organized in their resistance to teen-agers' demands and expectations.

Again, the reader will be able to supply many examples of both external and internal adaptations of the family which he or she regards as dysfunctional. The more difficult task is to make explicit: dysfunctional for whom and for what goal? To return the mental patient to the family may well serve the goals of reducing the inpatient load of mental hospitals, save the taxpayers money, and prove highly therapeutic for the patient. But will it also have some dysfunctional aspects for the family and society (28)? Will the family still be able to permit the emotional blow-offs and to provide the relaxation and the emotional input needed daily by its "well" members whose output, tight schedules, and emotional control will continue to be expected in the office, factory, and schoolroom?

The foregoing is not intended as an argument against the gradual return of emotionally disturbed, aged, or infirm persons to their families. It is simply a further attempt to illustrate: (a) that the adaptive function of the family system is crucial in any society characterized by rapid social change, (b) that the adaptive family system of our industrial era generally is *un*organized and unrepresented beyond each individual family, and (c) that it, therefore, is predisposed to being overloaded with or overadaptive to the demands and expectations with which it is confronted, internally and externally (29).

The author's thesis remains that an industrialized society characterized by rapid social change necessitates a highly adaptive family system. This adaptiveness of the family will be interpreted by some as evidence of weakness and by others as evidence of strength. Those who view it as weakness may point to the family's loss of power and authority, while those who interpret its adaptability as strength may see the dependence of the larger social system on the flexibility of the family and see the family's adaptive function as crucial to its socialization and mediation functions. The family's internal adaptiveness may well prove to be a key variable in socializing the child for the flexibility needed in future adult roles within a rapidly changing society.

References

1. William J. Goode, *The Family*. Englewood Cliffs, N.J.: Prentice-Hall, 1964.

2. See William F. Ogburn, *Social Change*, New York: Viking Press, 1922; and "The Changing Family," *Publications of the American Sociological Society*, vol. 23, 1929, pp. 124–33; and William F. Ogburn and Clark Tibbitts, "The Family and Its Functions," chap. 13 in *Recent Social Trends in the United States*, New York: McGraw-Hill, 1933.

3. William F. Ogburn, "The Changing Family," *The Family*, vol. 19, 1938, pp. 139–43; W. F. Ogburn and M. F. Nimkoff, *Technology and the Changing Family*, New York: Houghton-Mifflin, 1955, pp. 15, 45–48, 129–30, 244–47.

4. William J. Goode, *World Revolution and Family Patterns*. New York: Free Press, 1963, chap. 1.

5. See, for example, Franz C. Muller-Lyer, *The Family*, New York: Knopf, 1931; and Herbert Spencer, *Principles of Sociology*, New York: Appleton-Century, 1897, vol. 1, pp. 653, 681–83.

6. For a cogent review of the major "Traditionalists" and "Philosophical Conservatives" who reacted to the individualism of the French Revolution legislation with efforts to strengthen the family and reconstitute the *ancien régime*, see Robert A. Nisbet, *The Quest for Community: A Study in the Ethics of Freedom and Order*. New York: Oxford University Press, 1953.

7. Pitirim A. Sorokin: *Social and Cultural Dynamics*, 1937, vol. 5, p. 776; and *The Crisis of Our Age*, New York: Dutton, 1941, p. 187.

8. Carle C. Zimmerman, *Family and Civilization*, New York: Harper, 1947, pp. 782–83, 802 ff.; *The Family: Its Function and Destiny*, ed. Ruth N. Anshen, New York: Harper, 1949, p. 4.

9. Ernest W. Burgess and Harvey J. Locke, *The Family: From Institution to Companionship*. New York: American Book Co., 1945.

10. Murray A. Straus, "Conjugal Power Structure and Adolescent Personality," *Marriage and Family Living*, vol. 24, no. 1, February 1962, pp. 17–25.

11. Clifford Kirkpatrick, "Housewife and Woman: The Best of Both Worlds?" *Man and Civilization: The Family's Search for Survival*, ed. S. M. Farber, P. Mustacchi, and R. H . L. Wilson. New York: McGraw-Hill, 1963, pp. 135–152.

12. Hyman Rodman, "Introduction" to chap. 8, "The Changing American Family," in *Marriage, Family and Society: A Reader*, ed. Hyman Rodman. New York: Random House, 1965, pp. 249–58.

13. Daniel R. Miller and Guy E. Swanson, *The Changing American Parent*. New York: Wiley, 1958, pp. 198–202.

14. Bernard Farber, *Family: Organization and Interaction*. San Francisco: Chandler, 1964.

15. Hyman Rodman, "Talcott Parsons' View of the Changing American Family," in *Marriage, Family and Society*, pp. 262–86.

16. Talcott Parsons, "The Point of View of the Author," in *The Social Theories of Talcott Parsons*, ed. May Black. Englewood Cliffs, N.J.: Prentice-Hall, 1961, p. 129.

17. Talcott Parsons and Robert Bales, *Family, Socialization and Interaction Process*. Glencoe, Ill.: Free Press, 1955, pp. 10–11.

18. Goode, *The Family*, p. 2.

19. Karl Mannheim, *Ideology and Utopia*. New York: Harcourt, Brace, 1957, p. 86.

20. Goode, *World Revolution and Family Patterns*, pp. 10–26.

21. For a critical review of some of the historic, methodologic, and theoretic issues involved in "functionalism," see *Functionalism in the Social Sciences*, ed. Don Martindale. Philadelphia: American Academy of Political and Social Science, Monograph 5, 1965.

22. W. F. Cottrell, "Of Time and the Railroader," *American Sociological Review*, vol. 4, April 1939, pp. 190–98.

23. Willaim F. Whyte, Jr., "The Wives of Management," *Fortune*, October 1951; and "The Corporation and the Wife," *Fortune*, November 1951.

24. Alvin W. Gouldner, "Attitudes of 'Progressive' Trade Union Leaders," *American Journal of Sociology,* vol. 52, March 1947, pp. 389–92.

25. Goode, *World Revolution and Family Patterns*, p. 14.

26. Ibid. p. 1.

27. See Marvin B. Sussman and Lee Burchinal, "Kin Family Network: Unheralded Structure in Current Conceptualizations of Family Functioning," *Marriage and Family Living*, vol. 24, August 1962, pp. 231–40; Eugene Litwak, "Geographic Mobility and Extended Family Cohesion," *American Sociological Review,* vol. 25, June 1960, pp. 385–94; and "Occupational Mobility and Extended Family Cohesion," *American Sociological Review*, vol. 25, February 1960, pp. 9–21.

28. Some of the contraindications and possible dysfunctional aspects of returning the patient to the family are discussed and relevant literature is cited in Clark E. Vincent, "The Family in Health and Illness," *Annals of the American Academy of Political and Social Science*, vol. 346, March 1963, pp. 109–115.

29. Lee Rainwater has commented on this predisposition to overloading as a concomitant of individualism and of how the family unit is perceived as secondary to the roles of its individual members in other social systems. "The [family] internal adaptation process presumably is guided by the demands that family members bring home based on their involvement with other institutions — career, teen-age peer group, school, etc. The family is seen by each particular institution as an extension of the person who has a role in that institution — business sees the executive's family as an extension of him in his executive role, the school sees the pupil's family as an extension of the pupil in his learning role, etc. Because of the value placed on individual achievement and gratification, the individual often identifies more with the demands of the secondary institution in which he has a role than he does with a solitary family." (Personal communication to the author.)

PART IV

Old Age: An Introduction to the Determinants Affecting the Old Age Stage of the Life Cycle

There seems to be little disagreement on the decline of biological and physical capacities in the aging process. Many biological theories accounting for this process are concisely discussed in Busse's article "Theories of Aging" (1). Busse lists five major theories: exhaustion theory, stochastic theory, composite theory, error theory, and cross-linkage theory. Despite differences in theoretical stance, all who consider the phenomena of old age are confronted with the inescapable fact of the inevitable deterioration of the human organism.

Biochemical theories of aging detail the physical processes of aging, and are of course of great importance. In addition, there are a number of viewpoints on the psychosocial components of aging, although these views are limited in their predictive and explanatory power. In the absence of a comprehensive psychosocial theory on aging, we shall consider those most frequently encountered in recent gerontological literature.

The disengagement theory emerged from the Kansas City studies of Robert Havighurst, Bernice Neugarten, and Sheldon Tobin (2), who have advanced three basic assumptions: (*a*) as a result of physical decline, the aged individual must lessen his activities; (*b*) relinquishing his social roles is therefore an intrinsic part of natural aging processes; and (*c*) in view of increased psychological preoccupation with himself and decreased emotional investment in others, the older person who is well adjusted for his age benefits from the disengagement process.

In opposition to the disengagement theory, the activity theory (3) proposes that (*a*) until a person becomes incapacitated by senility, compulsory withdrawal of social activities such as total retirement and termination of other meaningful social roles is detrimental to the well-being of old people; and (*b*) to maintain a level of social activities according to one's past life style and current physical and mental condition is therefore essential to the betterment of the aging process.

These two divergent views on aging have stimulated those in the gerontology field to search for a unified theory of aging. On the personal front, one factor complicating theoretical construction is individual differences in the rate of deterioration of physical condition and psychosocial needs. Therefore, we do not know whether the compulsory retirement system can be considered beneficial to every sixty-five-year-old person. Individuals differ as to the age at which they can be considered old, and the extent to which social life has to be readjusted. In Western industrial society, being active is highly valued and inactivity or decrease in social interaction is considered a decrease in self-esteem and self-worth. The change from a high-value status to a low-value status can be a very difficult adjustment problem for the aged. Frequently the economic status of the aged shifts radically. The man who was formerly independent is now dependent on others. He is supported by social security, perhaps a retirement pension, and if he is lucky, by supplements from his adult children. This role has little or no social reward.

In other societies, such as traditional Oriental societies, old age is culturally valued, and very often is equated with wisdom. Old age carries a worthwhile social status in its own right and is revered. Once an aged person is retired from his vocational status, he is immediately eligible to be a member of a social position of equal value, and is awarded certain honors reserved exclusively for his age group. His leisure life style is respected by society and he does not have to demonstrate his physical prowess and youthfulness in order to assert his self-worth. He accepts his position as an old man and he enjoys the new status and privileges extended to him by the society and his family.

A third outlook views aging as an adaptation process. Erikson characterizes the eighth stage of man as focused on ego integrity vs. despair (4), and points out the necessity for old people to accept the inevitability of being old and deteriorating, and to meet their condition with mature ego qualities without despair. To achieve this resolution one must have adaptive ego capacity to cope with the various changes ultimately facing him. If one lacks these adaptive qualities, the sense of despair takes a heavy toll and stress accelerates the aging process, as suggested by Hans Selye (5). Like Erikson, Maxwell Gitelson (6), Caroline Ford (7), and Robert Peck (8) all address themselves to the importance of the adaptive processes and qualities of the elderly. Lacking these, malaise ensues.

The adaptation concept seems to have overcome the extreme theoretical divergence of the previous two views — disengagement vs. activity — and has taken into account individual differences and environmental circumstances.

In Chapter 37 Gitelson states that emotional problems of elderly people are problems of adaptation. In view of the difference in adaptive processes and resources of young and old, the elderly have greater odds against success. The aged individual experiences a decline in his powers, capacities, and resilience. This creates a maladaptive potential and in turn various physical and psychological symptoms emerge. On the basis of these assumptions, Gitelson explains the processes by which various symptoms are formed for the purpose of maintaining one's selfness and security in the declining years. In discussing normal adaptive processes, Gitelson observes that one has to be flexible within himself and in relation to his new environment to allow the best utilization of his energies for overcoming adverse circumstances, and at the same time to allow for generating and creating suitable gratification.

The aging process viewed as the rate of wear and tear on the body is discussed by Selye in Chapter 38. He states that there is a close relationship between aging and stress and between limited human vitality and the rate at which it is withdrawn. Depletable human energy has to be employed in a balanced manner without placing undue stress on a particular organ. According to Selye's observation, in most cases the death of an old person was not necessarily caused by the end of the natural human life span, but was the consequence of wear and tear on a vital organ. The excessive stress on the organ meant overdrawing the energy and vitality of that organ. Selye further argues that stress is unavoidable in our everyday life, but that one should be able to adapt to an optimum stress level that is best suited to his capacity, and not exceed the limits of certain fundamental biological laws.

Ford, in Chapter 39, gives a more detailed illustration of ego-adaptive mechanisms frequently employed by the aged. On the basis of her years of professional experience with the aged, she is able to demonstrate with case materials how old people face their fear and anxiety, and show certain types of defense mechanisms that have been successfully employed by the aged.

In Chapter 40, Peck, taking a developmental approach to the second half of life, illustrates the ways by which old people achieve psychosocial adjustment. He views the adaptive system of the individual as based on his ego and role adjustment to a state that best suits his current functional capacity. For instance, a retired man should be able to use his ego capacity to modify his long internalized work ethic or work-related ego identity into a new basis of self-valuation. Peck conceptualizes this as a compromise between ego differentiation vs. work-role preoccupation. Finding alternative roles that accommodate one's personal assets, such as skills, knowledge, and experience, is one of the several adaptive systems discussed in this chapter. Peck's conceptualization has provided a working relationship between psychic structure, which has developmental antecedence, and the environment, which can be manipulated by social intervention. Once the psychic and environmental factors are mutually harmonized, problems of aging will be lightened.

Searching for understanding of the late life period, Havinghurst discusses in Chapter 41 two basic questions on aging: How do people structure their lives after age 65? and Under what conditions do they achieve satisfaction? In tackling these questions, he views the satisfactory management of one's later life as a problem of adaptation consisting of three components — ego strength and resilience, the supportive social environment, and physical health. In the study of the adaptation process, Havighurst lists six variables for measuring the success of adaptation, ranging from assessments of the ego or personality, to societal provisions to assist adaptation.

Is it true that older people have little or no desire for sexual activity? Gustave Newman and Claude Nichols report on this question in Chapter 42. In their study, they found a noticeable decline in sexual activities after seventy-five years of age. The reason for curtailment of sexual activities is more often attributable to various physical illnesses, such as arthritis, arteriosclerotic heart disease, and so on, and the availability of one's customary sexual partner rather than age alone. There seems to be some variation among social groups in response to sexual drive. The Negroid old people have more sexual activity than Caucasoid; male has more than female; and lower socioeconomic class has more than high socioeconomic class.

Mental illness of the elderly can be examined, at least in part, in terms of role changes, patterns of group living, and value systems. In Chapter 43, Donald Kent asserts that there is a lack of provision for new roles once a man reaches retirement age. This lack affects the elderly man more than the elderly woman, since the man's lifetime role function is in the instrumental sphere and it is more difficult to substitute other roles of a similar character. Lack of social provision for smooth transition bridging role discontinuity has alienated the aged male group. Many are characterized by a state of anomie that is also a result of the gradual loss of group membership and participation affecting the norms of behavior that have functioned as guideposts of social relationships. Furthermore, the American value system, placing strong emphasis on youth, is certainly not helpful to the aged.

In Chapter 43, Kent distinguishes two basic types of role function — instrumental and socio-emotional — and comments that the aged female has a better chance to adapt to her new roles than does the aged male. In Chapter 44, Margaret Clark and Monique Mendelson present this same point by a vivid case illustration. They depict cultural influence on adaptation to aging by using a Mexican-American aged female in contrast to her Anglo-American counterpart. They demonstrate the cultural variation affecting the aged female in two different types of cultural system.

Robert Kleemeier, in Chapter 45, makes a comparative analysis of the disengagement theory and the activity theory in an attempt to present some arguments and implications on the issue of relinquishment of social relationship vs. maintenance of sufficient levels of activity. The effects of both theories on the aged are aptly discussed. In view of the individual differences in aging processes, it is desirable to avoid any excessive optimism or rigidity in recommending levels of disengagement or activity. Recommendations should be made with careful consideration given to factors such as physical condition and the degree of satisfaction or pleasure in work and in social relationships. Kleemeier attempts to use the concept of leisure to bridge the gap between these two divergent concepts of aging. He suggests that leisure interests can provide activity without undue pressure and can substitute for work needs and be engaged in solely for pleasure. He adds that Western culture builds into its members an incapacity for effective use of leisure.

The problems of getting old are manifold. Most frequently encountered are problems of physical health, emotional health, personal financial resources, and availability of social resources. Gordon Streib, in Chapter 46, points out that in an industrialized society, many aged families may need some assistance from the public and society. It is a false belief that the older generation will always get assistance from its younger members. In fact, many old people have no children or close relatives accessible to them. For this and other reasons discussed in the chapter, Streib suggests that some social provision should be made available to those who cannot meet their basic needs. Three types of intervention are mentioned. One is aid from the natural family or kinsmen; second is "true" social intervention, consisting of quasi-family intervention; and third, institutional intervention.

In Chapter 47, Wayne Davidson and Karl Kunze report their study of preparation for retirement by factory workers, noting the problem confronted by the prospective retiree. Our retirement process is rather an abrupt one. Unpreparedness for retirement may result from psychological denial or resistance to a conscious awareness of the inevitable. Or it may be due to lack of an adequate system within our social structure to ease the problem, such as retirement counseling. Whether as a result of either or both of these factors, retirement is a painful experience for many factory workers. In the study, four basic causes of the retirement dilemma are identified: (*a*) status maintenance, (*b*) financial circumstances, (*c*) resistance to change, and (*d*) intrinsic nature of work. The authors are hopeful that as we learn more about the implications of retirement in the future, better retirement systems can be instituted. They envision better social attitudes toward retirement, viewing retirement as self-actualization instead of self-depreciation, better social provision for economic security, and better counseling services providing early preparation and planning.

We are all consciously aware of ourselves as mortal. Yet this awareness alone has not provided sufficient channels to handle the disruptive effects following death of a fellow man. The social structure, as Robert Blaumer contends in Chapter 48, has to provide additional means to alleviate the impact of death. He indicates that the ways by which the social system deals with the death impact are varied in relation to the type of society and the stratification of the society. The death of a member of a society is sometimes not only an individual's concern, but also the concern of the society, depending on the position that the deceased had occupied in the social structure. For example, the fact of death and its consequences are viewed differently in the case of a young person as compared to an old person, in light of the social roles and responsibilities of the deceased. In order to normalize the death impact and quickly end mourning and depression, the responsibilities for handling illness and the disposal of the corpse are given to institutions such as the hospital and funeral home. Blaumer characterizes this phenomenon as bureaucratization of death control. This in turn depersonalizes death and minimizes emotional and disruptive consequences to the living and to society.

References

1. Ebward W. Busse and Eric Pfeiffer, eds. *Behavior and Adaptation in Late Life*. Boston: Little, Brown, 1969, pp. 11-32.

2. Robert J. Havighurst, Bernice L. Neugarten, and Sheldon S. Tobin. "Disengagement and Patterns of Aging." In *Middle Age and Aging*, ed. Bernice L. Neugarten. Chicago: University of Chicago Press, 1968, pp. 161-72.

3. E. Palmore. "The Effects of Aging on Activities and Attitudes." *Gerontologist*, vol. 8, 1968, pp. 259-63.

4. Erik H. Erikson. *Childhood and Society*. Norton, 1950, pp. 231-33.

5. Hans Selye, Chapter 38 of this volume.

6. Maxwell Gitelson, Chapter 37 of this volume.

7. Caroline Ford, Chapter 39 of this volume.

8. Robert Peck, Chapter 40 of this volume.

A. INTRAPERSONAL SYSTEM

CHAPTER 37

The Emotional Problems of Elderly People

MAXWELL GITELSON

The emotional problems of elderly people, like all other psychic problems, are problems of adaptation. Adaptation consists of the modifications or changes which occur in an organism in the general direction of more perfectly fitting it for successful existence under the conditions of its environment. The problem of adaptation, therefore, must be considered from two sides. We need to know what the organism's powers and capacities are, and we must know also what are the tasks which the organism needs to perform.

The infant organism of the human species has one outstanding task: it has to survive. Pending the maturation and development of those functions which ultimately enable the individual to exert himself actively to his own ends, vegetative adaptation is of the first importance. The infant's environment is constructed around that fact. It is modified to conform with the limited adaptive capacities of the infant. In the normal course of events we see this situation gradually changing. The successful survival of the infant goes hand in hand with developments which ultimately convert the human organism from a more or less completely vegetative adaptive state into one in which self-initiated active mastery of life is more or less the case.

From the standpoint of the individual, the aging organism of the human species has the same first task of survival. However, there are these differences in the problem: The infant is developing from helplessness to self-sufficiency; the elder is losing self-sufficiency and becoming helpless. As the child develops, his self respect and security increase progressively with his personal capacity for mastering the problems of his environment; as age advances there is declining power of mastery and with it declining security and self-respect. The child, until his own development towards independent survival is assured, has the continuing security of his dependence. The elder's declining powers of independent survival are not comparably associated with a supplementary secure dependency.

The healthy young child is on the make; his organs are developing; his integrations are becoming more perfect; his external adaptive powers are supported by adequate internal functioning. In all of these aspects of adaptation the aging individual is on the decline. Thus, the factors that maintain homeostasis operate within much narrower limits of variation in the old; we know that the learning curve begins to drop significantly in the early twenties and that reaction time begins to increase then. In terms of the highly refined adaptations of which we are capable the war experience has revealed to us that a man may be functionally old at thirty years.

If we state these considerations in psychobiological terms, we see the following: The helpless dependence of the infant is overlapped in time by the structural and functional maturation of the nervous system and the gradual acquisition of knowledge and experience. As the latter increases, helplessness and dependence decrease. In the meanwhile, the child lacks for nothing. Youth is growth and growth is hope. On the contrary, with advancing age not only is the person confronted with the experience of failing personal powers but the involution of the vegetative organs deprives him of the balancing compensations which the infant finds in his vegetative functions. Finally, while the helpless child may be coddled and protected by his mother pending his maturation and self-sufficiency, the helpless grandfather may meet with short shrift. The old man's memory may be full of past glories and his heart be empty of hope. It is this overlapping of the waning powers of maturity and the increasing helplessness of the second childhood that is the basis for the psychological picture with which old age presents us.

Character Patterns and Old Age

During the course of childhood and adolescence the individual matures physically and mentally. At the same time the character of the person is developed and consolidated into the form which it will have throughout the duration of his mature life. Character is the person's habitual mode of reaction to problems. It is his habitual mode of entering into and participating in relationships with other people. It is his habitual outlook on life. Finally, it is the person's habitual attitude towards himself, his picture of and evaluation of himself, his moral-ethical attitudes and his method of dealing with his primitive impulses and instincts. Character is, in short, the over-all adaptive pattern of the person.

Character is in part determined constitutionally. Thus the person's temperament is manifested by his prevalent mood, his relative liveliness or relative sluggishness of motor and intellectual behavior, his relative sensitiveness or dullness of reaction to stimulations. Such factors as these appear to be inborn and determined by the person's heredity. However, by far the larger part of the determination of a person's character comes from his experience, beginning with his infancy and depending on the characters of his parents. Thus, prevalent attitudes of security or insecurity, ascendancy or submission, optimism or pessimism, confidence in others or distrust of them, and many less tangible traits may distinguish the various persons we know. The normal individual who has reached old age will have lived his mature life on the basis of a character structure which is pretty well fixed in a pattern which has worked more or less successfully for the total life situation in which he has found himself. And in the end the psychological pattern of the aging person will be determined by his mature character.

Let us take the following ideal example: The person has been born into a family of good heredity. He has had no unusual crippling illnesses in childhood. He is of good intelligence. With the usual vicissitudes and the normal incidence of failure and successes in his development he has, by and large, come out on top and has no particular reason either to doubt his capacity or to have any delusional notions about being invulnerable. Consequently he is confident of himself without under-rating the problems of his life. His parents have demanded from him neither moral perfection nor superlative achievement. They have not whitewashed his failings nor condoned his delinquencies. On the other hand, neither have they depreciated his assets nor rejected him for his deficits. They have respected him and he respects himself. He knows his own assets and liabilities and is realistically aware of the assets and liabilities of his friends and associates. His life is successful as regards his work and his human relationships. He may be without distinction but he is good human material. His wife has been fond of him and his children care for him and are loyal without being dependent. We may summarize his life prior to old age as an example of successful adaptation under favorable environmental circumstances.

Now he is at last old. He is in good health but he gets tired rather easily. He is mildly forgetful. His wife has died of an acute infection. His children have married and live at a distance. He sees them and his grandchildren occasionally but they lead their own lives. He lives in a small apartment which is maintained for him by a housekeeper. He belongs to a club and he has his acquaintances with whom he plays gin and shares a drink. But old friends are dead and others are ailing or have moved away. He comes to his physician complaining of his fatigability. He thinks that it may be due to constipation, though this is not severe.

What is the doctor likely to find? There may be the beginnings of a malignant disease, but let us assume that serious illness is ruled out. The diastolic pressure may be a little high and there may be a mild senile type of glycosuria. Appropriate measures do not remedy the chief symptom of fatigue. Are the etiological findings complete? Organically they may be, but what about the patient as a person?

If the doctor were now to have an easy-going kind of a chat with the patient he might discover the following: When this patient's wife died a younger daughter had suggested that he retire and come to live with her. He didn't want that. He was a young man yet — there were still many years of good hard work left in him. Besides, his daughter had young children — he would only be in the way, much as he knew his daughter loved him. So he set up in the small apartment. There had followed a period during which this new routine made him rather irritable. His housekeeper didn't seem to get the hang of doing things as he had learned to expect to have them done. She was a good cook but somehow the food she served was different. These were all small matters, but things just weren't right. About a year after his wife's death members of his firm began to suggest that he take a vacation. It was years since he'd been away. He deserved a rest. The business was well established and he could afford to go.

His business associates were also friends. There was no doubt in his mind that they meant well. Yet he couldn't help thinking once in a while that they wanted him to retire. There had been that deal on which he'd been overruled. He'd been opposed to changing the policy under which the firm had been operating for years — thought the suggestion of a younger colleague was too risky. Anyway, he'd been overruled. And the deal had been successful. He still held his opinion about the deal for the long run but he'd been overruled and it looked as if his judgment had been wrong. Perhaps he did need a rest. Perhaps he would be able to get back in his stride with a little medical help. No surrender. Only a little doubt.

The doctor who can elicit this history will be able to make the diagnosis. Here is an organism which is wearing out; here is a human being who is getting old. To the man whom I have described this is an unwelcome and an inconceivable thing. He would like to believe that there is nothing wrong with him which a little tinkering won't fix up. For this man cannot really envisage his own death. As an individual he feels only in terms of his survival.

Patterns of Adjustment

What does this add up to? As men mature successfully they do so on the basis of techniques of mastery which have insured that success. In the upshot, that which has worked well for them in their personal relations, in their work, and, as you know, also in their politics, becomes a fixed and rigid habit of adjustment. Conservatism of outlook and action is the older man's reaction to a decreasing tolerance for change. With age there is increasing anxiety before the new and the untried. Problems of adaptation which have not been previously mastered successfully become too much of a burden for failing psychological and physical powers. The older person inwardly senses increasing actual inadequacies. Particularly disturbing to him are the even slight failures of memory which may be the first heralds of decline.

The reactions to this are several:

(1) It is a fact of common observation that the elderly gentleman will forget his hat but will remember the score and the details of the football game in which he made the winning point fifty years ago. The elderly lady may forget her nurse's name but will remember the names of her bridesmaids and their hair-dos, the wedding presents and their donors.

This dulling of recent memory and sharpening of the remembrance of things past is not due to organic changes alone. Psychologically these are to be seen as an actual turning away from the painfulness of the present. The present is lacking in both the dependent security of childhood and the independent powers for maintaining security of the mature years. The past carries forever and undeniably the record of life lived successfully, problems overcome, disaster survived. Memory turns backward to periods of highest capacity and greatest security as the elderly person's means of saving his self-esteem and as an attempt to find reassurance that the threats of the present will be as transient as those of the past. It is no inconsistency that this occurs even in the case of those aged persons who may not have much to look back upon according to external standards. The fact is that they had at least survived previous threats to life, that someone at least had once cared for them. And memory is aided by confabulation. Never were such heroic deeds performed as in the memory of an old man's youth.

(2) It is observable that older people may become more emphatically self-assertive, even domineering. These tendencies are compensatory reactions for the feelings of inferiority and inadequacy which have been engendered by actual physical and psychological decline. The loss of centrality of position in the family group or in the person's work, the sense of loss of social status through actual or relative decline in occupational status, general cultural attitudes towards old age as a period of necessary and actual failure — something which is pitied but without real sympathy by those younger — these are some of the factors that produce the feelings of insecurity and inadequacy that result in the reactive cantankerousness of some elderly people.

(3) Mild depressiveness is a common characteristic of elderly persons such as the one whom I have described. As a basis for this there is the inevitably increasing isolation and loneliness as friends and relatives die or become absorbed in their own lives. Essentially this produces the same kind

of desolation as is experienced by the child who feels or has actually been deserted or rejected. The elders are being left behind by life. They as individuals can experience this only as a desertion. Added to this is the loss of self-respect and self-esteem which goes with feelings of helplessness and worthlessness and which further feeds the depressive feelings.

(4) Gradually feeling themselves more and more isolated, older persons turn in more and more on themselves. In addition to their increasing preoccupation with a past in which they still felt themselves as having some significance, the following may be observed: There is increasing sensitiveness to slights. Old age may bolster itself by standing tremulously on prerogatives which may be traditional but are in fact little honored. Because of this there may be increasing querulousness. Sensitiveness may become tinged with paranoid attitudes. These do not make the person more comfortable but they may at least increase his feeling that he counts for something.

(5) The increasing incidence of illness and death among those in their own age group gradually tends to undermine the delusion of invulnerability which all harbor. At the same time, new thought, new techniques, new ways of living make old habits less reliable as bolsters of security. The consequence is an increase in what psychiatrists call "free floating anxiety." Psychological tensions become translated into somatic tensions. The tendency is for the organism to try to cope with vague general anxiety by attaching it to specific susceptible organs. The elderly person now has an apparently tangible problem which can be dealt with. Constipation or aching joints or paresthesias are matters which can be legitimately looked upon as being of interest and concern to relatives or to the doctor. They therefore provide some hope to the elder of restoring a little of the lost security. Older people therefore tend to become more and more sensitive to their bodily functions. Infantile body-interest is revived. Actual organic changes tend to become psychologically elaborated and thus become even more severe and incapacitating. Here we can see how the situation in old age is the exact psychological reverse of that in infancy and childhood. The normal child sloughs off his dependency attitudes as his own powers mature. The older person tends to resume dependency attitudes as his own powers become atrophied. The child extroverts more and more of his interest and energy. The older person becomes more and more introverted.

As we have seen, the basic function of that over-all adaptive process which we call character is to deal with the person's particular reality in such a way as to overcome its vicissitudes, minimize its dangers, and add to his security, pleasure, and self-esteem. Some individuals are generally recognized to be "flexible." Others we look upon as being rigid. Some are more or less flexible in certain aspects of their adaptive functions and more or less rigid in others of these functions. Thus a man may be able to adjust and change with the times in his vocation and become rigid and stereotyped in his social outlook. In another man the reverse may be the case.

Our discussion thus far has concerned itself with that which is more or less generally true of the psychology of the aging process insofar as this occurs in even normal individuals. Depending on the character structure of the particular person, the psychological phenomena which I have described will be seen more or less in the open. It may be stated that the more rigid the character adaptation has been, the more open will tend to be the signs of anxiety and insecurity as manifested by sensitiveness, irritability and querulousness, compensatory self-assertiveness and stubbornness, depressiveness and hypochondriasis. On the contrary, the more flexible the character structure has been the more likely are we to encounter that delightful person whom we speak of as never having grown old. In contrast to the latent envy of and hostility to younger up-and-comers manifested in the first type, we will observe in the second type the qualities of generous fatherliness and motherliness that goes with security and self-respect.

You will have noted that I have frequently resorted to the modifying phrase "more or less." I have done this deliberately in an effort to adhere as closely as possible to the facts. The adaptive functions are so complex; there are so many internal overlappings and contradictions in any character structure — even such as has been by and large "successful" — that in the upshot we are bound to observe both summations and cancellations in the final observable phenomena.

Examples of Old-Age Reactions

The following are some examples of old-age reactions in persons with various character organizations:

(1) In the case which I have previously described, we saw evidence of mild depressiveness manifested by irritability, fatigue and constipation as a reaction to the death of his wife, his increasing isolation, and the change of his status in his business.

(2) In the case of another patient who had been an insatiable go-getter; who had made an outstanding success against many obstacles, and who had been incapacitated by a coronary attack from which he had made a good physical recovery, the final reaction was one of angry resentment with paranoid ideas. He could not tolerate the fact that he had been ordered by his physicians to slow down and that during his absence his business had gone on quite successfully under the leadership of a younger associate. In an effort to preserve his need for status and his confidence in himself he had elaborated the delusion that his doctors and his business associate had plotted to get him out of the business and that he was not really as ill as he had been told he was.

This is the reaction in a person who was a go-getter because his life had been dominated by hostile competitiveness beginning early in a disturbed relationship to a string of younger brothers. Sensitive to his actual internal helplessness, he externalized his problems in a form with which he could feel he had a fighting chance.

(3) Another man reacted to a coronary attack with a refusal to believe the diagnosis. He rebelled against the management of his case and continued his activity in his affairs by way of telephones and visiting secretaries. He finally died in his boots.

In this case we see a man who had lived down a relationship with a mother who had rejected him. He could not believe that anyone ever gave anybody anything. After a lifetime of confidence in and dependence on himself only, he could not trust himself to sink into the actual dependence of illness.

(4) Still another coronary case, a man who had successfully operated an old family business, finally presented a clinical picture of unwarranted invalidism and querulous dependence on physicians and wife.

Here was seen the collapse of a character which had been largely a facade. Pampered as a child, this man had later in life assumed the habiliments of active responsibility only against inner protest. In the end he resumed, with all the more intensity, the dependent claims of his childhood.

In short, in the elderly person we will see a personality picture which will represent either an accentuation, with some distortion, of life-long character traits, or the breakdown of such traits and the exacerbation of latent tendencies.

Regression in Old Age

This last fact brings us to some less frequent, but not uncommon, features of old age. The maturation of the personality is to a large extent attended by the regulation and repression of primitive instinctual tendencies. Outstanding among these is the sexual drive and the infantile interest in the body and its functions. It is notable that some old people lose their former fastidiousness and others their moral scruples. Sometimes we encounter rather tragic forensic problems in this area. The general fact here is that the process of introversion which characterizes old age involves what we refer to as a regression of the emotions. Emotional energy, formerly directed externally in various ways such as work, friendships and creative activity, returns along the paths of the person's development to earlier levels of expression. Sexual interest which may have been more or less dormant may be revived in partial or distorted forms. Besides the regressive activation of infantile sexual interest, sexuality may be turned to, by elderly men particularly, as a kind of self-therapy — a sort of gesture in the direction of recapturing a sense of capacity and self-esteem.

However, we must distinguish clearly at this point between regressive sexuality, sometimes in the form of perversion or incest, and sexuality as it may survive in the old in its mature forms. It is a fact that sexual capacity and sexual enjoyment may remain effective until advanced ages in both sexes. Not only do elderly married couples confirm this but the marriages of widowed elderly persons may be successful emotionally and complete sexually. By and large, however, the involution of sexual capacity and sexual interest increases with age and its exacerbation is more likely to represent an anxiety-determined protest against old age.

In this connection we can again turn to earlier phases of life to see a complementary situation. Normally, sexual interest and, within the given cultural limitations, sexual activity are greatest during adolescence. As men and women direct more and more of their mature emotional energies towards their careers, their homes, their children or their creative work, the earlier intensity of sexuality subsides. It survives for a longer or shorter time as a harmonious overtone in the lives of normal couples but it is no longer the theme song. The recrudescence of sexuality as a leading preoccupation in an elderly person therefore provides an index to the degree of involution which has occurred in the mature aspects of that person's character and emotionality.

Interdependence of Elderly Couples

Elderly couples at the end of a long married life will have established a symbiosis on which each of them is dependent even if the relationship has been something less than ideal according to external standards. It is a matter of note that one of a couple may not long survive the death of the other without too much basis for this in the survivor's state of health. In other cases a fairly severe depression will ensue. In the cases of couples whose lives together have been marked by an ambivalent mixture of hostility and affection the survivor will tend to become neurotically ill with symptoms comparable to those involved in the cause of death of the spouse. These are all manifestations of the intensity of the identification which occurs. The dependence of such old people on each other is increased because they remain for each other the familiar landmarks in a life whose course has been changed by age. They lend each other support and security because there survives in at least this aspect of their lives something predictable and capable of being dealt with according to old familiar techniques.

Psychosomatic Complexities

One of the striking phenomena in old age is the inextricable relationship of emotional problems and actual organic disease. Illness is much more of a threat to older people, not only because of their decreased constitutional reserves but for such psychological reasons as we have already discussed. Every such blow to failing powers increases insecurity and tends to produce either over-compensatory reactions, which burden the failing organism still more, or invalid reactions which enforce the claim of the person for the care and attention which he has felt himself to be in danger of losing.

More striking, however, are those instances in which an apparently healthy person will succumb rapidly to some previously latent organic disease process under the impact of an emotional blow. Physicians are familiar with the cerebral apoplexy which follows the loss of a spouse or some injury to pride. Psychiatrists and neurologists not infrequently encounter cases with signs of rapidly advancing mental deterioration among elderly persons who have remained quite intact up to the time of some emotional blow. We are all familiar with the post mortem findings in the brains of elderly persons in which there will be considerable discrepancy between the organic pathology and the clinical picture of the person at the time of death. There is no one-to-one relationship between cerebral arteriosclerosis and the degree of mental failure which may be present. Elderly people with more or less well balanced personalities, such as that of the first case we have discussed, are able to withstand a considerable amount of cerebral damage, while less balanced personalities may produce a frank psychosis with a minimum of cerebral pathology. It seems to be the case that cerebral organic pathology is only a final precipitating factor in persons already excessively burdened by such emotional problems as we have been discussing. Even more complex psychosomatic relationships exist in the case of some of the chronic diseases of old age such as heart disease and arthritis. All such actual organic illness presents us with secondary elaborations based on the emotional problems of the person. Im some instances it is difficult to be certain that it is not the psychological problem which has primary etiologic significance.

Organic Mental Diseases

There are a number of specific mental diseases which occur characteristically during the declining years.

(1) Alzheimer's disease or presenile psychosis tends to develop in the fifties, though it may occur earlier. Beginning with marked failure of memory there is a rapid development of confusion and disorientation leading by way of aphasia and apraxia to complete organic dementia. Even in this undeniably organic disease of the brain, symptoms of emotional origin occur in its early phases and are representative of the person's

reaction to the overwhelming nature of the attack on his integrity. Irritability and anxious depression may be attended by compensatory psychic symptoms such as sexual deviations, grandiose business plans, paranoid ideas. Pick's disease, which tends to occur somewhat earlier than Alzheimer's disease, and in women more frequently than in men, is also characterized in its early phases by reactive and compensatory psychological phenomena. Here we often see euphoric overactivity and generally uninhibited behavior which is in part to be accounted for by the infantile regression induced by the atrophy of the cerebral cortex and in part a psychological denial of catastrophe.

(2) There are many cases of cerebral arteriosclerosis, with vertigo, tremors, unsteady gait, paresthesias, etc., in which psychotic symptoms never appear. The commonest precipitating factor in the onset of a psychosis with cerebral arteriosclerosis is a major or minor cerebral accident. The psychotic symptoms often date from such an episode though the onset may be gradual. These symptoms consist of organically determined disturbances of consciousness and mentation and of emotionally determined apprehension and panic, delusions of threats of bodily harm, nightmares, and depression. Restlessness and sleeplessness are common. There may occur episodes of violence against even those closest to the patient, these being dictated by delusions of threat against the patient's person. The emotional symptoms again demonstrate the reaction of the personality to the inwardly perceived threat to its integrity. The anxiety is a direct consequence of this. The violence which may appear is expressive of an attempt on the patient's part to externalize a problem and then to attempt to master it by brute force.

Functional Personality Disorders

Apart from the general psychological reactions of old age which we have discussed and the organic mental diseases which we have touched on, elderly people may present any of the various categories of functional disorders of the personality. In order to clarify this it is necessary to make a digression.

In our previous discussion of character we have spoken of it as the overall adaptive pattern of the personality. We have referred to its function in the repression and regulation of the instincts and drives. Every neurosis represents a breakdown of this function and is an emergency repair job. The neurotic symptom, be it a phobia, a compulsion, a conversion or a tic, represents the partial breaking through of a primitive impulse which the mature personality repudiates and encapsulates in the symptom. This occurs at a considerable cost of emotional energy, and with a consequent decrease in the resources of the personality for coping with its adaptive problems. Nevertheless, the integrity of the personality is thus preserved. Most persons are burdened to a certain extent with neurotic symptoms or traits. Very often these go unrecognized and for the most part they do not seriously damage the adaptive functions.

However, the failure of adaptive strength which goes with age is associated with a weakening of the defenses against the instinctual drives. Thus it can come about that minor neurotic symptoms which have sufficed to protect the individual against eruptions of the instincts during his maturity may become intensified and new symptoms may be added as a means of bolstering these defenses. Thus a compulsion neurosis, which first made its appearance in youth and then became latent when an emergent need for it had subsided, may exacerbate in the later years. This may be similarly true of a hysteria or a paranoid or depressive syndrome. We see then that the emotional reactions of elderly people not only are a reaction to the external reality of their adaptive problems but are also determined by the tendency of the internal defense to be weakened as the strength of the personality declines.

There are a number of gross functional disturbances which are considered to be characteristic of the later years.

(1) *Involutional melancholia* is mistakenly thought to be a disorder of the female menopause. In its classical form it is an anxious agitated depression with delusions of guilt, hypochondriasis, and nihilism. It tends to appear as a first attack of overt mental illness in women between the ages of 40 and 55, and in men between the ages of 50 and 65. I shall not go into the details of this illness other than to tie it in with the general formulation which I have given to the emotional problems of elderly people. The anxiety is again a reaction to both the internal and external dangers which threaten the involuting organism. The depression and the delusion of guilt are reactions to the increased pressure of unacceptable instinctual impulses. Often the delusions are concerned with sexual fantasies and ideas of irreparable injuries done to others. The hypochondriasis is a regressive return of infantile body interest and the fantasies of world destruction which I have called nihilism are simple projections of the sense of failure in the integrity of the personality itself.

(2) *Paranoid states* characterized by ideas of persecution may be representative of the resurgence of primitive hostile impulses and of the reactive resentment of the elder towards his unwelcome position in his present life. The unacceptable hostility is projected and becomes the delusion of persecution.

(3) Still other elderly people may be subject to *simple depressions,* with feelings of desolation connected with their actual life situation. Self punishing attitudes of guilt may be present as a reaction to the resentment and hostility which in the paranoid cases has been projected.

(4) The *senile psychosis,* distinguished from the psychosis with cerebral arteriosclerosis by the lack of, or minimal significance of, organic signs and symptoms is a final caricature of old age. Its onset is gradual, beginning with an insidious weakening of initiative, loss of interest in things normally of vital importance, reversal of the sleep rhythm, failure of memory and comprehension, general deterioration with final confusion and delusional formations. What we observe is the gradual breakdown of a personality into what may be compared to the marasmus of an emotionally and physically starved infant. In these cases life has become just too much and yet the spark lingers on.

Normal Adaptive Reactions

It should not be presumed on the basis of what has been said that there is no such thing as a normal old age. Persons whose life-long stability has been based on resiliency rather than rigidity will live out their years with dignity and decency though none will escape his share of anxiety and quiet desolation. The flexible person is one who respects himself but has no conceit, who has been guided by principles but has not been the slave of dogma, who has had steadiness of purpose without being hypnotized by an immutable goal, who has bent his energies to making the grade rather than getting to the top first, who has tolerated his weakness while employing his strengths, and who has respected his neighbor. Such a person is buffered throughout his life against both internal and external vicissitudes. Such a person has regulated and diverted his instinctual tendencies rather than attempted to smother them. In such a person the pressure of the instincts will at all times be less threatening. Such a person has applied his adaptive capacities to their proper task of mastering the environment to the end of overcoming its vicissitudes, minimizing its dangers, adding to its security, evolving from it the possibilities of creativeness, and deriving from it its legitimate gratifications. Such a person, having lived within the possibilities, will be less vulnerable to the limitations imposed by the external harsh reality of aging.

It must not be thought that I am advocating a belief in poetic justice. There are the notable exceptions. The tyrannical old man or old woman, whose throne goes to the grave together with its incumbent, is well known in fact and fiction. On the other hand, the external facts of the case may be too much for the least egocentric and best balanced elderly person. By and large, however, the principles of which I have been speaking do operate.

CHAPTER 38

The Stress of Life

HANS SELYE

The Wear and Tear of Life

For our scientific research in the laboratory we needed an operational definition of stress, that is, one which showed us what to do in order to see stress. It is only by the intensity of its manifestations — the adrenal enlargement, the increased corticoid concentration in the blood, the loss of weight, and so forth — that we can recognize the presence and gauge the intensity of stress. The fact that you cannot see it directly, as such, does not make stress less real. After all, as Robert Louis Stevenson put it:

> Who has seen the wind?
> Neither you nor I
> But when the trees bow down their heads
> The wind is passing by.

For the present discussion, however, our shorter, Aristotelian definition — which merely classifies stress as one aspect of wear and tear — is more satisfactory. We can look upon stress as the "rate of wear and tear in the body." When so defined, the close relationship between aging and stress becomes particularly evident. Stress is the sum of all the wear and tear caused by any kind of vital reaction throughout the body at any one time. That is why it can act as a common denominator of all the biologic changes which go on in the body; it is a kind of "speedometer of life."

Now, in discussing my experiments, I have often had occasion to point out that aging, at least true physiologic aging, is not determined by the time elapsed since birth, but by the total amount of wear and tear to which the body has been exposed. There is, indeed, a great *difference between physiologic and chronologic age*. One man may be much more senile in body and mind, and much closer to the grave, at forty than another person at sixty. True age depends largely on the rate of wear and tear, on the speed of self-consumption; for life is essentially a process which gradually spends the given amount of adaptation energy that we inherited from our parents. Vitality is like a special kind of bank account which you can use up by withdrawals but cannot increase by deposits. Your only control over this most precious fortune is the rate at which you make your withdrawals. The solution is evidently not to stop withdrawing, for this would be death. Nor is it to withdraw just enough for survival, for this would permit only a vegetative life, worse than death. The intelligent thing to do is to withdraw generously, but never expend wastefully.

Many people believe that, after they have exposed themselves to very stressful activities, a rest can restore them to where they were before. This is false. Experiments on animals have clearly shown that each exposure leaves an indelible scar, in that it uses up reserves of adaptability which cannot be replaced. It is true that immediately after some harassing experience, rest can restore us almost to the original level of fitness by eliminating acute fatigue. But the emphasis is on the word *almost*. Since we constantly go through periods of stress and rest during life, just a little deficit of adaptation energy every day adds up — it adds up to what we call *aging*.

Apparently, there are *two kinds of adaptation energy*: the superficial kind, which is ready to use, and the deeper kind, which acts as a sort of frozen reserve. When superficial adaptation energy is exhausted during exertion, it can slowly be restored from a deeper store during rest. This gives a certain plasticity to our resistance. It also protects us from wasting adaptation energy too lavishly in certain foolish moments, because acute fatigue automatically stops us. It is the restoration of the superficial adaptation energy from the deep reserves that tricks us into believing that the loss has been made good. Actually, it has only been covered from reserves — and at the cost of depleting reserves. We might compare this feeling of having suffered no loss to the careless optimism of a spendthrift who keeps forgetting that whenever he restores the vanishing supply of dollars in his wallet by withdrawing from the invisible stocks of his bank account, the loss has not really been made good: there was merely a transfer of money from a less accessible to a more accessible form.

I think, in this respect, the lesson of animal experimentation has a great practical bearing upon the way we should live; it helps us to translate knowledge into wisdom.

The lesson is a particularly timely one. Due to the great advances made by classic medicine during the last half century, premature death caused by specific disease-producers (microbes, malnutrition, etc.) has declined at a phenomenal rate. As a result of this, the *average human life span* increased in the United States from 48 years in 1900 to 69.8 years in 1956. But since everybody still has to die sometime, more and more people are killed by disease-producers which cannot be eliminated by the methods of classic medicine. An ever increasing proportion of the human population dies from the so-called wear-and-tear diseases, or degenerative diseases, which are primarily due to stress.

In other words, the more man learns about ways to combat external causes of death (germs, cold, hunger), the more likely is he to die from his own voluntary, suicidal actions. I am not competent to speak about wars — though these are also signs of maladaptation — but perhaps my experiments can teach us something about the way to conduct our personal lives in keeping with natural laws. Life is a continuous series of adaptations to our surroundings and, as far as we know, our reserve of adaptation energy is an inherited finite amount, which cannot be regenerated. On the other hand, I am sure we could still enormously lengthen the average human life span by living in better harmony with natural laws.

To Die of Old Age

What makes me so certain that the natural human life span is far in excess of the actual one is this:

Among all my autopsies (and I have performed quite a few), I have never seen a man who dies of old age. In fact, *I do not think anyone has ever died of old age yet*. To permit this would be the ideal accomplishment of medical research (if we disregard the unlikely event of someone discovering how to regenerate adaptation energy). To die of old age would mean that all the organs of the body would be worn out proportionately, merely by having been used too long. This is never the case. We invariably die because one vital part has worn out too early in proportion to the rest of the body. Life, the biologic chain that holds our parts together, is only as strong as its weakest vital link. When this breaks — no matter which vital link it be — our parts can no longer be held together as a single living being.

You will note I did not say "our parts die," because this is not necessarily so. In tissue cultures, isolated cells of a man can go on living for a long time after he, as a whole, has die. It is only the complex organization of all our cells into a single individual that necessarily dies when one vital part breaks down. An old man may die because one worn-out, hardened artery breaks in his brain, or because his kidneys can no longer wash out the metabolic wastes from his blood, or because his heart muscle is damaged by excessive work. But *there is always one part which wears out first and wrecks the whole human machinery*, merely because the other parts cannot function without it.

This is the price we pay for the evolution of the human body from a simple cell into a highly complex organization. *Unicellular animals never need to die*. They just divide, and the parts live on.

The lesson seems to be that, as far as man can regulate his life by voluntary actions, he should seek to equalize stress throughout his being, by what we have called *deviation*, the frequent shifting-over of work from one part to the other. The human body — like the tires on a car, or the rug on a floor — wears longest when it wears evenly. We can do ourselves a great deal of good in this respect by just yielding to our natural cravings for variety in everyday life. We must not forget that the more we vary our actions the less any one part suffers from attrition.

We have seen[1] through what mechanisms stress itself can act as an equalizer of biologic activities; but it is equally true that stress, perhaps precisely due to its equalizing effect, gives an excellent chance to develop innate potential talents, no matter where they may be slumbering in the mind or body. In fact, *it is only in the heat of stress that individuality can be perfectly molded.*

A Way of Life

Can the scientific study of stress help us to formulate a precise program of conduct? Can it teach us the wisdom to live a rich and meaningful life which satisfies our needs for self-expression and yet is not marred or cut short by the stresses of senseless struggles?

I have seen a great many books and articles of late, which tell you "how to . . .": how to achieve peace of mind, how to enjoy life, how to become a millionaire or a centenarian, and how to be a success in general. Can one really give definite and generally applicable directions on how to accomplish such complex aims? I could tell a stranger how to get to the station, without having to walk along with him to show the way, but I doubt that by using words only, I could really explain even such a comparatively simple thing as how to drive a car. The best I could do is to show how the car is made (at least as far as I understand its mechanism) and how I drive it. Usually, this kind of explanation is quite sufficient, because one learns things involving personal conduct by adding one's own experience to that of others. This is even more true when it comes to complex behavioral problems which must be solved differently by every person in agreement with his own particular personality traits. I believe that the mere fact of understanding the general rules about the way stress acts upon mind and body, with a few remarks about the way this has been used as a basis for one man's personal philosophy, can help others, better than fixed rules, to formulate their own solutions.

Perhaps the best thing to do, therefore, is to summarize here as precisely as possible the rules of conduct which I have found most practical in my

[1]See Hans Selye, *The Stress of Life*. New York: McGraw-Hill, 1956, p. 266.

own particular case, repeat briefly how I arrived at them from my research on stress, and then let the reader decide for himself what he should accept and how he should use it.

I found it helpful to subdivide my aspirations into three kinds: short-range aims, long-range aims, and the ultimate aim. So, even if this means some recapitulation, let us now arrange them this way.

Short-Range Aims

Man's short-range aims are designed for immediate gratification. They have comparatively little influence upon our well-being in the distant future. Most of these are readily accessible enjoyments, in that they require no prolonged preparation by learning or planning; you simply allow something pleasant to act upon you. The *pleasures of the flesh* brought to us by our senses are the best example of this type. Some short-range aims are not wholly passive; they imply some activity in the form of self-expression, as for instance creative activities and the various games. But, in all these, work and reward are virtually simultaneous. In other words, to achieve short-range aims, you merely let things act upon you which give you a pleasant feeling and you do things you like to do.

That these pleasures can give happiness is self-evident. But man's mind is capable of much *more lasting and profound passive satisfactions.* These are not so readily accessible, because they presuppose a carefully developed, acquired taste. They really form a transition between short- and long-range aims. We must plan these; they are not fully developed by our heredity. They are therefore more optional, and it is well to establish which ones are worth working for. Not everybody has an ear for music, an eye for painting, or a mind for the enjoyment of nature. Each person must seek the kind of pleasure he is fit to enjoy. The task of determining our special predispositions in this respect deserves to be taken very seriously. "To know thyself" from this point of view can do much more to bring happiness than such conventional ideals as money or position.

Personally, I have derived my greatest satisfaction from the mere contemplation of natural laws. But I want to make it very clear that these satisfactions are no farther removed from the general public than are those offered by art. Of course, the more you know about nature, the more she can give you; but there is no essential difference between the pleasure a child can get from watching a butterfly and that of the professional biologist who studies the microscopic structure of a cell. The great thing about these passive pleasures is that they are not practical. They are pure. I think that people in all practical walks of life should keep in mind that, deep in their hearts, they also need the pure enjoyment of impractical pleasures to live a balanced life.

Let me explain. Anybody who likes music can enjoy at least a simple folk song; no special schooling is needed for this. But a person who has a real understanding for the art of combining tones can appreciate really great music much better than the casual amateur. The same is true of literature and of painting. Many people who want to enjoy the finer things in life spend a great deal of time studying the structure and techniques of art or literature, precisely with this in mind — even if they have never wanted to become professionals. Anyone can enjoy the universe by looking at the stars or life by looking at flowers and animals. But there is no reason why one should have to be a professional biologist or a physician to penetrate deeper into the enjoyment of living things. The profound pleasure of studying nature comes from our understanding — however superficially — the lawfulness which governs it. The image of this understanding, the reflection of nature in our mind, somehow fits us more harmoniously into the world within and around us. *The source of our pleasure is the intimacy of this contact with nature;* and it is rewarding to cultivate our acquaintance with her and thus make this communion ever more intimate. Of course no scholar can fully understand even that little section of nature which he has selected for special study. Yet the more he learns about it, the more he can get out of it. The difference between the rewards of the most eminent expert and those of the complete ignoramus is not one of kind but only one of degree.

This second kind of aim — the appreciation of art, the admiration of nature — has another advantage over the simple pleasures of the flesh: its afterglow lasts longer. The more you learn to appreciate detail, the more you get in return. There are cumulative benefits to be derived from this which come quite close to what we shall have to say about the long-range aims. Most simple pleasures of the flesh tend to become stale as time goes by, but these more complex satisfactions of the mind become increasingly more gratifying as they are cultivated. To paraphrase the Bard: Age cannot wither them, nor custom stale their infinite variety.

Now, just a few words about the *relationship between short-range aims and stress.* Activity in any part of the body or mind apparently tends to go through three stages. Our adaptation energy is a finite, hereditarily determined amount of vitality which must necessarily be spent. Hence the inherent urge for man to express himself so as to achieve fulfillment and completion. In general, we must go through with what our bodily and mental structure is built to undertake and, in particular, we feel the urge to finish whatever task we start. Frustration and indecision are only types of incompletion.

Even what we call *failure* is not, as one might think, the most general, all-comprising class of biologic defeat; it is only one subdivision of incompletion. Failure implies merely the inability to complete a voluntary endeavor, but incompletion includes also the purely passive suffering, which may involve neither the mental humiliation of defeat in an undertaking nor physical pain. When you must listen throughout the night to the monotonous sound of drops falling at a regular intervals from a faulty faucet, it is the pure incompletion of the theme that hurts; and this hurt can be so great that it has actually been used as a form of severe torture. What we call *deviation* is only a way to simulate completion when true fulfillment is unattainable. Actually, it fools us, but even so, it helps.

I mention all this to underline the importance of completing the three phases of the stress-response in all our activities and all our passive sensations. To me this is the biologic basis of man's need to express himself and to fulfill his mission. In all our activities, we proceed through surprise (alarm reaction), to mastery (stage of resistance), to fatigue (stage of exhaustion), and hence either to rest (with a repetition of the cycle in the same or some other part of our being) or, eventually, to death. Man is constructed for this cycle. He should direct his life accordingly, neglecting no phase of it, and giving each manifestation of life the emphasis which fits his personal requirements.

The great practical lesson is to realize the deep-rooted biologic necessity for completion, the fulfillment of all our smallest needs and greatest aspirations, in harmony with our hereditary makeup.

Long-Range Aims

Man's long-range aims are designed to permit future gratification. They have comparatively little influence upon our well-being in the present; in fact, they are often in conflict with it. But whether he puts his faith in God or in creation, man realizes that his ultimate aim must outlast the moment and must usually be earned at the price of some momentary sacrifices. To achieve our long-range aims we have to act and we must learn how to choose between various optional modes of action. The difficulty here is to formulate these aims precisely and to develop a code of conduct to guide us in the perpetual dilemmas created by the competition between immediate and future happiness. The long-range aims are essentially social — or, at least, impersonal — in that their object is to create a favorable milieu for future happiness. They should lead us through a meaningful, happy, active, and long life, steering us clear of the unpleasant and unnecessary stresses of fights, frustrations, and insecurities.

Some people hope to find such aims in the acquisition of wealth, power, and social position; others in religion or philosophy. Yet others have instinctively realized that they are unable to solve this problem. They just give up and drift from day to day, trying to divert their attention from the future by some such sedative as compulsive promiscuity, frantic work, or simply alcohol.

That none of these things can assure lasting happiness is self-evident. Of course, there are many better guides to it: love, kindness, or simply the desire to do some good. These are often very successful; but it seems to me that they all have common roots in *man's innate, though often subconscious, desire to earn gratitude and to avoid being the target for revenge*. Why then should we not make this quite consciously our principal long-range social aim in life? Of course, it is an egotistic aim; but then as we have seen, egotism is an essential characteristic of life. In a sense, selfishness is the original sin, not only of man, but of all living beings. Why pretend that we can do without it? We cannot, and trying this just leads to frustration and self-incrimination. But if we follow the philosophy of gratitude, we necessarily make all our selfish impulses also altruistic without curtailing any of their egotistic, self-protecting values. From a scientific point of view, this strikes me as the most highly ethical among all possible natural guides to conduct. No one will blame us for hoarding the gratitude of our fellow men. This long-range aim is inextricably rooted in the natural laws governing man's actions. This is perhaps most clearly demonstrated by the fact that, rather than competing with ethical codes laid down by other philosophies and by religions, it actually finds support in one essential aspect common to them all.

After what we have said, the relationship between long-range aims and stress is so evident that it hardly justifies more than a fleeting comment. Mental tensions, frustrations, the sense of insecurity, and aimlessness are among the most important stressors. As psychosomatic studies have shown, they are also very common causes of physical disease. This is especially important, now that knowledge about microbes, vitamin deficiencies, and other specific disease-producers has given us such effective weapons to combat the maladies which, even at the beginning of this century, were still the major scourges of mankind. How often are migraine headache, gastric and duodenal ulcer, coronary thrombosis, arthritis, hypertension, insanity, suicide, or just hopeless unhappiness actually caused by the failure to find a satisfactory guide for conduct?

But neither short-range nor long-range aims are actually the ultimate aim of man: the objective which should furnish a basis for all our actions. We instinctively feel that one final aim should somehow coordinate and give unity to all our strivings.

The Ultimate Aim

As I see it, man's ultimate aim is *to express himself as fully as possible, according to his own lights*. Whether he seeks this by establishing harmony and communion with his Maker or with nature, he can do so only by finding that balance between long- and short-range aims, between sowing and harvesting, which best fits his own individuality.

The goal is certainly not to avoid stress. Stress is part of life. It is a natural by-product of all our activities; there is no more justification for avoiding stress than for shunning food, exercise, or love. But, in order to express yourself fully, you must first find your optimum stress-level, and then use your adaptation energy at a rate and in a direction adjusted to the innate structure of your mind and body.

The study of stress has shown that complete rest is not good, either for the body as a whole, or even for any organ within the body. Stress, applied in moderation, is necessary for life. Besides, enforced inactivity may be very harmful and cause more stress than normal activity.

I have always been against the advice of physicians who would send a high-strung, extremely active business executive to a long, enforced exile in some health resort, with the view of relieving him from stress by absolute inactivity. Naturally ambitious and active men often become much more tense when they feel frustrated by not being allowed to pursue their usual activities; if they cannot express themselves through actions, they spend the time worrying about what might be going on in their business during their absence.

At the risk of sounding facetious, let me present a little motto which I developed while analyzing stress in my experimental animals, in my colleagues, my friends, and myself. It may sound trivial and purely abstract, but it is based on solid biologic laws and — at least in my case — it works. Whatever happens during the day to threaten my equanimity or throw some doubt upon the value of my actions, I just think of this little jingle:

> Fight always for the highest attainable aim
> But never put up resistance in vain.

Everyone should *fight* for whatever seems really worthwhile to him. On the other hand, he should aim only for *things attainable* to him, for otherwise he will merely become frustrated. Finally, *resistance* should be put up whenever there is reasonable expectation of its succeeding, but never if we know it would be *in vain*.

It is not easy to live by this motto; it takes much practice and almost cmnstant self-analysis. Any time during the day, in discussions, at work and at play, when I begin to feel keyed up, I consciously stop to analyze the situation. I ask myself: "Is this really the best thing I could do now, and is it worth the trouble of putting up resistance against counterarguments, boredom, or fatigue?" If the answer is no, I just stop; or whenever this cannot be done gracefully I simply "float" and let things go on as they will, with a minimum of active participation (e.g., during most committee meetings, solemn academic ceremonies, and unavoidable interviews with crackpots).

Probably few people would be inclined to contest the soundness of this motto. The trick is to follow it! But that is where my assistance must stop. That is where you come in. This may sound like an anticlimatic ending, but it really should not. We all must live our own lives. No self-respecting person wants to go from cradle to grave sheepishly following the directives of another man.

Success Formula?

One of the main points of this whole discussion is that *there is no ready-made success formula which would suit everybody.* We are all different. The only thing we have in common is our obedience to certain fundamental biologic laws which govern all men. I think the best the professional investigator of stress can do is to explain the mechanism of stress as far as he can understand it; then, to outline the way he thinks this knoweldge could be applied to problems of daily life; and, finally, as a kind of laboratory demonstration, to describe the way he himself can apply it successfully to his own problems.

It was the dissection of stress and the analysis of its structure which helped me most with my own problems; and I do not think there is any other way to learn something that, of necessity, must be done differently by every person.

What is the use, for example, of dissecting a sentence and explaining its structure? In actual speech you would never have the time to apply the rules of syntax and grammar by conscious intellectual processes. Still, people who know something about syntax and grammar use better language, thanks to this knowledge. You cannot teach a man how to express himself because it is the first rule of the game that his speech must reflect his personality, not yours. Moreover, few of the rules of syntax and grammar are absolute; the most unpolished slang is often more effective and picturesque than the King's English. All formal teaching can do is to explain the basic elements of language, so as to make them available for translation from conscious intellectual appraisal — which is impersonal, slow, and cold — into instinct — which is personal, quick, and warm.

I intended to do, and could do , no more in these pages than to present the syntax and grammar of stress, illustrating its application to the philosophy of life by one example: my own. A single case does not prove much; but, against my laboratory background, one actual experiment proves a great deal more than volumes of pure speculation. In such an experiment the indices of success are purely subjective; therefore, I could not repeat the test on others and still vouch for the veracity of my findings. All I can say is that the philosophy of stress has helped me enormously in achieving equanimity and a personally staisfactory program for the way I want to go through life. I rather think if you tried it, it might help you too.

CHAPTER 39

Ego-Adaptive Mechanisms of Older Persons

CAROLINE S. FORD

Stress — the urgent pressure and severe strain produced by inner conflicts or the force of external circumstances — appears to have become a permanent characteristic of our society. Its presence is revealed in the sounds of children's anger, the taut faces of the unemployed, and the anxious grimace of the sick. Above all, stress is the constant companion of the aged; in many ways the older person seems singled out to be its special prey.

There appear to be three major areas — the social, the physical, and the emotional — in which unusual demands are made on him. First, the rude and obvious signs of his loss of social and economic status are everywhere. They can be found in compulsory retirement programs, in daily advertisements urging everyone to "wash away the gray," and in the development of retirement or "golden age" villages, perhaps a thinly disguised form of segregation.

In general, the wisdom of older persons is not sought or esteemed in many of our social institutions, although in certain professions, such as law and medicine, age and experience still appear to hold honor and value. Women seem to fare somewhat better than men, since only extreme illness or disability results in women's compulsory retirement from their accustomed household activities. For the most part, however, the policies of schools, churches, and government agencies result in downgrading the older teacher, minister, or civil servant. Even in the family itself, older family members are accorded low status; they are no longer revered and respected participants in major family decisions. If they are fortunate, their advice is set aside politely. In our society councils of elders no longer exist. The older person's concept of himself is subtly subverted by the impact of all these circumstances. In the end his respect for himself is eroded and he internalizes society's concept of him.

A second demand on the elderly person is the need to accept evidence of his diminishing physical stamina and, often, of a severe illness or disability. The exigencies of long-term chronic illness or disability are a major source of stress. Despite many years in which extensive medical research on chronic illness has been carried forward, such illnesses as Parkinsonism, arteriosclerosis, glaucoma, and arthritis still remain severely disabling so that long-term physical disability is a common problem among the aged.

A third demand made on the older person is that he become physically, financially, or emotionally dependent on others. These dependency demands with their many painful aspects may be the greatest source of stress to the older person. He may reluctantly decide he must relinquish the pleasures and freedom of driving a car. His mental lapses may force him to accept the distasteful necessity to procure legal protection in handling money. His limited finances may even force him to undergo the humiliation of applying for a public assistance grant. The rigidly conceived and often brutally applied residence requirements in a community, the "means" test, and the substandard payments of many OAA grants make this step particularly distasteful.

A final demand the dependent older person may have to face is that he leave his own home and enter some type of facility for the aged, which in turn brings forth its own particular brand of stress. The aged person feels a certain degree of stigma or loss of status when he enters even the best of group residences. Despite the most considerate and thoughtful arrangements in group living, there is some loss of privacy and freedom and a necessity to conform to institutional rules.

The older person may thus be bombarded by a rapid succession of blows to the ego singly or in various combinations that need to be dealt with. As his choices and decisions narrow and constrict, he may suffer some loss of control over the direction of his life. Various combinations of these blows and losses form a confluence of increasing pressure, compressing and restricting his personality and ego with a resultant, and at times seemingly unbearable, load of stress and anxiety.

Examination of the external signs of stress and anxiety is important for everyone dealing with older people but most particularly for the professional social worker. This initial step increases his understanding of them and improves his ability to evaluate behavior that may have been causing them difficulty in their relationships with family members and staff members of institutions. The words "childish," "contrary," and "impossible," which younger persons may use in describing older people, connote strong emotional reactions and suggest the need for an objective analysis of the older person's special stresses and the defensive mechanisms he is making use of to sustain his adjustment.

Signs of Stress

Fear and Anxiety

Fear is a prominent reaction of the older person and often reflects a kind of generalized feeling of uncertainty. He may express his fear verbally in the form of a general concern about the future or concern about his own future. Often his fear of rejection by family members may center around being put in a home or what may appear to be some other place of abandonment. Another common fear centers around finances so that discussion of budget and fee take on an air of unreality not ordinarily prevalent among younger persons.

> Mrs. W, aged 85, asked to discuss the Meals on Wheels service with the caseworker. After reviewing the simple procedure for obtaining the service, the worker mentioned the fee. At once Mrs. W commented forcefully that she could not afford it and requested a lower rate. The worker learned that Mrs. W was receiving a monthly income of $475 from investments, as well as a monthly social security benefit of $85. Her monthly expenses amounted to approximately $325. Thus, she could afford to pay the $2.00 per day service charge. Nevertheless, Mrs. W continued to express much fear about this added expense and finally withdrew her application despite the worker's efforts to help her make a more realistic appraisal of her situation.

Confusion and Immobility

Confusion and immobility in planning and decision-making often reflect the older person's stress and characterize many of his attempts to use various direct services.

> Mr. B, an 84-year-old widower, living alone, telephoned to inquire about the Meals on Wheels service. He liked to eat in a restaurant but his daughter had urged him to try the meal service during the winter months.
>
> After deciding to try the service, Mr. B soon called to say he had changed his mind. He was not sure he would like the food and thought it might not be hot. His final comment was that he seemed to be having trouble making up his mind.

Hostility

An indication of intense stress is seen in the older person's explosive outbursts of anger or in his steady outpouring of resentment and frustration. Such feelings may be triggered by relatively minor incidents and the extent of the anger may appear wholly inappropriate to the precipitating cause, which suggests his great inner turmoil.

Mr. and Mrs. M, aged 73 and 71 respectively, entered a small group residence after thoughtful and intensive planning with the caseworker over a period of several months. Both of them participated fully in the plans to move, as did their daughter. All seemed well for several months following their admission to the residence, until it was announced that the main meal of the day would be served at noon instead of in the evening. This decision, which was based on medical and administrative reasons, was interpreted in advance to all the residents.

Mr. and Mrs. M's responses to the change were extreme and violent. The resident manager was startled and amazed at Mr. M's loud expressions of anger and at Mrs. M's sharp, unceasing complaints about the food. The daughter, the physician, the minister, and the administrative staff all received complaints that the M's were being "starved" despite the fact that the couple's total caloric intake remained constant. These stormy reactions abated after Mr. M was able to comment to the caseworker somewhat sheepishly that his physician had told him to be more careful about overeating.

The intensity of this couple's anger suggests that their relatively placid admission to the residence was deceptive and, in fact, concealed a high degree of underlying stress.

Some older people may display this same sort of intense rage constantly rather than in periodic outbursts; it invades all areas of their life and may be directed toward family members, friends, or staff members of a group facility. Even with continued ventilation the rage does not appear to be dissipated or reduced. Instead, it seems to become incorporated as a permanent part of the personality. Such rage has elements of an infantile temper tantrum and of an emotional storm characteristic of teen-agers. Because it is a continuing phenomenon, it cannot be ignored. Family members and those in the immediate environment must attempt to understand and deal with it since the strength and depth of this feeling is indicative of the severe stress the older person is experiencing.

Mrs. L, aged 79, lived alone for many years following the death of her husband. The daughter of a physician and wife of a wealthy businessman, Mrs. L had traveled extensively and had lived surrounded with the comforts and luxuries commensurate with her economic status. Now she resided in a rapidly deteriorating home that she could neither afford to keep nor manage to sell. Her dominating ways and hostility had long since alienated her few remaining relatives. As Mrs. L's social isolation increased, her few social contacts were restricted to the paper boy, the grocery delivery man, and a few others like them. Even the family minister stopped visiting her following her angry attack on him for "neglecting" her.

Mrs. L's ankle was broken by a fall and she was taken to a hospital and thence to a local convalescent-care unit for the aged and chronically ill where the outpouring of her expressions of anger and frustration assumed monumental proportions from time to time.

Thus, to a greater or lesser degree, fear, confusion, immobility, and hostility are signs of the hidden emotional warfare being waged within the ego of the older person as he comes to grips with the destructive forces associated with old age. How does the aged person solve his conflicts? How does he ward off the effects of many simultaneous blows to the ego? What defenses does he employ? Are they similar to those used by younger persons? Do the defenses help him achieve a comfortable and satisfying adaptation to his reality situation?

At first glance, it may seem that only a small proportion of aged persons successfully adapt to the high degree of stress often present in old age and, further, that the ego defenses they utilize are somewhat rigid and nonadaptive.

However, the author's observation of a population of 675 persons known to a foundation providing multiple services to the aged clearly indicated that 75 percent of them had made a good adaptation to their particular life situation, displaying varying degrees of strength and flexibility.[1] After they came to the foundation, their symptoms of stress diminished, and their ability to maintain social relationships and to participate in the activities of daily living improved noticeably. Furthermore, their grace and readiness to adapt, along with their considerable wit and general good humor, served to stimulate the author's search for the details of the adaptive process.

Defenses Commonly Employed

Denial

A frequent comment heard in a group residence for the aged is "I don't see what I'm doing here with all these old people." Men seem to need to emphasize their physical strength and vigor. Often they flex their biceps like small boys and they attempt to walk briskly and to jump up from their chairs quickly. Their main intent apparently is to deny signs of their impaired physical stamina. This behavior often persists despite the presence of obvious evidence of disability, so that the disparity between the person's wish and reality is marked.

> Mr. J, aged 80, suffers from heart disease and cerebral arteriosclerosis. He is an ambulatory but somewhat feeble patient in a rehabilitation center for the aged, and he displays some mental confusion. Each day Mr. J tells a staff member stories of his former athletic prowess, comments on his present muscular vigor (even as the nurse is helping him to walk), and requests his discharge from the center, insisting in a rather desperate way that he feels well and strong.

[1]The study was made possible through a grant from the U.S. Public Health Service.

In general, clinicians view denial as an unhealthy defense, but it can also be viewed as protective and supportive, since it sometimes walls off, at least temporarily, the excessive anxiety and the unbearable knowledge that one is afflicted with partial paralysis, Parkinson tremor, or mental confusion. The vigorous use of denial appears to buy the individual time that enables him to work toward developing other, perhaps more realistic, defenses.

An older person's total denial of a problem over a long period shows a quality of undaunted strength that commands respect.

> Mrs. A, aged 83, entered a convalescent facility for the aged, following a hospitalization for coronary heart disease. Mrs. A suffered from a severe diabetic ulcer on her left foot. Prior to her hospitalization, she had attempted to remain at home, having employed a series of housekeepers and practical nurses, all of whom she eventually discharged as unsuitable or incompetent. She completely rejected her illness, her dependency needs, and indeed the facility itself. She preferred to think of it as a hotel and always referred to the staff members as maids and waitresses.
>
> Each night Mrs. A requested that she be wheeled to the entrance, where she sat, impatiently waiting to be taken home by her son and daughter. That this plan did not materialize never seemed to bother her. She always returned to her room at bedtime, commenting that undoubtedly her family would call for her later. Mrs. A endured great pain and eventually the amputation of her left leg. Through it all she remained reasonably sociable, had a good appetite, slept fairly well, and appeared relatively free from obvious stress.

The defense of denial clearly was of great value to Mrs. A, since it enabled her to adapt to a most painful life situation.

Regression into Dependency

In certain situations an individual may regress to an infantile level, making many excessive demands for help in daily living although there is no organic basis for the incapacities he claims to have. He may ask to be fed or bathed or he may ask for assistance in walking. He may insist on using a walker or a wheelchair, and he may become incontinent. Such a person may cling to anyone who is nearby, even another patient, with a high degree of desperation, and he may resist all efforts to help him assume any responsibility for his own care.

> Mrs. F, aged 81, and recently widowed, entered a group residence for the aged, in general good health except for a moderately severe glaucoma. She was able to see large objects and could distinguish light and darkness. She quickly developed a pattern of severe dependency, clinging to her daughter and to the nurse and making many frantic demands for help. She kept her eyes

> closed most of the time and described herself as blind. She requested assistance in all activities of daily living and felt that she could not feed or dress herself. Mrs. F frequently asked for a wheelchair, saying she felt weak and faint. Although she had asked for a private room at the time of her admission, she spent most of her time in another patient's room, explaining that she could not bear to be alone.

Regression in this instance must be viewed as a basically unsatisfactory defense since it failed to relieve Mrs. F's anxieties or provide her with a comfortable adaptation to reality.

Other Adaptive Mechanisms

Flight

Some aged persons take refuge from an apparently unbearable situation by physical flight, which is similar in some respects to the flight of teenage runaways. Their activity may be construed as an attempt to take some sort of step to cope with their painful situation.

Flight may result from an unplanned impulse and may be executed in plain view of witnesses. Or it may be planned and executed at a time when the aged person knows the staff members are elsewhere. The fugitive may wear adequate outdoor clothing and have a specific objective in mind, such as returning to his former residence. In other instances, flight seems aimless, a restless wandering without any specific objective or wish to escape from the group residence.

> Shortly after his arrival at a convalescent-care unit, Mr. D dashed from the bailding, barefoot, clad only in his pajama bottoms. After he was returned to his room, he subsequently made repeated attempts to leave the building, despite the severe winter weather and his own weakened state.
>
> Mr. O planned to wait for the nurses' supper hour before making his attempted getaway. He put on a coat, hat, and muffler and quietly slipped out by the side door. Soon located with the assistance of the neighbors, Mr. O was disappointed that he had not found his own house.
>
> Following a stroke Mrs. S wandered restlessly both in and out of the group residence, lost and rather helpless, searching for some relief from her confusion and physical disability.

For all three of these old people flight was not a satisfactory adaptive mechanism, since it provided them with only a temporary respite from their anxiety and did not lead to any kind of long-range adaptation to their basic problems. Indeed, flight may add to such anxieties and difficulties or create problems that result from accidents and overexposure to the elements. These complications are occasionally fatal.

Withdrawal

The recluse or neighborhood eccentric is a common figure in society. Isolated and cut off from others, he is much misunderstood and even feared by his neighbors. Like the recluse, the resident in a home for the aged or a nursing home also may retreat from the social life about him. His emotional affect is neutral or detached. He appears listless and passive, displaying reduced emotional energy in all his relationships, although he may continue to make adequate responses in his daily living. Quiet and undemanding, he is apt to stay in his room most of the day and participate only minimally in the life of the group.

> Mrs. R, a 75-year-old widow, entered a group residence for the aged, suffering from Parkinsonism, moderately advanced. She was alert, still retained good ambulation and gait, and showed only a mild hand tremor. Mrs. R appeared quiet and contented. She ate and slept well, and she performed her daily activities independently. She made no demands on the staff, her family, or anyone else. She remained alone in her room except at mealtimes. She had few visitors, received little mail, and made no complaints.
>
> Mrs. R had been a lively person. A mother of four, she had once been employed full time as a newspaper reporter. The staff considered her present withdrawal from social activities to be a protective mechanism that was quite different from those she had formerly made use of.

Excessive sleeping or resting in bed may be viewed as another form of withdrawal. The Victorian lady who "took to her bed" has her modern counterpart.

> Mrs. Z entered a convalescent-care facility following a life of travel and of residence in a series of apartment hotels in various sections of the country. She would rise from her bed only for meals; otherwise she was found there either sleeping or resting. She expressed no interest in any activities. Sleep appeared to shield her from having to face her current situation, which clearly represented to her a loss of status.

Persons suffering from a terminal illness sometimes display a more positive form of withdrawal or detachment. They often develop a total acceptance of the nature and the eventual result of the illness without any appearance of anxiety. Such persons convey a spiritual quality of strength and serenity to the staff and other old people.

> Mrs. E, aged 90, was suffering from cancer of the pancreas. Although eating and sleeping well, she appeared to be gradually withdrawing herself from involvement in the life around her. She was pleasant and cheerful, but somewhat distant and preoccupied. Mrs. E commented that she was thankful for her life and was prepared for death whenever her time should come.

In such instances withdrawal appears to be essentially ego-protective in nature, and it may be considered a relatively successful defense since the individual using it is able to make a fair adaptation and to maintain some degree of self-esteem.

Manipulation

Manipulation is a relatively healthy mechanism of ego defense since the person making use of it is still vitally involved with his environment and the people in it, even though he may use them to avoid facing reality. He uses any available tool or circumstance to help him gain his ends. He may cry, threaten, sulk, or pout. His "sinking spells," feelings of faintness, agitated telephone calls to his family, or his frantic calls to nurses are his means of attempting to control and change the rules of a facility or the behavior of the staff. His goal is to increase his status or prop up his feelings of self-esteem and retain some semblance of mastery over his affairs.

This particular narcissistic mechanism is an indication of vitality. Over a period of time it brings the individual considerable satisfaction despite the disturbing effect on others.

> Mrs. H, aged 70, entered a convalescent facility for the aged after a complete diagnostic work-up failed to reveal any organic basis for her feelings of weakness. She seemed to want to retreat to her bed and die. Mrs. H was cheerful and pleasant with members of the nursing staff, but with both her daughters she appeared tearful and agitated. In a long series of scenes she left them in various stages of emotional exhaustion. Mrs. H then appeared calm and satisfied once more, having carried the day.
>
> This situation continued without change throughout Mrs. H's stay and was clearly a source of great satisfaction to her. Her ability to control her adult children reduced her anxieties about her increasing dependency.

Integration and Sublimation

Despite illness, disability, or other limiting factors, the aging process apparently represents to certain older persons a challenge that they approach with vigor and even enthusiasm. To them the best defense is a good offense, and they rally their forces and handle their anxieties by integrating and channeling their energies into activity and accomplishment. These individuals are excellent candidates for rehabilitation. They show great determination in working to overcome physical handicaps, and they are creative in finding new kinds of satisfaction within their realistic restrictions.

> Miss T, aged 90, formerly a professional teacher of music, entered the

rehabilitation center following a stroke that left her aphasic but otherwise relatively intact. She suffered from diabetes and poor vision, but these difficulties appeared to present no obstacles to Miss T's adaptation to her situation. She continued to try to speak and was friendly and warm to everyone. She accepted her diet restrictions with little difficulty and she ate and slept well.

Miss T took great pride in her appearance and visited the beauty shop each week. She welcomed her many visitors with her customary graciousness. Clearly her feelings of self-esteem remained intact and her social responses unimpaired. Her apparent enjoyment of her daily activities and all that went on about her was contagious. Her presence in the residence was considered therapeutic for the other patients.

Thus, certain aged individuals who achieve a high level of integrative adaptation not only accept their restrictions but also sublimate their anxieties by reaching out to others with warmth and support. In addition, they often display a highly developed sense of humor and wit that further enlarges their perspective.

Implications

The clinician's recognition of ego defenses provides him with a view of the rich array of behavioral responses the older person may employ in countering the attacks on his ego. Many of these defenses are healthy and adaptive. Certain defenses that are frequently considered unsuitable and unhealthy at another stage of life are most useful in enabling the aged individual to attain a comfortable and adequate level of adaptation.

Surely the orientation of all programs for older persons, whether public or private, whether concerned with financial assistance, health needs, or counseling, should be ego-supportive, rather than ego-destructive as so many are today. Surely in all agencies dealing with aged clients highly skilled and mature workers should be employed rather than students, beginning caseworkers, or completely untrained workers.

Workers whose professional skills are used in behalf of older persons find that their clients like any other age group possess a wide variety of ego structures and make use of a wide variety of defenses. It is a challenge to understand their needs, to learn from them, and to provide them with the best possible professional casework services.

Social workers have taken a long time to arrive at a fully developed professional interest in the aged client. They now share in the responsibility for initiating and supporting programs and services in all areas touching the lives of older persons, so that these persons may have the chance to live out their appointed time in comfort, dignity, and grace.

B. INTERPERSONAL RELATIONS

CHAPTER 40

Psychological Developments in the Second Half of Life

ROBERT PECK

Many of the changes which take place in middle age and old age are decrements in physiological capacity and function. To apply the term "development" to such changes would be a misnomer. But do physiological powers, alone, adequately describe what it means to be a human being? Such a definition would make people nothing but animals — and that would be a "nothing but" explanation, it seems to me. It would ignore the very characteristics which uniquely distinguish *Homo sapiens.*

The Concept of Development in Later Life

It is precisely in the supplanting of sheer physical vigor and animal powers by the development of *mental powers,* and by the creation and transmission of a *social culture,* that we show our uniquely human capacities. If this be the case, it would seem worthwhile to look at the life cycle in terms of those mental and social capacities, as well as in terms of the physiological aspects of human living. Physiology may inevitably suffer in the second half of life. It is by no means a foregone conclusion that mental and social powers must or do show a parallel decline. On the contrary, there seems reason to believe that the second half of life may be necessarily the period when these human capacities can be most fully developed and used. Barring that minority of cases where physiological "senility" puts an end to learning, or to independent living, it is a matter of common observation that many people continue to increase their knowledge and their wisdom up to the day of their deaths. While constitutional factors probably play a significant part in permitting such steady development of wisdom and "know how" through life, experiential and motivational forces are probably essential to the realization of this potentiality.

The idea seems worth investigating that it is only in the years beyond the midpoint, in most cases, that human beings have the necessary experience to develop fully what we call *wisdom*. Sheer mental and perceptual acuity may well reach a final peak in the twenties; there is a good deal of evidence that this is true. The ability to interpret perceptions farsightedly, foresee complex consequences, and make wise decisions — this ability may require more lived-through experience than the years of youth permit.

To me, this suggests that there is a truely developmental process in the second half of life. It is not that *all* of later life consists of developmental processes. It is not how people "adjust" to those attributes which decline, but how they transcend their animal limitations — how they develop new, different, uniquely human powers, to the fullest extent — which may mark the truly developmental aspects of the years beyond thirty-five.

It seems that the developmental concepts can be justified in *either* or *both* of two ways: one "historical" and nonteleological; the other, teleological in nature. The "historical" kind of developmental approach is based on this reasoning: what a person is and does at sixty is importantly, though not exclusively, a function of what he was, did, and experienced between fifty and sixty. Similarly, what he was and did at fifty was importantly a function of what he was, experienced, and did from forty to fifty; and so on, back as far into childhood as data can go. Consequently, no real understanding or even half-complete explanation of the aging process can be achieved, it seems to me, unless historical development, in this sense, is considered as fully as all our data allow. Of itself, this seems reason enough to use a developmental approach, simply to allow one to find out how the past affects the present for people in later life. It has been pointed out, however, that the concept of development often connotes an end point — a teleological goal. If the "end point" be conceived not as some physiologically defined goal, but as an end-of-life state of mind whose vision shapes and colors all the actions of an older person, then might not the human end point be this: to achieve the ability to live so fully, so generously, so unselfishly that the prospect of personal death looks and feels less important than the secure knowledge that one has built for a broader future, for one's children and one's society, than one ego could even encompass.

To summarize, there appear to be three distinct reasons for looking at certain important aspects of human aging with a developmental view.

1. Since human beings are exceptionally time-binding creatures, the present life style, feelings, motives and actions of a mature person cannot be very well or fully understood unless the past which affects his present condition is seen in its unfolding, developmental perspective. This is "development" in the historical sense. It is possible, true enough, to restrict oneself to an analysis of

the present field of the person's inner and outer life; but *learning* and *change* cannot be understood without data on antecedent conditions.

2. Certain aspects of human potentialities appear to develop to maximum realization only the the second half of life. Even while decremental physical changes are occurring, increments in judgmental power and in social wisdom may take place. Consequently, not all, but some aspects of later life would seem to be *developmental* in every sense of the word, including the concept of incremental growth in mental and social powers.

3. If it be accepted that man is a goal-seeking being, by nature; and if it be accepted that a man who shows a purposeful drive to make life more than a closed, self-centered circle which death totally obliterates — if such men are both happier and more constructive in their impact on those about them, then an optimal end state for the late years of life could well be defined. This would be a goal toward which development could be seen to proceed — whether well or poorly, from case to case. It would define personal death not as an ultimate, irretrievable tragedy, but as a limiting fact of normal life which it is possible for man to transcend by vicariously building and enjoying a future he will not personally see. This conception of the life cycle would thus include an element of adaptive, goal-seeking change which extends into the latest years of life.

Any or all of these points may seem to idealize the later years of life. This is not intended; nor does it mean that one need assume that all or most people perfectly achieve these potential increments in their human powers as they grow old. Yet the observable fact of imperfect development in many people's lives is just as true of the first twenty years of life as it is of the last twenty. Each theorist who has attempted to describe the optimum goals of development, at any age, has stated them in their idealized, perfected form, as ultimate standards to aim for.

If not all men achieve such growth, if even a majority of old people "lead lives of quiet desperation," it has not yet been shown that this is innate or inevitable. On the contrary, there appears to be some reason to believe that imperfect development in the later periods of life may be just as dependent on experiential imperfections as we believe to be true of the devleopmental aberrations or imperfections of childhood and adolescence. Decay of psychological and social powers in later life *may* be as alterable by environmental changes as are the neuroses, infantilisms, and aberrancies which develop in many people during childhood. Niether is likely to be an easy change to effect, for both practical and theoretical reasons; but if the possiblity exists for old people, as we know it does for younger people, then it would seem important to include some such developmental hypotheses in any study of the aging process.

Stages in Psychological Development in the Second Half of Life

Erikson, like Freud before him, conceived his "stages" of early life by considering the psychological problems that must be universally met and mastered at specific developmental (age) periods. Thus, his first four stages are defined as "psychic developmental tasks" (one might say) which must be faced in infancy and childhood. The fifth, identity, seems to be defined as it is because adolescence (at least, in our society) uniquely poses the problem of developing a new kind of sense of identity. That is, this problem does not ordinarily arise acutely during middle childhood; and it cannot be perfectly resolved if deferred to the adult years. The sixth and seventh stages are defined and located as they are, it seems, because they describe tasks which are uniquely crucial issues in young adulthood That is, they do not arise until adolescent problems are behind, and they probably cannot be successfully deferred much beyond the age of thirty.

Erikson's eighth stage, however, ego-integrity vs. despair, seems to be intended to represent in a global, nonspecific way all of the psychological crises and crisis solutions of the last forty or fifty years of life. Clearly, his phrasing of it states a major issue of life after thirty. A closer look at the second half of life, however, suggests that it might be accurate and useful to divide it into several quite different kinds of psychological learnings and adjustments, at different stages in the latter half of life. If this is true, these stages, and the tasks they present, may be as worthy of distinct definition and study as Erikson has devoted to the stages of early life. For reasons which will be discussed later, the chief chronological division which seems sound is between a Middle Age period and an Old Age period. Within these periods, the stages may occur in different time sequences for different individuals.

Middle Age

1. *Valuing Wisdom vs. Valuing Physical Powers.* One of the inescapable consequences of aging, after the late twenties, is a decrease in physical strength, stamina, and attractiveness (if, as in America, "attractiveness" is usually defined as "young-looking"). On the other hand, the sheer experience which longer living brings can, if it is used, make the middle-aged person able to accomplish a good deal more than younger people, *though by a different means*. It might be summed up by the old adage that more can be achieved if you "use your head instead of your hands."

"Wisdom" seems to be a widely used work which may sum up this increment in judgmental powers that aging makes possible. Certainly, not everyone grows wiser with increasing age. It is not an automatic process. Indeed, it would be a worthwhile undertaking to investigate what inner psychodynamic patterns and what environmental experience predispose some people to appraise and use their accrued experience — achieving wisdom — whereas others seem to learn little through the years.

Wisdom is to be distinguished from intellectual capacity. It might be defined as the ability to make the most effective choices among the alternatives which intellectual perception and imagination present for one's decision. Such choice-making is affected by one's emotional stability, and one's unconflicted or conflicted motivation set, as well as by intellectual ability. Sheer life experience seems to be essential in giving one a chance to encounter a wide range of emotiona relationships, as a corrective to the overgeneralized perceptual-attitudinal set derived from one's necessarily limited experience in one family, and one subculture, during childhood and adolescence.

Judging from personality analysis of some thousands of business people in the middle range of life — mostly men — it is my impression that most reach a critical transition point somewhere between the late thirties and the late forties. Some people cling to physical powers both as their chief "tool" for coping with life, and as the most important element in their value hierarchy, especially in their self-definition. Since physical powers inevitably decline, such people tend to grow increasingly depressed, bitter, or otherwise unhappy as they grow older. Moreover, they may become increasingly ineffective in their work roles and social roles if they try to rely on physical powers which they no longer possess. (This appears to be a major etiological element in the "middle-age depression," particularly in men.)

Conversely, it has been my impression that those people who age most "successfully" in this stage, with little psychic discomfort and with no less effectiveness, are those who calmly invert their previous value heirarchy, now putting the use of their "heads" above the use of their "hands," both as their standard for self-evaluation and as their chief resource for solving life problems.

Thus, it might be conceived that the optimum course for people who reach this first stage of physical decline is to switch from physique-based values to wisdom-based — or mental-based — values, in their self-definition and in their behavior.

If wisdom is a "good thing," then such a transition would not necessarily represent a defeated surrender to physical aging. Rather, it would be replacing physique-based powers of youth with equally attractive, useful powers *of a different kind,* which aging has conferred — and which, indeed, may require about this much life experience.

2. *Socializing vs. Sexualizing in Human Relationships.* Allied to general physical decline, but partially separate from it, is the sexual climacteric. Upon analysis of the changes that necessarily follow this event, particularly changes in heterosexual relations, it appears that there may be some logic in viewing this transition point as a crossroads, too; not necessarily as a one-way, downward path. The opportunity the climacteric presents might

be this: that people can take on a new kind of value for one — or to a much more dominant degree — as individual personalities, rather than primarily as sex objects.

It appears to be a simple, basic fact of human life that sex is biologically and socially an active, primary part of the first half of life; and equally, that it is of decidedly secondary importance in the second half of life, both in the nature of human physiology at this second period, and by social definition. If this is true, especially if it is biologically true that older people experience many other drives more strongly than the sex drive, then why need they feel particularly frustrated about lessened sexual potency, *unless* they are still trying to work out unfulfilled needs from an earlier age? May it not be that the climacteric is felt to be a tragedy, or a gap-leaving loss, only by people who irrationally persist in valuing themselves primarily by their sexual powers, even though their actual impulses are no longer strong, or no longer call intensely for release?

If a person takes positive action at this point, redefining men and women as individuals and as companions, with the sexual element decreasingly significant, it would at least be understandable that interpersonal living *could* take on a depth of understanding which the earlier, perhaps inevitably more egocentric, sex-drive would have tended to prevent to some degree.

3. *Cathectic Flexibility vs. Cathectic Impoverishment.* The phenomenon for which this label is intended might equally well be described as "emotional flexibility": the capacity to shift emotional investment from one person to another, and from one activity to another. In some ways, this cross-cuts any and all adjustive shifts that are made throughout life. The reason for considering it as a distinct function, perhaps more crucial in middle age than at earlier ages, rests in the fact that this is the period, for most people, when their parents die, their children grow up and leave home, and their circle of friends and relatives of similar age begins to be broken by death.

On the other hand, for many people this is the time of life when they have the greatest range of potential cathexis objects. They have the widest circle of acquaintances in their community and vocational worlds. They have achieved informal and formal status as "mature" or "experienced" people, to whom others actively turn. In fact, this may give them contacts with people over the widest age range, from young to old, which they will ever encounter. Further, by contact with younger ages, it may be that experience with a greater variety of people, of roles, of relationships, can lead to a more complex set of more varied, differentiated relationships than is possible at younger ages. If so, this would be another positive benefit which aging confers.

Some people suffer an increasingly impoverished emotional life through the years, because as their cathexis objects disappear thay are unable to reinvest their emotions in other poeple, other pursuits, or other life settings. Hence this too looks like a crisis stage where positive adaptation requires new learning — not only of specific new cathexis, but of a generalized set toward *making* new cathexes (or redefining existing cathectic relationships, as in the case of grown-up children). Certainly, adaptation to the death of one's parents, and to the departure of one's children, is a totally new experience, never met in earlier years. These primary ties being so deeply important, the adaptive problem they pose is perhaps more crucial than any earlier experience which requires cathectic shifts.

4. *Mental Flexibility vs. Mental Rigidity.* One of the major issues in human growth and living seems to be the question which will dictate one's life — oneself, or the events and experiences one undergoes? Some people learn to master their experiences, achieve a degree of detached perspective on them, and make use of them as *provisional* guides to the solution of new issues. There are other people — more than a few — who seem to become dominated by their experiences. They take the patterns of events and actions which they happen to have encountered, as a set of fixed inflexible rules which almost automatically govern their subsequent behavior.

In any case, there appears to be a widespread feeling by a great many people that "too many" (other people, of course) tend to grow increasingly set in their ways, inflexible in their opinions and actions, and closed-minded to new ideas, as they go through the middle years. This is often said of elderly people, of course; but it seems that the *first* time when it becomes a critical issue for most people may well be during middle age, when they have peak status, have worked out a set of "answers" to life, and may be tempted to forgo further mental effort to envision new or different "answers."

Like cathectic flexibility,[2] this function cross-cuts all adaptive learning behavior. It is no doubt particularly related to "stage one," wisdom vs. physique; but insofar as it may be a generalized phenomenon, including that first choice point as a special case, it may be worthy of separate study.

Old Age

1. *Ego Differentiation vs. Work-Role Preoccupation.* The specific issue here, particularly for most men in our society, is created by the impact of vocational retirement, usually in the sixties. What this phrase is intended to

[2] The third and fourth stage adaptation, cathectic flexibility, and mental flexibility seem very closely related to Kuhlen's discussion of intolerance for ambiguity and Henry's concept of affective complexity.

represent is a general, crucial shift in the value system by which the retiring individual can reappraise and redefine his worth, and can take satisfaction in a broader range of role activities than just his long-time, specific work role. The chief issue might be put this way: "Am I a worthwhile person only insofar as I can do a full-time job; or can I be worthwhile in other, different ways — as a performer of several other roles, and also because of the kind of person I am?"

The process of ego differentiation into a complex, varied set of self-identifications begins in early childhood. There are reasons, however, for considering it a centrally important issue at the time of vocational retirement. For most men, the ability to find a sense of self-worth in activities beyond the "job" seems to make the most difference between a despairing loss of meaning in life and a continued, vital interest in living. (For many women, this stage may arrive when their "vocational" role as mother is removed by the departure of the grown children. In that case, this crisis stage might well come in middle age for many women.)

There is an even broader issue at stake, however. For most Americans, at least, vocational retirement means a sharp reduction in income. This means a reduced standard of living, with all of the sense of depreciation of status symbols and worth symbols this may entail. Another consequence is that many retired people must adjust to a new state of dependence on others, in sharp contrast with their decades-long experience of being self-sufficient and self-supporting. If this enforced shift is to be met successfully, it requires ego-differentiation of a different kind: not just differentiation among varied role activities, but among different attributes of personality and interpersonal relationship. Thus, the person who has built his self-respect primarily on the value of rugged independence may see nothing left to live for when he can no longer fulfill this part of his makeup. On the other hand, a man or woman who finds meaningful satisfaction in being a "good friend" to people, in sensual pleasures of sunning, swimming, or "just sitting," or in some other side of his nature, may positively welcome retirement as an opportunity to develop these other aspects of life more fully.

Thus, one critical requisite for successful adaptation to old age may be the establishment of a varied set of valued activities and valued self-attributes, so that any one of several alternatives can be pursued with a sense of satisfaction and worthwhileness. This, at any rate, is what the term ego differentiation is here intended to represent.

2. *Body Transcendence vs. Body Preoccupation.* Old age brings to almost everyone a marked decline in resistance to illness, a decline in recuperative powers, and increasing experience with bodily aches and pains. For people to whom pleasure and comfort mean predominantly physical well-being, this may be the gravest, most mortal of insults. There

are many such people whose elder years seem to move in a decreasing spiral, centered around their growing preoccupation with the state of their bodies.

There are other people, however, who suffer just as painful physical unease, yet who enjoy life greatly. It may be that these are people who have learned to define "happiness" and "comfort" more in terms of satisfying human relationships, or creative activities of a mental nature, which only sheer physical destruction could seriously interfere with. In their value system, social and mental sources of pleasure and self-respect may transcend physical comfort alone.

This is the hypothesis underlying the selection of this issue as a critical decision point of old age. While such an orientation must almost certainly be developed in its initial form by early adulthood, if it is to be achieved at all, old age may bring the most critical test of whether this kind of value system has been achieved. In the form in which this issue occurs in late life, it may thus be viewed as one of the goals of human development. It recognizes that physical decline occurs, but it also takes account of mental and social powers which may actually increase with age, for many people.

3. *Ego Transcendence vs. Ego Preoccupation.* One of the new and crucial facts of old age is the appearance of the certain prospect of personal death. In earlier years death comes unexpectedly, as it were; but elderly people know it must come. Chinese and Hindu philosophers, as well as Western thinkers, have suggested that a positivie adaptation is possible even to this most unwelcome of prospects. The constructive way of living the late years might be defined in this way: to live so generously and unselfishly that the prospect of personal death — the night of the ego, it might be called — looks and feels less important than the secure knowledge that one has built for a broader, longer future than any one ego ever could encompass. Through children, through contributions to the culture, through friendships — these are ways in which human beings can achieve enduring significance for their actions which goes beyond the limit of their own skins and their own lives. It may, indeed, be the only *knowable* kind of self-perpetuation after death.

Such an adaptation would not be a stage of passive resignation or of ego denial. On the contrary, it requires deep, active effort to make life more secure, more meaningful, or happier for the people who will go on after one dies. Since death is the one absolute certainty for all people, this kind of adaptation to its prospect may well be the most crucial achievement of the elder years. Success in this respect would probably be measurable, both in terms of the individual's inner state of contentment or stress, and in terms of his constructive or stress-inducing impact on those around him. It seems reasonable to suppose that one could find objective evidence that there are destructive effects from a narrowly ego-centered clinging to one's private, separate identity, at the expense of contributing to others' welfare or happiness. The "successful ager" at this final stage would be the person who is purposefully active as an ego-transcending perpetuation of that culture which, more than anything else, differentiates human living from animal living. Such a person would be experiencing a vital, gratifying absorption in the future. He would be doing all he could to make it a good world for his familial or cultural descendants. While in a sense this might be considered a vicarious source of satisfaction, actually as long as one lives this is a direct, active, emotionally significant involvement in the daily life around one. It might almost be seen as the most complete kind of ego-realization, even while it is focused on people and on issues which go far beyond immediate self-gratification in the narrow sense.

Use of Developmental Criteria Rather Than Age Criteria for Studying Stages in Later Life. If stages in later life are to be defined, certain special problems must be faced which do not pertain, or not as much, to the study of early life. For one thing, there is far greater variability in the chronological age at which a given psychic crisis arises in later life than is true of the crisis points of youth. For instance, one critical test of cathectic vitality occurs when one's children grow up and leave home for good. In one family, this may occur in the parents' late thirties. In another family, the parents may be close to sixty before this happens. Thus, if one practical criterion of mastery of a later-life psychological task is the person's handling of certain critical experiences, then older people who are equated for the *stage* they are "working on" may differ very widely in chronological *age*.

An even more complex situation exists, moreover. In studying children who are at the pre-pubertal stage, we can almost take it for granted that they are almost all working on the same *total set* of developmental tasks. With adults, the pattern of developmental tasks can vary more greatly, from one individual to another. For example, the man whose children are grown when he is forty may not yet have experienced the male climacteric; he may still be working "uphill" to master his vocational role; and he may just be entering a widened circle of social, political, or other activities, and a widened circle of friends. This makes "the departure of children" a much different thing for this man than for a man of sixty whose youngest child is just leaving home; who is nearing vocational retirement; whose family and friendship circle has been broken by several deaths; and whose interest or potency in sexual activity may be markedly less than in his earlier years.

One practical conclusion might be drawn from such reflections, with regard to the conceptualizing of stages in later life: they may have to be much more divorced from chronological age than is true of the childhood stages. There probably are still certain broadly delimitable periods, such as "middle age" and "old age," but these are apt to be statistical artifacts, describing "the average person" of forty to sixty, or some such span. There are bound to be some people of sixty-five who act, think, and feel like the "middle age" group, while other sixty-five-year-olds act, think, and feel very elderly. At least, observation indicates that this is likely to be found.

This leads to one conclusion about the design of future researches on aging: it may be that the best way to get samples which are homogeneous with respect to their "stage in life" will be to use some "stage" criterion and disregard chronological age, except as it proves to be similar for the members of a sample which is defined by a nonchronological criterion. To illustrate, it may be that most can be learned about the principles of psychological development and learning in later life if samples are drawn of women who are at the climacteric, regardless of age; of men who are at the point of retiring (probably less independent of age, for reasons of social policy); of parents whose youngest child has just left home; and so on.

Such considerations may be good reason for using a stage index to define samples, *when developmental processes are the subject for study*. It goes without saying that there are other kinds of reasons for making other kinds of studies, where the age criterion is the most sensible. If one wants to find out what the average forty-year-old man is like, or the average sixty-year-old woman, clearly the thing to do is to study age samples.

Comparison of groups a decade apart in age, as 45-55-65, should also allow some rough estimates of "normal" developmental changes. It merely seems that the *best* way to study such *developmental* changes is probably to use a developmental criterion; and that age is probably decreasingly correlated with developmental stages as age increases.

Factors in Successful Aging: A First Sketch for a Study of Personal and Social Determinants of Adjustment to Retirement

Of the fairly universal crises that older people encounter, it would probably be easiest, by far, to get a sample of people who are facing the issue of career retirement. Such people could be located through industrial and other organizations; whereas it would be more difficult, practically, to find a cross-sectional sample of people whose families have departed, who are currently faced with a severe illness, or who have other kinds of major adjustments to make. (It would be possible to make later studies, of similar design, on people at other crisis points in middle age and old age.)

Another separate outcome of the study could be a measure of the nature and frequency of the specific problems which people face at sixty-five in the community studied.

Deriving the Sample. A stratified sample could be drawn from a community (Kansas City, Mo., or Austin, Texas, for example), equally representing each of the sexes, and each of the socioeconomic levels and ethnic groups in the community . If this sample were located via retired-employee lists, and also via Old Age Assistance lists for the lower economic strata, it might be possible to get as much as a 5 percent sample of the people sixty-five and older. Use of existing lists would reduce or obviate the need to knock on many doors before finding suitable subjects. Stratifying the sample carefully could nonetheless make it possible to generalize from such a sample to the total population of people over sixty-five in the community.

Two kinds of data could be gathered: (1) a simple count of the various problems the subjects face: housing, economic, health, loneliness, sense of worthwhile activity, etc. — together with a description of their present circumstances in these respects; (2) data on the present attitudes, behavior pattern, and adjustment of the subjects; plus such data as can be obtained about their previous life history.

The basic approach would be a personal interview with each subject, covering the areas indicated above. Supplementing the record of this interview would be interviewer ratings on problem areas; possibly self-ratings by the subjects on the various aspects of adjustment; one or two brief projective instruments, including a sentence completion form which can be responded to either in writing or verbally; and possibly a short Q-sort, designed to get a some of the major psychological variables.

A detailed interview guide would be prepared, of the "focused interview" type, indicating the areas to be covered. Similarly, existing rating scales, and newly developed ones, would be employed. Especially adapted, brief projective techniques and a brief Q-sort would also be designed. The design of these instruments would build on the work we have done for the present Kansas City study.

A briefer, follow-up interview would be conducted a year later, to obtain evidence on longitudinal trends in the subjects' adjustment.

Analysis and Interpretation of the Data. In analyzing the data the following would be the procedure:

1. A statistical analysis would be made of the reported "problems," to give a picture of the number of "retiring" people who are faced with each kind of problem. A summary of these findings would inform the community quite precisely (by comparison with the present lack of information) about the incidence and nature of the adjustment problems its older people face.

2. One way to approach the personal data is as follows: On the basis of the interview data on role perfomance, and sense of personal adequacy and satisfaction, the subjects could be classified into three or four groups, on a scale of "successful aging." (The definition of this scale would be much more specific than this general description.) Thereafter, a set of judges, different (ideally) from those who make the "successful aging" ratings, would analyze the various instruments.

3. The data — e.g., Q-scores, sentence completion responses, etc. — would be grouped according to the "success" criterion, and analyzed to find what attributes characterize, and perhaps differentiate, the groups. Both qualitative and quantitative analysis of the projective data are possible. The Q-sort data would be treated by analysis of variance (having been designed for such analysis); they might also be subjected to correlational and factor analysis. Wherever possible, all interpretations of all data would be expressed in terms of either categorizations or ratings, so that statistical tests of significance could be applied. Cross-validation of different instruments would also be possible under this system.

4. Subsequently, the subjects could be de-grouped and individual case studies could be made, using all the data. Here, ratings would doubtless be the most efficient way to quantify interpretations.

5. Certain variables would be predetermined and measured, whatever analytic procedure was used. These include most of all of the characteristics listed below. In addition, some unforeseen variables might emerge from study of the data. These could be added to the research plan, and scaled, before interpretation of the data was concluded.

Here is a rough sketch of the way one instrument — the Sentence Completion — could be analyzed. Using the ratings on "success in aging," select high, middle, and low groups of subjects. Analyze each group's responses to S.C. item one; then item two; etc. Thema differentiating the groups could be identified by assessing predetermined research variables, and also any additional characteristics which the data themselves suggest. Thematic categories could be rated on scales, or by a present-absent criterion.

When all pertinent response categories have been identified and scored, significance tests can be applied to identify those characteristics which differentiate the "successful aging" group. Similarly, by the use of contingency measures, it can be found what characteristics tend to be associated with each other. Some form of cluster analysis would provide a first approximation to the "primary factors" at work in these data. (This general procedure is one we are currently working out for a research project on Mental Health in Teacher Education, at the University of Texas.)

Conceptual Framework. In selecting psychological areas to be studied, and variables to be measured, I would tend to draw most heavily on the work already in progess in our Kansas City study. There, a great debt is owed to the creative thinking of Robert Havighurst, William Henry, Bernice Neugarten, and Walter Gruen, to name only those most centrally contributing to the psychological aspects of the study. My own theoretical contributions to that project have drawn on previous work with some of these same people — as in the 1946–47 study of adolescent personality and social role behavior led by W. E. Henry and the late Caroline Tryon; and the study on the development of moral character which I had the privilege of conducting with Robert Havighurst and others. Other direct contributors of ideas notably include W. Lloyd Warner and Carson McGuire. Through published writings, a major debt is owed to Erik Erikson and to Carl Jung; not to mention the original, pervasive contributions to our personality theories of Signumd Freud. It will be evident, as soon as the framework below is perused, why such acknowledgments are in order.

Social Role Activity Pattern. Here we could assess (*a*) the subject's relative effectiveness in each role, (*b*) his degree of satisfaction with each role, (*c*) the degree of role differentiation, (*d*) the degree of balance in the time, effort, and emotion invested in the various roles.

The purpose would be to give us a picture of (*a*) the subject's overt behavior pattern, (*b*) its adjudged adequacy in terms of his society's expectations, (*c*) its adequacy in his own eyes, (*d*) the degree to which it seems to imply — or perhaps even produce — personal integration and psychological health. Thus we could study both the overt role performance and its inner significance to the individual. This might yield upward of thirty or forty distinct variables worth measuring, considering the number of activities, personal relationships, and attitudes about them, to be appraised.

Familial Roles: (1) provider, (2) mate, (3) parent, (4) child (if subject's parents are still alive), (5) sibling (if close relatives are alive and significant in subject's life). Here we could assess subject's attitudes and behavior toward the people close to him; his perception of their reactions to him; and how satisfied or dissatisfied all this makes him feel.

Occupational Role: Vocation subject followed until retirement; kind and intensity of satisfactions he found in it; how he feels about leaving it; any entire or partial replacement to occupy his time as a "worker" now.

Friendship Role: Number, relative closeness, attitudinal tone, duration, et al., of subject's friendships.

Community Role: kind and degree of involvement with formal social or civic groups; extent of "social horizon," etc.

Political Role: degree, nature, and personal meaning of subject's political activities, affiliations, and attitudes.

Religious Role: degree of religious activity; its social and personal significance for subjects, etc.

Leisure Role: nature and personal significance of subject's leisure pursuits; proportion of time spent in them.

Social Status Role: subject's life pattern in terms of objective social status (climber, static, decliner; social level; etc.); effects of aging, or retirement, on subject's social position; subject's feelings in this area.

Age-Grade Role: degree to which subject's behavior matches the pattern expected of his age group by those around him; any specific respects in which subject deviates from the "norm"; reasons (conscious and unconscious) for any deviations.

Personality Dynamics

Erickson's first seven stages (through early adulthood).
Rate on degree of subject's success in mastering the psychological developments represented by Erikson's stages 1, Trust, through 7, Generativity.

Suggested Stages of Middle Age and Old Age

Rate on degree of success in mastering the behavioral and attitudinal developments represented by (1) wisdom vs. physique, through (7) ego transcendence.

Such assessments might be a way of constructing a profile of subject's "psychological maturity," and of identifying any adaptive themes appropriate to earlier stages of life on which subject still is fixating significant parts of his affect and effort. They could also be considered as general evaluations of such things as ego functioning, body concept, psychological autonomy, and the like. While some overlapping definition might be entailed, it might be well worthwhile to take the additional step of assessing a number of smaller, discrete aspects of the latter attributes. For example:

Ego Functioning

1. Degree of accurate awareness of objective, outer events (perhaps divided into the several role areas).
2. Degree of accurate awareness of inner feelings, thoughts, wishes.
3. Psychological capacity to take effective steps to deal with external events.
4. Capacity to handle and express inner impulses in a balanced, controlled way.
5. Degree of psychological autonomy with respect to external pressures.
6. Degree of confidence or anxiety in the face of external events.
7. Degree of confidence or anxiety in the face of inner pressures.
8. Conflict tolerance.
9. Generalized aspiration level.

Affective Life

1. Emotional vitality: amount and intensity of affect available for

investment; degree of liveliness and vigor in reactions to events. (May be related to constitutional energy or temperamental surgency.)

2. Emotional perception of external world: good, benign, friendly, full of gratifying opportunities vs. bad, difficult, hostile, or thwarting.
3. Emotional perception of inner impulses: good, "natural," acceptable, enjoyable vs. bad, "unnatural," unacceptable, pain-arousing.
4. Variety and complexity of inner feelings and motivations.
5. Relative importance assigned to external events or demands vs. inner feelings, wishes, experiences.
6. Degree of sensual reactivity; relative degree of satisfaction gained in sensual experiences.
7. Intensity of sexual impulses (perhaps differentiated further, as to outlets sought).
8. Feeling about own sexuality, from positive to negative.
9. Intensity of aggressive impulses (with descriptive statement of modes of expression used).
10. Feelings about own aggressive impulses, from positive to negative.

Body Concept

1. Degree of body awareness and preoccupation.
2. Feeling of body adequacy and attractiveness, from pride and liking to shame or disgust. (Descriptively specifiying particular body parts of special pride or concern to subject.)
3. Degree of integration of body image.
4. Degree of body-self integration.

Finally, certain attitudes toward aging, per se, might repay study. Some of these are now being assessed in the Kansas City study:

1. Subject's acceptance of his actual age.
2. Degree of discrepancy between perceived present self and perceived past self (e.g., ten years ago).
3. Time orientation: proportions of thought and concern devoted to past, present, and future.

Interpersonal Relations

This area, which would doubtless subsume many specific variables, might be taken care of by variables included in the social roles area, above. It would include the nature of subject's relationships, the degree and accuracy of his awareness of them, and his evaluative feelings about them, specified for each significant person in his life.

These lists of variables undoubtedly do not exhaust the possiblities. They are more illustrative than completely definitive of the kinds of characteristics which might prove to have a bearing on successful adjustment to retirement, and to aging generally. The ultimate aim would be to discover how such psychological attributes affect adjustments to aging changes; and, if this can be separated in some way, how aging affects these attributes in various ways, with different people.

If it proves possible to assess such psychological factors in the aging process — and only practical problems of staff and funds would seem to stand in the way — it might be possible within a few years to know much more than we do now about "the psychology of successful aging."

References

Erikson, Erik H. *Childhood and Society.* New York: Norton, 1950.
Jung, C. G. *Modern Man in Search of a Soul.* New York: Harcourt, Brace, 1934.

CHAPTER 41

A Social-Psychological Perspective on Aging

ROBERT J. HAVIGHURST

One of the principal unanswered questions about the human life cycle is: How do people structure their lives after about age sixty-five? Under what conditions do they achieve satisfaction?

This is the central problem of the social psychology of aging. It is a problem worth studying for both its theoretical and its practical value. The problem is of practical importance because the commercial-industrial societies are gaining in longevity and in material affluence and thus are providing people with an increase of free time after the age of sixty-five and with a fairly adequate financial income. Therefore the material conditions for life satisfaction are present.

As I read these two paragraphs, I ask myself: How would I have described the problem of adjustment in old age twenty-two years ago when I first began to work in this field? I think the main difference lies in our present emphasis on the *initiative* of the person who is growing older. I would not then have used such an active expression as "structure their lives."

The widespread but superficial view of aging in the 1940's saw it as a period of declines, losses, stresses, with the society outside the family doing very little to help older people make a satisfactory adjustment. Professor Burgess and I were members of the Social Science Research Council's Committee on Adjustment to Retirement, and we helped to make a survey of research on Social Adjustment in Old Age. With Ruth Cavan and Herbert Goldhamer and later with Ethel Shanas, we studied the lives of several occupational groups at and after the age of their retirement. We had a series of first-class doctoral students who studied retired teachers, YMCA secretaries, Methodist ministers, old age assistance recipients. For these studies we elaborated an interview schedule entitled "Your Activities and

Attitudes." The Attitudes Inventory was a self-report measure of successful adjustment. Later we developed a rating scale for measuring personal-social adjustment, based on the Activities and Attitudes Inventory or on an interview.

We then studied the meanings of work and retirement to several occupational groups, including men and women generally in the age range from sixty to seventy. My associates in these studies were Eugene Friedmann, William Harlan, Janet Bower, Ralph Ireland, and Dolores Gruen, and we studied medical doctors, steel workers, coal miners, department store employees, and photoengravers.

Our next step was to study a sample of older people in a small city by means of interviews. Dr. Ruth Albrecht made this study, which was published in the book *Older People.* She took the very important step of defining a set of social roles which cover the ordinary social interaction of people and of working out a set of rating scales so that we could measure with a degree of accuracy the extent and intensity of social interaction. The morale and attitude toward life of these people was generally favorable, and we were coming to see old age more and more as a period with substantial rewards and satisfactions.

At this point we were convinced that we must study the *process* of growing old rather than the end product, and we moved to Kansas City for what was to be ten years of field work in two separate studies.

There were four characteristics of our Kansas City studies that we had learned were important in the social psychology of aging.

1. To study an adequate-sized sample of each of the major socioeconomic groups.
2. To study a group extending from middle age to old age.
3. To study the social interaction of people based on some kind of role analysis.
4. To study the personality structure of the members of our sample.

In the first of our two Kansas City studies the social role analysis was carried through in a fairly thorough manner, which was to serve as a basis for later social role studies. This was published in the monograph entitled *The Social Competence of Middle-Aged People.*

Up to this point in our Chicago studies I had generally taken the role of the psychologist while my colleagues were generally sociologists. However, with the advent of the Kansas City studies, we had several able psychologists on our research team, and I began to operate more as a sociologist.

The second Kansas City study, generally known as the Kansas City Study of Adult Life, started in 1956 and completed field work in 1961. This was a more sophisticated work, both in sociological and psychological terms, than any of our earlier work. The first major publication was the book *Growing Old* by Elaine Cumming and William E. Henry. Reporting on a sample of people from fifty to eighty, they showed systematically how persons change in their social interaction over time and also they began to explore the inner life or personality changes. They formulated the disengagement theory of successful aging, which was to become a stimulus for much discussion, writing, and further research. They proposed that social disengagement and "inner" or intrinsic disengagement were inevitable processes, and that the aged individual would be generally high in morale as he demonstrated in his own life these "natural" changes.

The second book had as its title *Personality in Middle and Later Life* and was organized by my colleague Bernice Neugarten, who had taken the lead in making the personality studies in the Kansas City projects. This book presented various ways of studying personality change in adults, elaborating upon the problem of ego-development, and the relations between personality and adjustment, and it gave a factor-analytic treatment of the personality data which served as a basis for the study of personality types in middle-aged and older people. Among those working with Dr. Neugarten on this book were: William J. Crotty, Walter Gruen, David L. Gutmann, Robert F. Peck, Jacqueline Rosen, and Sheldon S. Tobin.

The next step was taken in 1960, when the field work on the Kansas City study was nearing completion. Dr. Neugarten and I formed a clinical research group of graduate students to study intensively the seven rounds of Kansas City interviews. We needed better measures of social interaction and of personal adjustment (morale) than we had used in the past. At this time we developed the *Life Satisfaction Rating* which we have since used as a measure of personal-social adjustment or morale and we developed several measures of role performance, including role activity, satisfaction with role activity, ego-involvement in role activity, and change in role activity.

Our study of the relations of life satisfaction to social interaction in the Kansas City sample led us to a major modification of the disengagement theory. This was reported by us at the Copenhagen meeting of the International Gerontological Congress in 1963 and is about to appear in book form with the title *Patterns of Aging*. Dr. Neugarten, Dr. Tobin, and I are saying that both the activity theory and the disengagement theory of successful aging are insufficient to explain the patterns of aging to be seen in the Kansas City sample. There is no single manner in which older people combine social interaction and life satisfaction. Old people, like young people, are different. If anything, older people are more complex than younger people, because they have had such a wide variety of life histories.

Some of the complexity of life for older people is pointed out by Williams and Wirths in their book *Lives Through the Years*, which is based on the Kansas City sample and pays special attention to the *social systems* of friends, family members, and acquaintances which center around each person.

The Central Position of Personality

This twenty-year series of studies has brought us to the conclusion that personality organization and coping style is the major factor in the life adjustment of the individual as he grows older. It is the manner in which the individual deals with the various contingencies in his life, some of them social, some biological, which is the important fact. It is what one makes of the world that is the important thing.

The key concept is the concept of *adaptation*. The individual continually adapts himself to the conditions of his life. Adaptation is an active process, ruled by the ego. There are remarkably few people who refuse to try to adapt to changes in their circumstances. A very few give up the effort and commit suicide. Most people attempt to cope with whatever situation they face. For instance, a sixty-five-year-old widow who had recently lost her strong and dominant husband spent a half hour telling an interviewer that she could not go on living without "Daddy." She wept almost continuously during the interview, and never ceased talking about her wonderful husband and how she could not get along without him. But six months later she had adapted reasonably well to this loss and was finding sources of satisfaction. Again, an elderly brother-sister pair had lived together for years in a small house heated by a coal-burning stove. They were becoming too feeble to look after their house in the cold winter months, and then the brother broke his hip. When this was mended enough to permit him to return home, the county health authorities decided that the couple must go to the county infirmary where they could get competent care. They both insisted that they would rather die than leave their home; but they were gently moved to the infirmary. After three months they had adapted reasonably well to the new life, where they were separated in different wings of the building, and could see each other only during daytime hours.

The Process of Adaptation. In all segments of the life span, growing older means an adaptation to (*a*) changes in the structure and functions of the human body, and (*b*) changes in the social environment.

The adaptation process is ruled, more or less actively and autonomously, by the ego or personality. What we call personal adjustment, measured in various ways and called by such names as *life satisfaction*, is a product of the adaptation process.

At a given place in the life cycle, adaptation is focused on certain roles and activities. These adaptation areas shift from one age level to another.

For example, during early adulthood the adaptive emphasis is on marriage and procreation and career-building, with the roles of spouse, parent, and worker the most important ones.

Then during middle age the emphasis is on career performance, family life, and civic activity, with greatest importance given to the roles of worker, parent, and the complex of civic and associational activities.

Adaptation in the Sixties. Up to the age of sixty or sixty-five the great majority of people can count on fairly good health, so that the state of their physical organism does not generally count very much either in favor of or against their adaptation. Still, those who are blessed with abundant vigor generally make a better adaptation than the average, and those who are physically weak generally have difficulty with their adaptation.

By the time a person reaches the decade from sixty to seventy, he or she generally has to adapt to some marked changes in physical vigor and in the social environment. With respect to physical vigor, he cannot do as much heavy physical work as he has done formerly. This is a major problem especially for industrial workers, who often find it necessary to drop the worker's role or to change to a lesser job for reasons of health before they are ready to retire. Nobody escapes from the necessity to adapt his way of living to decreasing physical vigor and acuity if he lives to be seventy.

The social environment changes in several ways so as to require adaptive effort by the aging individual. The majority of women lose their husbands before they are seventy, and must adapt to widowhood. The great majority of employed men and women lose their roles as workers during this decade. People who have been active in social clubs and professional and civic organizations often find that they are relegated to positions of less importance than they have enjoyed during their middle years.

That part of the social environment which consists of the expectations that other people have of the individual as he moves through his sixties changes slowly but surely. He is expected to be less energetic, less autonomous, less creative than he was a decade earlier. At the same time he may have to adjust to a reduced income. The supportive environment of friends and colleagues thins out.

For the first time, for most persons, the adaptation process takes place against a set of negative changes in the body and the social environment. Some people have a kind of personality that accepts these negative changes in a passive-dependent manner. Others have a personality that seeks to replace lost roles by greater activity in other roles, especially those of grandparent, neighbor, friend, and church member.

Also, with time freed from earlier role-obligations, some people develop a set of free-time activities which give interest and enjoyment.

The end result of the adaptation process in the sixties is a reorganization of role structure which is accomplished by the ego in the face of the losses and gains of this decade of life.

Adaptation in the Seventies. Speaking in a general way about older people, one can say that the decade from seventy to eighty is one of *maintenance* of the reorganized role structure that resulted from the adaptations of the sixties. However, there is further role loss, such as widowhood for men as well as for women, and the loss of friends, with a general reduction in role activity caused by decreasing physical vigor. The *task* of the seventies for most people is maintenance of a structure of satisfactory activity which was developed by the reorganization process of the sixties.

Successful Aging. Successful aging may be defined in various ways. It may be defined from the point of view of the friends and relatives and neighbors of an aging person. Does he or she do the things that we expect a person to do while growing older? It may be defined on the basis of a set of assumptions about what is "natural" as people grow older. One set of assumptions may lead to a definition of successful aging as the maintenance of the activities of middle age with little or no reduction. Another set of assumptions may lead to a definition of successful aging as progressive disengagement from much of the social interaction of middle age.

The writer will assume that successful aging consists of successful adaptation.

When there is a close "fit" between the personality, the social environment, and the physical organism, the adaptation will be relatively easy and aging will be successful. In general, the goodness of fit is maximized when:

1. The personality is strong and flexible.
2. The social environment is supportive.
3. The body is vigorous.

Studying the Adaptation Process

To understand the adaptation process in the latter part of the life cycle and to be able to predict the degree of life satisfaction of a particular individual we need to describe systematically for this person his:

1. Personality.
2. Social interaction.
3. Norms and expectations of the subculture in which the person lives.
4. Economic security.

5. Health and vigor.
6. Societal provisions to assist adaptation.

The study of norms and expectations of the various subcultures and of the societal provisions to assist adaptation can best be made in a cross-cultural setting which amplifies the subcultural differences. For this reason we have taken part in a cross-national study that allows us to compare two occupational groups of retired men in six countries. In this research the social setting emerges clearly as a significant dimension of the adaptation process. For example, we have noted that older men living in Vienna are rated high on the extent of their individual expressive activities (attendance at theater and concert hall, summer excursions to the mountains, gardening, etc.) but relatively low on certain of the social interaction variables (especially participation in formal associations). This may be due to the liberal provision Vienna makes for expressive free-time activity. Again, we noted that retired Dutch teachers are much more active in church work than teachers from any of the other countries, but are rated relatively low on friendship activities. This may be due to the prominent place of the church in the social and political as well as religious life of Holland.

Another example of the influence of societal provisions was noted in the case of retired industrial workers of Milan. The members of our sample had all worked in a particular large industrial plant which provided housing for its workers and a social center for them. The retired workers continued to live in the apartment blocks maintained by the industry and they used the social center as a kind of clubhouse. Their ratings on friendship activities were higher than those of workers in any of the other national groups. This rating, as well as that on the acquaintance-colleague role, was higher than the ratings of retired teachers in these roles in Milan. Here the societal provision affected the forms of their adaptation to aging, through making it easy for them to interact practically every day for several hours a day with friends and former work colleagues. In contrast, the retired teachers lived widely dispersed and had little easy opportunity to associate with old friends and colleagues. Their social interaction was more intensive with their families, and less with friends and colleagues, than that of the industrial workers.

The cross-national study thus far has been only a pilot one, based upon only small samples, and therefore not sufficient for valid generalizations. Nevertheless, the findings point to the conclusion that, while particular patterns of role activity differ from society to society, the relationships between social interaction and life satisfaction are about the same as found in Kansas City. In other words, in various Western societies most people show greater life satisfaction if their general level of social activity remains medium or high, even though that activity takes place in a wide variety of behavior settings.

The Significance of Life History

Another line of inquiry, independent of our Kansas City studies and primarily concerned with a somewhat different set of problems, has been under investigation by Tobin, Lieberman, and their students. They have been concerned with the function of memory, the place of reminiscence, and ways in which aged people reconstruct their life histories.

The older person is continually integrating and reinterpreting his past, attempting to make of it a meaningful whole. Older persons always *adapt to the present in terms of a past history* — a past history which is, in a sense, an active component in present adaptation.

The individual is not only coping with a present biological state and a present social state but also with his past; and in a very active way, making sense of the present in terms of the past, and the past in terms of the present.

My colleague Bernice Neugarten has been especially concerned with the life history element in the adaptation process. From her we have learned to see adaptation in later maturity as a process in which the ego, or the personality, mediates between a set of biological processes, a set of social processes, and a past as represented in the individual's memory.

Patterns of Aging

The adaptation process can be studied empirically in groups of people when the several dimensions mentioned above are systemically measured and described. The actual forms of adaptation that result can be called *patterns of aging*. A pattern of aging is a coherent complex of behavior, including social interaction and use of free time, achieved by an individual through the interaction of his personality with his physical organism and with his social setting. There is a limited number of patterns which can be discovered empirically. Observers agree on assigning persons to these patterns, with a small number of exceptional persons who exhibit rare patterns that will only be observed repeatedly in relatively large samples.

The social setting and the physical organism give the ego or personality a set of possibilities, from which ego works out a pattern which is comfortable to him and is approved or at least tolerated by the society. The actual adaptation which the ego makes depends on his past experience, or his *life history*. The life history gives the ego a supply of experience and habits with which he makes up his own adaptation.

Next Steps

We have noted that there does not exist any study of the process of adaptation to aging which takes account of all of the dimensions of this process. Some useful studies have been made, but they are only partial. Generally the studies have been limited to people in reasonably good health and in a limited social class range, and in one country.

The most useful extensions of our present knowledge would probably come from extending the range of the health and vigor variable and extending the range of the social setting variable, combined with the best of our present personality and social interaction study techniques.

Through extension of cross-cultural studies as well as continued attention to comparative social-class studies we may be able to take into account the differing *group values* which are a part of the social setting. The system of values which is present in a subculture and which the individual has internalized acts to mediate between the personality and the tasks presented to the individual by his society.

Finally, we do not yet understand the *individual psychological process* by which individuals deal with the tasks of later maturity. We do have some understanding of the processes underlying the transition from adolescence to maturity, through close studies of individuals. We need analogous studies of the transition from middle age to later maturity. For this purpose it would be wise to make longitudinal studies, following persons for ten to fifteen years in the period from fifty to seventy.

References

Cumming, E., and W. E. Henry: *Growing old.* Basic Books, New York, 1961.

Friedmann, E., and R. J. Havighurst: *The meaning of work and retirement.* Univ. Chicago Press, Chicago, 1954.

Havighurst, R. J., R. Cavan, E. W. Burgess, and H. Goldhamer: *Personal adjustment in old age.* Science Research Associates, Chicago, 1949.

Havighurst, R. J., and R. Albrecht: *Older people.* Longmans, Green, New York, 1953.

Havighurst, R. J.: The social competence of middle-aged people. *Genet. Psychol. Monogr., 56:* 297-375, 1957.

Havighurst, R. J., B. L. Neugarten, and S. S. Tobin: Disengagement, personality, and life satisfaction. *Age with a future.* Munsgaard, Copenhagen, 1964.

Neugarten, B. L., R. J. Havighurst, and S. S. Tobin: The measurement of life satisfaction. *J. Geront., 16:* 134-43, 1961.

Neugarten, B. L.: *Personality in middle and later life.* Atherton Press, New York, 1964.

Neugarten, B. L.: Personality and patterns of aging. *Gawein, 13:* 249-56, 1965.

Neugarten, B. L.: Personality development in adulthood. Paper presented at Amer. Psychol. Ass. meeting, Sept., 1966.

Williams, R. H., and C. Wirths: *Lives through the years.* Atherton Press, New York, 1965.

CHAPTER 42

Sexual Activities and Attitudes in Older Persons

GUSTAVE NEWMAN and CLAUDE R. NICHOLS

Misconceptions of an Asexual Adjustment

Recent emphasis in medical education has stressed the need for greater understanding of human behavior. This broadening scope of knowledge is of practical use to the physician in making possible more comprehensive care of his patients. With a rapidly growing geriatric population to attend, the physician of today needs more knowledge of what is ordinary or usual in the life of the older person. Little has been reported concerning the sexual activity and attitudes of older people. There have been many misconceptions, certainly, about the role of sex in the lives of older people in our society. One commonly recognized belief among younger people in our society is that older persons, especially grandparents, have no sexual feelings. As we now recognize, the feelings and attitudes of people are directly related to the expectations of the society in which they live. Thus, it is common for the physician to see in his daily practice older persons who feel guilty about having sexual feelings; often these feelings are not acceptable to the older person, to the physician or to other people in the environment in which the older person is living. Guilt or anxiety over sexual feelings may interfere with the adjustment of the older person and with his interpersonal relations, among which is the doctor-patient relationship. Guilt or anxiety on the part of the patient may thwart the therapeutic efforts of the physician.

While recognizing that the attitudes of society are slow in changing, we believe that more knowledge of this area of human behavior would serve both the physician and his patient, as well as strengthening their relationship to each other. Kinsey and co-workers (1) devoted only two pages in their study of male sexuality to the specific area of sexual activity in the aging man. Similarly a single table summarizes their data on the aging woman (2). Much more information is needed in this area in order to determine the role that sexual activity plays in the life of older persons. It is our purpose, then, to present some normative data on sexual activity and attitudes in a geriatric population.

Findings on Sexual Activity and Attitudes

Materials and Methods

Since 1953, a continuous interdisciplinary study of geriatric subjects living in the Piedmont area of North Carolina has been conducted by members of the Duke University Department of Psychiatry. Each subject has been studied for two days, or about sixteen hours. The table contains a complete list of the studies performed on each person. The data on sexual activity and attitudes are but one part of the medical and psychiatric history in this study and, as such, have always been obtained by a psychiatrist member of the research team. All subjects in this study were volunteers from the community in and around Durham, North Carolina. None of these subjects was hospitalized or living in nursing homes or homes for the aged, and they represent, as a group, a generally successful adaptation to aging. Subjects ranged from sixty to ninety-three years of age, with an average of seventy years. The study included both caucasoid and negroid men and women. A total of 250 subjects were included in this study.

Studies Performed on Persons in Population Studied

Social history
Medical history
Psychiatric evaluation
Physical examination
Neurological examination
Ophthalmological examination, including fundus photographs
Audiometry
Electroencephalogram
Electrocardiogram
Ballistocardiogram
Chest x-ray
Microscopic vascular study of bulbar conjuctiva
Laboratory studies
  Urinalysis
  Complete blood counts

Nonprotein nitrogen determination
Blood sugar determination
Serologic test for syphilis
Serum cholesterol level determination
Full-length photograph against grid
Psychological test data
Wechsler Adult Intelligence Scale
Roschach

Results

One hundred one subjects of the original 250 were either single, divorced, or widowed, the the greatest proportion of these being widowed. Of these 101 subjects, only 7 reported any sexual activity; thus about 7% of this group were sexually active. The remaining 149 subjects were still married and living with their spouses, and of these, 81 or 54% indicated that they were still sexually active to some degree. Subjects with any degree of

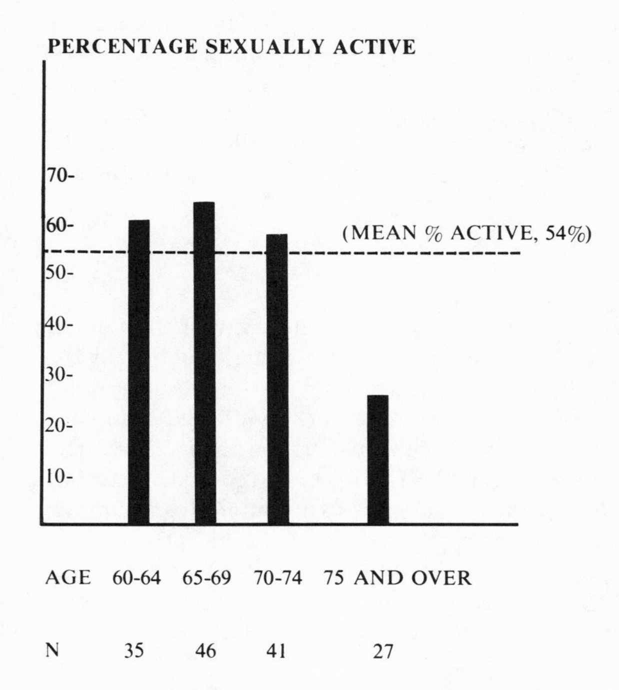

Fig. 1.— Sexual activity related to advancing age in 149 married subjects.

regularity and periodicity of sexual intercourse were included in the group called "active." Frequency of sexual relations for the persons in this group ranged from once every other month to three times weekly. Figure 1 indicates correlation of sexual activity with age. Statistical tests of significant differences (chi-square test) indicate that only the fourth group, consisting of those persons 75 years of age or older, showed a significantly lower level of sexual activity.

In figure 2 the sample of married subjects is divided by race, sex, and socioeconomic status. Inspection of these data shows that in this sample, the negroid subjects were sexually more active than the caucasoid, men more active than women, and persons of lower socioeconomic status more active than those of high socioeconomic status. The differences shown in these three bar graphs were valid differences by the chi-square test. We would like to point out, however, that there may be "cross-contamination," so that one variable may influence another; e.i., most of our negroid subjects were also of lower socioeconomic status. Thus the graphs are descriptive of our total sample but should not be taken as independent variables related to sexual activity.

During the interview, each older person was asked to compare the subjectively felt strength of the sexual urge at the present time with its strength in youth. Without exception, every subject rated the strength of the sexual urge as lower in old age than in youth. However, those who rated their sexual urges as strongest in youth tended to rate them as moderate in old age; most persons who described their sexual feelings as weak to moderate in youth described themselves as being without sexual feelings in old age.

Comment

The great difference between 7% of subjects sexually active in the goup of single persons as compared to 54% of subjects sexually active in the married group confirms the great influence that marital status has on sexual activity, as pointed out by Finkle and co-workers (3). This difference is indicative of the gradual lessening of the sexual drive with increasing age; in the absence of a socially sanctioned or legally approved sexual partner, i.e., spouse, the sexual drive, although present, is not often strong enough to cause the elderly person to seek an extramarital sexual partner.

The abrupt lessening of sexual activity in the group of persons aged seventy-five years and older deserves comment. Many subjects in this segment of our sample had chronic illnesses such as arthritis, arteriosclerotic heart disease, and diabetes. Persons in this group also frequently report such illness in their spouses. We have consistently found that chronic illness places limitations on all the activities of these persons, although there is an obvious variation in the degree of limitation which depends on the severity of the illness and the subject's attitude toward it. Thus, it was not uncommon for a person of this age group to indicate the presence of sexual feelings in the absence of any sexual activity. These subjects usually stated that sexual activity was stopped because of poor health either of themselves or of their spouses.

The fact that the women in our sample reported less sexual activity than the men also deserves comment, since all persons in this group were still living with their spouses and therefore had adequate opportunity for sexual expression. The possibility exists that the women were somewhat more

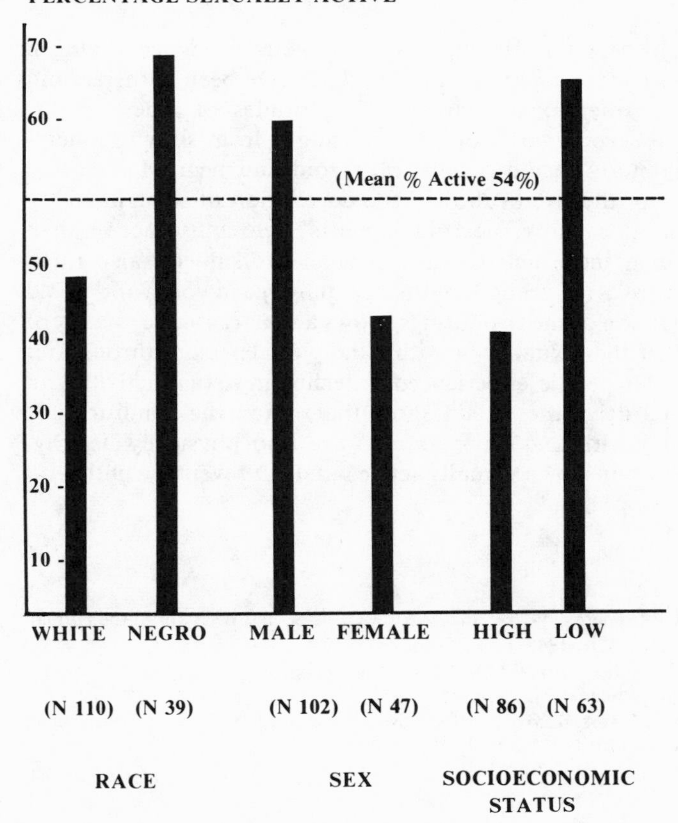

Fig. 2. — Sexual activity related to race, sex, and socioeconomic status in 149 married subjects aged over 60 years.

reluctant to give factual data about their sexual activity; however, we have generally felt that the data obtained in the face-to-face interviews have been reliable. We can point to one factor which may tend to explain this difference, which is that, while the average age of both men and women in this study was about seventy years, the men, by virtue of a trend to marry younger women, and conversely the women, marrying older men, were subjects who were individually reporting on sexual activity in marriages of significantly different age groups. For both males and females in our study, husbands were an average of four years older than wives. Another way of saying this is that our "average" male subject aged seventy had a sixty-six-year-old-wife, while our "average" female subject aged seventy had a seventy-four-year-old husband.

Conclusions

In a comprehensive, interdisciplinary study of geriatric subjects living in the Durham, North Carolina, community, data have been gathered and assessed regarding the sexual activity and attitudes of older people. Subjects averaged seventy years of age and ranged from sixty to ninety-three years. The study included both caucasoid and negroid men and women. Analysis of these data shows little correlation of sexual activity with age, but in this study, negroid subjects were more active than caucasoid and men more active than women. The subjects also rated themselves on the relative strength of their sexual urge in youth and in old age, and a comparison of the two ratings shows a remarkable constancy of the experiencing of the sexual drive within individual persons throughout life. Although older people experienced a decline in sexual activity and strength of sexual drive these data show that, given the conditions of reasonable good health and partners who are also physically healthy, elderly persons continue to be sexually active into their seventh, eighth, and ninth decades.

References

1. Kinsey, A. C.; Pomeroy, W. B.; and Martin, C. R; Sexual Behavior in the Human Male, Philadelphia, Saunders, 1948, pp. 235–36.

2. Kinsey, A. C.; Pomeroy, W. B.; Martin, C. R.; and Gebhard, P. H.: Sexual Behavior in the Human Female, Philadelphia, Saunders, 1953, p. 548.

3. Finkel, A. L.; Moyers, T. G.; Tobenkin, M. I.; and Karg, S. J.: Sexual Potency in Aging Males, *J.A.M.A.* 170: 1391–93 (July 18), 1959.

C. CULTURAL VARIATIONS

CHAPTER 43

Social and Cultural Factors Influencing the Mental Health of the Aged

DONALD P. KENT

The Influence of Social Structure

The complexity of any field of study is directly related to the number of variables with which the investigator must contend. By such standards the study of aging must rank among the most complex. All the variables ordinarily found in the behavioral sciences are present with the addition of a whole series of age-related ones. Consequently, any brief paper is necessarily an oversimplification; this is no exception.

This paper purports to limn certain aspects of the social life of the aged by focusing upon a few of the many aspects of the sociocultural system of American society. It will be necessary to paint with a broad brush. Factors as important as social class, different biological capacities and varying physiographic conditions will be ignored as well as less obvious but significant aspects which would be necessary for individual portraiture.

The starting point of this essay is the social system. This in its simplest terms is a

> . . . plurality of individual actors interacting with each other in a situation which has at least a physical or environmental aspect, actors who are motivated in terms of a tendency to the "optimization of gratification" and whose relation to their situations, including each other, is defined and mediated in terms of a system of culturally structured and shared symbols. Thus conceived, a social system is only one of three aspects of the structuring of a completely concrete system of social action. The other two are the personality systems of the individual actors and the cultural system which is built into their actions. Each of the three must be considered to be an independent focus of the organization of the elements of action system in the sense that no one of them is theoretically reducible to terms of one or a combination of the other two. Each is indispensable to the other two in the sense that without personalities and culture there would be no social system and so on around the roster of logical possibilities [1].

It is obvious that aspects of a social system and of the actors in it are perceived differently from different vantage points. Thus we have Jones retiring from the presidency of a large corporation. His successor may perceive this retirement very differently from Jones. And the junior executive employed as a consequence of the general promotions that follow Jones' retirement will have yet a different view. Stockholders may in turn think not in terms of personalities at all but in terms of the infusion of "new blood" and hoped-for higher dividends. From the vantage of the social observer the act may be viewed as merely a means by which a social organization recognizing the increasing probability of death seeks transfer of power that will be orderly and minimally disruptive.

Our view will at times be from the vantage of the actor, at times from that of the total society. Neither view is more accurate or more important but it is incumbent upon any investigator to make clear his vantage. This we shall try to keep in mind.

The interactions of individuals usually take place within a structured situation, i.e. the individuals have an understood position, or status, and they follow somewhat scripted behavior patterns, or roles. Throughout life the ascription of statuses and roles is fairly well marked except for old age. The positions of son or daughter, brother or sister, student and teenager are well established, as are the expected behavioral patterns. With adulthood come equally clearly marked, yet more differentiated, statuses and roles. However, with old age comes a number of changes.

Many observers have noted the marked reduction in roles that accompanies old age. This disengagement has been described as a function of societal pressure and individual preference. To the extent that it is the latter there are no personal problems involved save perhaps "regrets" that it is necessary. It has also been noted that a drop in status frequently accompanies the loss of roles.

However, a reduction in the number of roles played or a decline in status in themselves may not be as personally upsetting as is the necessity to *change* roles. The aged, and particularly aged men, find it necessary not only to slough off some former roles but also to add some new ones. It is perhaps the latter that is the source of much personal distress.

One of the social functions of any status or role is that it gives an individual a measure of security. He knows that if he follows the script appropriate to the position, he will be playing the part expected; and he can reasonably expect an appropriate reciprocal response. After once learning the role, he has neither to think about how to behave nor to wonder about the response.

Throughout life there are changes in role and status. However, there are major differences between those which occur in old age and those in other stages of the life cycle.

The changes that an individual faces in early life occur over a protracted period. The changes from infancy to childhood, to adolescence, to young adulthood, to the middle years are gradual ones. The retirement from gainful employmsnt, with its corresponding social definition of old age, is abrupt. Partly because of this there is considerable preparation for each of the primary changes. Few in our society become teenagers without a pretty thorough schooling in proper behavior. This is less true in the case of retirement.

Two added differences are present. One is the social support given individuals in transition. Our society has set up nursery schools and kindergartens to ease the transition to the school years. We have set up intermediate and junior high schools to ease the transition to adolescence; and in addition we have augmented both with a series of clubs and organizations serving a similar function. Part-time work experiences, practicums, apprenticeships and intensive counseling are devices that we use to cushion the transition to work; and in-service training serves the same function in preparing one for added responsibilities. The transition to retirement and old age, however, lacks the same cushioning supports.

Secondly, each of the trends and transitions preceding old age has built into it social pressures and social rewards. While school places greater demands upon the individual than those of infancy, the pleasures also are greater. If the demands of work are somewhat greater than of school, so also are its freedom and the monetary rewards. The social pressures to retire are present but the social rewards are less apparent.

However, these aspects of role change seem to me to be less significant then the discontinuity of the role change in old age. Each of the other changes is linked in that there is a progressive increase in responsibility, in individual freedom, and in status. Even more important, at least for the male, each change tends toward increasing role specificity, increasing role specialization, increasing focus upon the instrumental role of occupation.

Dr. Cumming points out:

> Socialization insures that everyone learns to play the two basic kinds of roles that are known as instrumental and socio-emotional. The instrumental roles in any given social system are those primarily concerned with active adaptation to the world outside the system during the pursuit of system goals. Socio-emotional roles are concerned with the inner integration of the social system and the maintenance of the value patterns that inform its goals.
>
> Men, for reasons at once too obvious and too complex to consider here, must perform instrumental roles on behalf of their families, and this, for most men, means working at an occupation. Although men play socio-emotional roles in business and elsewhere, they tend to assign the integrative tasks to women when they are present [2].

In our society the roles of men become increasingly instrumental until retirement, at which point the roles abruptly change to integrated ones: those that are ordinarily associated with the female. This change is reflected in the better adjustment to retirement on the part of women that has long been noted and described by Dr. Cumming in these words:

> . . . lifelong training to a role that is primarily socio-emotional but nevertheless includes adaptive skills leaves her more diffusely adaptable than a man's working career leaves him, because he does not automatically need integrative skills. Integrative skills are, in a sense, the *lingua franca* wherever people interact with one another. Adaptive skills, in contrast, tend to be more functionally specific and less easily transferred. The disposition toward the intrumental role can remain after retirement, but the specific skills lose relevance. Only rarely does a woman find herself with no membership group that can use her integrative contribution [3].

Two other potential consequences of this role shift should at least be mentioned. In being forced to assume a role type previously indentified with the feminine, the male may suffer an additional blow to his ego. The unfavorable self-image frequently held by retired men may partly stem from this.

The second aspect centers about role conflict between husband and wife that may emerge from the role change. Many families develop considerable stability arising from the complementary nature of the respective roles. The new roles in many cases necessitate new adjustments.

Thus, with little institutional support and slight prior preparation, the aged male is forced to assume a role that is in distinct contrast to those toward which all his prior life has been pointing. It is not surprising that for many, old age and retirement are a crisis.

Dr. Ruth Benedict, in speaking of discontinuity in other culturally conditioned roles, notes:

> It is not surprising that in such a society many individuals fear to use behavior which has up to that time been under a ban and trust instead, though at great psychic cost, to attitudes that have been exercised with approval during their formative years. Insofar as we invoke a physiological scheme to account for these neurotic adjustments, we are led to overlook the possibility of developing social institutions which would lessen the social cost we now pay [4].

The impact of these role changes is compounded by another change that accompanies old age and retirement: a change in the nature of group living.

Over the past century the social functions of the group have been spelled out in detail by sociologists. Bennett and Tumin note that "the existence of groups makes it easier for society and its members to solve the crucial problems associated with man as a biologic and social animal" (5).

The group gives support, defines behavior and bolsters morale when one is "low" and keeps one in bounds when "high."

With old age comes a very significant decrease in the number of groups of which one is a member. The aged are often unbelievably socially isolated, frequently no longer being an active member of work, club, neighborhood or church. Often the family is the sole group to which the individual belongs, and even this is diminished in size in most cases and in many others nonexistent.

A number of things follow from this social isolation. We know that group membership and interaction lead to feelings of kinship and liking. Dr. Homans puts it this way:

> Stated more precisely, the hypothesis is that a decrease in the frequency of interaction between the members of a group and outsiders, accompanied by an increase in the strength of their negative sentiments toward outsiders, will increase the frequency of interaction and the strength of positive sentiments among the members of the group, and vice versa [6].

With social isolation may come not only indifference to people on the outside but a distrust or even hostility toward these "outsiders" who become increasingly more numerous as one's group contacts diminish.

As has been noted, the group exercises a normative influence on the individual. A norm, as Dr. Homans points out, is ". . . an idea in the minds of the members of a group specifying what the members or other men should do, ought to do, are expected to do under given circumstances" (7).

With decreasing group influence comes decreasing normative control. Idiosyncratic behavior is thus made more possible, and even finds some support in society, which does tolerate such behavior on the part of children, the mentally deranged and elderly people.

This deviation from norms, however, has other consequences relevant to a discussion here. It further adds to the social isolation of the aged and further increases the separation of the generations. Ours has been characterized as a society where the generations are probably more separate than most. However, there is one big difference between the separate social world of the aged and that of the others. There are definite institutional links between the worlds of each of the other groups, but less so with the world of the aged. Young children and parents are functionally integrated, as are young adults and the middle-aged. The elderly are separated ideologically and functionally.

There is yet another consequence in the change in group living that has implications for mental health and social participation. Group living provides a means of collectively sharing knowledge and experience, of gaining insights into the problems of others and of oneself. The absence of group living further impedes the older person in his personal efforts to orient himself to a new position and its demands.

Prevailing Values Becoming Alienated

Underlying social roles and social group life are a set of values. While there is no precise way to measure values, or even to identify them, they can be inferred from our formal statements, our behavior and our rewards. There is some agreement that any listing of American values would include these (8):

1. *Achievement and success.* "American culture is marked by a central stress upon personal achievement, especially secular, occupational achievement." This value is manifest from the stories of Horatio Alger to the career-centeredness of our educational and social institutions.
2. *Activity and work.* From the time of the first colonists, who announced, "If you don't work, you don't eat," to the present, work has been judged to be good and desirable. In all societies one has had to work to achieve ends; but usually it is the end product that is "good" rather than the process of work itself. Throughout our history we have had no leisure class, and idleness leads, at least in our mythology, only to mischief.
3. *Efficiency and practicality.* The ingenuity of the frontier man, the assembly line, the machine and automation all are a part of the American penchant for the "efficient" and the "practical."
4. *Progress.* Deeply ingrained in American life is a belief in both the inevitability and the desirability of "progress." Life gets better; the new is better than the old. We move onward and upward.
5. *External conformity.* Writing in the early years of the American republic, Tocqueville warned of the "tyranny of the majority," of the tendency in America to make all conform. Recent studies of the "organization man" and suburbia reveal that the pressures to conform are still strong and may even have become intensified.
6. *Science and rationality.* No society has been as wedded to science, or at least to applied science and engineering, as our own.

Every one of these values is antithetical to old age. The roles assigned to the aged do not mesh with these values. Nor does the group life of the aged give even vicarious involvement with them. Thus to his social disorientation stemming from a discontinuity in a role and his social isolation arising from the change in his group living comes also to the older person a social alienation. He becomes a stranger in a society that is both his and of his making. Yet he no longer belongs nor can he participate. In my view it is this social alienation of the aged that is our greatest problem in gerontology. In its most extreme forms it leads to suicide, or its counterpart of social withdrawal. In other forms it is perceived in hostility, aggression, and depression.

The personal anguish that comes with social alienation has been well documented by psychiatrists and artists. The social consequences are less dramatic but nonetheless real. The picture painted here has been unrelieved. This is deliberate for the contention is that the formal institutional structure of our society offers little psychological support to the aged. This does not mean, however, that all elderly persons are overwhelmed. Obviously this is not the case. The informal structures of our society combined with the strength of character and personality built over several decades of living often more than compensate for lack of an ideology favorable to aging. However, this does not obviate the need for the development of social structures congenial to the aged. The promotion of the common welfare is an explicit goal of American society. Our work will not be complete until it includes a meaningful place for all ages and a value system oriented not only to work but also to leisure and retirement.

References

1. Parsons, T. 1951. The Social System. Free Press, Glencoe, Ill., pp. 5–6.
2. Cumming, M. E. 1964. New thoughts on the theory of disengagement. *In* New Thoughts on Old Age. Robert Kastenbaum, ed. Springer Publishing Co., New York. p. 11.
3. Ibid., pp. 13–14.
4. Benedict, R. 1939. Continuities and discontinuities in cultural conditioning. Psychiatry. May, p. 167.
5. Bennett, J. W., and M. M. Tumin, 1948. Social Life. Alfred A. Knopf, New York, p. 157.
6. Homans, G. C. 1950. The Human Group. Harcourt, Brace, New York, p. 113.
7. Ibid., p. 123.
8. Williams, R. M., Jr. 1951. American Society: A Sociological Interpretation. Alfred A. Knopf, New York. p. 390.

CHAPTER 44

Mexican-American Aged in San Francisco

MARGARET CLARK and MONIQUE MENDELSON

Cultural Factors in Aging Patterns

Through the work of people in the field of culture and personality, it has become increasingly well documented that, while the need for ego strength and self-esteem is a universal human requirement, constitution of the self is a symbolic process, largely if not wholly dependent on culture for its content. Self-esteem is based on the performance of behaviors or the embodiment of characteristics that have been learned by the individual through the medium of cultural values to be "good" and "desirable." Great pride may be felt if the individual can attach to himself the value-laden symbolism of certain roles which his society prizes. He may take even greater comfort in manifesting the characterological values of his culture, as modified for his age-sex role. He feels worthy when he can point to his self-reliance, his economic independence, his Yankee cleverness in business affairs, his efficiency in scheduling his time, his wisdom in family planning, and his youthful appearance.

Any anthropologist would have no difficulty in citing cases of very different societies in which any one of these characteristics would be regarded as a weakness rather than a virtue, with the possible exception of performance of certain kinship roles, such as that of parent and grandparent. In other words, the fabric of self-esteem is inevitably supplied by the cultural matrix in which the individual learns to define himself. In some cultures there is a much greater continuity of self-referred values through the life span than is possible in other cultures. Earlier work done with American *S*s in San Francisco (Clark, 1967) has demonstrated the fact that the accession to the status of old age in this society represents a dramatic cultural discontinuity, in that some of the most basic value orientations we hold — those relating to time perspective, competitiveness and cooperation, aggression and passivity, doing and being — must be changed at that stage of the life cycle if adaptation is to occur. This conclusion emerged out of a study of the values underlying self-esteem in two samples of aged people — one a group of community *S*s and the other a group of geriatric psychiatric first admission patients. We found that the hospitalized *S*s based their self-evaluation on the embodiment of such values as achievement and success, aggressiveness, acquisition (rather than possession) of money and goods, activity and work, individualism, control, progress, and orientation toward the future. Community aged, on the other hand, more often based their self-esteem, not on these values, which are modal for young Americans, but on a contrasting profile, including congeniality, conservation, resilience, harmoniousness, cooperation, continuity, and an orientation on the present.

We further postulated that the requirement in American culture that the aged make such dramatic shifts in basic philosophy — the hard cultural imperatives to develop entirely new orientations in later life — posed critical stresses on the aging individual, and left him more vulnerable to emotional disorder.

Even among aged Americans in the community, with no history of psychiatric disturbance, we found that there were some serious problems in adaptation arising from difficulties in changing one's basis of self-esteem. And in this task, the aged individuals we studied are hindered rather than helped by the cultural emphases of contemporary American culture (Clark and Anderson, 1967).

A natural extension of this inquiry, it seemed to us, was to examine values, self-esteem, and aging patterns in some very different cultural or subcultural settings, particularly in societies where the values our aged are pressed to adopt in later life are already the core values of the culture. Would we find comparable discontinuities and stresses impinging upon the elderly in groups where continuity, conservation, and an orientation toward the present are modal values for even the young and middle-aged?

We therefore have begun a study of three contrasting subcultures in the San Francisco area — one a sample of Mexican-American aged, the second Japanese-Americans, and the third a group of Euro-American elderly. These investigations comprise the Subcultural Studies Project of the Adult Development Research Program at Langley Porter Neuropsychiatric Institute. Although the studies are still under way, I would like to present some case material from our early interviewing in the Mexican-American group that may illustrate some ethnic variations in aging patterns.

In California and the other states of the American Southwest — Colorado, New Mexico, Texas, and Arizona — the largest minority ethnic group is the Latin-American, which outnumbers even the Negro. (In fact, Los Angeles now has the second-largest concentration of people of Mexican descent in the world — only Mexico City itself contains more of them.) According to the 1960 census there were over 51,000 persons of Spanish surname residing in San Francisco. This represented an increase of 20,000 over the number reported ten years previously. It is estimated that between 30 and 40% of the Spanish-speaking colony in San Francisco are of Mexican-American descent; the remainder came mainly from El Salvador, Nicaragua, Puerto Rico, Guatemala, and other parts of Central and South America. Persons of Latin-American origin are centered in the Mission District of San Francisco, where they tend to live in ethnic neighborhoods. People with Spanish surnames form from 10 to 40% of the populations of the census tracts within this area. In spite of the fact that the Mission District of San Francisco contains persons of diverse Latin-American origin, the study described here is limited to elderly Mexican-Americans and their families.

Most elderly persons of Mexican descent who now live in San Francisco came from villages in Mexico. A study by Gamio, published in 1930 (and covering that period of time when most Mexican-Americans now aged sixty and over entered the United States), indicated that most of the Mexican immigrants to the United States during the early part of this century came from the central and northern plateau, with more than half coming from the three states of Michoacán, Guanajuato, and Jalisco. A 1950 check by Saunders (1954), which used records of the Immigration and Naturalization Service, confirmed this finding.

It must be remembered that the bulk of elderly Mexican-Americans in California came to this country from a rural or semirural environment, often a partially Indian background. The immigration pattern of Mexican-Americans has tended to create a dominant group whose older members came to this country before the social revolution which has been going on in Mexico for the last thirty years (Grebler, 1966). These immigrants came from a limited area of Mexico, were from an economy which was nonindustrialized and nonmechanized, and from a region where the feudal characteristics of society and church were (and to some extent still are) the most persistent in all of Mexico.

> Although many were originally small agriculturalists, artisans, and small tradespeople (rather than peons on haciendas), they mainly entered this country illegally and began their life here as railway or migrant agricultural laborers. Few were of the upper class, and at that time there was virtually no true Mexican middle-class [Beals, 1951].

Because of this pattern of immigration, older persons of Mexican descent now living in San Francisco have been exposed to the combined processes of acculturation and urbanization. Both of these factors have had a significant impact on the cultural heritage of Mexican-Americans in San Francisco, and understanding of the cultural dynamics of aging in this group will be enhanced by a later study of an urban Mexican population geographically situated in an area from which a large segment of the San Francisco Mexican-American population has migrated. This is a part of our long-range goal.

A Case Description of Mexican-American Aged

We are now investigating the specific social and cultural traits that may be germane to aging adjustment among Mexican-Americans and exploring some of the hypotheses and research questions that have been formulated for the study of this group. Some of these may be illustrated by the case of one of our elderly *S*s from the Mexican-American community in the Mission District of San Francisco, a seventy-one-year-old grandmother named Beatriz Chávez.

Senora Chávez was born in the state of Jalisco, Mexico, the only girl in a family of eight children. Her father was a rancher, and although he was never rich, the children never went hungry. She remembers her mother as a kind and generous person who taught her children that they must share whatever they had with those who lacked, and, according to Senora Chávez, she set a good example of this. Her mother would say, "I cannot sit down to eat knowing that others nearby are hungry," and would send the children with parcels of food to old, sick, or needy neighbors. Senora Chávez has certainly followed her mother's teaching, for she is generous to what her children consider a fault.

She was married in Mexico and came to this country around 1920 with her husband and their one-year-old son. Her husband came under contract to the Southern Pacific Railway Company, and they first lived in Oregon. She does not remember this very vividly, other than that it rained a great deal and the streets would become flooded. However, to her (in her retelling, at any rate) this was not a hardship, but added charm to the area, for, she says, "There would be water everywhere, things floating around, and people had to use little boats to get from place to place — so pretty, just like a little Venice." After they had lived in Oregon for about six or seven months, they moved to Southern California, and later after her husband had had an accident and needed hospitalization, to San Francisco. This was about twenty years ago, and by this time she had had eleven children, nine of whom were still alive. Shortly after the family's move to San Francisco her husband left her, and, although he infrequently came to visit the children, she has not exchanged a word with him. Although understandably bitter over this, she never presents herself as a martyr. This year she will have been married fifty years, but when asked if she intended celebrating her golden wedding anniversary, she replies, "Why? I have no husband; he might as well be dead so why should I celebrate?"

During the years that her children were growing up and she was without the support of her husband, times were very hard. She worked in various places, including a nearby sugar factory, canneries, and an almond packing company. She enjoyed, and still marvels at, the complicated mechanism of the sugar factory, but said that she was laid off after nine years, thereby being deprived of employee benefits — one must work for ten years to be eligible.

Today Senora Chávez, who is seventy-one, owns her own home in the Mission District of San Francisco. Were one to judge by her looks, seventy-one seems plausible. She has grey hair, cut short, sparkling brown eyes, a beautifully wrinkled forehead, and five front teeth, all lowers, three of which are gold. But when one considers her energy, her age is difficult to believe.

The house she owns consists of one lower and one upper apartment and a small bedroom and bathroom in the basement. This basement room is used, rent free, by a middle-aged Mexican man, Ramón. Ramón was a cabinetmaker in Mexico, does odd jobs in this country, and, in return for his room, does any carpentry, plus numerous errands, for Senora Chávez. At present he is completely renovating her kitchen and doing an excellent job.

The lower apartment is occupied by her 35-year-old daughter, Angie, Angie's husband, and their two children, one five years old and the other eleven months. Upstairs lives Senora Chávez, one daughter who is separated from her husband, two sons, one recently married and divorced (both events took place within two weeks) and the other divorced. The second is the father of two children, eleven and eight, both of whom also live with Senora Chávez. Recently, for a period of about six months, a niece of hers also lived there.

Senora Chávez' two sons and daughter work, and so she assumes responsibility for all the housekeeping and cooking, as well as care of the two grandchildren. In addition, she frequently looks after Angie's children, particularly the baby, and the downstairs family members frequently take their meals upstairs. It is not uncommon for one of her other children in San Francisco to phone and ask if they may come for dinner, usually with their spouse and children, since they are tired, have not shopped, or do not feel like cooking. This evidence of her continued importance in the family pleases her, and she seems to be in her element with a house full of family and friends and she busily cooking in the midst of it all.

In addition to her rather full schedule of household activities, Senora

Cha'vez carries on an active trade (that some might refer to as smuggling) between this country and Guadalajara. She lives less than a block from a large Salvation Army thrift shop, which she visits at least four times a week, and where she purchases large quantities of used clothing. About three times a year she makes bundles and boxes of these, plus other household and luxury items (such as radios, TV sets, and food blenders) she procures from somewhat suspicious sources, and travels to Mexico, where she sells everything, paying for her trip from the profits. Her children laughingly tease her about this activity, saying that not only does she look like a gypsy when she travels, but the immigration authorities are going to catch up with her one of these days. But that is no problem. She has a good friend who works at the border and with whom she has established a friendship over the years. She gives him gifts and he allows her to go through Mexican customs untouched. In fact, to ensure an unmolested passage, she writes to this man before leaving, telling him what day she intends crossing the border, and he either gives her the O.K., or tells her to wait awhile since he cannot be on duty that particular day.

This brings up the subject of Senora Cha'vez' interaction with people. One does not have a fleeting encounter with Senora Cha'vez; Senora Cha'vez involves herself with people, is genuinely interested in them, and thoroughly enjoys establishing ties. She is by no means unaware of the benefit she can reap from these friendships, but this is far from her first motivation. For example, on her frequent trips to the Salvation Army store she is waited on by the same person — according to her, a very nice old Anglo lady who works so hard. If the weather is particularly cold, she often takes a thermos of hot coffee to this lady, simply because it is a nice thing to do. Or if she has prepared some special dish, she takes lunch to the lady. Now they are close friends, and in return for Senora Cha'vez' kindness, the lady gives her significant reductions on what she buys at the store.

Also, there is an old Chinese lady who runs a shop where Senora Cha'vez trades. Through a stroke this lady has lost practically all use of one arm and hand, but with enthusiastic encouragement from Senora Cha'vez, she attempts to move the limb and exercise the fingers. It is beautiful to watch these two, one speaking only Spanish and the other only Chinese, engage in meaningful and affectionate interaction.

Second only to Senora Cha'vez' delight in her trading trips to Mexico is her penchant for gambling. Whenever she can accumulate $30 to $40 she takes off for Reno or Lake Tahoe for an evening of slot-machine playing. For birthdays and Christmas her children give her gifts of money, and this she saves and uses for her gambling. Her children all seem to enjoy gambling also, and so at least one of them usually accompanies her. A typical trip consists of leaving San Francisco by bus around seven P.M. and arriving at Lake Tahoe around midnight. Without wasting a minute,

Senora Chávez heads for the nickel slot machines, and these she plays (usually three at a time) until prevailed upon to have something to eat and to sit for a while. But this interruption (unnecessary, in her view) does not detain her long. She keeps this up until the departure of the bus at nine A.M., and, napping a little on the bus, carries on a usual day's activities at her home upon her arrival there around two P.M. Whoever has accompanied her collapses, but not Senora Chávez, who says she does not understand what is wrong with the young people of today — they seem to be so tired all the time.

There is no doubt that Senora Chávez' main satisfactions derive from her family, and even in her extrafamilial activities, i.e., trading and gambling, her companions are family members. Six of her seven brothers are in this country, and four of them frequently visit her — in fact, one brother who lives nearby drops in every weekday merely to visit for awhile, smoke a cigarette or two, and have a cup of coffee. It is one of her brothers who usually accompanies her to and from Mexico. But if her satisfactions spring from her family, her dissatisfactions also come from this source. She is accepting, although somewhat resentful, of the fact that even at the age of seventy-one, she is still involved with care of the two grandchildren who live with her. As she says, "I raised nine of my own, and even now, at this age, I cannot rest — but what can I do? The poor children have no mother and Alfredo [her son] could never control them."

Senora Chávez' excellent health probably has a great deal to do with her vitality. She complains of nothing, and, other than having an occasional cold, is never ill. This she attributes to the fact that she has always eaten well and led an active life. Her prescription for a happy life is to work hard when necessary, enjoy people, be generous, and live every day to its fullest. "You are going to die; there is nothing you can do about it, and there is nothing material you can take with you when you go."

Two precautions should be taken in interpreting the case of Senora Chávez. First of all, this family is relatively unacculturated. They manifest traditional Mexican patterns to a somewhat greater extent than many more Anglicized families. Second, the Chávez family are working-class people; whether or not the elderly in middle-class or professional Mexican-American families retain such important family roles remains to be explored in our research.

In the earlier Langley Porter Studies of Aging that I have referred to, we found quite a few old people in an Anglo-American sample who were as involved, hardy, and psychologically indestructible as Senora Chávez. There was, however, one major difference between the two samples — the "tough old birds" in the Anglo-American group represented a minority pattern, while Senora Chávez, from our preliminary observations, seems to be quite modal for the healthy aged in her ethnic community.

Lest, however, I give the impression that all anthropologists are romantic archaeists still in search of a Rousseauian paradise, let me add these speculative comments. There seems to be a perversity in human affairs, in Western cultures at least, which decrees that one group of people in a society shall be guarded, rewarded, and enriched only at the expense of another group. In the case of the Chávez family, Senora Chávez' pride, power, and freedom as a matriarch are purchased at some psychological cost to her children. They often continue, according to Anglo-American perceptions, to be forty- and fifty-year-old dependent children, and they may suffer the conflicts and ambivalences that we usually associate with adolescence in our own society. Madsen (1964) has described one form that such conflicts may take. It is possibly they, rather than the aged, who are most emotionally vulnerable — who have doubts concerning their own worth and competence to influence their own destinies. When we have completed our study of these adult children, I hope to have more to say along these lines. However, this idea is suggested by a study by Murphy (1959). This noted psychiatric epidemiologist, in a study of rates of schizophrenia among four ethnic groups in Singapore, found that age-sex rates of diagnosed mental illness varied enormously, depending upon the cultural background of the patients. Among the Chinese group, age rates for both sexes rose from adolescence to young adulthood, but declined sharply during the last decades of life. Malaysian rates for both sexes were highest in adolescence and followed a generally declining curve to reach a minimum after the age of sixty. Conversely, for British and American *S*s, the shape of this curve was absolutely reversed — there were fairly low rates in the teens, rising sharply in the twenties, leveling off somewhat in the thirties and forties, then resuming an even sharper upward slope to reach maximum rates after the age of sixty. Rates by both age and sex showed even more suggestive patterns. Among East Indian *S*s, rates for both sexes were quite high in adolescence. The rates for both sexes dropped somewhat in young adulthood (up to about age thirty). After that, however, the curves are reversed in slope, with male rates dropping sharply through subsequent decades and the female rates rising steadily until after the age of sixty. In old age, the curves for both sexes decline. It may be that the patrilocal clan system with separation of women from their own kindred represents a stress for Indian women that is not paralleled among males. The drop in female rates during the child-bearing years may reflect the strong positive sanctions of maternal status. The rise in late middle age among females may perhaps indicate something about the low status of widows and the fear of widowhood in traditional Hindu society.

All of these data suggest that human societies, as I have said, are somewhat perverse in that the sense of well-being in one social group seems so often dependent upon the helplessness and defamation of another. In our own society, the aged (and perhaps the children) pay for the relative competence and well-being of the mature adults. Negroes pay for the social and economic ascendency of whites. And, if we can believe those strange bedfellows, Phillip Wylie and Ashley Montague, men may, in some covert way, have paid for the "emancipation" of American women.

It seems to me that the major task in mental health today, and one upon which the survival of man as a species may rest, is to seek a way of helping people — *all* people — to establish and maintain a sense of self, meaning, and worth, without recourse to arbitrary hierarchical arrangements in which one may assure his own competence and value by the devaluation of others. It seems an impossible task, but our lives and the future of our species may depend upon it.

References

Beals, R. L. Culture patterns of Mexican-American life. In *Proceedings of the Fifth Annual Conference on the Education of Spanish-Speaking People.* Los Angeles, January 1951.

Clark, M. The anthropology of aging; a new area for studies of culture and personality. *Gerontologist,* 1967, 7, 55–64.

Clark, M., & Anderson, B. G. *Culture and aging: an anthropological study of older Americans.* Springfield, Ill.: Charles C. Thomas, 1967.

Gamio, M. *Mexican immigration to the United States.* Chicago, University of Chicago Press, 1930.

Grebler, L. *Mexican immigration to the United States: the record and its implications.* Mexican-American Study Project, University of California, Los Angeles, Advance Report 2, January 1966 (mimeographed).

Madsen, W. The alcoholic Agringado. *American Anthropology,* 1964. **66**, 355–61.

Murphy, H. B. M. Culture and mental disorder in Singapore. In M. Opler, ed., *Culture and mental health.* New York: Macmillan, 1959.

Saunders, L. *Cultural difference and medical care: the case of the Spanish-speaking people of the Southwest.* New York: Russell Sage Foundation, 1954.

CHAPTER 45

Leisure and Disengagement in Retirement

ROBERT W. KLEEMEIER

> To transform the lead of free time into the gold of leisure, one must first be free of the clock. And that is just the start.
>
> Sebastian de Grazia (1962, p. 328)

Leisure is a complex idea. It starts with the simple concept of freedom to do as one pleases, and ends, before one is completely aware of the subtle transition, with a serious consideration of a basic philosophical position. Society views leisure with suspicion and rejects it summarily as a way of life. Western culture is better geared to and better understands work. The almost total exclusion of leisure from the higher levels of our value system was epitomized by the late Anton Carlson, who declared on his seventy-fifth birthday that his lifelong guide had been the three W's and the three D's — "Work, work, work from diaper days 'til death."

Carlson offered this dogma as the proper retirement formula. As such it is undeniably in tune with the thinking of our society and congruent with its tradition; but it begs the question. It says in effect that man was made to work, and that the ideal prescription for retirement living is not to retire. But today millions *must* retire, and the vast majority of these will remain outside of the work force without the possibility of further gainful employment during the remainder of their lives. To those who, like Carlson, believe in the three W's and the three D's this prospect of retirement is very bleak indeed, unless they can find an acceptable substitute for work.

It may be argued, however, that retirement spent pursuing an activity which serves primarily as a substitute for work is the denial of retirement or of leisure. For many this denial would be justified, because they are hostile to leisure and to what it may do to the character, the moral fiber, and the physical being of man.

But what is leisure that it should deserve such denial? The dictionary reveals the generally accepted meaning of the term, but leaves its implications unexplored. In this formal sense leisure has three characteristics: (1) freedom or lack of constraint; (2) unoccupied time; and (3) the doing of something. While this third meaning is technically obsolete, it is nevertheless implicit in the modern usage of the term. Indeed the relationship between leisure and activity is an interesting one.

Leisure: Activity vs. Inactivity

It should be clear that leisure and inactivity are not synonymous terms, yet some of the negative attitudes toward leisure can undoubtedly be traced to the feeling that leisure is associated with indolence, laziness, and do-nothingness. Yet, because leisure is characterized by freedom and lack of restraint, the possibility of doing nothing as a full-time leisure activity must be seriously considered. Indeed it has been.

Disengagement. Within the past decade the term disengagement has become a familiar one in gerontology. It stems from a series of studies based upon long interviews with a panel of older men and women drawn from the population of Kansas City. Its most precise statement is found in a book by Elaine Cumming and William Henry (1961). These authors, noting the reduction in the number and quality of social interrelations in successive age groups of older persons, contend that with increasing age beyond maturity there is an increasing tendency for the individual to withdraw from social relationships, to conserve his energies, and to reduce his life space.

This they contend is both a social and a personal process. It is a social process in that society, its institutions and organizations, tends to disengage or to exclude the older person in order to make room for vigorous, younger people and thus to avoid its own obsolescence. On the other hand, the individual, feeling the strain and difficulty in keeping up with the pace of society, tends himself to withdraw. Thus, the disengagement process is a mutual one, and voluntary retirement from work is the ideal example of it.

But society and the person do not always agree about the time that disengagement should take place. If society decides to disengage the person before he is ready, it does so. The compulsory retirement of a worker who wishes to remain on the job is a common example of this. The aging person, on the other hand, may himself wish to disengage before society is ready to let him initiate the process. In both of these latter cases the individual is put in stress, which, under ideal circumstances, Cumming and Henry believe might better be avoided.

While this theoretical position has obvious implications for the issue of compulsory retirement, it is more intimately related to the question of leisure and activity in retirement. From the disengagement theory it is easily deduced that the foisting of activities upon the older, retired person, who is in the process of disengagement, may not be wholly desirable. If disengagement is an inherent characteristic of aging, why not let it proceed at its own rate, rather than to devise activities designed to contravene this tendency.

"Activity" theory. This issue has been rather sharply drawn, and an "activity" theory has been developed, somewhat informally, to serve as a counterbalance to the disengagement position. As you would expect, this theory holds simply that the maintenance of high activity levels is necessary in order to inhibit deteriorative age trends in the behavioral potential of the individual and to increase his satisfaction with life.

These two positions or theories have marked significance for society. If we were to pursue the obvious injunction of the disengagement position, the social prescription for retirement would be, I am sure, much different from what most social planners now advocate. Following this line of thought, society would question seriously the intent to involve the older person in any activities to a greater extent than his present state of disengagement would seem to indicate. Certainly we would hesitate to re-engage the individual who gave clear evidence of low activity and interest level, for to do so might cause him stress, might disturb him, and thereby lessen his personal feelings of happiness. Activity theorists, on the other hand, would be dedicated to the idea that increases in activity levels are of necessity beneficial and tend to increase the life satisfaction of the person.

The inherent danger of the disengagement theory is, of course, its encouragement of the idea that no action, no intervention in the lives of older persons to increase the richness of their experiences, is the desirable course to take. But in many ways society has already intervened. It has withdrawn support and it has withdrawn personal interaction, and that with which it so often provides the person is bland and sterile when compared to its provisions for other age groups.

The vital question is whether this blandness and sterility of society's offerings is not more of the cause of apparent disengagement than is the hypothetical inherent disengagement process within the individual. The answer to this question is not immediately apparent, and certainly research must be done upon it; however, there is ample evidence to indicate that the disengagement hypothesis, as originally proposed, does not adequately handle the apparent facts. Indeed, William Henry in a paper presented at the recent International Gerontological Research Seminar in Markaryd, Sweden, stated that neither the activity theory nor the disengagement theory ". . . is sufficient to account for the facts we now know. And neither deals explicitly with the question of the effects upon the persons of specific plans for increasing or decreasing their activity, or for plans for providing one kind of activity as opposed to another."

Activity Level and Behavior

This statement is clearly an invitation to explore further in order to find answers to these important questions. Surely activity levels in the adult go down with increasing age, but at what rate and for what reasons we can only surmise. In human beings, it is an extremely difficult problem to investigate; in animals, it is less so but still not easy. Animal studies do show us, however, that spontaneous activity increases rapidly during the early part of life and reaches its peak, probably sometime within the first third of the life span, and thereafter diminishes regularly until death. Physical activity curves for human beings would probably reveal similar declines, but, as in animals, persistent individual differences in activity levels would undoubtedly be found. Furthermore, the amount of activity would unquestionably be related to many factors other than age, and indeed activity level maybe determined in only small measure by the complex of variables we call age or aging.

For example, it is well known that amount of activity is related to amount of stimulation, although the relationship is not a simple one. University of Minnesota's Professor John Anderson insistently calls attention to this fact in his discussion of behavioral changes with age. He says (1959, p. 77l):

> Attention should go to the recent work of Hebb . . . and his associates and Heron . . . , who restricted the amount of stimulation received by humans and animals to minimal levels and found that behavior and activity quickly deteriorated. They believe that a certain massive amount of arousal stimulation is necessary to keep the organism functioning and that, above this stimulation level, there is the cue stimulation which enables us to perceive objects and relations and build specific responses. Thus there is a kind of psychological tonus, not too different from the concept of physiological tonus, which is maintained by arousal stimuli. When these drop below a certain level, all activities are affected.

The implication of Professor Anderson's observation is that, regardless of age, the individual must be subjected to some minimal amount of general stimulation and if this is not provided serious psychological dysfunction will occur.

It is not then simply a matter of choice between activity theory and disengagement theory. The presumptive test for the validity of these theories is whether the aging person is happier and more satisfied when he engages in many activities than when he is relatively disengaged.

Anderson's conceptualization suggests that lack of stimulation leads to a diminution of activity, which circularly results in still less stimulation and ultimately in deterioration of behavior. This deterioration can happen at any age and to anyone, but the aged are particularly vulnerable to it. Lessened sensory capacities, diminished energy, and frail physical condition predispose the aging individual to stimulus deprivation with its deleterious behavioral consequences.

We must conclude from this that maintenance of activity, regardless of kind and regardless of the attitude of the person toward it, is highly important to forestall quick deterioration. Unquestionably optimum activity levels vary between individuals and with differing physical conditions. Undoubtedly excessive or stress-producing activities may also be deleterious. How such levels that are beneficial to the individual can be selected is more a matter of art than of science at this time. Probably, however, the optimum amount of activity is somewhat above the spontaneous level adopted by the person. The answer to this question, however, remains speculative until more adequate research is directed toward this problem.

Characteristics of Leisure Activities

Although, as I have tried to indicate, leisure may be characterized by activity, not all activity is leisure. The very practical questions which concern us here are: (1) what distinguishes leisure activities from other activities, and (2) which leisure activities are most appropriate for older persons?

Max Kaplan (1961) has attempted the difficult job of distinguishing leisure from other activities by establishing a set of leisure criteria. The first of these is that leisure must not be considered work, particularly in the economic sense. This is not always as easy a distinction to make as one would think. By this criterion, however, Kaplan would exclude activities done for pay rather than for the pure pleasure of doing them. This is really a distinction in motivation and not in kind of activity. In effect, Kaplan says that a leisure activity is not defined by what one does, but by why one does it.

In addition Kaplan states that leisure activities must be characterized by pleasant expectation and recollection. We must look forward to them with pleasure and recall them in the same way. In addition we must feel that we are perfectly free in our choice of these activities, that we are not forced or constrained to engage in them. All of these characterizations of leisure activities are subjective and cannot possibly be determined or ascertained by the outside observer. Indeed, the only external stricture that Kaplan applies to the leisure concept is that for an activity to be so classified it must be congruent with cultural values; that is, illegal and delinquent activities, no matter how much pleasure they give a person, are not considered as proper leisure activities. Apart from this there are no additional external limitations, for he believes that leisure activities can involve anything from the most trivial to the most weighty in significance.

It is clear from this kind of classification that leisure is a state of mind rather than a particular kind of activity. One may plant a garden, milk a cow, or build a house for pleasure or for profit. If for profit, it is work; if for pleasure, it is leisure. It is obvious, however, that this distinction is not an easy one to maintain. In a sense, the housewife washes dishes neither for pleasure nor for profit, but most often for some reason more related to profit than to pleasure. She wants to enjoy a clean house. She needs to have utensils ready for the preparation of the next meal. Thus if she is not paid for washing the dishes, the act is instrumental in achieving some end which she wants to enjoy. Thus the dishwashing is as instrumental to the achievement of that end as is her husband's job by which he earns his weekly pay check. These instrumental acts, carried on to reach some related goal, have come to be considered work, or non-leisure; and acts which are carried on for the sheer pleasure of doing them become identified with leisure.

It is at precisely this point that the weight of our cultural tradition makes the facts of retirement a profound moral confrontation in the life of the individual.

For many centuries Western culture has believed that to work is good, and not to work is bad. Hannah Arendt (1959) has carefully traced the evolution of this cultural ideal from the times of the ancient Greeks, who *could* appreciate a leisure society, to modern Western man, to whom work is the sole *raison d'etre*, and leisure is only the absence of work.

Arendt looks bleakly at Western civilization and says that we are a society of laborers, in which all of us must have a job whether we be president, intellectual, king, or peasant. We are alarmed because modern technological achievement is gradually freeing us from the fetters of labor, and we may become ". . . a society of laborers without labor, that is, without the only activity left to [us]. Surely, nothing could be worse" (Arendt, 1959, p.5).

There is much to indicate that for older workers this time of freedom has indeed arrived, nor is there any prospect that they will return to the labor force in significant numbers. A. J. Jaffe, Director of the Manpower and Population Program of Columbia University, after careful study of the available labor supply from the present to 1980, states flatly that ". . . the United States' economy does not 'need' the labors of . . . persons . . . aged 65 and over, unless very unusual conditions should arise" (Jaffe, 1963, p. 37).

Jaffe's incisive conclusion does not allow us seriously to entertain the expectation that a return to work movement can be mounted to save us from a "life of leisure." We must look to other alternatives and these seem to me to be reasonably clear-cut. If actual continuation of labor is impossible for the great majority of workers during the later years of their life, then the retirement years are likely to be spent in activities which may be characterized in one of three ways: (1) in work substitute activities, (2) in leisure activities, or (3) in doing nothing.

Work Substitute Activities

There is no question but that work substitute activities, congruent with the physical and mental capabilities of the individual, are highly successful in creating good morale and high life satisfaction in the later years. This is the practical solution to the void filled by the loss of job. It is simply to have an activity which fills the same personal needs that were formally satisfied by paid employment.

Robert Havighurst (1961), after extensive study of the ways in which older people use their time, concluded that the satisfactions of work and of free time activity are essentially the same for most people. This is particularly so when one compares work with instrumental, "non-work" activities. Both offer opportunities for pleasure, to be creative, to be with friends, to have self-respect, to make time pass, to be of service to others, and to give the person prestige and popularity.

My own experience in a large fraternal home for the aged has also convinced me that the provision of work opportunities for institutional residents has many values and can bring significance into the lives of those who might otherwise be desolate (Kleemeier, 1951). Certainly many of these quasi-work or work substitute activities are beneficial. If they help to maintain at an optimum the activity level of the individual, they are justified; and if in addition the person enjoys them, no further justification is required. The question remains, however, whether this is truly the life of leisure about which we are concerned. In one sense it may be considered as a perpetuation of the life of work of the individual, hopefully carried on at an equal or higher level of satisfaction.

Leisure Activities

I would be in error, however, if I were to conclude that work substitute and instrumental activities in general were the only admissible ways in which time could be employed when the call of occupation is no longer heard. William Henry, in his Markaryd paper to which I have already referred, has associated instrumental activities to engagement and non-instrumental activities to disengagement. He is inclined to believe that, in spite of irregularities, there is a tendency for the aging person to shift from the instrumental to the non-instrumental type of activity. Since non-instrumental activities are engaged in solely for the purpose of doing the activity, the private, personal, and idiosyncratic nature of such activities becomes apparent.

Josef Pieper (1963), the Roman Catholic philosopher, and Hannah Arendt (1959) both concerned themselves with this issue. Arendt decried the failure of modern Western civilization to comprehend the values of the contemplative life as exemplified in ancient Greece. Pieper's thesis is similar and is specifically related to the concept of leisure. He would have no question that a life of leisure could be lived, and that it would be a good life. But this would be a contemplative life, a life not concerned basically with utilitarian or instrumental objectives. For him the salient characteristic of leisure is neither time nor activity, but freedom — complete and absolute without pressure from necessity or need. Leisure would be occupied with the *bona non utilia sed honesta,* which is best typified by philosophy or the philosophical act. This, the philosophical act, he says, is the fundamental relation to reality. It requires

> an attitude which presupposes silence, a contemplative attention to things, in which man begins to see how worthy of veneration they really are . . . [and this] . . . can only be preserved and realized within the sphere of leisure, and leisure, in its turn, is free because of its relation to worship, to the *cultus* [Pieper, 1963, p. 18].

In Pieper's sense of the word, our culture has trained us away from leisure, and, therefore, we cannot at any age know how to live the life of leisure. But, if by chance, by personality or by experience, we have learned the use of leisure in this sense, Pieper offers it as a fulfilling way of life.

Doing Nothing

If neither work substitute activities nor Pieper's contemplative leisure are available to us, then we are left with the third alternative — doing nothing; a dismal prospect, indeed. Doing nothing means just that — a vegetative existence characterized by minimal overt and covert behavior, a part of the senile cycle of inactivity, low morale, depression, deterioration. Obviously this is an end to be avoided. Certainly it cannot be considered as the ultimate adjustment to social and personal disengagement.

The choice is not between activity and inactivity, but between kinds and levels of activity. And the choice is not one which must be made only in the later years or in retirement. It is faced at all ages and the course of life itself is determined by the decisions made.

References

Anderson, J. E.: The use of time and energy. *In*: J. E. Birren (editor), *Handbook of Aging and the Individual.* University of Chicago Press, Chicago, Ill., 1959, pp. 769-96.

Arendt, Hannah: *The Human Condition.* Doubleday Anchor Books, Garden City, New York, 1959, p. 385.

Cumming, Elaine, and W. E. Henry: *Growing old.* Basic Books, New York, 1961, p. 293.

de Grazia, S.: *Of Time, Work and Leisure.* The 20th Century Fund, New York, 1962, p. 559.

Havighurst, R. J.: The nature and values of meaningful free time. *In*: R. W. Kleemeier (editor), *Aging and Leisure, a Research Perspective into the Meaningful Use of Time.* Oxford University Press, New York, 1961, pp. 309-44.

Jaffe, A. J.: Population, needs, production, and older manpower requirements. *In*: H. L. Orbach and C. Tibbitts (editors), *Aging and the Economy.* University of Michigan Press, Ann Arbor, Mich., 1963, pp. 31-39.

Kaplan, M.: Toward a theory of leisure for social gerontology. *In:* R. W. Kleemeier (editor), *Aging and Leisure, a Research Perspective into the Meaningful Use of Time.* Oxford University Press, New York, 1961, pp. 389-412.

Kleemeier, R. W.: The effect of a work program on adjustment attitudes in an aged population. *J: Gerontol.,* 6:373-79, 1951.

Pieper, J.: *Leisure, the Basis of Culture.* Pantheon Books, Inc. Mentor-Omega Books, New York, 1963, p. 127.

D. SOCIETAL STRUCTURE

CHAPTER 46

Older Families and Their Troubles: Familial and Social Responses

GORDON F. STREIB

The assertion is frequently made that the American family is declining. One piece of evidence offered is that the American family is not meeting its traditional obligations, particularly in the care of its older members. The idea that the family alone should be primarily responsible for the health and welfare of older persons is a vestigial attitude from an earlier historical period. The conception of the family as an autonomous social unit which is supposed to solve all of the basic problems of living is, in some ways, a carry-over of thinking about the family retained from the agrarian and early phases of industrialization.

Many families do have considerable autonomy, and this is a goal desired by most family units. However, when we consider the family in the latter part of the life cycle, there is a need to rethink the notion of the autonomous family — independent, self-regulated, and able to take care of itself as a group. Family autonomy and independence are very congruent with American goals and values in the spheres of the political, the economic, and the educational. But family autonomy must be more than a shibboleth, particularly when one studies families in trouble at any stage of the life cycle. Independence and autonomy may be desired by families, but in reality they are sometimes hard for some families to attain, particularly under conditions of economic or medical crisis at the end of life. It is during the latter phases of the life cycle that some notion of shared function must be brought into the analysis of the situation of older families and into the thinking of persons involved as practitioners with older families.

What is the theory of shared functions? It is the notion that formal organizations and families must coordinate their efforts if they are to achieve their goals. The idea of shared functions was developed by Eugene Litwak, who with several colleagues explicated and tested the implications of the theory in a series of stimulating papers (Litwak, 1965; Litwak and Meyer, 1966; Litwak and Figueira, 1968; Litwak and Szelenyi, 1969).

The ways in which bureaucratic structures and family groups are articulated is a crucial matter in urban-industrialized societies. Although the theoretical analysis is still incomplete and many of the applications must be worked out, the basic idea of shared functions is sound.[1]

This paper will explore some of the ways in which various kinds of primary groups and bureaucratic organizations may share functions in meeting problems or crises which older families face. Bureaucracies are social structures which have an instrumental basis for operation, emphasize impersonality, are organized on the basis of formal rules, and stress professional expertise. On the other hand, a family as a prototype of primary groups is characterized by face-to-face contacts, employs affective bases for judgments, stresses diffuse demands and expectations, and so on.

One primary assumption which needs to be emphasized is that family groups and bureaucratic organizations are not to be considered in competition with each other. The two types of structures have multiple goals and tasks which in many instances may overlap. Furthermore, it is assumed that the family has not "failed" when it utilizes formal organizations for assistance. There is a range or a continuum of groups from a "pure" primary group — the nuclear family — to the monocratic bureaucracy, and there is a variety of mixed types of groups or organizations which may be located on the continuum.

There are writers who describe the United States as a post-industrial society (Bell, 1967), and there are others who write about American society as being in the mature — not the developmental — phase, of industrialization (Litwak and Figueira, 1968). Clearly a society whose urban-industrial structure is as developed as is that of the contemporary United States is markedly different from the society emerging at the turn of the century or even the society which America's senior citizens lived in just a generation ago at the beginning of World War II. Technology has changed; some phases of private and governmental bureaucracies have changed, and primary groups — such as the family — are changing. But as Litwak and Szelenyi state: "It would be an error to say that, because a primary group structure changes from one stage of historical development to another, it is moving to destruction" (Litwak and Szelenyi, 1969, 480).

[1]The theory of coordination of groups has been developed by Litwak and Meyer (1966). Other writers have also been concerned with the subject. For example, John Mogey (1964) reviewed the literature and provided a valuable discussion. From a different persepctive Elaine Cumming (1968) discusses systems of social regulation in one city, Syracuse, New York. Rosow has criticized Litwak's early work on the articulation of functions between family and bureaucracy (Rosow, 1965).

Two Kinds of Families: Residential and Extended

One of the first steps which is necessary in order to understand the notion of shared functions is to clarify what is meant by the family. A major distinction must be made between: (1) the residential family or the family in which one lives, and (2) the extended family or the kin network to which a person belongs by blood ties or marriage.

Residential families are diverse because of their intrinsic structure and also because of two other important considerations, age and sex. Whether a person is 65 or 85 makes a great deal of difference in terms of family relations and the need for assistance. Sex is a factor because of the differential death rates; the proportion of widowhood is much greater among women than among men. About 10 percent of the men in the age category 65 to 69 are widowed, in comparison to 38 percent of the women (Riley et al., 1968). Considering all persons over 65, about a fifth of the males are widowers and about half of the women are widows. This means that a man is much more likely to have another person in his household to help in times of trouble or crisis, while a woman is more likely to need to turn to outsiders or bureaucratic organizations for assistance.

Most older people — 70 percent of men and women over 65 — live in families (living arrangements) with two or more members. Twenty-one percent of older persons over 65 live alone and only four percent live in institutions or other kinds of group quarters (Riley et al., 1968).

A more detailed examination of the kinds of persons who live in a household of one person accentuates the differentiation by sex: only 16 percent of the males compared to 32 percent of the women live in households entirely alone or with nonrelatives (Riley et al., 1968). These proportions increase for subcategories of the older persons, for living alone is a characteristic of the later rather than the earlier phases of the life cycle.

An examination of the facts about kin networks in Table 1 shows that the great majority of older persons in the United States belong to kin networks of three or four generations. A nationwide survey conducted by Shanas

TABLE 1. KIN NETWORKS OF OLDER PERSONS IN THE U.S.

Number of Generations in Family	Percentage of Older Persons
One	18
Two	6
Three	44
Four	32
Total number of families	*2,436*

Source: Shanas et al., 1968, 143.

and her associates revealed that about one older person in five was a part of a one generation kin network and almost one-third of the older persons in the United States were members of three or four generation networks.

Thus the analysis of the older family in situations of trouble and stress encompasses a tremendous variety of family structures based on the variation in residential types of families and the way they are or are not tied in with kin networks of varying complexity. There are, for example, three generation families living in one household, older couples with young children, single-person families, and newly-weds in post-retirement families, to mention a few of the variations. Thus it is important to stress: there is no "typical family" in later life, for there are many patterns.

This great variation in family structures and relations is the result of a complicated set of factors and of decisions which go back much earlier in the life cycle. To understand those families which are most vulnerable, have the fewest resources to draw upon, and probably require the most attention from health and other helping agencies, one must realize that many of these families have been limping along through most of life. In addition, there are some families which are in trouble for the first time in later life. Many families which have met the vicissitudes of child rearing, have dealt successfully with intermittent economic distress, and have been able to solve previous medical emergencies now find the dilemmas of later maturity too much of a challenge to cope with.

Family Resources and Family Problems

Attention to the major factors that create problem families in later maturity follows. There are four major resources which contribute to the strength of older families. They are:

1. Physical health
2. Emotional health
3. Economic resources
4. Social resources — family, kin, friends

TABLE 2. FAMILY RESOURCES AND PROBLEM PRONENESS IN OLDER FAMILIES

	Resources			
Family Types	Physical Health	Emotional Health	Economic Resources	Social Resources
I "Strong" — "Organized"	+	+	+	+
II	-	+	+	+
III	-	-	+	+
IV	-	-	-	+
V "Weak" — "Disorganized"	-	-	-	-

The interrelationship between these four factors is very complex, as one can see in Table 2, and the absence of one or more can bring about severe dislocation in the life of the older person. On the other hand, an abundance of any of the resources can greatly alleviate the stress caused by the absence of the others. For example, a warm, supportive family can help to ease the crises caused by failing physical and emotional helath, and assist in supplying economic resources.

An examination of five types of families in terms of these major resources follows. Type I, the Golden Sunset Family, obviously is not the concern of this paper. Fortunately, they constitute a significant number of older families in this society. These are the people in good health and spirits with a comfortable pension, warm family relations and friends.

At the other extreme are the "unfortunate families" who end their years in misery — the Totally Deprived. These families are the ones who present the most serious crisis in old age — indeed, many of them are in institutions for they lack *any* of the basic resources needed to meet their needs. When people talk about the family and a care crisis, it is this type of "family" that they have in mind, and from their knowledge or acquaintance of such tragic types, they may assume that there are more of these families than really exist. Between these two polar types there is a variety of combinations; in all there are sixteen different combinations considering the four resources.

An important and interesting question is how many families are there of the various types referred to in Table 2? At the present time only a rough estimate can be made of the numbers and proportions of the various types by fitting together items of information on single variables and isolated characteristics. A crude estimate is that about a quarter of the older population are "priviledged aged" and about a half are "typical aged" and the remaining proportion, roughly 25 percent, are the "needy aged."[2]

These estimates are based upon facts of the following kind: About 95 percent of the United States' older population lives in the community, and of those who do about nine out of ten are ambulatory and perhaps 2 percent are bedfast (Shanas et al., 1968, 24). Moreover, among the older persons living at home almost two-thirds scored zero on an incapacity scale; that is, they could perform six basic tasks without difficulty.

The economic resource variable is probably the one that has been surveyed more often and with greater precision than the others. For example, a recent U.S. Senate Committee reported that one out of every four individuals 65 and older lives in poverty (Special Committee on Aging, U.S. Senate, 1971, 3). Poverty, like other characteristics, is

[2]Bernice Neugarten made a distinction between the "needy aged" (a minority) and the "typical aged." See *Aging,* no. 127, May 1965, 6.

differentially distributed in the population, for the same Senate document reported that among the elderly women who live alone about one-half fall below the poverty line; and the non-white older females are especially disadvantaged.

Another way to index economic resources is by the ownership of property. About two-thirds of the aged own their own homes, and of these thirteen million plus older American homeowners about 80 percent own their homes free and clear. Moreover, over six and a half million (about 50 percent of the homeowners) have an equity of $25,000 or more in their homes (Special Committee on Aging, U.S. Senate, 1971, 29).

Social resources are more difficult to ascertain than financial resources. What indices of social resources should one utilize? Presence or absence of children? Siblings? Neighbors? Membership in organizations? The presence of facilities like the telephone, radio, or television which may increase one's social contacts with the wider world? For example, about a third of the aged never married or have no living children.

The psychic or emotional resources are probably the most difficult resource to estimate. Gurin and associates in a nationwide survey of America's mental health, using five questions, found that among persons 55 and over about 27 percent reported themselved as "very happy," 55 percent as "pretty happy," and 18 percent as "not too happy" (Gurin et al., 1960). The same national survey found that if one employs "worry" as an index of emotional health, the elderly report themselves as not worrying any more than younger categories of the population. About a third of the persons over 55 years of age say they "worry a lot" or "worry all the time."

How can these complicated variables be grouped together into clusters or types? In a nationwide urban survey of males 60 and over conducted some years ago, four variables were grouped together in order to determine how health, socioeconomic status, and whether a person was working or retired were related to morale (Streib, 1956). The three independent variables had a cumulative effect upon morale. For example, among those who were retired, in poor health, and had a low socioeconomic status, 71 percent had low morale. Conversely, among the men who were employed, in good health, and had a high socioeconomic status, only 25 percent had low morale. And between the extreme categories low morale varied according to whether a postivie or negative factor was present.

Some persons have argued that the way to solve the problems of families in old age is to give them more money. Yet they forget that some people can be rich in economic resources and poor in some of the other factors that make life meaningful. Some readers will recall the news accounts of the "richest girl in the world" who spent a recent Christmas sick in a luxury hotel suite in San Francisco and alone except for an entourage of servants. If one is to believe news reports, her circle of friends and relatives is small, she is in poor physical health, and she is unhappy. Yet she has tremendous economic resources at her command.

The most elusive and most difficult resource to identify and measure is the emotional or psychological health of the individual. Just as one person may have immense financial resources and lack the other three, there are some old persons who are lacking in health, financial and social resources and yet have sufficient psychological and emotional resources. These are the kinds of older people whom Dr. George Reader of the Cornell Medical School has described as performing the Indian rope trick for they seem to be able to function without any visible means of support (Reader, 1969, 312). These people who sometimes come to the attention of public and private agencies are in very poor physical health, exist in dire poverty, have no family or friends — in fact, seem to be totally lacking in resources. Yet they seem to be able to remain cheerful and can cope with life — in short, they possess an abundance of the elusive factor of emotional resource which somehow enables them to integrate their lives. Dr. Reader said it was a source of amazement to some medical students that such people could even survive. Just as the Indian *fakir* seems to defy the laws of gravity, so these deprived persons seem to defy all the rules and generalizations of social workers and sociologists about the basic resources needed to maintain an integrated life, for somehow they are able to cope.

This schematic presentation obviously leaves out the variations in resource level which must be considered in analyzing the diverse kinds of families. The easiest resource to measure in an objective sense is money; however, the way in which a given set of financial resources are perceived and utilized by older people varies considerably. Streib and Schneider summarized the situation in their longitudinal study of 2000 retirees: "Thus a given amount of income might be considered plentiful by one retiree while the same amount would represent dire poverty to another. The amount of income an individual receives may not be as important as whether he thinks it is "enough' " (Streib and Schneider, 1971, 82).

Profiles of Families in Trouble

There is another aspect of families in stress situations which requires some analysis, namely, the way in which families respond to difficulties which confront them. This complicated subject is the central focus of a book, *Families in Trouble*, by the sociologist Earl L. Koos (1946).

In 1946 Koos published an imaginative study, based on his field work with lower-class families in New York City. He lived and worked closely for two years with 62 families who resided in one tenement block north of the Cornell Medical Center. Although Koos did not specifically focus his attention on gerontological phenomena, there are several aspects of his study which are very pertinent to understanding older families. In the course of his work, Koos made a judgment of what he called the "level of organization" of families. He emphasized that he tried to utilize an internal standard used by the families themselves; that is, he did not set down criteria of what was normative for families in American society. The organization or adequacy of the family were considered in terms of the following:

> (1) There must be a consciousness of an acceptance by each member of his and of the complementary roles in the family; (2) family members must have a willingness to accept the common good of the family over their own good; (3) the family must provide satisfactions with the family unit; (4) the family must have a sense of direction and be moving in keeping with this, to however small a degree [Koos, 1946, 33].

Koos classified his families into three rough categories according to these organizational criteria with about half the families classified as "average" in their adequacy and about equal proportions of his sample above or below average.

The 109 troubles which these low income families faced comprise a familiar catalogue of difficulties; about a third were financial and about a third of the troubles resulted from illness (Koos, 1946, 63). There were some problems which resulted from pregnancies (a source of difficulty which most geronotologists do not have to be concerned about very often). Educational problems, alcohol, sexual incompatibilities, and family conflicts were other sources of difficulty.

One aspect of Koos' study which is pertinent to this analysis is the graphic means by which Koos depicted the way in which families coped with their troubles (Koos, 1946;[3] see diagram 1).

On Profile A the normal interaction and operation of the family is shown roughly by the line a — b in the diagram. The descent into difficulty is depicted by the line b — c and the recovery period is represented by the angle of recovery e. The plateau of recovery is depicted by the line d to f.

The families can be divided into four profile types according to their response to problems (Koos, 1946). Profile A reprsents those families which, after the trouble, returned to the same level of organization they had attained before the onset of the difficulty. Profile B was a kind observed in only one case, in which the family attained a higher level of family cohesiveness after the trouble than they had experienced before the

[3]The utility of Koos' profiles is shown by the research of Hill and his associates. They studied the adjustment of families which were separated by the father's absence due to war service and they developed a larger number of profiles to describe the 116 Iowa families. See Hill et al., 1949, 74–99.

trouble. Profile C is the kind observed in about 60 percent of the troubles. In this profile the families never recovered fully after the trouble. Finally about 20 percent of the families exhibited profile D, in which the families experienced what Koos called "incomplete recovery" in that a second or third trouble eroded the family's level of organization so that it suffered additional assaults from which it was not able to recoup its earlier level of integration.

Diagram 1

Family Recovery Profiles

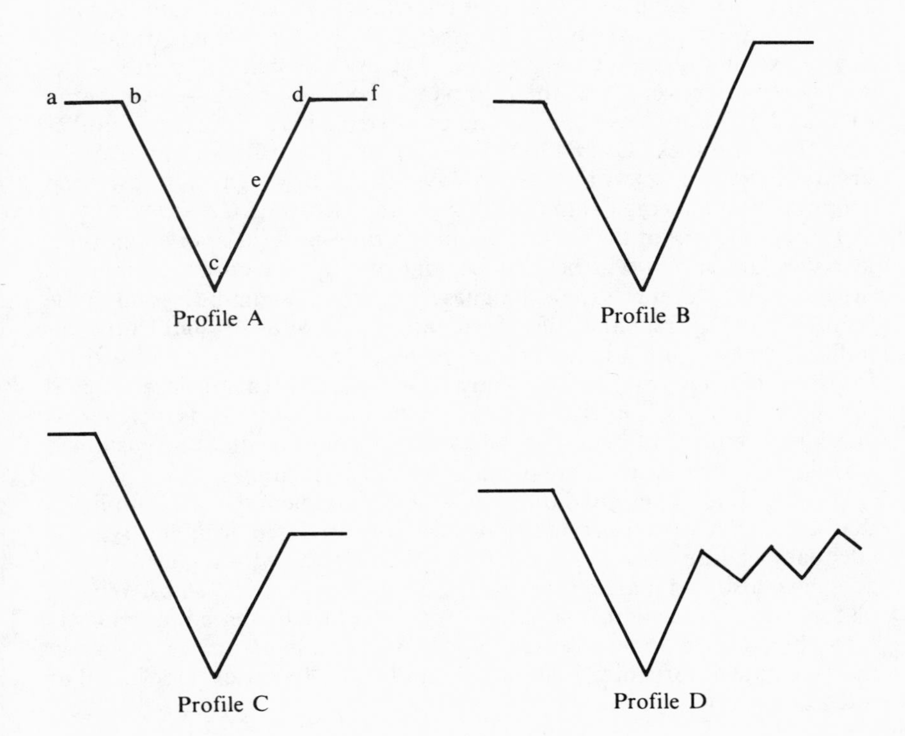

How does the work of Koos shed light upon the study of older families? First, it should be reiterated that many of the troubles of these lower-class families originated in health and financial matters. This is a familiar story to those who study or assist families in the later part of the life cycle. Secondly, it should be noted that many of the families had erratic recovery patterns. These profiles are the kind which can be observed in many older families, because the later years are often characterized by a continuing series of troubles: economic, health, and social. Important role losses result from the death or incapacity of friends and relatives. Declining health is the situation for many persons. And financial difficulties resulting from sharp cuts in income upon retirement and the persistent, glaciallike erosion caused by inflation increase economic insecurity and make life a period of trial and tribulation for many. Thirdly, the way in which families cope with difficulties in later life are like the families in Koos' study in that the recovery profile can be best understood as a process and not as an event. Families do not miraculously find ways to deal with their difficulties. They encounter troubles because of their environmental pressures and also because of the resources which they can marshal and which have been a part of the family's capital for some time.

Finally, the way in which families recover from crises is related to their previous level of organization. The previous level of family integration is a key to explaining the recovery period. The families that had profile A, in which they recovered to their previous level, were those who were "average" or "above average" in the pre-trouble period, and the families who had profiles C and D were often those whose pre-trouble organizational level was "below average," although a substantial proportion of "average" families also exhibited profile C.

The significance of the pre-trouble organizational level is very important in older families. It will be recalled that one dimension of the family's organizational level was the willingness to accept the common good of the family over the personal wishes of individuals. Therefore it would be rather unlikely that a family which has been pulling in several different directions for thirty or forty years will suddenly become integrated and accept a goal of the common good in later life. Adversity does bring friends, neighbors, and kin together, but momentary help at a point of trouble and sustained working for the common good are quite different things.

Another finding reported by Koos which is pertinent to older families is that the average number of friendships is correlated with the type of adequacy or integration of the family. The families with higher levels of adequacy also had a greater than average number of friendships. What is also of significance and is congruent with the idea of shared functions is that the "better than average" families had almost twice as many memberships in organizations as did families of "below average" level of integration.

Forms of Intervention and Older Families

To this point the analysis has centered on trying to clarify what is the nature of the family in relation to trouble and stress situations and to specify family types and the resources whose absence tends to create problems. The trajectories of the crisis process have also been described. In the following sections the discussion will focus on the forms of intervention and the way in which resources are employed in meeting troubles. The analysis of the intervention process illustrates how the theory of shared functions works out in practice. The articulation of family and other groups and organizations is very intricate and thus the following discussion will only highlight a few aspects as they pertain to older families.

The forms of intervention can be classified into three broad categories: (1) help and intervention by family and kin; (2) quasi-family or pseudo-family forms of intervention; (3) institutional or bureaucratic forms of intervention.[4] These categories are not mutually exclusive, but there is utility in considering them separately. The principal criterion employed to distinguish the three forms of intervention is the nature of the relationship between the person offering help and the recipient. Does the intervention agent have an ascribed kin relationship to the recipient? Or does he assume a pseudo-family role in the relationship? Or is the intervention agent acting as a bureaucrat?

The kinds of resources which the three forms of intervention may offer can encompass all of the four resources referred to earlier in the paper, namely, (1) health and medical, (2) emotional and psychological, (3) economic and financial, (4) social or interactional. For example, kinsmen may render nursing care; they may give financial help; and they may also be a source of social interaction and emotional support. Similarly, a government agency or several agencies can and do intervene and offer help which involves one or more and sometimes all of the four resources.

Family Help and Intervention

There is a great deal of popular writing and public discussion about the inadequacy and breakdown of the American family and its failure to meet the needs of older people in an industrialized society. An examination of documented studies presents a picture that is not as bleak or as dysfunctional as some persons assert (Adams, 1968; Sussman, 1965).

For example, many Americans were moved by the studies made by Ralph Nader's investigators who reported the sad plight of many old

[4]A fourth form of intervention is that initiated by the person himself. Psychotherapy and counseling with the individual person is, of course, an important kind of intervention.

people who seemed to be adandoned and forgotten in nursing homes, and perhaps the more so because the investigators were young females. But it would be inaccurate for Americans to hold a stereotyped picture that many older Americans are abandoned by their children. It will be recalled that 8 percent of persons over 65 have never married (Riley et al., 1968). Not all married persons have surviving children, for of the noninstitutional population over 65 about 25 percent do not have any living children (Stehouwer, 1965).

What do research studies show concerning interaction of older families and their adult children? One almost universal finding from family studies in Western society is summarized in the phrase offered by Rosenmayr and Kockeis (1963), "intimacy at a distance." Again and again in surveys, case studies, and from clinical observations one learns that old people wish to have continuous meaningful contact with their children and other kin, but they do not wish to reside in the same household. This is what is meant by "intimacy at a distance."

The form in which intimacy at a distance manifests itself has been clearly demonstrated in numerous studies which show that the great majority of older Americans who have children live within one-half hour driving distance of at least one child. Moreover, they see one another quite frequently. In the nationwide study conducted by Ethel Shanas and her associates 65 percent of the elderly in the sample had seen at least one child in the twenty-four-hour period prior to the interview[5] (Shanas et al., 1968).

George Rosenberg (1970) found in his recently published book about Philadelphians that 90 percent of his respondents had at least one primary or secondary relative living outside their household in metropolitan Philadelphia. And over 80 percent received a visit from at least one such relative in the seven days preceding the interview.

One aspect of the kin network interaction patterns of older persons is the well-documented observation that there are mutual help patterns present in many older families (Sussman, 1965). In the Cornell Study of Retirement, for example, it was found that more help in the form of baby sitting, home repairs, and help during illness flowed from the older generation to the adult child than vice versa (Streib, 1958).

Another aspect of parent-child relations in old age which is rarely brought to public attention is the way in which *some* older parents exploit and tyrannize their adult children. An example is a case study known to the author. Mr. Jones, father of five children, is 82 and comfortably situated financially. As a businessman, he was accustomed to giving orders. When his wife died five years ago, he demanded that his 45-year-old bachelor son

[5]It should be noted that visiting and assistance patterns tell very little about the emotional or affective dimensions of the interactions. See, for example, Adams, 1968, for a discussion of the subjective aspects of kin interaction and the significance of filial obligation.

give up his job and move in to take care of him (he is in a wheelchair). This seemed a reasonable solution to all the children as Buddy did not like his job anyway. Thus he because Dad's nurse, being given an allowance of $50 a week, a new car, and free board and room. Now, five years later, the arrangement is still continuing. Dad is completely satisfied and refuses to consider any other arrangement. *But what about Buddy?* When his father does finally die, what chance will he have to get a job again? To some old people this may sound like an ideal arrangement for Mr. Jones, but the cost to Buddy must be considered also.[6]

The crucial variable in this case is that the older person has enough money to buy this life style. A combination of an authoritarian personality and sufficient money enables him to enforce his wishes. Also, it must be emphasized that this example reveals as much about Buddy's personality as Dad's. Gerontological studies have tended to overlook the rights and needs of the adult children and have seemed to blame them if they do not work out an arrangement that keeps the aged parent happy and comfortable. There has not been enough attention to the needs of the adult children when a "crisis" extends for ten or fifteen years. But as medical advances extend life, there will be more and more persons living for several years when they are not necessarily "happy." And no amount of care and sensitive arrangements by a loving and concerned family can really solve this problem (Goldfarb, 1965).

Quasi-Family Involvement

A further illustration of how the theory of shared functions operates in regard to older families is shown by the development of quasi- or pseudo-family types of intervention. Some of these programs may be supported by private or governmental agencies or organizations, but the nature of the relationship between the persons involved — usually older and younger — is that of a quasi-family relationship.[7] The persons act toward one another and assume the role of family members during the period of social contact. But they are not genuine family members. The name of one of the best known of these programs indicated the quasi-family nature — Foster Grandparents (Nash, 1968).[8] This program employs persons sixty and

[6]A side issue to be mentioned in the kinship patterns is that research has shown that daughters give much more attention and support to their parents than do sons. Hence, the burden of care in crisis situations in most instances falls to the daughter in families with children of both sexes (Blenkner, 1965, 51).

[7] For a list of a variety of research and community action projects, see Wray, 1969.

[8] For a descritpive account of the Foster Grandparent program, see *Aging*, no. 130, August 1965; no. 136, February 1966; no. 137, March 1966; no. 139, May 1966; no. 141, July 1966; no. 142, August 1966; no. 144 October 1966; no. 163, May 1968; nos. 180–81, October–November 1969.

older to work in close association, on a one-to-one basis, with children in orphanages, schools for mentally retarded persons, and other insitutional settings. There are other volunteer programs, such as SERVE, which are similar in their operation (Sainer and Zander, 1969). The important aspect to stress here is the dual benefits through the creation of a warm family kind of relationship between the older persons and the children and young people they serve.[9] The program gives the older person a role in the community as a quasi-family member and at the same time the child receives the kind of love and attention that is essential for development.

There are a number of other programs which have been initiated in the last few years in which younger persons — usually teenagers — take on a quasi-family role in relation to the older person.[10] Some of these have been started by Girl Scout troops. One program in New York State is operated in connection with a state training school for girls. In this instance the girls from the training schhool visit as "granddaughters" the residents of a nursing home who become "grandparents" to the girls. These programs, like those in which the initiative was taken by older persons, seem to have a very positive and enriching benefit for all who take part.

There are also a number of programs in some states which are operated in connection with the discharge of patients from mental hospitals. While these programs do not have a specific concern for the elderly, they often help older persons (Crown, 1970). In these programs a person may become a quasi-relative for the former patient so he can make a more satisfactory adjustment upon his return to his home community.[11]

Bureaucratic Intervention and Help

The theory of shared functions involves the notion that both primary groups and bureaucratic organizations are important for achieving a variety of tasks in American society. There are thousands of bureaucratic groups which are concerned with the health and welfare of older families in the United States. It would be a formidable task to even list the many public and private bureaucracies which offer services to older Americans. The Social Security Administration is the one bureaucracy which contacts the

[9] See *Aging*, no. 165, July 1968, and no. 170, December 1968.

[10] For examples of these kinds of programs, see *Aging*, no. 134, December 1965; no. 145, November 1966; no. 156, October 1967; no. 162, April 1968; no. 195 , January 1971. There is also a number of programs which involve home visitors who help with shopping and transportation, act as interpreters, etc. See *Aging*, no. 125, March 1965; no. 144, October 1966; no. 146, December 1966; no. 164, June 1968. For a more detailed account of one program, see Byron, 1961. Other examples of pseudo-family kinds of intervention are foster homes for the aged (*Aging*, no. 126, April 1965) and Meals-on-Wheels (*Aging*, no. 127, May 1965).

[11] In one program in California a "family agent" acts as a knowledgeable friend who cuts through red tape, gets in touch with agencies, etc. See Crowne, 1969.

most older persons; the Veterans Administration, the Social and Rehabilitation Service, and the numerous health and welfare agencies which are in contact with thousands of older persons at the local, state, and national levels of government also offer a wide variety of services. In addition, the bureaucracies affiliated with educational, religious, health, and private charitable organizations provide hundreds of programs. It is clear why the problem of referral and coordination of bureaucratic services is a major need of older persons which must receive more attention in the near future.

In this section of the paper it is proposed to focus upon the ways in which the mass media of communication, particularly television and radio, may play a greater role in coordinating the work of bureaucracies by serving as information and referral agencies, and also by assuming a new role of offering emotional and interactional resources to older persons.

Why should the mass media be singled out as possible means of increasing the social resources and particularly the emotional health of older people? It is because these powerful omnipresent communications media of modern society already permeate the residential units of almost all Americans. A variety of statistical information points to the high degree of participation of persons over 65 years of age in the use of television and radio. For example, in a typical twenty-four-hour day in which an older person spends approximately nine hours in sleep and six to seven in obligated time (eating meals, housekeeping, personal care) the next largest segment of time is devoted to leisure. Of this 6.5 hourse of leisure, 2.8 is devoted to television (Riley et al., 1968).[12]

Another way to obtain a picture of the leisure activities of older people is by a survey of five thousand OASI beneficiaries in four urban areas. In this survey it was found that 70 percent of the five thousand persons had watched TV a median number of three hours on the day preceding the survey (Riley et al., 1968). In terms of the total number of hours involved, this activity engaged more time than any other leisure activity reported. No other leisure activity was reported by such a large percentage of older persons.

The attitude of many gerontologists is to deplore the fact that housebound older people spend so much time with the mass media. The author thinks that more attention should be given instead to making the media a positive mechanism for enriching their lives.

[12]It is important to point out that these figures were reported by Riley and Foner and their associates in their monumental study, *Aging and Society*, published in 1968. This thorough and carefully documented handbook of facts concerning the aged had to rely on surveys conducted in 1957 and 1958. These figures are an underestimation of the present utilization of television-radio by America's older families. A nation-wide survey of 1500 viewers recently conducted for *Newsweek* by the Gallup organization reported the average household watches TV about six hours per day (*Newsweek*, May 31, 1971, 73–74).

From many surveys that were conducted the last few months preceding the Second White House Conference on Aging, it was found that one of the prime needs of older citizens is the provision of transportation so they can attend meetings, religious services, go visiting, etc. Indeed it is important to develop cheap public transportation for old and young in this society. Given the fact, however, that local, state, and national legislative bodies are reluctant and even recalcitrant to take steps to deal with the complex problems in urban transportation, an auxiliary approach can be the more imaginative use of mass media — particularly local stations— as a means of informing and involving older people in programs and activities which are specifically geared to their needs and stage of the life cycle.

What kinds of programs might be developed which would offer a meaningful and realistic fare for America's older families?[13] These are suggestions based on limited present-day knowledge of the older population and which may be modified with increasing knowledge of wishes of older viewers and a more precise understanding of the impact of various kinds of programs and services transmitted by television. For example, the content of more programs could be devoted to health and nutrition of the aged; the problems of inflation as it relates to social security, medical insurance, taxation — local, state, and national. There is a great range of programs which could be specifically geared to older ethnic, regional, and language audiences. Hobby shows and displays of arts and crafts could be offered. Reminiscing — what Dr. Robert Butler has called the "life review" — could be attempted. Interviews and life stories of a wide variety of older persons could be more widely presented and also "meet the traveler" and historical programs. These are a few things which an amateur suggests. If the vast talents of the mass media were utilized, many other kinds of subjects could be discovered and developed into attractive programs. Local programs which involve the concept of a "help column," now found in many newspapers, might be used on television. Closely related to this kind of program would be some kind of ombudsman program which would mediate grievances of the elderly. There should be more attempts to involve the viewer into the local programs. In many places a large percentage of the aged have telephones so they could call in and easily participate in various kinds of programs.

It is essential to stress the nature of the electronic communication media in American society in their broadest social context — in their structure,

[13]Some years ago a series of twenty educational TV programs sponsored by the State University of New York was developed. The series was called "Living for the Sixties," and topics such as diet, exercise, social security benefits, etc., were included. The series was shown by Station WPSX of University Park, Pennsylvania, and a few other educational TV stations. See *Aging*, no. 143, September 1966. For a description of a local radio program for older persons produced by a retired person, see *Aging*, nos. 176–177, June–July 1969.

their control, and their operations. In the minds of most people, the mass media are viewed only as an entertainment device and the main emphasis is upon programming to maximize the audience of viewers and listeners. It is necessary to broaden our expectation of what television can provide.

At the time when licenses for stations come up for review by the Federal Communications Commission, the owners of stations and the commission engage in a public minuet showing how they operate "in the public interest." However, this is sometime a charades which masks completely the true operation of the electronic media in their singular neglect of the needs and interests of America's 20 million older persons. Local stations may point with pride to occasional interviews with the director of the local senior citizens' center or an occasional conversation with a prominent retiree or older person. Less frequently the networks may produce a documentary that bemoans the plight of one segment of America's older population. The impact of such programs in the short run is probably beneficial by focusing public attention for an instant on a severe, neglected social problem. However, the sustained concern and attention by the electronic media to the needs and interests of approximately 10 percent of the population is minimal and niggardly.

Perhaps the primary reason for the studied neglect of America's older viewing families is the obvious fact that television and radio are primarily geared to selling goods and gadgets which America's older population either cannot afford or do not want. This tremendous concern with the huckstering of products to the maximum number of the youthful audience is a misuse of the airwaves which belong to all Americans, including its 20 million senior citizens. The very fact that older people do not have the money to purchase the products does not mean that their needs should be neglected. Older people are less mobile and thus cannot have the social contacts enjoyed by other segments of the population. This makes it even more imperative that relevant and useful materials are offered them — not merely to entertain them but to inform, to educate, and to give them some feeling of importance, worth, and dignity. The mass media can be their lifeline to the outside world and its more imaginative use can serve to inform them, inspire them, raise their spirits, and perhaps develop more feeling of community with other older persons. It can also direct them to the available bureaucratic services which are already provided but which, too often, they do not know about.

Summary

Both the analyst and the practitioner who is concerned with older families must be cognizant of the diversity of older families and their varying access to four major resources: physical health, emotional health, economic resources, and social resources. The accessibility and management of these resources may involve family members, kinsmen, and outsiders and for this reason the theory of shared functions is a useful analytical tool.

Generally speaking, older families are resourceful in coping with troubles. Available evidence indicates that younger family members also provide a great deal of help and assistance for older families. The picture of abandonment and neglect is not as bleak as is sometime portrayed. However, a realistic assessment requires an understanding of the complexity of the troubles and their long-term nature and that individuals and family groups can only deal with *some* of the problems. For those who have no immediate kin or whose family members are too far away to help, the possiblity for raising the level of family integration can be increased by quasi-family members. There are a number of exciting programs of this kind, exemplified by Foster Grandparents, Foster Granchildren, SERVE, and other kinds of quasi-family projects which are being developed in many communities. The present programs could be greatly expanded, for they are still very limited in the number of persons who are involved, and their financial support is minimal.

Other developments which might contribute new psychic and social resources to older families involve better utilization of bureaucratic structures. One of the most challenging possibilities is to find new ways for the mass media of communication, particularly television and radio, to enrich the lives of older people who have less mobility and decreased social contacts than formerly. It can also inform them of the services available from the many agencies and organizations already designed to serve them.

Part of the neglect of 10 percent of America's population, its older citizens, is due to the fact that commercial television is oriented to maximizing the sales of youth-oriented products and services. Older populations are neglected. There is little doubt but that the commercial networks and their affiliated stations should be made aware of their responsibilities to serve older families. The great amount of time and talent which is devoted to promoting products to young buyers can be rechanneled to serving older persons. Governmental bodies share some of the responsibility for the continued wasteland nature of television as it relates to older Americans.

There are also increasing opportunities — presently untapped — for educational television and for cable television to be more sensitive to the listening and viewing needs of older Americans. More attention should be given to using technology to serve *all* citizens — not just the young. There is little doubt that many older people are in trouble and their burdens can be eased and their lives enriched — psychologically and socially — if the imagination, resources, and skills of the society are shifted to helping more older families.

There is now greater understanding that care of older families in trouble is not merely a private family burden. Other persons — non-family, quasi-family members — can visit and help on a volunteer bases. Finally, there must be greater recognition and utilization of the many services of bureaucratic organizations by older families if the United States is to become a more humane industrialized society.

References:

Adams, Bert N. *Kinship in an Urban Setting.* Chicago: Markham, 1968.

Aging, Washington, D.C.: Administration on Aging, Department of Health, Education and Welfare.

Bell, Daniel. Notes on the Post-Industrial Society. *The Public Interest*, 1967, no. 7, 102-118.

Blenkner, Margaret. Social Work and Family Relationships in Later Life with Some Thoughts on Filial Maturity. In Ethel Shanas and Gordon F. Streib (eds.), *Social Structure and the Family: Generational Relations.* Englewood Cliffs, N.J.: Prentice-Hall, 1965, 46-59.

Byron, Evelyn S. A Friendly Visiting Program. Case Study No. 13. Washington, D.C.: Special Staff on Aging, Department of Health, Education and Welfare, 1961.

Cowne, Leslie J. Approaches to the Mental Health Manpower Problem: A Review of the Literature. *Mental Hygiene*, 1969, 53, 176-87.

Cowne, Leslie J. Case Studies of Volunteer Programs in Mental Health. *Mental Hygiene*, 1970, 54, 337-46.

Cumming, Elaine. *Systems of Social Regulation* New York: Atherton, 1968.

Goldfarb, Alvin I. Psychodynamics and the Three-Generation Family. In Ethel Shanas and Gordon F. Strieb (eds)., *Social Structure and the Family: Generational Relations.* Englewood Cliffs, N.J.: Prentice-Hall, 1965, 10-45.

Gurin, Gerald, et al. *Americans View Their Mental Health: A Nationwide Interview Study*, New York: Basic Books, 1960.

Hill Reuben, et al., *Families Under Stress.* New York: Harper, 1949.

Koos, Earl L. *Families in Trouble.* New York: King's Crown Press, 1946.

Litwak, Eugene. Extended Kin Relations in an Industrial Society. In Ethen Shanas and Gordon F. Streib (eds)., *Social Structure and the Family: Generational Relations.* Englewood Cliffs, N.J.: Prentice-Hall 1965, 290-323.

Litwak, Eugene, and Henry J. Meyer. A Balance Theory of Coordination Between Bureaucratic Organizations and Community Primary Groups. *Administrative Science Quarterly*, 1966, 11, 31-58.

Litwak, Eugene, and Josefina Figueira. Technological Innovation and Theoretical Functions of Primary Groups and Bureaucratic Structures. *American Journal of Sociology*, 1968, 73, 468-81.

Litwak, Eugene, and Ivan Szelanyi. Primary Group Structures and Their Functions: Kin, Neighbors, and Friends. *American Sociological Review,* 1969, 34, 465-81.

Mogey, John. Family and Communities in Urban-Industrial Societies. In Harold T. Christensen (ed)., *Handbook of Marriage and the Family.* Chicago: Rand McNally, 1964, 501-29.

Nash, Bernard E. Foster Grandparents in Child-Care Settings. *Public Welfare*, 1968, 26, 272-80.

Reader, George. Group Discussion. Seminar: Geriatrics and the Medical School Curriculum. *Journal of the American Geriatrics Society,* 1969, 17, 312.

Riley, Matilda W., et al. *Aging in American Society*. New York: Russell Sage Foundation, 1968.

Rosenberg, George S. *The Worker Grows Old.* San Francisco: Jossey-Bass, 1970.

Rosenmayr, Leopold, and Eva Kockeis. Propositions for a Sociological Theory of Aging and the Family. *International Social Science Journal*, 1963, 15, 410–26.

Rosow, Irving. Intergenerational Relationships: Problems and Proposals. In Ethel Shanas and Gordon F. Streib eds., *Social Structure and the Family. Generational Relations.* Englewood Cliffs, N.J.: Prentice-Hall, 1965, 341–78.

Sanier, Janet, and Mary Zander. Guidelines for Conducting a Viable Volunteer Program for Older Persons. Paper presented at the Eighth International Gerontological Congress, Washington, D.C., August 1969 (mimeographed).

Shanas, Ethel, et al. *Old People in Three Industrialized Societies*. London: Routledge & Kegan Paul, 1968.

Special Committee on Aging, United States Senate, *Developments in Aging — 1970*. Washington: U.S. Government Printing Office, 1971.

Stehouwer, Jan. Relations Between Generations and the Three-Generation Household in Denmark. In Ethel Shanas and Gordon F. Streib (eds)., *Social Structure and the Family: Generational Relations.* Englewood Cliffs, N.J.: Prentice-Hall, 1965, 142–62.

Streib, Gordon F. Morale of the Retired. *Social Problems,* 1956, 3, 270–76.

Streib, Gordon F. Family Patterns in Retirement *Journal of Social Issues,* 1958, 14, 46–60.

Streib, Gordon F., and Clement J. Schneider. *Retirement in American Society*. Ithaca, N.Y.: Cornell University Press, 1971.

Sussman, Marvin B. Relationships of Adult Children with their Parents in the United States. In Ethel Shanas and Gordon F. Streib (eds)., *Social Structure and the Family: Generational Relations* Englewood Cliffs, N.J.: Prentice-Hall, 1965, 62–92.

Wray, Robert P. Projects in Gerontology. Appendix 3. In Rosamonde R. Boyd and Charles G. Oakes (eds)., *Foundations of Practical Gerontology*. Columbia, S.C.: University of South Carolina Press, 1969, 255–61.

What's Ahead for Television. *Newsweek*, May 31, 1971, 72–79.

CHAPTER 47

Psychological, Social, and Economic Meanings of Work in Modern Society: Their Effects on the Worker Facing Retirement

WAYNE R. DAVIDSON and KARL R. KUNZE

As he progresses along the age continuum, the worker is the recipient of many influences which ultimately have a bearing on his readiness for retirement. The authors have been fortunate to have had the opportunity to observe some of these influences, to discuss experiences and problems with pre-retirees, and to be a part of a retirement counseling program. Because of this front-row position, we have decided to pay most attention in this article to working people of the company with which we are associated — the Lockheed-California Company.

First, to give you better insight into the circumstances of those we will discuss, a few words about the Lockheed-California Company. This company was an airplane manufacturer; and it now manufactures products and provides services in the aircraft, spacecraft, ocean systems, and research fields. Over the last two decades great changes have taken place in the kinds of jobs and the kinds of people needed in the company's work; formerly, few scientists, engineers, technicians, and large numbers of semi-skilled workers characterized the work force, and now just the opposite is the case. As people grow old in this company (and in so many other present-day companies) they must adjust to new jobs and to changing circumstances.

One other bit of background information: our retirement income plan is non-contributory, has a provision permitting an early retirement at 55, and provides for normal retirement at 65 for salaried and 68 for hourly people. The company's retirement plan was placed into effect in 1942 and its retirement counseling plan had its beginnings ten years later.

Characteristics of the Worker Facing Retirement

We have evidence that some people face retirement from very remote vantage points (that is, if we consider "facing retirement" as any deliberate consideration by an individual of his future status as a retiree). Our employment interviewers report to us that it is not uncommon for applicants in their twenties to ask, freely and without timidity, about the company's retirement plan. Interviewers report that applicants have become more sophisticated about this subject; they take for granted that the company has a retirement plan and ask questions about specific provisions of the plan. One interviewer reported that, in his opinion, the better the applicant and the more professional his status, the more apt he is to ask these questions. Although most of our retirement counseling is with older people, a few employees request counseling at a fairly early age. Our experience suggests that younger interviewees usually discuss retirement with frankness and detachment. They may show little insight into the meaning of old age, and their ideas about retirement may be unrealistic, but it is heartening to know that some young adults are looking into the subject. Older people seem to discuss the subject with greater reluctance, or if pressed into a disucssion, are apt to be superficial. The table below gives the ages of our employees who were interviewed about retirement during 1963.

Table 1. Age of Counselees

Under 40	3
40–50	65
50–60	218
60–70	768
Over 70	3

It may now be well to describe the characteristics of the employee using our counseling service. As we have said, he could possibly be quite young; however, the usual worker we counsel is perhaps from 50 to 68 years of age. He has probably had several, perhaps a half dozen, different but related jobs at Lockheed. His career conflicts are probably behind him but because of the fast pace of developments in his occupation, a gap might be growing between his qualifications and his job requirements. Therefore, although he has long decided on the kind of career to follow, he may have decisions to make: whether to broaden, specialize, or update his job capabilities; or whether to convert his skills to a related field. He is probably at the peak of his earning power and job status. Because the majority of our employees are on an hourly rate, we can assume that this person's income ranges somewhere between $110 and $150 a week. His family and social life are still quite active; he has children, some of whom are probably married; he has a relatively large number of friends — an accumulation in adult life from both work and family sources.

Attitudes and Motivations Relating to Work and Retirement

There is evidence that attitude toward work is associated with age. Many studies with which the authors are familiar show a consistent relationship between work attitudes and age. In one study of salaried people conducted at Lockheed by Kunze and Stockford (1950), personal morale (or attitude toward oneself) was found to be high in the early twenties, low in the middle thirties, and high from age fifty on. In a well-known survey by Benge and Copell (1947), this same curvilinear relationship was found to exist among workers in a confectionery plant, with the exception that the point of low morale was in the late twenties. In the Kunze-Stockford study a close relationship was found between personal and company morale (the latter including measures of job satisfaction), although company morale seemed to lag behind personal morale by about five years.

All of the studies referred to suggest that the morale of older people at work is high — in fact, higher than those in their early twenties. It has been conjectured that low morale in the middle period of employment is due to a first awareness of competition, low seniority, feelings of insecurity, disappointment with progress; and that high morale later on is attributable to the tendency of dissatisfied people with low morale to remove themselves or to be removed from the work situation and to a resolution of earlier problems and concerns by those who remain.

We cannot infer from these studies that older people are more satisfied than others with the work they are performing. They may be or they may be simply reconciled to the work and have good feelings about themselves in relation to many things, including their work. However, we have noted from our interviews that, as employees come close to retirement, work may become recognized as a means to an end rather than an end in itself. This may be because of an awareness that their work lacks permanence and that retirement is the next period in their life sequence.

In spite of an awareness that retirement is imminent, many employees facing retirement have no conception of what retirement means, of what it will consist, or whether or not they will be prepared for it financially and psychologically. They are vague and apprehensive about what retirement will bring. Some believe hopefully that the government and the company collectively will provide for them.

Fear of loss of group medical and life insurance coverage causes some to want to stay employed as long as possible (medical and life insurance policies sometimes can be converted to individual policies at retirement, but the coverage invariably is not as liberal). There are numerous other motivations for continuing work:

1. *Status maintenance.* Among retirees may be found an attitude that working for compensation has some inherent good in it and carries with it certain status significance. Even among retirees who seem to have adequate finances to retire, some are disinclined to make the transition because of values placed on working for money.

2. *Financial circumstances.* We find that many employees facing retirement are uncertain about their financial ability to retire, and others know that their modest income sources would not meet the demands of life in retirement. They may have substantial bills to pay, dependents to support, or imminent medical costs. They may be cognizant of the increased costs of insurance — but whatever the specific reason, financial motives play a part in the reluctance of employees to relinquish work in favor of retirement.

3. *Resistance to change.* As has been mentioned, the worker facing retirement may be enjoying the highest state of his vocational career. In comparison with this status, retirement may be an undesirable alternative. This switch to a less attractive mode of life may be thwarted further by a resistance to, or at times a fear of, change.

4. *Intrinsic nature of work.* People in scientific, engineering, managerial, technical, or other kinds of more sophisticated work are apt to be highly motivated by ego satisfactions derived from the work itself. However, there are satisfying elements to most jobs, and these can cause or reinforce a hesitancy to enter retirement.

Physical and Health Aspects

No doubt the state of health plays a paramount role in the determination of an older person's job success and preparation for retirement. Many of the early researches on the performance of the older worker were conducted in Great Britain during the first and second world wars. The use of older people in industry was necessary because younger men were required by the military. These early studies revealed that some physiological changes (such as reduced visual and auditory acuity, deterioration in eye-hand coordination) became manifest at a relatively early age — sometimes during the mid-twenties. These reduced abilities of older workers were found to be correctable through glasses and similar devices; and, furthermore, were found to correlate poorly with work performance.

Our experience is that, from the mid-twenties on up to 55 or even beyond, with the exception of specific acute illnesses, our employees maintain a reasonably good state of health. Beyond this point, however, the incidence of illness increases, slightly at first and more markedly later on. We find this evidence in the number of visits of employees to our Medical Department and in the disproportionate number of employees 55 years of age and over who are on prolonged absence. Table 2 gives the percentage of employees over and those under 55 who are on prolonged absence. These percentages are compared with the percentages of all employees in both age categories, with a further breakdown by sex.

Table 2. Employees on Prolonged Absence for Medical Reasons

Males		Females		Total	
On Prolonged Absence	In the Work Force	On Prolonged Absence	In the Work Force	On Prolonged Absence	In the Work Force
Under 55 Years of Age					
31%	70%	37% [a]	13%	68%	83%
55 Years of Age and Over					
21%	14%	11%	3%	32%	17%

[a] 15% of these employees are on prolonged absence because of pregnancy.

Table 2 indicates that, although the younger employees (under 55) constitute 83% of the company's work force, they account for only 68% of the prolonged absences, and that the older workers represent 17% of the work force to the 1% level of confidence. The percentage differences between the younger and older employees are even more striking if the comparison is limited to males alone. Males 55 and over represent 14% of the work force but 21% of those on prolonged absence; whereas males 55 and under represent 70% of the work force but only 31% of those on

Table 3. Distribution of Deaths by Age Groups, 1963

Age	Number	% of Deaths by Age	Distribution All Employees
20-29	1	1.0	15.5
30-39	7	7.0	20.3
40-49	24	24.0	33.2
50-59	49	49.0	24.2
60-69	19	19.0	6.8

prolonged absence. There is additional evidence that the illness suffered by those on prolonged absence on which the sick leave is based is likely to be more chronic and to result in longer periods of absence. Further, chronic illnesses of the older employee are more likely to cause residual disabilities which later may affect adversely his employability.

Another indication of the influence of age on health can be found in a study (Davidson, 1964) of the distribution of deaths by age groups at the Lockheed-California Company.

Table 3 reveals that, although those employees 50 and over constitute only 31% of the total Lockheed-California Company work force, 68% of all deaths were of people 50 years of age or more.

Economic Aspects

Although economic and other aspects will be referred to in more detail in the next section when we discuss the employee about to retire, a few points should be raised at this time. Often when an older worker changes jobs, because of technological developments or for any smilar reason, a reduction in status and income may take place. Of course a reduction of income at a time when an employee is making final preparations for retirement can have serious consequences. A few decades ago, an older person, let us say in his sixties, could in fact live adequately on reduced income. Now this is less apt to be the case. Things that were formerly luxuries are now necessities. It is now quite necessary for older people to have some kind of automobile for everyday transportation, to have books, to go to the theater or, on occasion, to go out to dinner. Although we believe that the often discussed influence of inflation on the retired person is to some extent exaggerated, it certainly is another factor that can create economic problems for the retiree.

The Employee About to Retire

Our typical retirees are married (there is a greater likelihood that a woman retiree is widowed or divorced). Retirees are apt to be from 62 to 68 years of age (women retire nearer the earlier age). Certain circumstances sometimes make early retirement either quite attractive or, on the other hand, quite necessary. Chief reasons for early retirement are illness, the nature of the work being performed (its unpleasantness, the necessity of accepting a downgrading, improbability of advancement, etc.), persuasion of a spouse or relative, attractiveness of retirement, or simply adequate finances for retirement.

The usual retiree has two or three children, most of whom live within a 50-mile radius of the parent's residence. If he married late in life or remarried, he may still have extraordinary educational expenses for the youngest of the children. The children meet on occasion with the parents, but because of diverging interests do not usually engage in much collective recreational, educational, or family activities. The decreasing contact with the children may be a source of concern. The wife especially may be discontented if she cannot see children and grandchildren frequently. The wife usually has more contact with the daughter than with the son, and more contact than does the husband with either the son or daughter.

For a majority of retirees, the annual income ranges from $2,400 to $3,000. This includes income from social security, the company retirement income plan, and from savings and other investments. About one-third receive veterans' or other pensions, and a few qualify for Old Age Assistance. Most own their own homes. Their other assets are usually not in excess of $10,000. Personal property ownership often prevents them from securing Old Age Assistance.

The majority of our retirees are free of individual debts of over $500. The larger debts consist of those for the home, home improvements, a car, and medical bills. Medical bills increase considerably during old age. Most of our retirees seem to have health problems. Heart or circulatory disease, arthritis, lung conditions, emphysema, and overweight seem to occupy top place among the chief problems. The wife's health usually is better than the husband's.

Most of our retirees want to live in their own homes. Increasing numbers are going into trailers, and of course this trend may be attributable to our Southern California living. On the other hand, very few have yet cast their lot with the senior citizen communities.

The changing relationship of the male retiree to the spouse warrants discussion. As a couple grows older, the wife seems to take on a role of pronounced leadership, tends to dominate and to make decisions to a greater extent. Because this may not have been the traditional role of the wife, she, because of lack of experience and an unrealistic conception of what is needed in the family, may make unwise decisions.

From our impressions, the usual retiree has fewer community contacts than he had previously — let us say, in his mid-fifties. The male retiree is greatly influenced by his wife's desires about church membership and attendance. His interest in church and politics may increase. Dining out, attending movies or church functions, taking long and short trips seem to be popular forms of activity. Some time usually is spent in some repair and gardening, playing golf, and attending sports events.

By this time the man has dropped some of his club and other organizational memberships, and is becoming less active socially. However, there are differences in inclinations to maintain memberships, and these seem to be associated with the occupation held by the retiree (or with the personal characteristics that made him qualified for his occupation). Those engaged in so-called white-collar jobs tend to hold on to memberships and to engage in leadership roles in church, professional, or other similar activities.

In one sense, then, there may be more similar elements in the pre- and post-retirement activities of the white-collar worker than there are of the blue-collar counterpart. To extend this subject, the blue-collar worker is apt to believe that his education has long since been completed, may read mass media periodicals rather than more selective literature, and might spend much of his time watching television. It is unlikely that he would patronize plays, musical comedies, or concerts with any degree of frequency.

Discussion

It can be assumed that those who visit our quarters at Lockheed for counseling are at least somewhat motivated to plan and prepare for retirement. The service operates wholly on a voluntary basis. Most employees hear of the service from those who have been counseled. We can speculate, then, that as a group, our counselees are more prone to think about and do somethiong about retirement than other company employees (or industrial employees generally, for that matter).

Prior studies suggest that if older workers are in good health, their personal morale may be high, yet they may or may not have favorable sentiments about their work. And seemingly regardless of their job feelings, they usually are apprehensive about retirement and would prefer to continue working rather than to enter what should be a rewarding phase of their lives.

Why is this? Is it because, even if someone dislikes his job, a dislike about a real thing is more acceptable than an apprehension about an imaginary thing? Or is it that the retiree (for a number of reasons — many beyond the control of the individual himself) is unprepared for the transition?

If he is unprepared, is it because people have needs, like the needs for recognition, achievement, competition, prestige, belonging, etc., and these needs are at least partially fulfilled on the job? If so, then, are people hesitant to leave the work situation feeling that retirement will not provide these same satisfactions? Is this a problem that society can do something about?

Is the potential retiree unprepared because he is unwilling to accept something that represents to him a downgrading — unwilling to be reduced in status, for example, from a purchasing agent to a retired purchasing agent? Pursuing this thought further, has modern society established a value in which the connotations of retirement and "putting out to pasture" are similar?

Is he unprepared because he thinks of relaxation as something only permissible if engaged in between periods of strenuous work, or of fun and play as something reserved for children, or education as something that stops with the onset of maturity? Is he unprepared because society hasn't yet defined retirement, or clarified its content?

Is he unprepared for financial reasons — because of rising prices, the trap of easy credit, or the effective persuasion of the friendly salesman? Has the urge to save been stifled? Or is he unprepared because of an ignorance of some of the economic facts of life that have a direct bearing on his future?

To continue this subject of financial preparedness, only rarely does published information about Social Security and company pensions tell of the limitations and inadequacies of these income sources, or warn of the costs these plans do not provide for. Only rarely do we find an employee who is aware of the need for long-range planning if his future is to be provided for adequately. The absence of precautionary advice to people of all ages concerning the need for pre-retirement planning could not be more complete if society had contrived a deliberate scheme to withhold the information. Under present society, employees can grow old realizing that sick leave, vacation, group insurance, school tuition reimbursement and severance pay are theirs when the occasion requires. Why then should they not assume that adequate retirement money will be forthcoming, if they receive no information to the contrary? The void of information about the responsibilities of a pension plan vs. those of the employee can and often does place the employee in a truly hazardous position.

Let us conclude with a final query: Is our potential retiree not ready for his transition because of the unavailability of retirement counseling?

It is true that some churches, schools, government, and nonprofit organizations and industries have retirement counseling service available to workers. In spite of this, it seems apparent that existing opportunities are insufficient, considering the great number who might benefit from this kind of help. Further, there is evidence that preparation for retirement should take place over an extended period of time. Planning for retirement has many elements in common with planning for a career: the long-range aspect; the need to take into account interests, abilities, personality factors; and the necessity for making decisions and carrying out certain preparatory steps. One need is then, as we see it, for developing an early interest in pre-retirement counseling, coupled with a better availability of counseling. How can this be accomplished? And how does one go about persuading government and industry to give more of the total retirement picture so that whenever money does become available through pension plans, this will be merged with self-generated funds as pension designers had intended?

What is in store for the retiree of the future? In the first place, this retiree will have a better education. Not only is the formal educational level of adults increasing; learning as a lifelong experience is gaining acceptance.

The retiree of the future will have benefited by more experienced counseling and, hopefully, by better planning than in the past. He will be motivated to work diligently toward retirement because retirement will have been accepted by society as a means of self-actualization rather than as an anti-climax to a career. Retirement, better defined, will present a positive attraction — a life objective — that will provide a motivation for concluding work responsibilities that now exists only in embryonic form. The present trend toward an acceptance of a respectability of leisure for adults may continue, thus making retirement more inviting.

We predict that retirement will be on a somewhat more voluntary basis than it now is. More attention is being paid to health, yearly physical examinations are becoming accepted as a necessity, and the average life span is now over 70 years of age. Work will become less strenuous physically, and the work week of the future may contain fewer hours. It is true that retirement dates based on chronological age help insurance companies, federal social security agencies, and industry to plan various operating requirements. On the other hand, this criterion of chronological age has been only a fair indicator of who should retire, because it fails to identify the younger adult who is psychologically and economically ready for early retirement and the vigorous, spry, older person who in his and his company's judgment is capable of continuing to work. It is our prediction that the average age at which people retire will reduce slightly but continuously over the years.

Why is this optimistic view held? The last few decades have brought about an increased awareness of the potential of retirement as a pleasant, self-actualizing, and creative period of life. It can be speculated that social security measures and private plans will be more adequate. Perhaps through some form of education people can be convinced of the great value of early retirement planning.

When will these universally beneficial changes take place? Changes in social processes occur slowly, in fact so slowly that these predictions can be made without fear of contradiction for many years to come.

References

Benge, E. J., and D. F. Copell. Employee moral survey. *Mod. Mgmt., 7:* 19–22, 1947.

Davidson, Louise. Mortality survey of Lockheed-California Company employees (unpublished), 1964.

Kunze, K. R., and L. O. Stockford. Preliminary results from a study of supervisors' attitudes (unpublished), 1950.

CHAPTER 48

Death and Social Structure

ROBERT BLAUNER

Bureaucratization of Modern Death Control

Since there is no death without a body — except in mystery thrillers — the corpse is another consequence of mortality that contributes to its disruptiveness, tending to produce fear, generalized anxiety, and disgust (1). Since families and work groups must eventually return to some kind of normal life, the time they are exposed to corpses must be limited. Some form of disposal (earth or sea burial, cremation, exposure to the elements) is the core of mortuary institutions everywhere. A disaster that brings about massive and unregulated exposure to the dead, such as that experienced by the survivors of Hiroshima and also at various times by survivors of great plagues, famines, and death camps, appears to produce a profound identification with the dead and a consequent depressive state (2).

The disruptive impact of a death is greater to the extent that its consequences spill over onto the larger social territory and affect large numbers of people. This depends not only on the frequency and massiveness of mortality, but also on the physical and social settings of death. These vary in different societies, as does also the specialization of responsibility for the care of the dying and the preparation of the body for disposal. In premodern societies, many deaths take place amid the hubbub of life, in the central social territory of the tribe, clan, or other familial group. In modern societies, where the majority of deaths are now predictably in the older age brackets, disengagement from family and economic function has permitted the segregation of death settings from the more workaday social territory. Probably in small towns and rural communities, more people die at home than do so in urban areas. But the proportion of people who die at home, on the job, and in public places must have declined consistently over the past generations with the growing importance of specialized dying institutions — hospitals, old people's homes, and nursing homes (3).

Modern societies control death through bureaucratization, our characteristic form of social structure. Max Weber has described how bureaucratization in the West proceeded by removing social functions from the family and the houshold and implanting them in specialized institutions autonomous of kinship considerations. Early manufacturing and entrepreneurship took place in or close to the home; modern industry and corporate bureaucracies are based on the separation of the work place from the houshold (4). Similarly, only a few generations ago most people in the United States either died at home or were brought into the home if they had died elsewhere. It was the responsibility of the family to lay out the corpse — that is, to prepare the body for the funeral (5). Today, of course, the hospital cares for the terminally ill and manages the crisis of dying; the mortuary industry (whose establishments are usually called "homes" in deference to past tradition) prepares the body for burial and makes many of the funeral arrangements. A study in Philadelphia found that about 90 percent of funerals started out from the funeral parlor, rather than from the home, as was customary in the past (6). This separation of the handling of illness and death from the family minimizes the average person's exposure to death and its disruption of the social process. When the dying are segregated among specialists for whom contact with death has become routine and even somewhat impersonal, neither their presence while alive nor as corpses interferes greatly with the mainstream of life.

Another principle of bureaucracy is the ordering of regularly occurring as well as extraordinary events into predictable and routinized procedures. In addition to treating the ill and isolating them from the rest of society, the modern hospital as an organization is committed to the routinization of the handling of death. Its distinctive competence is to contain through isolation, and reduce through orderly procedures, the disturbance and disruption that are associated with the death crisis. The decline in the authority of religion as well as shifts in the functions of the family underlies this fact. With the growth of the secular and rational outlook, hegemony in the affairs of death has been transferred from the church to science and its representatives, the medical profession and the rationally organized hospital.

Death in the modern hospital has been the subject of two recent sociological studies: Sudnow has focused on the handling of death and the dead in a county hospital catering to charity patients; and Glaser and Strauss have concentrated on the dying situation in a number of hospitals of varying status (7). The county hospital well illustrates various trends in modern death. Three-quarters of its patients are over 60 years old. Of the 250 deaths Sudnow observed, only a handful involved people younger than

40 (8). This hospital is a setting for the concentration of death. There are 1,000 deaths a year; thus approximately three die daily, of the 330 patients typically in residence. But death is even more concentrated in the four wards of the critically ill; here roughly 75 percent of all mortality occurs, and one in 25 patients will die each day (9).

Hospitals are organized to hide the facts of dying and death from patients as well as visitors. Sudnow quotes a major text in hospital administration: "The hospital morgue is best located on the ground floor and placed in an area inaccessible to the general public. It is important that the unit have a suitable exit leading onto a private loading platform which is concealed from hospital patients and the public" (10). Personnel in the high-mortality wards use a number of techniques to render death invisible. To protect relatives, bodies are not to be removed during visiting hours. To protect other inmates, the patient is moved to a private room when the end is foreseen. But some deaths are unexpected and may be noticed by roommates before the hospital staff is aware of them. These are considered troublesome because elaborate procedures are required to remove the corpse without offending the living.

The rationalization of death in the hospital takes place through standard procedures of covering the corpse, removing the body, identifying the deceased, informing relatives, and completing the death certificate and autopsy permit. Within the value hierarchy of the hospital, handling the corpse is "dirty work," and when possible attendants will leave a body to be processed by the next work shift. As with so many of the unpleasant jobs in our society, hospital morgue attendants and orderlies are often Negroes. Personnel become routinized to death and are easily able to pass from mention of the daily toll to other topics; new staff members stop counting after the first half-dozen deaths witnessed (11).

Standard operating procedures have even routinized the most charismatic and personal of relations, that between the priest and the dying patient. It is not that the church neglects charity patients. The chaplain at the county hospital daily goes through a file of the critically ill for the names of all known Catholic patients, then enters their rooms and administers extreme unction. After completing his round on each ward, he stamps the index card of the patient with a rubber stamp which reads: "Last Rites Administered. Date_____ Clergyman_____." Each day he consults the files to see if new patients have been admitted or put on the critical list. As Sudnow notes, this rubber stamp prevents him from performing the rites twice on the same patient (12). This example highlights the trend toward the depersonalization of modern death, and is certainly the antithesis of the historic Catholic notion of "the good death."

In the hospitals studied by Glaser and Strauss, depersonalization is less advanced. Fewer of the dying are comatose, and as paying patients with higher social status they are in a better position to negotiate certain aspects of their terminal situation. Yet nurses and doctors view death as an inconvenience, and manage interaction so as to minimize emotional reactions and fuss. They attempt to avoid announcing unexpected deaths because relatives break down too emotionally; they prefer to let the family members know that the patient has taken "a turn for the worse," so that they will be able to modulate their response in keeping with the hospital's need for order (13). And drugs are sometimes administered to a dying patient to minimize the disruptiveness of his passing — even when there is no reason for this in terms of treatment or the reduction of pain.

The dying patient in the hospital is subject to the kinds of alienation experienced by persons in other situations in bureaucratic organizations. Because doctors avoid the terminally ill, and nurses and relatives are rarely able to talk about death, he suffers psychic isolation (14). He experiences a sense of meaninglessness because he is typically kept unaware of the course of his disease and his impending fate, and is not in a position to understand the medical and other routines carried out in his behalf (15). He is powerless in that the medical staff and the hospital organization tend to program his death in keeping with their organizational and professional needs; control over one's death seems to be even more difficult to achieve than control over one's life in our society (16). Thus the modern hospital, devoted to the preservation of life and the reduction of pain, tends to become a "mass reduction" system, undermining the subjecthood of its dying patients.

The rationalization of modern death control cannot be fully achieved, however, because of an inevitable tension between death — as an event, a crisis, an experience laden with great emotionality — and bureaucracy, which must deal with routines rather than events and is committed to the smoothing out of affect and emotion. Although there was almost no interaction between dying patients and the staff in the county hospital studied by Sudnow, many nurses in the other hospitals became personally involved with their patients and experienced grief when they died. Despite these limits to the general trend, our society has gone far in containing the disruptive possibilities of mortality through its bureaucratized death control.

The Decline of the Funeral in Modern Society

Death creates a further problem because of the contradiction between society's need to push the dead away, and its need "to keep the dead alive" (17). The social distance between the living and the dead must be increased after death, so that the group first, and the most affected grievers later, can reestablish their normal activity without a paralyzing attachment to the corpse. Yet the deceased cannot simply be buried as a dead body: the prospect the total exclusion from the social world would be too anxiety-laden for the living, aware of their own eventual fate. The need to keep the dead alive directs societies to construct rituals that celebrate and insure a transition to a new social status, that of spirit, a being now believed to participate in a different realm (18). Thus, a funeral that combines this status transformation with the act of physical disposal is universal to all societies, and has justly been considered one of the crucial *rites de passage* (19).

Because the funeral has been typically employed to handle death's manifold disruptions, its character, importance, and frequency may be viewed as indicators of the place of mortality in society. The contrasting impact of death in primitive and modern societies, and the diversity in their modes of control, is suggested by the striking difference in the centrality of mortuary ceremonies in the collective life. Because death is so disruptive in simple societies, much "work" must be done to restore the social system's functioning. Funerals are not "mere rituals," but significant adaptive structures, as can be seen by considering the tasks that make up the funeral work among the LoDagaa of West Africa. The dead body must be buried with the appropriate ritual so as to give the dead man a new status that separates him from the living; he must be given the material goods and symbolic invocations that will help guarantee his safe journey to the final destination and at the same time protect the survivors againt his potentially dangerous intervention in their affairs (such as appearing in dreams, "walking," or attempting to drag others with him); his qualities, lifework, and accomplishments must be summed up and given appropriate recognition; his property, roles, rights, and privileges must be distributed so that social and economic life can continue; and, finally, the social units — family, clan, and community as a whole — whose very existence and functioning his death has threatened, must have a chance to vigorously reaffirm their identity and solidarity through participation in ritual ceremony (20).

Such complicated readjustments take time, and therefore the death of a mature person in many primitive societies is followed by not one, but a series of funerals (usually two or three) that may take place over a period ranging from a few months to two years, and in which the entire society, rather than just relatives and friends, participates (21). The duration of the funeral and the fine elaboration of its ceremonies suggest the great destructive possibilities of death in these societies. Mortuary institutions loom large in the daily life of the community, and the frequent occurrence of funerals may be no small element in maintaining societal continuity under the precarious conditions of high mortality (22).

In Western antiquity and the middle ages, funerals were important events in the life of city-states and rural communities (23). Though not so central as in high-mortality and sacred primitive cultures (reductions in mortality rates and secularism both antedate the industrial revolution in the West), they were still frequent and meaningful ceremonies in the life of small-town, agrarian America several generations ago. But in the modern context they have become relatively unimportant events for the life of the larger society. Formal mortuary observances are completed in a short time. Because of the segregation and disengagement of the aged and the gap between generations, much of the social distance to which funerals generally contribute has already been created before death. The deceased rarely have important roles or rights that the society must be concerned about allocating, and the transfer of property has become the responsibility of individuals in cooperation with legal functionaries. With the weakening of beliefs in the existence and malignancy of ghosts, the absence of "realistic" concern about the dead man's trials in his initiation to spirithood, and the lowered intensity of conventional beliefs in an afterlife, there is less demand for both magical precautions and religious ritual. In a society where disbelief or doubt is more common than a firm acceptance of the reality of a life after death (24), the funeral's classic function of status transformation becomes attenuated.

The recent attacks on modern funeral practices by social critics focus on alleged commercial exploitation by the mortuary industry and the vulgar ostentatiousness of its service. But at bottom this criticism reflects this crisis in the function of the funeral as a social institution. On the one hand, the religious and ritual meanings of the ceremony have lost significance for many people. But the crisis is not only due to the erosion of the sacred spirit by rational, scientific world views (25). The social substructure of the funeral is weakened when those who die tend to be irrelevant for the ongoing social life of the community and when the disruptive potentials of death are already controlled by compartmentalization into isolated spheres where bureaucratic routinization is the rule. Thus participation and interest in funerals are restricted to family members and friends rather than involving the larger community, unless an important leader has died (26). Since only individuals and families are affected, adaptation and bereavement have become their private responsibility, and there is little need for a transition period to permit society as a whole to adjust to the fact of a single death. Karl Marx was proved wrong about "the withering away of the state," but with the near disappearance of death as a public event in modern society, the withering away of the funeral may become a reality.

In modern societies, the bereaved person suffers from a paucity of ritualistic conventions in the mourning period. He experiences grief less frequently, but more intensely, since his emotional involvements are not diffused over an entire community, but are usually concentrated on one or a few people (27). Since mourning and a sense of loss are not widely shared, as in premodern communities, the individualization and deritualization of bereavement make for serious problems in adjustment. There are many who never fully recover and "get back to normal," in contrast to the frequently observed capacity of the bereaved in primitive societies to smile, laugh, and go about their ordinary pursuits the moment the official mourning period is ended (28). The lack of conventionalized stages in the mourning process results in an ambiguity as to when the bereaved person has grieved enough and thus can legitimately and guiltlessly feel free for new attachments and interests (29). Thus at the same time that death becomes less disruptive to the society, its prospects and consequences become more serious for the bereaved individual.

Some Consequences of Modern Death Control

I shall now consider some larger consequences that appear to follow from the demographic, organizational, and cultural trends in modern society that have diminished the presence of death in public life and have reduced most persons' experience of mortality to a minimum through the middle years (30).

The Place of the Dead in Modern Society

With the diminished visibility of death, the perceived reality and the effective status and power of the dead have also declined in modern societies. A central factor here is the rise of science: Eissler suggests that "the intensity of service to the dead and the proneness for scientific discovery are in reverse proportion" (31). But the weakening of religious imagery is not the sole cause; there is again a functional sociological basis. When those who die are not important to the life of society, the dead as a collective category will not be of major significance in the concerns of the living.

Compare the situation in high-mortality primitive and peasant societies. The living have not liberated themselves emotionally from many of the recently deceased and therefore need to maintain symbolic interpersonal relations with them. This can take place only when the life of the spirits and their world is conceived in well-structured form, and so, as Goode has phrased it, "practically every primitive religious system imputes both power and interest to the dead" (32).

Their spheres of influence in preindustrial societies are many: spirits watch over and guide economic activities and may determine the fate of trading exchanges, hunting and fishing expeditions, and harvests. Their most important realm of authority is probably that of social control: they are concerned with the general morality of society and the specific actions of individuals (usually kin or clansmen) under their jurisdiction. It is generally believed that the dead have the power to bring about both economic and personal misfortunes (including illness and death) to serve their own interests, to express their general capriciousness, or to specifically punish the sins and errors of the living. The fact that a man as spirit often receives more deference from, and exerts greater power over, people than while living may explain the apparent absence of the fear of death that has been observed in some primitive and ancestor-worship societies (33).

In modern societies the influence of the dead is indirect and is rarely experienced in personified form. Every cultural heritage is in the main the contribution of dead generations to the present society (34), and the living are confronted with problems that come from the sins of the past (for example, our heritage of Negro slavery). There are people who extend their control over others after death through wills, trust funds, and other arrangements. Certain exceptional figures such as John Kennedy and Malcolm X become legendary or almost sainted and retain influence as national symbols or role models. But, for the most part, the dead have little status or power in modern society, and the living tend to be liberated from their direct, personified influence (35). We do not attribute to the dead the range of material and ideal interests that adheres to their symbolic existence in other societies, such as property and possessions, the desire to recreate networks of close personal relationships, the concern for tradition and the morality of the society. Our concept of the inner life of spirits is most shadowy. In primitive societies a full range of attitudes and feelings is imputed to them, whereas a scientific culture has emptied out specific mental and emotional contents from its vague image of spirit life (36).

Generational Continuity and the Status of the Aged

The decline in the authority of the dead, and the widening social distance between them and the living, are both conditions and consequences of the youthful orientation, receptivity to innovation, and dynamic social change that characterize modern society. In most preindustrial societies, symbolic contacts with the spirits and ghosts of the dead were frequent, intimate, and often long-lasting. Such communion in modern society is associated with spiritualism and other deviant belief systems; "normal" relations with the dead seem to have come under increasing discipline and control. Except for observing Catholics perhaps, contact is limited to very specific spatial boundaries, primarily cemeteries, and is restricted to a brief time period following a death and possibly a periodic memorial (37). Otherwise the dead and their concerns are simply not relevant to the living in a society that feels liberated from the authority of the past and orients its energies toward immediate preoccupations and future possibilities.

Perhaps it is the irrelevance of the dead that is the clue to the status of old people in modern industrial societies. In a low-mortality society, most deaths occur in old age, and since the aged predominate among those who die, the association between old age and death is intensified (38). Industrial societies value people in terms of their present functions and their future prospects; the aged have not only become disengaged from significant family, economic, and community responsibilities in the present, but their future status (politely never referred to in our humane culture) is among the company of the powerless, anonymous, and virtually ignored dead (39). In societies where the dead continue to play an influential role in the community of the living, there is no period of the life span that marks the end of a person's connection to society, and the aged before death begin to receive some of the awe and authority that is conferred on the spirit world.

The social costs of these developments fall most heavily on our old people, but they also affect the integrity of the larger culture and the interests of the young and middle-aged. The traditional values that the dead and older generations represent lose significance, and the result is a fragmentation of each generation from a sense of belonging to and identity with a lineal stream of kinship and community. In modern societies where mobility and social change have eliminated the age-old sense of closeness to "roots," this alienation from the past — expressed in the distance between living and dead generations — may be an important source of tenuous personal identities.

These tendencies help to produce another contradiction. The very society that has so greatly controlled death has made it more difficult to die with dignity. The irrelevance of the dead, as well as other social and cultural trends, brings about a crisis in our sense of what is an appropriate death. Most societies, including our own past, have a notion of the ideal conditions under which the good man leaves the life of this world: for some primitives it is the influential grandfather; for classical antiquity, the hero's death in battle; in the Middle Ages, the Catholic idea of "holy dying." There is a clear relationship between the notion of appropriate death and the basic value emphases of the society, whether familial, warlike, or religious. I suggest that American culture is faced with a crisis of death because the changed demographic and structural conditions do not fit the traditional concepts of appropriate death, and no new ideal has arisen to take their place. Our nineteenth-century ideal was that of the patriarch, dying in his own home in ripe old age but in the full possession of his faculties, surrounded by family, heirs, and material symbols of a life of hard work and acquisition. Death was additionally appropriate because of the power of religious belief, which did not regard the event as a final ending. Today people characteristically die at an age when their physical, social, and mental powers are at an ebb, or even absent, typically in the hospital, and often separated from family and other meaningful surroundings. Thus "dying alone" is not only a symbolic theme of existential philosophers; it more and more epitomizes the inappropriateness of how people die under modern conditions.

I have said little about another modern prototype of mortality, mass violence. Despite its statistical infrequency in "normal times," violent death cannot be dismissed as an unimportant theme, since it looms so large in our recent past and in our anxieties about the future. The major forms, prosaic and bizarre, in which violent death occurs, or has occurred, in the present period are: (1) automobile and airplane accidents; (2) the concentration camp; and (3) nuclear disaster. All these expressions of modern violence result in a most inappropriate way of dying. In a brilliant treatment of the preponderance of death by violence in modern literature, Frederick Hoffman points out its inherent ambiguities. The fact that many people die at once, in most of these situations, makes it impossible to mitigate the effects on the survivors through ceremonies of respect. While these deaths are caused by human agents, the impersonality of the assailant and the distance between him and his victim makes it impossible to assign responsibility to understandable causes. Because of the suddenness of impact, the death that is died cannot be fitted into the life that has been lived. And finally, society experiences a crisis of meaning when the threat of death pervades the atmosphere, yet cannot be incorporated into a religious or philosophical context (40).

A Final Theoretical Note: Death and Social Institutions

Mortality implies that population is in a constant (though usually a gradual) state of turnover. Society's groups are fractured by the deaths of their members and must therefore maintain their identities through symbols that are external to and outlast individual persons. The social roles through which the functions of major societal institutions are carried out cannot be limited to particular individuals and their unique interpretations of the needs of social action; they must partake of general and transferable prescriptions and expectations. The order and stability required by a social system are threatened by the eventual deaths of members of small units such as families, as well as political, religious, and economic leaders. There is, therefore, a need for more permanent institutions embedding "impersonal" social roles, universal norms, and transcendent values.

The frequent presence of death in high-mortality societies is important in shaping their characteristic institutional structure. To the extent that death imperils the continuity of a society, its major institutions will be occupied with providing that sense of identity and integrity made precarious by its severity. In societies with high death rates, the kinship system and religion tend to be the major social institutions.

Kinship systems organized around the clan or the extended family are well suited to high-mortality societies because they provide a relative permanence and stability lacking in the smaller nuclear group. Both the totem of the clan and the extended family's ties to the past and the future are institutionalized representations of continuity. Thus, the differential impact of mortality on social structure explains the apparent paradox that the smaller the scale of a community, the larger in general is its ideal family unit (41). The very size of these kinship units provides a protection against the disintegrating potential of mortality, making possible within the family the varied resources in relational ties, age statuses, and cultural experience that guarantee the socialization of all its young, even if their natural parents should die before they have become adults.

In primitive and peasant societies the centrality of magic and religion are related to the dominant presence of death. If the extended family provides for the society's physical survival, magic and countermagic are weapons used by individuals to protect themselves from death's uncontrolled and erratic occurrence. And religion makes possible the moral survival of the society and the individual in an environment fraught with fear, anxiety, and uncertainty. As Malinowski and others have shown, religion owes its persistence and power (if not necessarily its origin) to its unique capacity to solve the societal and personal problems that death calls forth (42). Its rituals and beliefs impart to the funeral ceremonies those qualities of the sacred and the serious that help the stricken group reestablish and reintegrate itself through the collective reaffirmation of shared cultural assumptions. In all known societies it serves to reassure the individual against possible anxieties concerning destruction, nonbeing, and finitude by providing beliefs that make death meaningful, afterlife plausible, and the miseries and injustices of earthly existence endurable.

In complex modern societies there is a proliferation and differentiation of social institutions that have become autonomous in relationship to kinship and religion, as Durkheim pointed out (43). In a sense these institutions take on a permanence and autonomy that makes them effectively independent of the individuals who carry out the roles within them. The economic corporation is the prototype of a modern institution. Sociologically it is a bureaucracy and therefore relatively unconnected to family and kinship; constitutionally it has been graced with the legal fiction of immortality. Thus the major agencies that organize productive work (as well as other activities) are relatively invulnerable to the depletion of their personnel by death, for their offices and functions are impersonal and transferable from one role incumbent to another. The situation is very different in traditional societies. There family ties and kinship groups tend to be the basis of economic, religious, and other activities; social institutions interpenetrate one another around the kinship core. Deaths that strike the family therefore reverberate through the entire social structure. This type of social integration (which Durkheim termed "mechanical solidarity") makes premodern societies additionally vulnerable to death's disruptive potential — regardless of its quantitative frequency and age distribution.

On the broadest level, the relationship between death and society is a dialectic one. Mortality threatens the continuity of society and in so doing contributes to the strengthening of social structure and the development of culture. Death weakens the social group and calls forth personal anxieties; in response, members of a society cling closer together. Specific deaths disrupt the functioning of the social system and thereby encourage responses in the group that restore social equilibrium and become customary practices that strengthen the social fabric. Death's sword in time cuts down each individual, but with respect to the social order it is double-edged. The very sharpness of its disintegrating potential demands adaptations that can bring higher levels of cohesion and continuity. In the developmental course of an individual life, death always conquers; but, as I have attempted to demonstrate throughout this essay, the social system seems to have greatly contained mortality in the broad span of societal and historical development.

References

1. Many early anthropologists, including Malinowski, attributed human funerary customs to an alleged instinctive aversion to the corpse. Although there is no evidence for such an instinct, aversion to the corpse remains a widespread, if not universal, human reaction. See the extended discussion of the early theories in Goody, *Death, Property, and the Ancestors*, Stanford U. Press, 1962, pp. 20–30; and for some exceptions to the general rule, Robert W. Habenstein, "The Social Organization of Death," *International Encyclopedia of the Social Sciences*, forthcoming.

2. Robert J. Lifton, "Phychological Effects of the Atomic Bomb in Hiroshima: The Theme of Death," *Daedalus*, vol. 92, 1963, pp. 462–97. Among other things the dead body is too stark a reminder of man's mortal condition. Although man is the one species that knows he will eventually die, most people in most societies cannot live too successfully when constantly reminded of this truth. On the other hand, the exposure to the corpse has positive consequences for psychic functioning, as it contributes to the acceptance of the reality of a death on the part of the survivors. A study of deaths in military action during World War II found that the bereaved kin had particularly great difficulty in believing in and accepting the reality of their loss because they did not see the body and witness its disposal. T. D. Eliot, "Of the Shadow of Death," *Annals of the American Academy of Political and Social Science*, 1943, vol. 229, pp. 87–99.

3. Statistics on the settings of death are not readily available. Robert Fulton reports that 53 percent of all deaths in the United States take place in hospitals, but he does not give any

source for this figure. See Fulton, *Death and Identity.* New York: Wiley, 1965, pp. 81–82. Two recent English studies are also suggestive. In the case of the deaths of 72 working-class husbands, primarily in the middle years, 46 died in the hospital; 22 at home; and 4 at work or in the street. See Peter Marris, *Widows and Their Families.* London: Routledge & Kegan Paul, 1958, p. 146. Of 359 Britishers who had experienced a recent bereavement, 50 percent report that the death took place in a hospital; 44 percent at home; and 6 percent elsewhere. See Geoffrey Gorer, *Death, Grief, and Mourning.* London: Cresset, 1965, p. 149.

4. Max Weber, *Essays in Sociology*, trans. and ed. H.H. Gerth and C. Wright Mills. New York: Oxford University Press, 1953, pp. 196–98. See also Max Weber, *General Economic History,* trans. Frank H. Knight. Glencoe, Ill.: Free Press, 1950.

5. Leroy Bowman reports that aversion to the corpse made this preparation an unpleasant task. Although sometimes farmed out to experienced relatives or neighbors, the task was still considered the family's responsibility. See Bowman, *The American Funeral: A Study in Guilt, Extravagance, and Sublimity.* Washington, D.C.: Public Affairs Press, 1959, p. 71.

6. William K. Kephart, "Status After Death," *American Sociological Review,* vol. 15, 1950, pp. 635–43.

7. David N. Sudnow, "Passing On: The Social Organization of Dying in the County Hospital," unpublished Ph.D. thesis, University of California, Berkely, 1965. Sudnow also includes comparative materials from a more well-to-do Jewish-sponsored hospital where he did additional fieldwork, but most of his statements are based on the county institution. Barney G. Glaser and Anselm L. Strauss, *Awareness of Dying.* Chicago: Aldine, 1965.

8. See Sudnow, "Passing On," pp. 107, 109. This is even fewer than would be expected by the age-composition of mortality, because children's and teaching hospitals in the city were likely to care for many terminally ill children and younger adults.

9. Ibid., pp. 49, 50.

10. J. K. Owen, *Modern Concepts of Hospital Administration.* Philadelphia: Saunders, 1962, p. 304; cited in Sudnow, "Passing On," p. 80. Such practice attests to the accuracy of Edgar Morin's rather melodramatic statement: "Man hides his death as he hides his sex, as he hides his excrements." See E. Morin, *L'Homme et la mort dans l'histoire.* Paris: Correa, 1951, p. 331.

11. See Sudnow, "Passing On," pp. 20–40, 49–50.

12. Ibid., p. 114.

13. See Glaser and Strauss, *Awareness of Dying*, pp. 142–43, 151–52.

14. On the doctor's attitudes toward death and the dying, see August M. Kasper, "The Doctor and Death," pp. 259–70. in *The Meaning of Death*, ed. Herman Feifel. New York: McGraw-Hill, 1959. Many writers have commented on the tendency of relatives to avoid the subject of death with the terminally ill; see, for example, Herman Feifel's "Attitudes Toward Death in Some Normal and Mentally Ill Populations," pp. 114–32 in *The Meaning of Death.*.

15. The most favorable situation for reducing isolation and meaninglessness would seem to be "where personnel and patient both are aware that he is dying, and where they act on this awareness relatively openly." This atmosphere, which Glaser and Strauss term an "open awareness context," did not typically predominate in the hospitals they studied. More common were one of three other awareness contexts they distinguished: "the situation where the patient does not recognize his impending death even though everyone else does" (closed awareness); "the situation where the patient suspects what the others know and therefore attempts to confirm or invalidate his suspicion" (suspected awareness); and "the situation where each party defines the patient as dying, but each pretends that the other has not done so" (mutual pretense awareness). See Glaser and Strauss, *Awareness of Dying*, p. 11.

16. See Glaser and Strauss, *Awareness of Dying*, p. 129. Some patients, however, put up a struggle to control the pace and style of their dying, and some prefer to leave the hospital and

end their days at home for this reason (see Glaser and Strauss, *Awareness of Dying*, pp. 95, 181–83). For a classic and moving account of a cancer victim who struggled to achieve control over the conditions of his death, see Lael T. Wertenbaker, Death of a Man. New York: Random House, 1957.

For discussion of isolation, meaninglessness, and powerlessness as dimensions of alienation, See Melvin Seeman, "On the Meaning of Alienation," *American Sociological Review*, vol. 24, 1959, pp. 783–91; and Robert Blauner, *Alienation and Freedom: The Factory Worker and His Industry*. Chicago: University of Chicago Press, 1964.

17. Franz Borkenau, "The Concept of Death," *Twentieth Century*, vol. 157, 1955, pp. 313–29; reprinted in Fulton, *Death and Identity*. pp. 42–56 New York: Wiley, 1965.

18. The need to redefine the status of the departed is intensified because of tendencies to act toward him as if he were alive. There is a status discongruity inherent in the often abrupt change from a more or less responsive person to an inactive, non-responding one. This confusion makes it difficult for the living to shift their mode of interaction toward the neomort. Glaser and Strauss report that relatives in the hospital often speak to the newly deceased and caress him as if he were alive; they act as if he knows what they are saying and doing. Nurses who had become emotionally involved with the patient sometimes back away from postmortem care because of a "mystic illusion" that the deceased is still sentient. See Glaser and Strauss, *Awareness of Dying*, pp. 113–14. We are all familiar with the expression of "doing the right thing" for the deceased, probably the most common conscious motivation underlying the bereaved's funeral preparations. This whole situation is sensitively depicted in Jules Romain's novel *The Death of a Nobody*, New York: Knopf, 1944.

19. Arnold Van Gennep, *The Rites of Passage*. London: Routledge & Kegan Paul., 1960 (first published in 1909). See also W. L. Warner, *The Living and the Dead*, New Haven: Yale University Press, 1959, especially chap. 9; and Habenstein, "Social Organization of Death," for a discussion of funerals as "dramas of disposal."

20. See Goody, *Death, Property, and the Ancestors*, for for the specific material on the LoDagaa. For the general theoretical treatment, see Hertz, "Collective Representation of Death," and also Emile Durkheim, *The Elementary Forms of Religious Life*. Glencoe, Ill.: Free Press, 1947, especially p. 447.

21. Robert Hertz, in "The Collective Representation of Death," in *Death and the Right Hand* (trans. Rodney and Claudia Needham; Aberdeen: Cohen & West, 1960), took the multiple funerals of primitive societies as the strategic starting point for his analysis of mortality and social structure. See Goody, *Death, Property, and the Ancestors*, for a discussion of Hertz (pp. 26–27), and the entire book for an investigation of multiple funerals among the LoDagaa.

22. I have been unable to locate precise statistics on the comparative frequency of funerals. The following data are suggestive. In a year and a half, Goody attended 30 among the LoDagaa, a people numbering some 4,000 (see Goody, *Death, Property, and the Ancestors*). Of the Barra people, a Roman Catholic peasant folk culture in the Scottish Outer Hebrides, it is reported that "most men and women participate in some ten to fifteen funerals in their neighborhood every year." See D. Mandelbaum, "Social Uses of Funeral Rites," in *The Meaning of Death*, ed. Herman Feifel, p. 206. Considering the life expectancy in our society today, it is probable that only a minority of people would attend one funeral or more per year. Probably most people during the first forty (or even fifty) years of life attend only one or two funerals a decade. In old age, the deaths of the spouse, collateral relations, and friends become more common; thus funeral attendance in modern societies tends to become more age-specific. For a discussion of the loss of intimates in later years, See J. Moreno, "The Social Atom and Death," pp. 62–66, in *The Sociometry Reader*, ed. J. Moreno. Glencoe, Ill.: Free Press, 1960.

23. For a discussion of funerals among the Romans and early Christians, see Alfred C.

Rush, *Death and Burial in Christian Antiquity*. Washington, D.C.: Catholic University of America Press, 1941, especially pt. 3, pp. 187–273. On funerals in the medieval and preindustrial West, see Bertram S. Puckle, *Funeral Customs*. London: T. Werner Laurie, 1926.

24. See Eissler, in *The Psychiatrist and the Dying Patient*, p. 144: "The religious dogma is, with relatively rare exceptions, not an essential help to the psychiatrist since the belief in the immortality of the soul, although deeply rooted in man's unconscious, is only rarely encountered nowadays as a well-integrated idea from which the ego could draw strength." On the basis of a sociological survey, Gorer confirms the psychiatrist's judgment: "... how small a role dogmatic Christian beliefs play ... " (see Gorer, *Death, Grief, and Mourning*, p. 39). Forty-nine percent of his sample affirmed a belief in an afterlife; 25 percent disbelieved; 26 percent were uncertain or would not answer (see Gorer, p. 166).

25. The problem of sacred institutions in an essentially secular society has been well analyzed by Robert Fulton. See Fulton and Gilbert Geis, "Death and Social Values," pp. 67–75, and Fulton, "The Sacred and the Secular," pp. 89–105, in Fulton, *Death and Identity*,

26. Leroy Bowman interprets the decline of the American funeral primarily in terms of urbanization. When communities were made up of closely knit, geographically isolated groups of families, the death of an individual was a deprivation of the customary social give and take, a distinctly felt diminution of the total community. It made sense for the community as a whole to participate in a funeral. But in cities, individual families are in a much more limited relationship to other families, and the population loses its unity of social and religious ideals. For ethical and religious reasons, Bowman is unwilling to accept "a bitter deduction from this line of thought that the death of one person is not so important as once it would have been, at least to the communiy in which he has lived." But that is the logical implication of his perceptive sociological analysis. See Bowman, *The American Funeral*, pp. 9, 113–15, 126–28.

27. Edmund Volkart, "Bereavement and Mental Health," pp. 281–307 in *Explorations in Social Psychiatry*, ed. Alexander H. Leighton, John A. Clausen, and Robert N. Wilson. New York: Basic Books, 1957. Volkart suggests that bereavement is a greater crisis in modern American society than in similar cultures because our family system develops selves in which people relate to others as persons rather than in terms of roles (see pp. 293–95).

28. In a study of bereavement reactions in England, Geoffrey Gorer found that 30 of a group of 80 persons who had lost a close relative were mourning in a style he characterized as unlimited. He attributes the inability to get over one's grief "to the absence of any ritual, either individual or social, lay or religious, to guide them and the people they come in contact with." The study also attests to the virtual disappearance of traditional mourning conventions. See Gorer, *Death, Grief, and Mourning*, pp. 78–83.

29. See Marris, *Widows and Their Families*, pp. 39–40.

30. Irwin W. Goffman suggests that "a decline in the significance of death has occurred in our recent history." See "Suicide Motives and Categorization of the Living and the Dead in the United States," Syracuse, N.Y.: Mental Health Research Unit, February 1966, unpublished manuscript, p. 140.

31. See Eissler, *The Psychiatrist and the Dying Patient*, p. 44.

32. See Goode *Religion Among the Primitives*, p. 185. Perhaps the fulllest treatment is by Frazer in *Fear of the Dead*, especially vol. 1.

33. See Simmons, *Role of the Aged*, pp. 223–24. See also Effie Bendann, *Death Customs*. New York: Knopf, 1930, p. 180. However, there are primitive societies, such as the Hopi, that attribute little power and authority to dead spirits; in some cultures, the period of the dead man's influence is relatively limited; and in other cases only a minority of ghosts are reported to be the object of deference and awe. The general point holds despite these reservations.

34. See Warner, *The Living and the Dead*, pp. 4–5.

35. The novel *Death of a Nobody* (see footnote 18) is a sensitive treatment of how its protagonist, Jacques Godard, affects people after his death; his influence is extremely short-

lived, and his memory in the minds of the living vanishes after a brief period. Goffman suggests that "parents are much less likely today of tell stories of the dead, of their qualities, hardships, accomplishments and adventures than was true a hundred years ago" ("Suicide Motives," p. 30).

36. In an interesting treatment of the problem from a different theoretical framework, Goffman has concluded that the sense of contrast between what is living and what is dead in modern society has become attenuated, in large part becauseof the decline in exposure to death. He has assembled evidence on social differences within our society: for example, women, lower-class people, and Catholics tend to have closer and more frequent contact with death or images of the dead than men, middle-class persons, and Protestants. (See footnote 30.) The question of what is the representative American imagery of afterlife existence would be a fruitful one for research. Clear and well-developed imageries are probably typical only among Catholics, fundamentalists, and certain ethnic groups. The dominant attitude (if there is one) is likely quite nebulous. For some, the dead may be remembered as an "absent presence," never to be seen again; for others as "a loved one with whom I expect (or hope) to be reunited in some form someday." Yet the background of afterlife existences is only vaguely sketched, and expectation and belief probably alternate with hope, doubt, and fear in a striking ambiguity about the prospect and context of reunion.

37. In primitive societies ghosts and spirits of the dead range over the entire social territory or occupy central areas of the group's social space. In ancestor-worship civilizations such as Rome and China, spirits dwell in shrines that are located in the homes or family burial plots. In these preindustrial societies symbolic contact with the dead may be a daily occurrence. Likewise, in the middle ages, cemeteries were not on the periphery of the societal terrain but were central institutions in the community; regularly visited, they were even the sites for feasts and other celebrations, since it was believed that the dead were gladdened by sounds of merrymaking. (See Puckle, *Funeral Customs,* pp. 145–46.) The most trenchant analysis of the cemetery as a spatial territory marking the social boundaries between the "sacred dead and the secular world of the profane living" in a small modern community is found in Warner, *The Living and the Dead*, chap. 9. Yet Warner also notes that people tend to disregard cemeteries as a "collective representation" in rapidly changing and growing communities, in contrast to the situation in small, stable communities. Goffman (see footnote 30; p. 29) notes that "increasingly the remains of the dead are to be found in huge distant cemeteries that are not passed or frequented as part of everyday routines [or] in cities in which our very mobile population *used* to live."

38. Feifel has suggested that American society's rejection of (and even revulsion to) the old may be because they remind us unconsciously of death. See Feifel, ed., *The Meaning of Death*, p. 122.

39. According to Kastenbaum, the tendency of psychiatrists to eschew psychotherapy with the aged and to treat them, if at all, with supportive (rather than more prestigious depth) techniques may be a reflection of our society's future orientation, which results in an implicit devaluing of old people because of their limited time prospects. See Kastenbaum, "The Reluctant Therapist," pp. 139–45 in *New Thoughts on Old Age.* The research of Butler, a psychiatrist who presents evidence for significant personality change in old age despite the common contrary assumption, would seem to support Kastenbaum's view. (See Robert N. Butler, "The Life Review: An Interpretation of Reminiscence in the Aged," *Psychiatry,* vol. 26, 1963, pp. 65–76.) Sudnow contributes additional evidence of the devaluation of old people. Ambulance drivers bringing critical or "dead-on-arrival" cases to the county hospital's emergency entrance blow their horns more furiously and act more frantic when the patient is young than when he is old. A certain proportion of "dead-on-arrival" cases can be saved through mouth-to-mouth resuscitation, heart massages, or other unusual efforts. These measures were attempted with children and young people but not with the old; one intern

admitted being repulsed by the idea of such close contact with them. See Sudnow, "Passing On," pp. 160–63.

40. Frederick J. Hoffman, *The Mortal No: Death and the Modern Imagination*. Princeton: Princeton University Press, 1964; see especially pt. 2. In a second paper on Hiroshima, Robert J. Lifton also notes the tendency for the threat of mass death to undermine the meaning systems of society, and the absence of a clear sense of appropriate death in modern cultures. See "On Death and Death Symbolism: The Hiroshima Disaster," *Psychiatry*, vol. 27, 1964, pp. 191–210. Gorer has argued that our culture's repression of death as a natural event is the cause of the obsessive focus on fantasies of violence that are so prominent in the mass media. See Geoffrey Gorer, "The Pornography of Death," pp. 402–407 in *Identity and Anxiety*, ed. Maurice Stein and Arthur Vidich. Glencoe, Ill.: Free Press, 1960; also reprinted in Gorer's *Death, Grief, and Mourning*.

The inappropriateness inherent in the automobile accident, in which a man dies outside a communal and religious setting, is poignantly captured in the verse and chorus of the country and western song "Wreck on the Highway," popularized by Roy Acuff: "Who did you say it was, brother?/ Who was it fell by the way?/ When whiskey and blood run together,/ Did you hear anyone pray?" Chorus: "I didn't hear nobody pray, dear brother,/ I didn't hear nobody pray./ I heard the crash on the highway,/ But I didn't hear nobody pray."

41. The important distinction between ideal family structures and actual patterns of size of household, kinship composition, and authority relations has been stressed recently by William Goode in *World Revolution and Family Patterns*, New York: Free Press of Glencoe, 1963; and by Marion Levy, "Aspects of the Analysis of Family Structure," pp. 1–63 in A. J. Coale and Marion Levy, *Aspects of the Analysis of Family Structure*. Princeton: Princeton University Press, 1965.

42. Bronislaw Malinowski, *Magic, Science and Religion*. Garden City, N.Y.: Doubleday Anchor, 1955; see pp.. 47–53.

43. Emile Durkheim, *Division of Labor*. Glencoe, Ill.: Free Press, 1949.

Bibliography

Childhood

Asubel, D. P. *Theory and Problems of Child Development*. New York: Grune & Stratton, 1958.

Baldwin, A. L. *Behavior and Development in Childhood*. New York: Dryden Press, 1955.

———. *Theories of Child Development*. New York: Wiley, 1967.

Bassard, J. H. S., and E. S. Boll. *The Sociology of Child Development*. New York: Harper, 1948.

Baughman, E. E., and W. G. Dahlstrom. *Negro and White Children: A Psychological Study in the Rural South*. New York: Academic Press, 1968.

Bennett, I., and I. Hellman. "Psychoanalytic Material Related to Observations in Early Development." *In Psychoanalytic Study of the Child*. New York: International Universities Press, vol. 6, 1951, pp. 307–24.

Beres, D., and S. J. Obers. "The Effects of Extreme Deprivation in Infancy on Psychic Structure in Adolescence: A Study of Ego Development." *In Psychoanalytic Study of the Child*. New York: International Universities Press, vol. 5, 1950, pp. 212–35.

Bronfenbrenner, U., ed. *Influences on Human Development*. Hinsdale, Ill. : Dryden Press, 1972.

Cole, M., and J. S. Bruner. "Cultural Differences and Inferences About Psychological Processes." *American Psychologist*, vol. 26, 1971, pp. 867–76.

Dunbar, F. H. *Your Child's Mind and Body*. New York: Random House, 1951.

Elkin, F., and G. Handel. *The Child and Society*. New York: Random House, 1951.

Elkind, D., and J. H. Flavell, eds. *Studies in Cognitive Development*. New York: Oxford University Press, 1969.

English, H. B. *Dynamics of Child Development*. New York: Holt, Rinehart & Winston, 1962.

Erikson, E. H. *Childhood and Society*. New York: Norton, 1950.

Flavell, J. H. *The Developmental Psychology of Jean Piaget*. New York: Van Nostrand, 1963.

Fraiberg, S. *The Magic Years*. New York: Scribner, 1959.

Freud, A. *The Ego and the Mechanisms of Defense*. New York: International Universities Press, 1936.

Frost, J. L., and G. R. Hawkes, eds. *The Disadvantaged Child: Issues and Innovations*. Boston: Houghton Mifflin, 1966.

Gesell, A., et al. *The First Five Years of Life*. New York: Harper, 1940.

——— and C. S. Amatruda. *Developmental Diagnosis*. New York: Hoeber, 1941.

——— and F. L. Ilg. *Infant and Child in the Culture of Today: The Guidance of Development in Home and Nursery School.* New York: Harper, 1943.

——— and ———. *The Child from Five to Ten.* New York: Harper, 1946.

Glass, D. C., ed. *Environmental Influences.* New York: Rockefeller University Press, 1968.

Haas, M. B., and I. E. Harms. "Social Interaction Between Infants." *Child Development,* vol. 34, 1963, pp. 79–97.

Hoch, P., and J. Zubin, eds. *Psychopathology of Childhood.* New York: Grune & Stratton, 1955.

Hoffman, L. W. "Mother's Enjoyment of Work and Effects on the Child." *Child Development*, vol. 32, 1961, pp. 187–97.

Hoffman, M. L., and L. W. Hoffman, eds. *Review of Child Development Research,* vol. 1 and 2. New York: Russell Sage Foundation, 1964.

Hunt, M. J. "The Psychological Bases for Using Preschool Enrichment as an Antidote for Cultural Deprivation." *Merrill-Palmer Quarterly*, vol. 10, 1964, pp. 209–248.

Langer, J. *Theories of Development.* New York: Holt, Rinehart & Winston, 1969.

Lichtenberg, P., and D. G. Norton. *Cognitive and Mental Development in the First Five Years of Life: A Review of Recent Research.* Maryland: National Institute of Mental Health, H.E.W., 1970.

Lidz, T. *The Person.* New York: Basic Books, 1968.

Lipsitt, L. P., and C. C. Spikes, eds. *Advances in Child Development and Behavior.* New York: Academic Press, 1967.

Lustman, S. L. "Cultural Deprivation." In *Psychoanalytic Study of the Child*, vol. 25, pp. 483–502. New York: International Universities Press, 1970.

Maccoby, E. E., ed. *The Development of Sex Differences.* Stanford: Stanford University Press, 1966.

Maier, H. W. *Three Theories of Child Development.* New York: Harper & Row, 1965.

Martin, W. E., and C. B. Stendler, eds. *Readings in Child Development.* New York: Harcourt, Brace, 1954.

McCall, G. O., and J. L. Simmons. *Identities and Interactions.* New York: Free Press, 1966.

Murphy, L. B. *The Widening World of Childhood: The Paths Toward Mastery.* New York: Basic Books, 1962.

Mussen, P. H., J. J. Conger, and J. Kagan. *Child Development and Personality,* 3rd ed. New York: Harper & Row, 1956.

———, ed. *Carmichael's Manual of Child Psychology*, vol. 1 & 2. New York: Wiley, 1970.

Myers, J. K., and B. H. Roberts. *Family and Class Dynamics in Mental Illness.* New York: Wiley, 1958.

Piaget, J. *Play, Dreams, and Imitation in Childhood.* New York: Norton, 1951.

Pollard, M. B., and B. Geoghegan. *The Growing Child in Contemporary Society.* Milwaukee: Bruce Publishing Co., 1969.

Redl, F. *When We Deal with Children.* New York: Free Press, 1966.

Ribble, M. A. *The rights of Infants.* New York: Columbia University Press, 1943.

Riessman, F. *The Culturally Deprived Child.* New York: Harper & Row, 1962.

Rosen, B. C. "Family Structure and Value Transmission," *Merrill-Palmer Quarterly,* vol. 10, 1964, pp. 59–76.

Sarnoff, I. *Personality Dynamics and Development.* New York: Wiley, 1962.

Sewell, W., and A. O. Haller. "Factors in the Relationship Between Social Status and the Personality Adjustment of the Child." *American Sociological Review*, vol. 24, 1959, pp. 511–20.

Solomon, J. C. *A Synthesis of Human Behavior*. New York: Grune & Stratton, 1954.

Spitz, R. A. *The First Year of Life*. New York: International Universities Press, 1965.

Stone, J. "A Critique of Studies of Infant Isolation." *Child Development*, vol. 25, 1954, pp. 9–20.

Stott, L. H. "The Persisting Effects of Early Family Experiences upon Personality Development." *Merrill-Palmer Quarterly*, vol. 3, 1957, pp. 145–59.

Stuart, H. C., and D. G. Prugh, eds. *The Healthy Child*. Cambridge: Harvard University Press, 1960.

The Psychoanalytic Study of the Child. New York: International Universities Press, vol. 1 to 25, and New York Times Co., vol. 26 and 27.

U.S. Department of Health, Education and Welfare, The National Institute of Child Health and Human Development. *Perspectives on Human Deprivation: Biological, Psychological, and Sociological*. Washington, D.C., 1968.

Wortis, H., et al. "Child-Rearing Practices in a Low Socioeconomic Group." *Pediatrics*, vol. 32, 1963, pp. 298–307.

Yarrow, L. J. "Separation from Parents During Early Childhood." In *Review of Child Development Research*, ed. M. L. Hoffman and L. W. Hoffman, pp. 89–130. New York: Russell Sage Foundation, 1964.

Adolescence

Adams, J. F., ed. *Understanding Adolescence: Current Developments in Adolescent Psychology*. Boston: Allyn & Bacon, 1968.

Ausubel, D. P. *Theory and Problems of Adolescent Development*. New York: Grune & Stratton, 1954.

Bandura, A., and R. H. Walters. *Adolescent Aggression*. New York: Ronald Press, 1959.

Benedict, R. "Continuities and Discontinuities in Cultural Conditioning." In *Social Perspectives on Behavior*, ed. H. D. Stein and R. A. Cloward, pp. 240–47. New York: Free Press of Glencoe, 1958.

Blos, P. *On Adolescence: A Psychoanalytic Interpretation*. New York: Free Press, 1962.

Caplan, G., and S. Lebovici, eds. *Adolescence: Psychosocial Perspectives*. New York: Basic Books, 1969.

Cloward, R. A., and L. E. Ohlin. *Delinquency and Opportunity*. New York: Free Press, 1960.

Coleman, J. S. *The Adolescent Society*. New York: Free Press, 1961.

Douvan, E., and J. Adelson. *The Adolescent Experience*. New York: Wiley, 1966.

Farber, S., and R. H. L. Wilson. *Teenage Marriage and Divorce*. San Francisco: Diablo Press, 1967.

Freud, A. "Adolescence." In *Psychoanalytic Study of the Child*, vol. 13, pp. 255–78. New York: International Universities Press, 1958.

Friendenberg, E. J. *The Vanishing Adolescent*. Boston: Beacon Hill, 1959.

Goethals, G. W., and D. S. Klos. *Experiencing Youth*. Boston: Little, Brown, 1970.

Goodman, N. "Adolescent Norms and Behavior: Organization and Conformity." *Merrill-Palmer Quarterly*, vol. 15, 1969, pp. 199–211.

Gross, H. J. "Conceptual Systems Theory: Application to Some Problems of Adolescents." *Adolescence*, vol. 2, Summer 1967, pp. 153–66.

Hager, D. L., A. M. Vener, and C. S. Stewart. "Patterns of Adolescent Drug Use in Middle America." *Journal of Counseling Psychology*, vol. 18, 1971, pp. 292–97.

Hall, G. S. *Adolescence*, vols. 1 and 2. New York: Appleton, 1904.

Havighurst, R. J., and H. Taba, eds. *Adolescent Character and Personality*. New York: Wiley, 1949.

Hechinger, G., and F. Hechinger. *Teenage Tyranny.* New York: Morrow, 1962.

Horrocks, J. E. *The Psychology of Adolescence.* Boston: Houghton Mifflin, 1962.

Hurlock, E. B. "American Adolescents of Today: A New Species." *Adolescence,* vol. 1, 1966, pp. 7–21.

———. *Adolescent Development,* 4th ed. New York: McGraw-Hill, 1973.

Inhelder, B., and J. Piaget. *The Growth of Logical Thinking from Childhood to Adolescence.* New York: Basic Books, 1958.

Jacobson, E. "Adolescent Moods and the Remodeling of the Psychic Structure." In *Psychoanalytic Study of the Child,* vol. 16, pp. 164–83. New York: International Universities Press, 1961.

Keniston, K. *The Uncommitted: Alienated Youth in American Society.* New York: Harcourt, Brace & World, 1965.

Lief, H. I., and W. C. Thompson. "The Prediction of Behavior from Adolescence to Adulthood." *Psychiatry,* vol. 24, 1961, pp. 32–38.

Louria, D. *Nightmare Drugs.* New York: Pocket Books, 1966.

Martin, J. M., J. P. Fitzpatrick, and R. E. Gould. *The Analysis of Delinquent Behavior: A Structural Approach.* New York: Random House, 1968.

Masterson, J. F., and A. Washburne, "The Symptomatic Adolescent: Psychiatric Illness or Adolescent Turmoil?" *American Journal of Psychiatry,* vol. 122, 1966, pp. 1240–48.

McClelland, D. *The Achieving Society.* Princeton, N.J.: Van Nostrand, 1961.

Morse, B. *Adolescent Sexual Behavior.* Derby, Conn.: Monarch Books, 1964.

Muuss, R. E. *Theories of Adolescence.* New York: Random House, 1962.

Nye, I. *Family Relationships and Delinquent Behavior.* New York: Wiley, 1958.

Osofsky, H. J. *The Pregnant Teen-Ager.* Springfield, Ill.: Charles C. Thomas, 1968.

Packard, V. *The Pyramid Climbers.* New York: McGraw-Hill, 1962.

Phelps, H. R., and J. E. Horrocks. "Factors Influencing Informal Groups of Adolescents." *Child Development,* vol. 29, 1958, pp. 69–86.

Reiss, A. J., and A. L. Rhodes. "The Distribution of Juvenile Delinquency in the Social Class Structure." *American Sociological Review,* vol. 26, 1961, pp. 720–32.

Remmers, H. H. "Cross-Cultural Studies of Teenagers' Problems." *Journal of Educational Psychology,* vol. 53, 1962, pp. 254–61.

Resnick, H., M., L. Fauble, and S. H. Osipow. "Vocational Crystallization and Self-Esteem in College Students." *Journal of Counseling Psychology,* vol. 17, 1970, pp. 465–67.

Rogers, D., ed. *Issues in Adolescent Psychology.* New York: Appleton-Century-Crofts, 1969.

Rosenberg, M. *Society and the Adolescent Self-Image.* Princeton, N.J.: Princeton University Press, 1965.

Rubin, L., and L. A. Kirkendall. *Sex in the Adolescent Years.* New York: Association Press, 1968.

Rushing, W. A. "Adolescent-Parent Relationship and Mobility Aspirations." *Social Forces,* vol. 43, 1964, pp. 157–66.

Salisbury, H. E. *The Shook-Up Generation.* Greenwich, Conn.: Fawcett Publications, 1958.

Shapiro, R. "Adolescence and the Psychology of the Ego." *Psychiatry,* vol. 26, 1963, pp. 77–87.

Smith, E. A. *American Youth Culture.* New York: Free Press, 1962.

Stone, J. L., and J. Church. *Childhood and Adolescence,* 3rd ed. New York: Random House, 1973.

Symonds, P. M. *Adolescent Phantasy.* New York: Columbia University Press, 1949.

Walters, P. A. "Promiscuity in Adolescence." *American Journal of Orthopsychiatry,* vol. 35, 1965, pp. 670–75.

Weiner, I. B. *Psychological Disturbances in Adolescence.* New York: Wiley, 1970.

Early and Middle Adulthood

Becker, H., and R. Hill, eds. *Family, Marriage, and Parenthood.* Boston: D. C. Heath, 1955.

Bergler, E. *The Revolt of the Middle-Aged Man.* New York A. A. Wyn, 1954.

Billingsley, A. *Black Families in White America.* Englewood Cliffs, N.J.: Prentice-Hall, 1968.

Borow, H., ed. *Man in a World at Work.* Boston: Houghton Mifflin, 1964.

Brecher, R., and E. Brecher, eds. *An Analysis of Human Sexual Response.* New York: New American Library, 1966.

Caudill, H. M. *Night Comes to the Cumberlands.* Boston: Little, Brown, 1963.

Christensen, H. T., ed. *Handbook of Marriage and the Family.* Chicago: Rand McNally, 1964.

Crosby, J. *Illusion and Disillusion.* Belmont Calif.: Wadsworth Publishing Co., 1973.

Dublin, R. "Industrial Workers' Worlds: A Study of the Central Life Interests of Industrial Workers." *Social Problems,* vol. 3, January 1955, pp. 131–42.

Duvall, E. M. *Family Development.* Philadelphia: Lippincott, 1962.

Farber, B., ed. *Family Organization and Interaction.* San Francisco: Chandler, 1964.

Farber, S. M., and R. H. L. Wilson, eds. *The Potential of Women.* New York: McGraw-Hill, 1963.

Frazier, F. E. *On Race Relations.* Chicago: University of Chicago Press, 1968.

Fried, B. *The Middle-Age Crisis.* New York: Harper & Row, 1967.

Fromm, E. *The Art of Loving.* New York: Harper & Row, 1956.

Ginzberg, E., et al. *Occupational Choice.* New York: Columbia University Press, 1951.

Glasser, P. H., and L. N. Glasser, eds. *Families in Crisis.* New York: Harper & Row, 1970.

Goode, W. J. *After Divorce.* New York: Free Press, 1956.

———. *World Revolution and Family Patterns.* New York: Free Press, 1963.

Gordon, A. I. *Intermarriage.* Boston: Beacon Press, 1964.

Gornick, V., and B. K. Moran, eds. *Woman in Sexist Society.* New York: Basic Books, 1971.

Harrington, M. *The Other America: Poverty in the United States.* New York: Macmillan, 1962.

Hobart, C. W. "Commitment, Value Conflict, and the Future of the American Family." *Journal of Marriage and the Family,* vol. 25, November 1963, pp. 405–12.

Hoffman, M. "Homosexual." *Psychology Today,* vol. 3, July 1969, pp. 43–45, 70–71.

Horwitz, E. L. *Communes in America.* Philadelphia: Lippincott, 1972.

Huizinga, J. *The Waning of the Middle Ages.* New York: Doubleday, 1956.

Klemer, R. H. *Marriage and Family Relationships.* New York: Harper & Row, 1970.

Kolko, G. *Wealth and Power in the United States.* New York: Praeger, 1962.

Komarovsky, M. *The Unemployed Man and His Family.* New York: Holt, Rinehart & Winston, 1940.

Larry, L., and J. M. Constantine. *Group Marriage.* New York: Macmillan, 1973.

Lederer. W. J., and D. D. Jackson. *The Mirages of Marriage.* New York: Norton, 1968.

Le Masters, E. E. *Parents in Modern America.* Homewood, Ill.: Dorsey Press, 1970.

Lewis, Oscar. *La Vida.* New York: Vintage Books, 1965.

Lidz, T. *The Family and Human Adaptation.* New York: International Universities Press, 1963.

Lipset, S. M., and R. Bendix. *Social Mobility in Industrial Society.* Berkeley: University of California Press, 1959.

Masters, W. H., and V. E. Johnson. *Human Sexual Response.* Boston: Little, Brown, 1966.

Montagu, M. F. A., ed. *Man and Aggression.* New York: Oxford, 1968.

Moynihan, D. "Employment, Income and the Ordeal of the Negro Family," *Daedalus,* Fall 1965, pp. 745–69.

Neff, W. S. *Work and Human Behavior.* New York: Atherton Press, 1968.

Nimkoff, M. *Comparative Family Systems.* Boston: Houghton Mifflin, 1965.

Nye, I. F., and F. M. Berardo. *Emerging Conceptual Frameworks in Family Analysis.* New York: Macmillan, 1967.

Perlman, H. H. *Persona.* Chicago: University of Chicago Press, 1968.

Reiner, B. S., and I. Kaufman. *Character Disorders in Parents of Delinquents.* New York: Family Service Association of America, 1959.

Rogers, C. R. *On Becoming a Person.*Boston: Houghton Mifflin, 1961.

Rover, C. *Love, Morals, and the Feminists.*London: Routledge & Kegan Paul, 1970.

Saxton, L. *The Individual, Marriage, and the Family: Current Perspectives.* Belmont, Calif.: Wadsworth Publishing Co., 1970.

Shostak, A. B., and W. Gomberg, eds. *Blue-Collar World.* Englewood Cliffs, N.J.: Prentice-Hall, 1964.

Soddy, K. *Men in Middle Life.* London: Tavistock, 1967.

Stein, M. R., A. J. Vidick, and D. M. White, eds. *Identity and Anxiety.* Glencoe, Ill.: Free Press, 1960.

Szasz, T. *The Myth of Mental Illness.* New York: Hoeber-Harper, 1961.

Vedder, C. B. *Problems of the Middle-Aged.* Springfield, Ill.: Charles C. Thomas, 1965.

Weiss, R. S., and D. Riesman. "Work and Automation: Problems and Prospects." In *Contemporary Social Problems,* 2nd ed. R. K. Merton and R. A. Nisbet, eds. New York: Harcourt, Brace & World, 1966, pp. 553–618.

Wheelis, A. *The Quest for Identity.* New York: Norton, 1953.

Whyte, W. H. *The Organization Man.* New York : Doubleday Anchor Books, 1957.

Wilcock, R., and W. Franke. *Unwanted Workers.* New York: Free Press, 1963.

Wilensky, H. L. "Work, Careers, and Social Integration." *International Social Science Journal,* vol. 12, 1960, pp. 543–60.

Winch, R. F. "Another Look at the Theory of Complementary Needs in Mate Selection," *Journal of Marriage and Family Living,* vol, 29, 1967, pp. 756–62.

Yorburg, B. *The Changing Family.* New York: Columbia University Press, 1973.

Zweig, F. *The Worker in an Affluent Society.* New York: Free Press, 1961.

Old Age

Anderson, J. E., ed. *Psychological Aspects of Aging.* Washington, D.C.: American Psychological Association, 1956.

Birren, J. E., et al., eds. *Human Aging.* Public Health Service Publication no. 986. Washington, D. C.: U.S. Government Printing Office.

———, ed. *Handbook of Aging and the Individual: Psychological and Biological Aspects.* Chicago: University of Chicago Press, 1959.

———, ed. *Relations of Development and Aging.* Springfield, Ill.: Charles C. Thomas, 1964.

Busse, E. W. "Problems Affecting Psychiatric Care of the Aging." *Geriatrics,* vol. 15, 1960, pp. 673–80.

——— and E. Pfeiffer, eds. *Behavior and Adaptation in Late Life.* Boston: Little, Brown, 1969.

Carp, F. M., ed. *Retirement.* New York: Behavioral Publications, 1972.

Cavan, R. S., et al. *Personal Adjustment in Old Age.* Chicago: Science Research Associates, 1949.

Cumming, E., and W. E. Henry. *Growing Old.* New York: Basic Books, 1961.

Kalish, R. A. "The Aged and the Dying Process: The Inevitable Decisions." *Journal of Social Issues,* vol. 21, 1965, pp. 87–96.

Kubler-Ross, E. *On Death and Dying.* New York: Macmillan, 1969.

Lawton, G. *Aging Successfully.* New York: Columbia University Press, 1946.

Levin, S., and R. J. Kahana, eds. *Psychodynamic Studies on Aging: Creativity, Reminiscing, and Dying.* New York: International Universities Press, 1967.

Maddox, G. L. "Disengagement Theory: A Critical Evaluation." *Gerontologist,* vol. 4, 1964, pp. 80–83.

Neugarten, B. L., ed. *Middle Age and Aging.* Chicago: The University of Chicago Press, 1968.

Reichard, S., F. Livson, and P. G. Petterson. *Aging and Personality.* New York: Wiley, 1962.

Shrut, S. D. "Attitudes Toward Old Age and Death." *Mental Hygiene,* vol. 42, 1958, pp. 259–66.

Simpson, I. H., and J. G. McKinney. *Social Aspects of Aging.* Durham, N.C.: Duke University Press, 1966.

Vernon, G. M. *Sociology of Death: An Analysis of Death-Related Behavior.* New York: Ronald Press, 1970.

Wahl, C. W. "The Fear of Death." *Bulletin of the Menninger Clinic,* vol. 22, 1958, pp. 214–23.

ACKNOWLEDGMENTS

Chapter 1 — Reprinted with permission from *International Journal of Psycho-Analysis,* Vol. 47, 1966, pp. 218-229. The article was subsequently expanded into several chapters in *Anxiety and Ego Formation in Infancy* by Sylvia Brody and Sidney Axelrad, International Universities Press, New York, 1970.

Chapter 2 — From *Psychoanalytic Study of the Child,* Vol. 1, New York: International Universities Press, 1945, pp. 53-74. Reprinted with permission (Abridged).

Chapter 3 — From *The Developmental Psychology of Jean Piaget* by John H. Flavell © 1963. Reprinted by permission of D. Van Nostrand Company (Abridged).

Chapter 4 — Reprinted from Harry Stack Sullivan, "Beginnings of the Self-System," *The Interpersonal Theory of Psychiatry,* Vol. 1, edited by Helen S. Perry and Mary L. Gawel. New York: W.W. Norton and Co., pp. 158-171, 1953. By permission of W.W. Norton and Co., Inc. Copyright, 1953, by the William Alanson White Psychiatric Foundation.

Chapter 5 — Reprinted with permission of the National Association of Social Workers, from *Social Work,* Vol. 10, No. 3 (July 1965), pp. 47-50.

Chapter 6 — Reprinted with permission. From *American Journal of Orthopsychiatry,* Vol. 37, No. 1 (January 1967), pp. 8-21. Copyright, 1967, the American Orthopsychiatric Association.

Chapter 7 — Reprinted by permission of *Daedalus,* Journal of the American Academy of Arts and Sciences, Boston, Massachusetts, Winter 1966, *Crucible of Identity* (Abridged).

Chapter 8 — From *Welfare in Review,* Vol. 3, January 1965, pp. 9-19. Reprinted with permission.

Chapter 9 — From *American Journal of Psychiatry,* Vol. 120, No. 4 (October 1963), pp. 332-344. Copyright, 1963, the American Psychiatric Association. Reprinted with permission.

Chapter 10 Reprinted with permission. From *Saturday Review,* October 7, 1967, pp. 60-66.

Chapter 11 Reprinted with permission. From *Child Welfare,* Vol. 50 No. 3 (March 1971), pp. 132-142.

Chapter 12 Reprinted with permission. From *Mental Hygiene,* Vol. 55, No. 4 (October 1971), pp. 437-443.

Chapter 13 Reprinted with permission of Macmillian Publishing Co., Inc., from *On Adolescence,* Chapter 5, by Peter Blos. Copyright © 1962 by The Free Press (Abridged).

Chapter 14 Reprinted with permission. From *Adolescence,* Vol. 6, No. 25 (Spring 1972), pp. 121-127.

Chapter 15 Reprinted with permission from Chapter 1 of *Adolescence: Psychosocial Perspectives,* edited by Gerald Caplan and Serge Lebovici, © by Basic Books, Inc., publishers, New York.

Chapter 16 From *Smith College Studies in Social Work,* Vol. 38, No. 1 (November 1967), pp. 1-15. Reprinted with permission.

Chapter 17 Reprinted with permission of Fritz Redl and Child Study Press. From *Child Study,* Vol. 21, No. 2 (Winter 1943-44), pp. 44-48 and 58-59.

Chapter 18 From *Family Process,* Vol. 1, No. 2 (September 1962), pp. 202-213. Reprinted with permission of the copyright owner.

Chapter 19 Reprinted with permission. From *Adolescence,* Vol. 4, No. 15 (1969), pp. 333-360.

Chapter 20 From *American Sociological Review,* Vol. 20, No. 6 (December 1955), pp. 680-684. Reprinted with permission.

Chapter 21 Published by permission of Transaction, Inc., from *Transaction,* Vol. 7, No. 4 Transaction, Vol 7, No. 4 (February 1970). Copyright © 1970, by Transaction, Inc.

Chapter 22 © 1970 by Kenneth Keniston. Reprinted from his volume, *Youth and Dissent* by permission of Harcourt Brace Jovanovich, Inc.

Chapter 23 Reprinted by permission of *Daedalus,* Journal of the American Academy of Arts and Sciences, Boston, Massachusetts, Summer 1967, *Memorandum on Youth.*

Chapter 24 From *Mental Hygiene,* Vol. 52, No. 3 (July 1968), pp. 323-329. Reprinted with permission.

Chapter 25 Reprinted with permission of American Book Company from *Readings in General Psychology,* edited by W. Edgar Vinacke, New York: American Book Co., 1968, pp. 332-343.

Chapter 26 From *Psychological Review,* Vol. 66, No. 5 (1959), pp. 297-333. Copyright, 1959, by the American Psychological Association. Reprinted by permission (Abridged).

Chapter 27 From *Bulletin of the Menninger Clinic,* Vol. 6, No. 6 (November 1942), pp. 170-182. Reprinted with permission.

Chapter 28 From *Family Process,* Vol. 9, No. 3 (September 1970), pp. 259-278. Reprinted with permission.

Chapter 29 Reprinted with permission of Macmillan Publishing Co., Inc., from *The Patient and the Mental Hospital* by Milton Greenblatt, Daniel J. Levinson, and Richard H. Williams. © Copyright 1957 by the Free Press.

Chapter 30 From William H. Masters and Virginia E. Johnson, Counselling on Marital and Sexual Problems (Richard H. Klemer, ed.) Copyright, 1965, the William & Wilkins Co. Reproduced by permission.

Chapter 31 From *Journal of Marriage and the Family,* Vol. 26, No. 4 (November 1964), pp. 457-466. Reprinted with permission.

Chapter 32 From *The Family Coordinator,* Vol. 20, No. 4 (October 1971), pp. 363-368. Reprinted with permission.

Chapter 33 From *Journal of Marriage and the Family,* Vol. 30, No. 1 (February 1968), pp. 26-39. Reprinted with permission.

Chapter 34 From *The Family Coordinator,* Vol. 19, No. 4 (October 1970), pp. 317-323. Reprinted with permission.

Chapter 35 From *American Journal of Sociology,* Vol. 68, No. 4 (January 1963), pp. 471-480. Reprinted with permission of the University of Chicago Press, and the author.

Chapter 36 From *Journal of Marriage and the Family,* Vol. 28, No. 1 (February 1966), pp. 29-36. Reprinted with permission.

Chapter 37 Reprinted with permission from *Geriatrics,* Vol. 3, No. 3 (May-June 1948), pp. 135-150, Copyright The New York Times Media Company, Inc. (Abridged).

Chapter 38 From *The Stress of Life* by Hans Selye. Copyright © 1956 by McGraw-Hill, Inc. Used with permission of McGraw-Hill Book Company (Abridged).

Chapter 39 Reprinted with permission of Family Service Association of America. From *Social Casework,* Vol. 46, No. 1 (January 1965), pp. 16-21.

Chapter 40 Reprinted by permission. From *Psychological Aspects of Aging,* edited by John E. Anderson. Washington, D.C.: American Psychological Association, 1956, pp. 42-53. Copyright, 1956, by the American Psychological Association.

Chapter 41 From *The Gerontologist,* Vol. 8, No. 2 (Summer 1968), pp. 67-71. Reprinted with permission.

Chapter 42 From *Journal of the American Medical Association,* Vol. 173, No. 1 (May 1960), pp. 33-35. Reprinted by permission.

Chapter 43 Reprinted by permission. From *American Journal of Orthopsychiatry,* Vol. 36, No. 4 (July 1966), pp. 680-685. Copyright, 1966, the American Orthopsychiatric Association.

Chapter 44 From *The Gerontologist,* Vol. 9, No. 2 (Summer 1969), pp. 90-95. Reprinted with permission.

Chapter 45 From *The Gerontologist,* Vol. 4, No. 4 (December 1964), pp. 180-184. Reprinted with permission.

Chapter 46 From *The Family Coordinator,* Vol. 21, No. 1 (January 1972), pp. 5-19. Reprinted with permission.

Chapter 47 From *The Gerontologist,* Vol. 5, No 3 (September 1965), pp. 129-136 and 159. Reprinted with permission.

Chapter 48 Reprinted from Robert Blauner, "Death and Social Structure," *Psychiatry,* Vol. 29, No. 4 (November 1966), pp. 378-394. Copyright, 1966, by the William Alanson White Psychiatric Foundation (Abridged).

INDEX